American Environmental Leaders

Volume I

American Environmental Leaders

From Colonial Times to the Present

Volume I
A–K

Anne Becher

with Kyle McClure,
Rachel White Scheuering,
and Julia Willis

ABC-CLIO

Santa Barbara, California
Denver, Colorado
Oxford, England

Library of Congress Cataloging-in-Publication Data
Becher, Anne.
American environmental leaders / Anne Becher with Kyle McClure, Rachel White Scheuering, and Julia Willis.
p. cm.
Includes bibliographical references and index.
ISBN 1-57607-162-6 (acid-free paper)
1. Environmentalists—United States—Biography. I. Title.
GE55 .B43 2000
363.7'0092'273—dc21

00-011296

06 05 04 03 02 01 00 10 9 8 7 6 5 4 3 2 1

ABC-CLIO, Inc.
130 Cremona Drive, P.O. Box 1911
Santa Barbara, California 93116-1911

This book is printed on acid-free paper ♾.
Manufactured in the United States of America.

Contents

List of American Environmental Leaders

Volume I

Volume II

Preface

What Americans now refer to as the "Environmental Movement" is actually a multifaceted confluence of many different movements throughout history. These specialized environmental missions focus on knowing, celebrating, conserving, or cleaning up the environment, with *environment* defined to encompass wilderness, urban, suburban, or rural areas inhabited by humans and/or other species.

This book honors almost 350 "American environmental leaders": well-known founders and leaders of the diversified strands of the environmental movement as well as some less famous regional or behind-the-scenes participants. I have included historical figures active as many as three hundred years ago, but the majority are alive today, the youngest still in their twenties. Some subjects profiled in this book were selected for their indelible impact on what we know and appreciate about the environment, or how intact or healthy it is today. Others were selected to illustrate some of the tremendously diverse starting points and orientations from which one can work for a better environment. Throughout the selection process, an effort was made to include not only the most visible movers and shakers but also those who collaborated on important efforts from behind the scenes. More than half of the subjects still living have read their own profiles to check facts and to approve the way in which their life and work have been portrayed. (For a list of all of those who collaborated in this way, see the Acknowledgments section.)

The struggles to preserve or conserve spectacular tracts of wilderness have produced many heroes and heroines. Among the best-known protagonists of these battles have been John Muir, the voice of Yosemite and founding president of the Sierra Club, and Stephen Mather, the dynamic first director of the National Park Service, who with staff members Robert Sterling Yard and Horace Albright convinced Congress and the American public of the value of national parks. Another group of people have dedicated themselves to preserving endangered species and habitats. This first issue became a concern during the late nineteenth century when bison and passenger pigeons, both of which had once occurred in dense populations, were exterminated by hunters. Key to these early efforts were big-game hunters, including Theodore Roosevelt and George Bird Grinnell, and the first Audubon Society president, William Dutcher, who fought for the earliest state and federal bird protection laws. Efforts to preserve endangered species and their habitats continue today in the hands of such people as Jasper Carlton of the Biodiversity Legal Foundation and Kierán Suckling of the Center for Biological Diversity.

Another angle on environmental conservation that has steadily gained importance during the past two centuries has been discouraging the overuse of natural resources such as forests, water, minerals, and soil and to promote more efficient and sustainable use of them. Nineteenth-century scholar George Perkins Marsh first wrote about how human activity could damage the environment in his *Man and Nature, or Physical Geography as Modified by Human Action* (1864). Since then, there have been full-scale efforts to reduce the negative impact of agriculture and the extractive industries on the health of the environment. Agricultural scientist George Washington Carver taught impoverished Southern farmers to grow such soil-enriching crops as peanuts, sweet potatoes, and black-eyed peas, showed them how to prepare these foods, and convinced the U.S. Congress in the 1920s to provide economic incentives for their production. This book contains profiles of many subsequent promoters of sustainable agriculture, including Soil Conservation Service director Hugh Hammond Bennett; organic gardening guru Robert Rodale; Diane and Kent Whealy, who help conserve heirloom varieties of vegetables and fruits through Seeds of Change; philosoper-farmer Wes Jackson; and many others.

Seeing the potential of industry to help solve global energy and environmental problems, contemporary visionaries Paul Hawken, author of *The Ecology of Commerce* (1993), and Amory and Hunter Lovins, cofounders of the Rocky Mountain Institute, have helped private businesses—including some of the largest in the United States—to become more efficient and sustainable in their practices. One of their clients, Ray Anderson of Interface, Inc., the largest commercial carpet, tile, and interior fabrics company in the world, has vowed to convert his company into a model of sustainability and environmental responsibility.

Unfortunately, most industry leaders are not as environmentally conscientious as Anderson, and many of the environmental leaders featured in this book are activists who feel called to fight the industrial pollution that ruins their homes. Grace Thorpe, Oklahoma Sauk and Fox, founded the national Environmental Council of Native Americans in 1993 to support Indian nations that refused the persistent requests by nuclear industry and the federal government to store nuclear waste on their land. Aurora Castillo and Juana Gutiérrez are two of the dynamic leaders of Mothers of East Los Angeles, a group that has protested the siting of incinerators, oil pipelines, and toxic waste sites in their neighborhoods and those of other poor people of color. What's more, they have initiated a water conservation program for residents of their neighborhood through which they distribute water-conserving showerheads and toilets.

The subjects included in this book are diverse, and their diversity is what strengthens and enhances the environmental movement. Poets, lawyers, economists, farmers, grassroots activists, scientists, theologians, sanitary engineers, and musicians—their stories are lessons in determination and commitment. Many are guided by strong convictions and an apparent inner knowledge of how to make the world a healthier place. Fortunately for all of us, beneficiaries of their concern and hard work, they have the courage, intelligence, endurance, and strategic ability to be effective leaders.

Use this book as a springboard from which to learn more about these inspiring individuals and their work. Follow up by reading the sources listed in the bibliography of each profile: books, magazine articles, and organization websites are listed so that you may delve deeper. This book is meant not only as a source of information about the great leaders it profiles but also as a tool with which readers may be inspired to initiate their own careers as environmental leaders.

Acknowledgments

Working on this book was very rewarding. Over the phone, on line, and occasionally in person I met many wonderful and generous people: those who helped me compile the subject list, others who researched and wrote profiles, and many of the book's subjects themselves, who carved out time to tell me more about themselves and their work and also read the profiles for accuracy.

The following people helped me compile my subject list:

Cain Allen
David Augeri
Richard Ayres
David Barsamian
Ingrid Becher
Amos Bien
Stewart Brandborg
Charlotte Caldwell
Bill Chaloupka
Paul Dresman
Elizabeth Dubrulle
Michael Egan
Willie Fontenot
Tom Goldtooth
Bob Golten
Robert Gottlieb
Linda Irvine
Frank Joyce
Diane Jukofsky
Gwyn Kirk
Kamala Kempadoo
Meg Knox
Ron Mader
Bob McFarland
John Opie
Paula Palmer
Jane Perkins
Lori Lea Pourier
Gretchen Reinhardt
Brian Russell
Rachel White Scheuering
Chet Tchozewski
Will Toor
Jeff Wagenheim
Marty Walter
Chris Wille

The following subjects provided information for and/or reviewed their own entries:

John Adams
Elias Amidon and Elizabeth Roberts
Adrienne Anderson
Ray Anderson (with Lisa Cape Lilienthal)
Richard Ayres
Betty and Gary Ball
Albert Bartlett
Michael Bean
Ed Begley, Jr.
Peter Berg
Peter Berle
Amos Bien
Eula Bingham
Murray Bookchin (with Janet Biehl)
Barbara Bramble
Stewart Brandborg
Tom Cade
Lynton Caldwell
Baird Callicott
Jasper Carlton
Anita Clark
Theo Colborn (with Rose Baeur)
Jack Collom
Ellen and Paul Connett
Lisa Conte (with Gail Jones)
Paul Cox
John Craighead, John Willis Craighead, Charlie Craighead

Richard Dawson
Jeff DeBonis
Chris Desser
Bill Devall
Hank Dittmar
Richard Donovan
Mark Dowie
Alan Durning
Polly Dyer
Anne and Paul Ehrlich
William Fontenot
Jerry Franklin
Michael Frome
Kathryn Fuller (with Lisa Clark)
J. William Futrell
Ross Gelbspan
Lois Marie Gibbs
Lou Gold
Bob Golten
Robert Gottlieb
Pete Grogan
Joan Gussow
David Harrison
Debra Harry
Dorothy Harvey
Paul Hawken
Randy Hayes
Samuel P. Hays
Tim Hermach
Donna House
Daniel Janzen
Glenn S. Johnson
Hazel Johnson
Diane Jukowski
Hal Kane
Daniel Katz
Hugh Kaufman
Stephen Kellert
Robert F. Kennedy, Jr.
Samuel LaBudde
Alicia Littletree
Jackie Lockett
Amory and Hunter Lovins
Ron Mader
Richard Manning
Ed and Betsy Marston
Dennis Martínez
Carolyn Merchant
Doug Moss
Carlos Nagel
Reed Noss
Eugene Odum
Molly Olson
David Orr
Steve Packard
Paula Palmer
Jane Perkins
Frank and Deborah Popper
Sandra Postel
Paul Pritchard
Laura Pulido
John Robbins (with Deo Robbins)
Vicki Robin
Holmes Rolston
Rosemary Radford Ruether
Carl Safina
Bill Sanjour
Rocky Smith
Rosemary Reuther
Holmes Rolston III
Stephen Schneider (with Sue Evans)
Danny Seo
Philip Shabecoff
Chris Shuey
David Sive
Rocky Smith
Gary Snyder
Michael Soulé
James Gustave Speth (with Denise Schlener)
Wilma Subra
Kieran Suckling
Terri Swearingen
Chet Tchozewski
Debby Tewa
Oakleigh Thorne
Grace Thorpe
Brian Tokar

Will Toor
Martin Walter
Kent and Diane Whealy
Gilbert White
Louisa Willcox
Chris Wille
Howie Wolke
Donald Worster

The profiles for the following subjects were reviewed by the persons listed with them:

Dana Alston: Larry Kressley and Adisa Douglas
Carl Anthony: Martha Olson
Judi Bari: Alicia Littletree
Walt Bresette: Rick Whaley and Tom Goldtooth
Robert Bullard: Glenn S. Johnson
Kate and Tom Chappell: Kathleen Taggersall
Peter Montague: Maria Pellerano
Melissa Poe: Patricia Poe
Rodger Schlickeisen: Aimee Delach
Barbara Warburton: María Alma Solís and Larry Lof
Adam Werbach: Kelly Braucht
Howard Zahniser: Ed Zahniser

Further help was offered by Carol Taylor at the Boulder *Daily Camera* archives and Jacky Morales-Ferrand at the City of Boulder.

I would never have been able to complete this book during this lifetime without the help of these twenty-eight talented contributors. They submitted timely drafts and cheerful revisions and sent notes of encouragement along the way. The profiles that each wrote or contributed to are listed under his or her name.

Cain Allen
Chief Sealth (Seattle)
William Steel
Chief Tommy Thompson

Eleanor Crandall
Paul Pritchard

Kris Daehler
Jerry Mander

Kevin Dahl
Gary Nabhan

Michael Egan:
John James Audubon
Marston Bates
Peter Berg
J. Baird Callicott
William Colby
William Cronon
Bill Devall
William O. Douglas
Newton Drury
Alan Durning
Albert Gore, Jr.
Stephen Jay Gould
Asa Gray
Alice Hamilton
Samuel P. Hays
Chris and Martin Kratt
Edmund Meany
Olaus and Margaret Murie
John D. Rockefeller, Jr.
Holmes Rolston, III
Richard Evans Schultes
George Sessions
Michael Soulé
Morris Udall
Stewart Udall
Walt Whitman
Donald Worster

Joan Erben
Samuel LaBudde
Donella Meadows

Erica Ferg
Karen Silkwood
Carol Browner

T. L. Freeman-Toole
Fred Krupp
Wes Jackson
Owen Lammers
McDonough, Bill
Roderick Nash
Sigurd Olson

Harah Frost
Hazel Johnson

Jennifer Heath
Dolores Huerta
Bill McKibben
Lewis Mumford

Sarah W. Heim-Jonson
John Hamilton Adams
Richard Ayres
Michael Bean
Barbara Bramble
William Dutcher

Cassandra Kircher
Frederick Jackson Turner

Meghan McCarthy
Rodger Schlickeisen
Luther Standing Bear
Peter Warshall

Kyle McClure:
Ansel Adams
Henry Adams
Cecil Andrus
Mary Austin
Bruce Babbit
Peter Berle
Albert Bierstadt
Lester Brown
John Burroughs
Lynton Caldwell
Jimmy Carter
John Chafee
Marion Clawson
Thomas Cole
Douglas Costle
Herman Daly
Jeff DeBonis
Christina Desser
Bernard Devoto
John Dingell
Richard Grossman
Juana Gutiérrez
Jay Hair
Paul Hawken
Hazel Henderson
Edward Hoagland
Michael McCloskey
Ian McHarg
John McPhee
Stephanie Mills
Edmund Muskie
Helen and Scott Nearing
Marc Reisner
Theodore Roszak
John Sawhill
Stephen Schneider
JoAnn Tall
Wallace Stegner
Ted Turner
Howie Wolke

Bob Macfarland
Ernest Seton Thompson
Oakleigh Thorne
Gilbert White

Susan Muldowney
Debra Harry

Laura Pulido

Katherine Noble-Goodman
Eugene Odum
David Sive

Joe Richey
Elias Amidon and Elizabeth Roberts
Wendell Berry
Jack Collom
Ralph Waldo Emerson
Barry Lopez
Peter Matthiessen
Gary Snyder
Walt Whitman

Brian Russell
Dorothy Harvey

Rachel White Scheuering
John Quincy Adams
Jane Addams
Peter Bahouth
William Beebe
Stewart Brand
Michael Brown
Tom Cade
Archie Carr
Marjorie Carr
Harry Caudill
James Fenimore Cooper
Mark Dowie
Silvia Earle
David Ehrenfeld
Buckminster Fuller
Theodore Geisel
Ross Gelbspan
Joan Gussow
Julia Butterfly Hill
Helen Ingram
Stephen Kellert
Francis Moore Lappé
Lindbergh, Anne Morrow and Charles
Richard Manning
Carolyn Merchant
Carlos Nagel
Margaret Owings
Steve Packard
Roger Tory Peterson
Russell Peterson
Deborah and Frank Popper
George Powell
Peter Raven
Jeremy Rifkin
John Robbins
Robert Rodale
Carl Safina
Catherine Sneed
Sandra Steingraber
Terry Tempest Williams
Diane and KentWhealy
Louisa Willcox
Paul Winter
Hazel Wolf
George Woodwell

Noelle Sullivan
Laurance Rockefeller
Kirkpatrick Sale
Philip Shabecoff
Enos Mills
Sandra Postel
George Mitchell

Horace Voice
Henry David Thoreau

Deena Wade
Ed Begley
Woody Harrelson
Carl Sagan

Jeff Wagenheim
Vicky Robin

Ray Watt
Christopher Shuey
Henry Waxman

Julia Willis
Catherine Baeur
Mollie Beattie
Eula Bingham
Janet Brown
Aurora Castillo
Kate and Tom Chappell
John B. Cobb, Jr.
Henry Cowles
Dian Fossey
Henry Jackson
Lady Bird Johnson
Jim Jontz
Hugh Kaufman
Henry Kendall
Oren Lyons
Mary McDowell
Marion Moses
Herbert Needleman
Bryan Norton
Francis Parkman
William K. Reilly
Rosemary Radford Reuther
Bill Sanjour
Carl Schurz
Irving Selikoff
Paolo Soleri
Christopher Stone
Brian Tokar
William Toor
William Vogt
Lynn White, Jr.

Hilary Wood
Diane Ackerman
Archie Carr

Josh Zaffos
Reed Noss

Thanks too to the editors at ABC-CLIO who helped envision and mold this book: Kristi Ward, Elizabeth Dubrulle, Kevin Downing, Connie Oehring, Libby Barstow, Anna Kaltenbach, and Jennifer Loehr.

It indeed takes a village to raise a book like this. I would like to acknowledge the loving encouragement and support of my family and friends, especially Mary Long; Christy Jespersen; Kathleen Flynn; Judy, David, Ingrid, and Ted Becher; and the Richeys: my children, Flora and Jacob, and my husband, Joe.

This book is dedicated to those profiled within, in awe and gratitude for their devotion to making the world a healthier, more beautiful place for all species.

A

Abbey, Edward

(January 29, 1927–March 14, 1989)
Writer

Edward Abbey, wild man of the American West and the author of 22 books, defies literary definitions. He is known for his exquisite descriptions of his beloved Southwestern desert, for his bitter diatribes against those who defile such pristine areas (ranchers, loggers, even dumb tourists), and for the unruly characters—some autobiographical—who people his novels.

Edward Paul Abbey was born on January 29, 1927, in Home, Pennsylvania, a rural Allegheny community. His mother, Mildred, was an activist for the Woman's Christian Temperance Union, and his father, Paul Revere Abbey, was a Socialist union organizer who earned his livelihood cutting hickory fence posts. As a child Abbey wrote comic books, and he became a journalist while in high school (though he flunked his journalism class). During the summer of his seventeenth year, Abbey hitchhiked and rode buses and trains on an exploratory tour of the West. He fell in love with the deserts and canyons. And at the age of 19, after one year in the Army and another at Indiana State Teacher College, Abbey moved west, where he was to stay except for a few brief periods of his life. Abbey studied philosophy and English at the University of New Mexico, earning his B.A. in 1951 and his M.A. in 1956. His master's degree thesis, entitled "Anarchism and the Morality of Violence," examined political situations in which violence could be justified. His conclusion was that it was most justifiable when used in self-defense.

While working at varied jobs after completing his M.A., including inspecting roads for the U.S. Forest Service and being a ranger for the National Park Service, Abbey wrote several novels. His first widely acclaimed work was *Desert Solitaire* (1968), a compilation of his journals from the time he was working as a seasonal ranger in Arches National Monument in Utah. Many critics call this book his best. It is a medley of crystalline nature writing and enraged rants against the incursion of civilization into the pristine Southwest deserts. The author's note in *Desert Solitaire* warns the reader:

> Do not jump into your automobile next June and rush out to the canyon country hoping to see some of that which I have attempted to evoke in these pages. In the first place you can't see *anything* from a car; you've got to get out of the goddamned contraption and walk, better yet crawl, on hands and knees, over the sandstone and through the thornbush and cactus. When traces of blood begin to mark your trail you'll see something, maybe. Probably not.

The direct action arm of the environmental movement remembers Abbey best for *The Monkey Wrench Gang* (1975), an account of the exploits of a group of iconoclasts who specialized in what was later termed *monkeywrenching*, the deliberate damaging of equipment used to destroy nature. The Monkey Wrench Gang practiced by pouring sugar and dirt into the gas tanks of bulldozers and tractors at desert construc-

Edward Abbey at his typewriter during his work with the National Park Service (Buddy Mays/Corbis)

tion projects, but their ultimate goal was to blow up the Glen Canyon Dam. True to Abbey's master's thesis conclusions, the Monkey Wrench Gang's violence was undertaken in self-defense, for the gang identified so closely with the desert that the development was an attack on their very beings. This book inspired DAVE FOREMAN, HOWIE WOLKE, and three other friends to found Earth First! in 1979. Their first public action was at a Glen Canyon Dam protest during which Abbey spoke. From that time on, Abbey served as an elder adviser and shaman to the group.

During his lifetime, Abbey wrote 22 books. The literary establishment pegged him as a Western environmentalist writer. Abbey himself said in an interview published in *Resist Much, Obey Little* (1996) that he was content to remain in that pigeonhole because it assured him easy access to publishers and earned him a comfortable living. However environmentalists who wanted to see him as a spokesman for their causes were often disturbed by some of his assertions. He spouted brash, disturbing opinions in some of his books. He insulted literary critics who did not like his work; called on the U.S. Border Patrol to turn back all Mexican immigrants, hand them guns, and tell them to finish their revolution; and criticized mainstream environmentalist organizations for their compromises.

Abbey's friends and fans admired his dedication to the truth—about the world and his own life—even if his words were sometimes difficult to digest or undiplomatic. Abbey was continually outraged, wrote Wendell Berry in his contribution to *Resist Much, Obey Little*, but Abbey's humor made his outrage tolerable to his readers. During his life Abbey married five times and fathered six children.

Abbey died of internal bleeding on March 19, 1989, shortly after being informed that he had a terminal circulatory disorder. His death and burial have achieved the same mythological status that was given his life while he was alive. Two days before he died, he asked his friends to take him out of the hospital, into the desert where he enjoyed one last campfire circle. He died in a sleeping bag on the floor of his writing cabin, and his friends followed the instructions he had left them to drive his body as far as possible into the desert and bury him under a pile of rocks.

Bibliography

Bishop, James Jr., *Epitaph for a Desert Anarchist*, 1994; Hepworth, James R., and Gregory McNamee, *Resist Much, Obey Little: Remembering Ed Abbey*, 1996; Hoagland, Edward, "Edward Abbey: Standing Tough in the Desert," *New York Times Book Review*, 1989; Peterson, David, "Where Phantoms Brood and Mourn," *Backpacker*, 1993.

Ackerman, Diane

(October 7, 1948–)
Poet, Nature Writer

Writer Diane Ackerman is best known for her ability to combine the seemingly disparate disciplines of art and science. Her writing, which includes poetry, nonfiction, stories for children, and plays, is unique in its ambition to use creative expression to describe the natural world. In 1990, Ackerman's book, *A Natural History of the Senses*, became a national bestseller and inspired a miniseries that aired as part of public television's *Nova* in 1995.

Diane Ackerman was born in Waukegan, Illinois, on October 7, 1948. Eight years later, her family moved to the more rural location of Allentown, Pennsylvania. The change suited the young writer, who remembers herself as an outgoing tomboy who spent most of her time outside. When she was indoors, Ackerman spent her time reading and writing, the latter becoming a major passion that resulted in limericks, stories, and articles for the local newspaper. After spending her freshman year at Boston University, Ackerman transferred to Pennsylvania State University in 1968. It was there that she met British novelist Paul West, a professor who later became her life companion.

Ackerman graduated in 1970 with a B.A. degree in English. During the same year she began studying for her M.F.A. at Cornell University in Ithaca, New York. Eight years later, Ackerman had not only completed her intended degree but had also earned an M.A. and a Ph.D. in English literature. During this period she won two prestigious awards, the Academy of American Poets Prize and the Corson Bishop Prize for Poetry and published her first book of poetry, *The Planets: A Cosmic Pastoral*, in 1976. Even at this early point in her career, Ackerman showed signs of her future ambition to bring science and poetic writing into the same medium. *The Planets*, written entirely about astronomy, introduced readers to the author's unusual articulation of the natural world, a perception that did not distinguish hard data from passionate observation. This unique blend of disciplines was to be Ackerman's trademark, a result of what she terms her "nomadic curiosity."

Ackerman's fascination with the universe is reflected in the diversity of subjects she investigates in her writing. In 1980, *Twilight of the Tenderfoot*, a book detailing her experiences as a ranch hand in New Mexico, was published to wide acclaim. After earning her pilot's license, she wrote *On Extended Wings* (1985), a book exploring the implications of learning to fly, which was later adapted to the stage. During this time, Ackerman continued to write poetry. *Lady Faustus*, a collection of poems published in 1983, covered such diverse subjects as soccer, flying, and meditations on amphibians. Despite the wide spectrum of subject matter, Ackerman's work remains infused with a central theme. Whether branding cattle or staring at the stars, there is always a fascination and a deep enthusiasm for the natural world. Though this point of view received criticism from those who believed that poetry must be free of science, Ackerman's approach shows that the two can be combined gracefully.

In 1988, Ackerman published *Reverse Thunder: A Dramatic Poem*, detailing the life and times of Sor Juana de la Cruz, a nun who lived in seventeenth-century Spain. A play written in verse, *Reverse Thunder* is a tribute to a woman Ackerman deeply admires. Despite the restrictions of the times, Sor Juana was both a poet and a scientist, a woman who shared Ackerman's awe at the complexities of creation. Her meditations reveal a creative mind unafraid to combine religion with science or poetry with data, an unpopular concept in her day.

Ackerman's most popular book, *A Natural History of the Senses*, became a bestseller shortly after its publication in 1990. A celebration of the five senses, the book uses essays, vignettes, and observations to investigate the ways in which humans perceive the natural world. The sweeping success of *Senses* was a surprise, as very few scientifically based works become bestsellers. No doubt the success of the book can be attributed to Ackerman's skill as a writer and her ability to mix potentially dry material with humor, myth, and poetic description. In February 1995, the television series *Nova* invited Ackerman to host a five-part miniseries entitled "Mystery of the Senses," which received some of the highest ratings of the season. *Senses* is Ackerman's crowning achievement to date, a work that introduced countless readers to her unique genius.

In 1991, Ackerman returned to poetry with *Jaguar of Sweet Laughter: New and Selected Poems*, which was nominated for the National Book Critics Circle Award. Another publication, *The Moon by Whale Light: And Other Adventures among Bats, Penguins, Crocodilians, and Whales*, expanded upon a series of articles that had been previously published in the *New Yorker*. In 1994, Ackerman's *A Natural History of Love,* which was modeled upon her earlier book, received mixed reviews from critics who believed that the subject of love, unlike that of our senses, was broader than could be covered in a single volume. Despite these hesitations, *Love* was widely enjoyed by the public, many of whom savored the beauty of Ackerman's prose in its investigation of the many different concepts of love.

Ackerman's 1995 *The Rarest of the Rare: Vanishing Animals, Timeless Worlds* resulted from a series of pilgrimages to the world's rarest ecosystems, during which she paid homage to rare and unusual species. The book contains chapters on the monk seal, the short-tailed albatross, the golden lion tamarin, and other endangered species. Ackerman mixes descriptions of habitats and animals with biographical data about her human companions, biologists who study and strive to protect their chosen species.

Diane Ackerman resides in upstate New York and has received many awards for her writing, including the Black Warrior Review Poetry Prize in 1981, the Pushcart Prize VIII in 1984, the Lowell Thomas Award from the Society of American Travel Writers in 1990, the Wordsmith Award in 1992, and the New and Noteworthy Book of the Year from the *New York Times Book Review,* also in 1992, for *The Moon by Whale Light.*

BIBLIOGRAPHY

Brainard, Dulay, "Diane Ackerman," *Publisher's Weekly,* 1991; Dowd, Maureen, "Diane Ackerman," Vogue, 1991; Elder, John, ed., *American Nature Writers Volume 1,* 1996; Shelton, Pamela, ed., *Contemporary Woman Poets,* 1998.

Adams, Ansel

(February 20, 1902–April 22, 1984)
Photographer, Preservation Activist

Ansel Adams was a photographer and a preservationist. His pictures played an important role in defining how Americans think about wilderness, and they have been vital in supporting its preservation. Adams served on the board of directors of the Sierra Club for nearly 40 years.

Ansel Easton Adams was born on February 20, 1902, in San Francisco, California, the only child of Charles Hitchcock Adams and Olive Bray Adams. His father was a businessman whose enterprises included an insurance agency and a chemical plant. Adams grew up among the sand dunes on the westernmost edge of the San Francisco Peninsula, where from his early childhood he was surrounded by natural beauty. His education was erratic and largely self-acquired. He attended both public and private schools, and he also received some instruction at home from his father.

A significant event in Adams's early life was a vacation his family took to Yosemite National Park in 1916, when he was

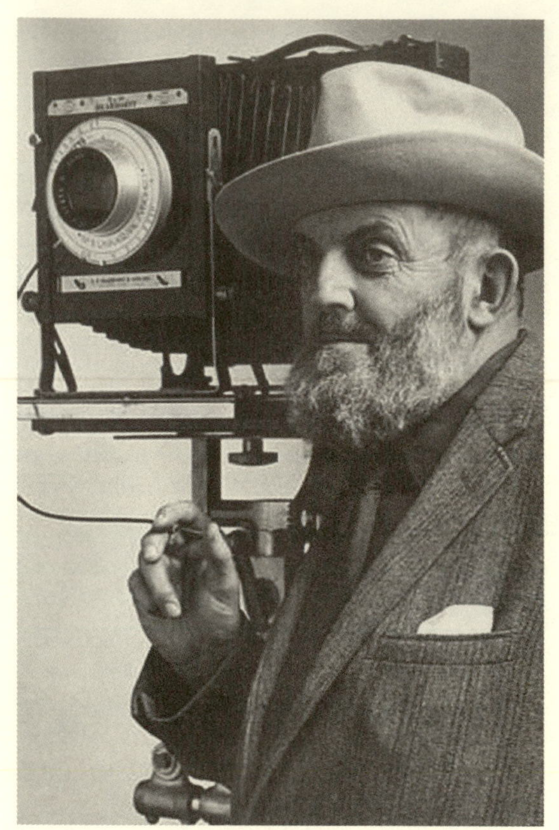
Ansel Adams (Bettmann/Corbis)

14 years old. He brought along a Kodak Box Brownie camera that he used to take his first photographs. As a result of this experience, Adams persuaded Frank Dittmman, the owner of a San Francisco photo finishing plant, to take him on as an apprentice in developing and other techniques in the darkroom. While Adams's enthusiasm for photography was growing, music, at this time, was his main concern. He was receiving four hours of instruction every day from Fredrich Zech, who had studied under Hans von Bulow, one of the great German pianists and conductors of the nineteenth century. By the time he was 18, Adams was convinced that he would be a professional pianist. It was not until 1930 that he decided to abandon his music studies and devote all of his time to photography. The study of technique and the attention to detail he acquired in his musical training, however, never left him. In a 1977 interview he stated that there is a very definite relationship between music and photography, all art being "essentially the same thing," and that he really benefited from two factors, a sense of discipline and a sense of aesthetics.

It was in 1930, two years after he had married Virginia Best, that Adams met Paul Strand, a photographer whom Adams credited with opening his eyes to the artistic possibilities of the stark, crisp photographic image. Most photographers of the time were practicing hand-tinted, soft-focus photography, creating images more like paintings than photographs. Adams helped to found a group called f/64 (the name f/64 being a reference to the small lens opening of the camera) in opposition to the general photographic practices of the time. The members of this group sought, through "straight photography," to create images with sharp focus, and great depth of field, to "create photographs which actually looked like photographs."

Adams is best known for his black and white photographs of nature and the American landscape. He pioneered the Zone System, a technique designed to enable the photographer to anticipate and control the tonal range of the print. He believed that a photograph is made, not taken. And in his photographs he sought to capture the spiritual excitement he felt about the subject.

Adams once said that "everybody needs something to believe in. Conservation is my point of focus." His involvement with conservation dates back to his first photographs. He joined the Sierra Club in 1919, and he worked for four summers as caretaker of the club's headquarters in Yosemite Valley. He later sat on the board of directors of the club for nearly 40 years, strongly influencing the club's philosophy and activities and encouraging activism and a national focus, in what until then had been primarily a California-based organization. In 1936, when the Sierra Club was lobbying for the creation of a national park in the King's Canyon area of California, Adams sent copies of his book, *Sierra Nevada: The John Muir Trail*, which contains many pictures of this area, to Pres. FRANKLIN D. ROOSEVELT and to other important political figures. He also made a trip to Washington to personally advocate the designation of this national park.

Adams was also active in protesting the widening of Yosemite's Tioga road as a part of the National Park Service's Mission 66 in 1952. He even went so far as to resign from the board of directors because of the Sierra Club's unwillingness to take a stance on the issue. As a private citizen, he denounced the plan as a violation of the National Park Service Act, bordering on criminal negligence. The Sierra Club board was not willing to accept his resignation, and its members persuaded him to return, stating that his "purist voice was needed to keep the club true to its ideals." Throughout his time on the board of directors, Adams wielded a powerful voice for cooperation and compromise. This would eventually estrange him somewhat, as younger members of the Sierra Club moved toward more aggressive and antagonistic means of promoting wilderness preservation.

Adams's interest in the national parks and wilderness was an abiding one. And, while he never specifically made pictures for environmental purposes, his pictures have been invaluable to environmental work. Indeed, his images of national parks, Yosemite especially, have virtually defined the way people see and experience these places. For many, Yosemite *is* what Adams captured with his camera.

Adams left a legacy in two parts, one being the result of his conservation work, his advocacy of wilderness, and national parks. The other was his photographs that record the beauty and majesty of these places for the enjoyment and education of future generations. He died on April 22, 1984, of heart failure in San Francisco. He left his wife, his son, Michael, and daughter, Anne Helms.

BIBLIOGRAPHY

"Ansel Adams," *Current Biography Yearbook*, 1977; Callahan, Harry M., ed., *Ansel Adams in Color*, 1993; Spaulding, Jonathan, *Ansel Adams and the American Landscape*, 1995.

Adams, Henry

(February 16, 1838–March 27, 1918)
Writer, Historian

Henry Adams was a historian, social critic, and travel writer. He was an incessant traveler, and he used his experiences as a tourist and outsider not only to understand the changing human society of the 1800s but also to understand the human soul. In his writings he examined the meaning and importance of wilderness as an idea and as a geographical location. He believed that the rapid technological advances and social changes occurring at the turn of the nineteenth century could only lead to economic and social collapse.

Henry Brooks Adams was born February 16, 1838, into the famous and successful Adams family that included presidents John Adams and JOHN QUINCY ADAMS, respectively his great-grandfather and grandfather. Adams's father, Charles Francis Adams, was also a powerful figure in national and international politics, serving as a congressman, a vice presidential candidate, and minister to Great Britain at various points in his career. As a member of such a prominent family, Adams grew up believing that he too would enter into public service and would eventually take his own turn as president of the United States. This, however, was not to be. Throughout his life, and despite all of his great accomplishments, Adams carried with him a certain sense of failure at not having lived up to the example set by his forebears.

Adams grew up in Boston, Massachusetts, attending the Dixwell School and graduating from Harvard University in 1858. After graduating from college, he spent two years in Europe visiting Belgium, Holland, Italy, Sicily, France, Germany, and England. While abroad, he acted as a newspaper correspondent to the *Boston Courier*. He returned to the United States in 1860 and took a position as his father's private secretary. His father was then a congressman. Later that year, when Abraham Lincoln appointed Adams's father minister to Great Britain, Adams accompanied him in his capacity as private secretary and remained in England until 1868. During this time, he came into contact with many distinguished artists and intellectuals, including Robert Browning, Charles Dickens, and one of Britain's foremost geologists, Sir Charles Lyell.

Upon returning to Washington, D.C., in 1868, Adams began working as a freelance journalist, contributing articles to various U.S. and British periodicals. One of his main topics was the corruption of the Ulysses S. Grant presidential administration. In 1870, Adams was offered a position as an assistant professor of medieval history at Harvard. He did not originally want to accept the position, being a disbeliever in the effectiveness of formal education. His family, however, eventually convinced him to take the position. He spent seven years at Harvard, during which time he edited the *North American Review* and developed a very popular seminar style of classroom instruction. In 1872, Adams married Marian Hooper. He gave up his teaching post in 1877 to concentrate on researching and writing history, settling with his wife in

Washington, D.C., where he could make use of the government archives for his historical research. During the period between 1877 and 1885, he wrote and took many trips to Europe with his wife. He published two biographies, one of Albert Gallatin (1879) and one of John Randolph (1882), and two novels, *Democracy* (1880) and *Esther* (1884), while also working on his nine-volume work, *The History of the United States during the Administrations of Thomas Jefferson and James Madison* (1889–1891). Adams published *Esther* under the pseudonymn Frances Snow Compton, and many believe that its main character, who grapples with the disharmonies inherent in science and religion, was modeled after his wife. The character, Esther, nearly commits suicide in the book. Adams's own wife, Marian, actually did kill herself in 1885 after a period of deep depression, a loss from which Adams would never fully recover.

Democracy was published anonymously. It was a satire about the political system in the United States and included many characters that had real-life counterparts. In this book, Adams writes about "wilderness" but in a way that completely reverses the original usage of the term. On the first page of the novel, referring to New York City, he writes, "What was it all worth, this wilderness of men and women as monotonous as the brown stone houses they lived in?" This "urban wilderness" became a theme of Adams's as he reflected on the rapidly changing Western world through his writings. He saw civilization, as it was developing during his life, as being just as bewildering and intimidating as any wilderness of trees and animals. He believed that society was too organized and that this organization stripped individuals of their autonomy, eventually leading to social and economic collapse. Which, from one perspective, it eventually did, in the form of two global wars and the Great Depression of the 1930s in the United States. These ideas, which were shared by others such as Oswald Spengler and Arnold Toynbee, led to an intellectual movement that believed that the shrinking natural wilderness should be preserved as an escape for humans from the growing oppressive urban social wilderness.

For the remainder of his life, Adams was an avid traveler and a prolific writer. He published a total of 18 books. He visited Japan and Russia and began spending half of every year in France until the onset of World War I. In his most famous book, *The Education of Henry Adams* (1918), for which he received a posthumous Pulitzer Prize for autobiography, he tracks his education through his life. It was travel, not formal education, that truly provided Adams with an education, an education that was usually "accidental" and adventurous. This book also provides an in-depth look at the changes occurring in the society of the 1800s. Adams attributed these changes to the increasing complications in life, coupled with increasing attempts by humans to control the forces of nature, compounded by the rapid technological advances of the time. He found himself in a state of amazement at the emerging "modern world." In a letter quoted by Viola Winner for an *American Heritage* magazine article, he writes, "Out of a medieval, primitive, crawling infant of 1838, to find oneself a howling, steaming, exploding, Marconing, radiummating, automobiling maniac of 1904 exceeds belief."

Winner also points out that Adams was concerned with environmental issues. He wrote, "Cities are now growing uninhabitable everywhere. The noise and wear are impossible. Paris is as bad as any. New York is already abandoned by everyone who can escape." "Nature," he believed simply, "has got to turn on us."

Adams suffered a cerebral thrombosis on April 24, 1912. He never entirely recovered but continued to travel and write. His health declined gradually over the next few years, and he died at his home in Washington on March 27, 1918.

BIBLIOGRAPHY

Adams, Henry, *Democracy*, 1880; Adams, Henry, *The Education of Henry Adams*, 1907; Blackmur, R. P., *Henry Adams*, 1980; Chalfant, Edward, *Both Sides of the Ocean*, 1982; Nash, Roderick, *Wilderness and the American Mind*, 1967; Winner, Viola Hopkins, "The Virgin and the Carburetor," *American Heritage*, 1995

Adams, John Hamilton

(February 15, 1936–)
Founder of Natural Resources Defense Council (NRDC)

John Hamilton Adams cofounded the Natural Resources Defense Council (NRDC) in 1970, an organization of public interest lawyers focused on the formation and enforcement of emerging environmental laws. As its executive director from 1970 to 1998, Adams created an influential nonprofit organization of lawyers and scientists with a membership that has grown to 475,000 people nationwide. As president of NRDC since October 1998, Adams advises politicians and members of industry on the need to create environmental policy that will help protect the nation's natural resources for future generations.

Adams was born on February 15, 1936, in New York City and grew up on a farm in the Catskills of New York State. He graduated from Michigan State University in 1959 with a B.A. in history. After college he attended law school at Duke University, graduating in 1962 with his LL.B. He returned to New York and worked as an associate at Cadwalader, Wickersham, and Taft on Wall Street from 1962 to 1965. From 1965 to 1970, Adams worked as the assistant U.S. attorney for the Southern District of New York.

At the U.S. attorney's office in Manhattan, Adams met a number of other young attorneys who were attracted to environmentally oriented public interest law. They cofounded the NRDC in 1970, and Adams became its first paid staff member. NRDC's first case was to represent a coalition of environmental groups that had been working throughout the 1960s against the construction of the Storm King Mountain pump storage plant on the Hudson River. In addition to marring one of the most spectacular sites on the river—Storm Mountain was a popular subject for THOMAS COLE and other painters of the Hudson River School—the plant would have resulted in devastating fish kills. NRDC's lawyers eventually won the Storm King case, which is still significant because it provided a historical precedent for environmental lawsuits brought on by citizen plaintiffs.

NRDC has won a reputation as one of the nation's most influential environmental organizations, and today claims a membership of 475,000. NRDC seeks to safeguard the health of humans and the natural world, attempting to persuade governments, business, and other institutions to adopt more environmentally friendly policies. NRDC is credited with having major impacts on national and international policies in many areas, including air and water pollution, toxic chemicals, energy, transportation and urban development, protection of old-growth forests and other terrestrial and marine habitat, limitation of nuclear proliferation, and global warming.

In 1972, Adams joined the adjunct faculty of New York University's School of Law, where he taught Clinical Environmental Law for 26 years. Each semester, six to ten law students participate in the environmental clinic. They attend lectures and discussions on a range of environmental subjects and work on projects at NRDC throughout the semester.

In 1973, Adams joined the board of the Open Space Institute, a conservancy devoted to the protection of open space lands in the Northeast. Since 1973, Adams has served as the chairman of the board. The Institute has purchased thousands of acres of land in the Hudson Valley, the Adirondacks, and the Catskills. One of its most successful projects was ensuring the protection of Sterling Forest, an area of 20,000 acres between New York and New Jersey. He is also on the boards of the Catskill Center for Conservation, the League of Conservation Voters, and the Woods Hole Research Center. He served on the President's Council on Sustainable Development and participated in the Environmental Protection Agency's Common Sense Initiative.

Over the years, Adams has received several notable honors and environmental awards, including the Judge Lumbard Cup for public service from the U.S. attorney's Southern District of New York, the National Conservation Achievement Award from the National Wildlife Federation (1999), and the Francis K. Hutchinson Conservation Award from the Garden Club of America (1990). In 1998, Adams was named one of the National Audubon Society's 100 Champions of Conservation.

Adams lives in New York City with his wife. They and their three grown children enjoy time spent at their home in the Catskills on the Beaverkill River.

BIBLIOGRAPHY

Adams, John, "Are Higher Fuel-Efficiency Standards a Good Idea?" *Insight*, 1996; Adams, John, "The Future of Environmental Management," *Environmental Management*, 2000; Adams, John, "Leadership on the Right Track," *Conservation Voice*, 1997; Adams, John, "On Environmental Cost-Benefit Scale, Prevention Scores High," *New York Times*, 1993.

Adams, John Quincy

(July 11, 1767–February 23, 1848)
President of the United States

John Quincy Adams was far ahead of his time in recognizing the need for conservation. While most of America's attention in the early nineteenth century was focused on expansion, he understood the problem of overusing the land. He was the first president to show any interest in setting land aside to be protected from the unchecked degradation that came with building a civilization. He was interested in natural history and did much to promote scientific study and discovery, at a time when these ideas were politically unpopular.

Born on July 11, 1767, in Braintree (now Quincy) Massachusetts, John Quincy Adams grew up as a child of the American Revolution, which began when he was seven years old. His father, John Adams, would later become the second president, and his mother, Abigail Smith Adams, was an accomplished and outspoken woman. He was educated in the village school and was urged along in his studies through letters from his absent father, who was serving in the Continental Congress in Philadelphia. He occasionally accompanied his father on diplomatic missions to Europe, and there was no question that his parents were grooming him for the presidency later in his life. After graduating from Harvard College in 1787, he became a lawyer and established a practice in Boston. On July 26, 1797, he married Louisa Catherine Johnson, with whom he would have a long and passionate marriage and four children.

His career as a politician began in 1802, when he was elected as a state senator from Suffolk County, Massachusetts. He then served in the United States Senate from Massachusetts, from 1803 to 1808. For the next few years he taught at Harvard, practiced law, and served as foreign minister to Russia. From 1817 to 1825, he served as President Monroe's secretary of state. And in 1825 he became president himself.

During his only term as president, he showed himself to be remarkably conscious of the need to protect the nation's forests. He was up against the frenzy to expand the nation, and very few people considered the cost to the environment. There seemed to exist an endless supply of timber, and there was an insatiable demand for it—which created an opportunity for great profit. Thus there was heavy pressure on politicians to provide policies favorable to the continued plundering of the forests.

Instead, Adams acted in the opposite direction. Understanding the need to preserve for the future, he set aside live oak lands along the Gulf Coast to be used as naval timber reserves. Santa Rosa Island off Pensacola was one of the reserves he established. Unfortunately his term in office was only four years, and his successor, Andrew Jackson, was opposed to all of Adams's conservation efforts. During Adams's terms in the House of Representatives following his presidency, an act decreed that it was unlawful to cut timber on public lands, but President Jackson saw to it that it was never implemented.

Adams also had a high regard for science and learning. He believed that the quest for knowledge was the greatest thing a person could aspire to, saying that "it prolongs life itself and enlarges the sphere of existence." He was a great supporter in Congress of scientific expeditions and led the movement for establishing the Smithsonian Institution in Washington, D.C., one of the nation's foremost centers of learning. In an age when most of his country cared more about productivity and progress, he had the foresight to advocate greater scientific knowledge of the land. In the early nineteenth century, his cry for knowledge was a precursor to today's concern for the environment. Indeed, as science progressed in that century, it produced sobering data about the consequences of human exploitation of the environment.

After his presidential term, Adams returned to politics in 1831, serving in the U.S. House of Representatives for the rest of his life. He died on February 23, 1848, in Washington, D.C., and was buried at First Parish Church in Quincy, Massachusetts.

Bibliography

Bartlett, Richard A., *The New Country*, 1974; Goetzmann, William H., *New Lands, New Men: America and the Second Great Age of Discovery*, 1986; Lipsky, George A., *John Quincy Adams: His Theory and Ideas*, 1950; Nagel, Paul C., *John Quincy Adams: A Public Life, A Private Life*, 1997; Shabecoff, Philip, *A Fierce Green Fire: The American Environmental Movement*, 1993.

Addams, Jane

(September 6, 1860–May 21, 1935)
Social Reformer, Peace Activist

A relentless reformer and a humanitarian activist, Jane Addams confronted the urban and industrial environmental realities of the Progressive era in the late nineteenth and early twentieth centuries. She founded Hull House in Chicago, an advocacy and education center run largely by women that provided cultural, social, and medical services to those who needed them. She centered her efforts on the quality of life of industrial workers and sought to abolish the urban miseries she saw daily, such as polluted and unsanitary neighborhoods, overcrowded tenements, inadequate schools, and unsafe working conditions. Hull House became a meeting ground for reformers and activists of all kinds, and Addams's work paved the way for better conditions in urban work environments.

Laura Jane Addams was born on September 6, 1860, in Cedarville, Illinois, to John Huy Addams and Sarah (Weber) Addams. She was the eighth of nine children, though only she and three others lived past childhood. When she was two years old her mother died, and as she grew up she became very close to her father. She was a bright and ambitious child but was plagued by constant back pain, caused by tuberculosis of the spine.

Jane Addams (Library of Congress)

Also, she sometimes felt she was an ugly duckling and was ashamed of the overwhelming adoration she felt for her father. He was a successful banker, miller, and later an admired state senator. He was strongly antislavery and was a friend and correspondent of Abraham Lincoln. His abolitionist views and his benevolent outlook were strong forces in Jane's life and influenced her own perception of the world. Though she admired and loved her father, she also felt occasionally restricted by him. After graduating from high school, she hoped to attend Smith College in Massachusetts, one of the few schools that offered women the same caliber of education as men. But her father objected, saying Smith was too far from home and that she did not need such a rigorous education. Though frustrated, she relented and ended up going to Rockford Female Seminary in Illinois, a religious school that did not even offer degrees. With her determination and vision, though, she quickly became an outstanding student and leader and worked hard in the hopes that her school would eventually decide to confer college degrees. In 1881 she graduated at the head of her class, and a year later Rockford Seminary became a degree-granting institution and she was granted a B.A.

In the same year that she completed her studies, her father died. She was devastated by this and had difficulty getting a career in order. In the fall of 1881 she enrolled in the Women's Medical College in Pennsylvania but dropped out a few months later, realizing she did not want to be a doctor. Her own health was deteriorating, and she soon had to have surgery on her spine, which confined her to bed for six months. Increasingly, she felt aimless and hindered by the fact that there were few options at this time for aspiring women. In the end, her circumstances combined forces against her, and she fell into a depression that lasted for years.

During this period, she occasionally traveled with her stepmother overseas. On one visit to London, Addams was shocked by the poverty and filth in the slums there. It had a profound impact on her and provoked her to action, and she finally began to feel as if she had a purpose. She and her close friend Ellen Gates Starr began making plans to buy a house in a poor neighborhood in Chicago where they and other interested women could attempt to reverse some of the abuses of industrialization. In 1889, they moved into Hull House in the nineteenth ward, a poor district in west Chicago inhabited by thousands of immigrants. Their goal was to establish a "settlement," or a center for ideas and reform.

Cities in the nineteenth and early twentieth centuries were experiencing industrial expansion, economic growth, and urban restructuring—all of which contributed to new forms of environmental degradation. Some of the most widespread problems included solid and hazardous waste disposal, inadequate sewerage and sanitation systems, loss of water quality, and other public health issues. For Addams, the boom in industry offered an opportunity to expand the nation's sense of civic responsibility. When the Hull House settlement was established, it became a major force for social reform in industrial cities. Addams worked aggressively to provide programs that addressed the immediate needs of the community. Hull House included a library, a nursery, an art studio, a cooperative boardinghouse for working women, a community kitchen, and a music school. With the help of her growing numbers of settlement activists, including Mary McDowell and physician Alice Hamilton, Addams also fought for other social and environmental reforms, such as abolishing sweatshops, child labor, unsafe factories, and overcrowded tenements; and for the establishment of effective urban infrastructure that would provide such services as adequate garbage collection. The garbage problem was a major symptom of the environmental downfall of the times and created a serious health threat in the urban neighborhoods—and Addams undertook a large-scale investigation of the city's garbage collection system. Conditions improved little as a result of her complaints, so in sheer desperation she submitted her own bid to collect the garbage in the nineteenth ward. Her application was thrown out on a technicality, but nevertheless, she started rising early every morning to follow the garbage trucks on their rounds and in general made enough of a ruckus that the city decided to restructure the garbage collection system.

In an era when very few women had significant public status, Addams became an international celebrity through her work. She wrote several books that became best sellers and traveled around the world attending conferences of the Women's International League for Peace and Freedom. Her pacifist stance brought her much criticism during World War I, but she never wavered. In 1931 she became the first woman in the United States to receive the Nobel Peace Prize. She died of cancer on May 21, 1935, in Chicago and is buried in Cedarville, Illinois.

Bibliography

Addams, Jane, *Twenty Years at Hull House*, 1910; Farrell, John C., *Beloved Lady: A History of Jane Addams' Ideas on Reform and Peace*, 1967; Gottlieb, Robert, *Forcing the Spring: The Transformation of the American Environmental Movement*, 1993.

Albright, Horace

(January 6, 1890–March 28, 1987)
National Park Service Director

Responsible for drafting the National Park Service Act, lobbying Congress to pass it, serving as assistant director of the National Park Service under its first director, Stephen Mather, and eventually becoming director himself, Horace Albright played the roles of midwife and nanny to the U.S. National Parks movement.

Horace Marden Albright was born on January 6, 1890, in Bishop, California, just north of Sequoia National Park, east of Kings Canyon, and south of Yosemite. He graduated from the University of California at Berkeley in 1912 and received his law degree from Georgetown University in 1914. At the age of 24, when he thought he was finishing up a temporary

Horace Albright (Library of Congress)

job assisting Interior Secretary Franklin Lane and preparing to return to California to marry and begin practicing law, Albright met the man who was to influence the course of the rest of his life. Secretary Lane was courting wealthy industrialist Stephen Mather for a near-volunteer job directing the national parks, and Lane asked Albright to tell Mather about the job's duties. By the end of the evening, Albright had convinced Mather to take the job, and Mather, in turn, had persuaded Albright to become his assistant.

Several national parks already existed in 1915 when the Mather-Albright team began its work: Crater Lake in Oregon; Glacier in Montana; Rocky Mountain in Colorado; Yellowstone straddling Wyoming, Montana, and Idaho; Yosemite in California—13 in total. But they lacked infrastructure for visitors, their administration was underfunded and uncoordinated, and most alarming, they were threatened by logging or mining at their borders and by unscrupulous profiteers catering to the few visitors that made their way there.

Mather and Albright made a list of priorities when they began work. Foremost was to draft and pass a National Park Service Act that would give them the authority they needed to reorganize park management and implement protection laws. They drafted the National Park Service Act, and thanks to publicity assistant ROBERT STERLING YARD's prolific and effusive articles about the national parks and to Mather's personal charisma and generosity in inviting the most influential congressmen and their wives on spectacular park tours, a majority of Congress voted for it. Albright's Capitol Hill connections got the bill to Pres. Woodrow Wilson ahead of schedule, and the bill became law August 25, 1916.

From 1916 to 1919, Assistant Director Albright continued to work at the National Park Service (NPS) in Washington, D.C., and to accompany Mather and groups of elite politicians or private donors on tours designed to convert them into committed conservationists. In 1919 he felt ready to move on to the private sector, where he could finally practice law and probably earn more money than he did at NPS, even with Mather's supplement. But once again Mather stepped in and offered Albright the superintendency of Yellowstone National Park combined with a position as field assistant to the director. Albright accepted it and moved his family—his wife, Grace, and young children Robert and Marian—to Yellowstone. They remained there for ten years.

Mather died in 1929, and Albright was invited to succeed him as National Park Service director. Albright could not boast the wealth, charisma, or sheer physical energy of Mather, but his time in Washington, D.C., had garnered him the personal acquaintance of 200 congressmen. During the four years that he served as director, he established three new national parks (Grand Teton, Carlsbad Cavern, Great Smoky Mountains) and several national monuments, including Death Valley and Canyon de Chelly. He was able to enlarge nine national parks, some of them significantly, in a way that would make their borders follow natural frontiers such as rivers or watersheds.

When FRANKLIN D. ROOSEVELT was elected president in 1932, he installed HAROLD ICKES as secretary of the interior. Ickes and Albright collaborated to find work in the national parks for the thousands of young men who formed Roosevelt's Civilian Conservation Corps. They built roads, trails, campsites, and bathrooms, strung phone lines, and fought forest fires.

By 1933, Albright had decided to leave the public sector, and he resigned his post with NPS to become president of the U.S. Potash mining company. For the rest of his life, Albright served as adviser to NPS directors. With philanthropist and conservationist JOHN D. ROCKEFELLER, JR., Albright lobbied to make Jackson Hole a national park and colonial Williamsburg a national monument. Albright was awarded the Sierra Club's John Muir Award, the Interior Department's Conservation Service Award in 1953, and the Presidential Medal of Freedom in 1980.

Albright died at the age of 97 in a nursing home in Los Angeles, California, on March 28, 1987.

BIBLIOGRAPHY

Albright, Horace, and Robert Cahn, *The Birth of the National Park Service*, 1985; McPherson, Stephen Mather, "Horace Albright," *National Parks*, 1987; Shankland, Robert, *Steve Mather of the National Parks*, 1970; Sidey, Hugh, "Present at the Preservation," *Time*, 1985.

Alston, Dana

(December 18, 1951–August 7, 1999)
Environmental Justice Funder and Activist

As one of the founders of the environmental justice movement, Dana Alston coconvened the First National People of Color Environmental Leadership Summit in 1991, which brought 600 environmental justice activists of color together for the first time and at which the influential Principles of Environmental Justice were formulated. Alston's particular niche in the movement was to help grassroots environmental justice groups attain the funding they needed, and she spent the last years of her life working to assure their continued financial support.

Dana Ann Alston was born in New York on December 18, 1951, to Garlan and Betty Alston. She attended Wheelock

College in Boston, graduating with a B.S. in 1973. At Wheelock, Alston served as president of the Black Student Organization and led fellow Black students to demand more African American courses and faculty members. She completed a master's degree from the School of Public Health at Columbia University in 1979.

During the 1970s and 1980s Alston worked with organizations devoted to improving the standard and quality of life for poor, rural people of color. She was a founding board member of the Southern Rural Women's Network and worked with farm workers on pesticide issues for Rural America. She aided the solidarity campaign for the South African anti-apartheid movement and helped organize the U.S. visit of South African president and former political prisoner Nelson Mandela.

Committed to increasing funding for people of color and grassroots organizing for social justice, Alston became president of the National Black United Fund, where she oversaw completion of a lawsuit that resulted in workers, for the first time, being able to contribute to Black-led and -organized charitable funds. Alston also worked on development and grant making with the National Committee for Responsive Philanthropy and worked for TransAfrica Forum.

In 1990, Alston developed the Environment, Community Development, and Race project for the Panos Institute, an independent policy studies organization that works to raise public understanding of sustainable development. One of her first tasks was to edit *We Speak for Ourselves: Social Justice, Race And Environment*, a collection of articles and interviews about the environmental justice movement. In her introduction to the book, Alston contrasted the environmental justice movement with the better-known, conservation-oriented mainstream environmental movement: "Communities of color have often taken a more holistic approach than the mainstream environmental movement, integrating 'environmental' concerns into a broader agenda that emphasizes social, racial and economic justice."

Alston and other environmental justice leaders, including Patrick Bryant, ROBERT BULLARD, BENJAMIN CHAVIS, Donna Chavis, Charles Lee, and Richard Moore, convened the First National People of Color Environmental Leadership Summit, a four-day conference held in Washington, D.C., in October 1991. This event, which attracted 600 representatives of environmental justice organizations led by people of color to address the environmental problems faced by people of color, allowed participants to share experiences and network among themselves. The attendees adopted seventeen Principles of Environmental Justice, which have endured as a reminder of the ultimate goals of the environmental justice movement.

Representatives of mainstream environmental organizations had been invited to attend the second half of the Summit, and Alston was chosen to deliver a welcome and explain the environmental justice movement to them. Her speech continues to be widely quoted because of its directness and its eloquence.

> Our vision of the environment is woven into an overall framework of social, racial and economic justice. It is deeply rooted in our cultures and our spirituality. It is based in a long tradition and understanding and respect for the natural world. . . . The environment affords us the platform to address the critical issues of

> our time: questions of militarism and defense policy; religious freedom; cultural survival; energy and sustainable development; the future of our cities; transportation; housing; land and sovereignty rights; self-determination; and employment.

Alston led a delegation of environmental justice activists to the Earth Summit in Río de Janeiro in 1992 and made an international call against racism and its effects on the public health in communities of color. That year, she was awarded the Charles Bannerman Memorial Fellowship, a sabbatical given to social justice activists and organizers in recognition of their outstanding activist work. She spent her sabbatical visiting South Africa and resting and reflecting on her work.

In late 1992, Alston joined the staff of the Public Welfare Foundation (PWF), whose mission is to help organizations that seek to remove the barriers that disadvantaged people face to full participation in society. She was senior program officer for PWF's Environmental Initiative, a premier environmental justice grant-making program that prioritizes projects focused on the impact of environmental degradation and pollution on public health, particularly in poor communities lacking resources from other funding sources. PWF now awards more than $3 million per year to projects fitting these criteria. Alston also encouraged other foundations to take a greater interest in and support the environmental justice movement. She contributed to an influential 1994 open letter to funders that defined the movement, explained its origins and significance, and described various activists and groups and the concrete work they were engaged in.

Alston died on August 7, 1999, in San Francisco, California, while being treated for kidney disease and the consequences of a stroke she had suffered two years earlier. She left a young son, Khalil. Alston's friends and family honored her memory by establishing the Dana A. Alston Fund for the Bannerman Memorial Fellowship. Recognizing that working for social change usually means long hours at low pay, with few tangible rewards and few escapes from the day-to-day pressures, the fund provides outstanding activists of color with sabbaticals for reflection and renewal.

Bibliography

Alston, Dana, "Black, Brown, Poor and Poisoned: Minority Grassroots Environmentalism and the Quest for Eco-Justice," *Kansas Journal of Law & Public Policy*, 1991; Alston, Dana, *Taking Back Our Lives: Environment, Community Development and Race in the U.S.*, 1990; Alston, Dana, and Nicole Brown, "Global Threats to People of Color," *Confronting Environmental Racism: Voices from the Grassroots*, Robert Bullard, ed., 1993; Douglas, Adisa, "A Tribute . . . Dana Ann Alston: Sister-Friend," *Sister Ink, National Black Women's Health Project*, 1999; Douglas, Adisa, "Dana A. Alston: Activist and Funder," *Networker News, National Network of Grantmakers*, 1999; "Environmental Justice Resource Center: Unsung Heroes And Sheroes on the Front Line for Environmental Justice," http://www.ejrc.cau.edu/(s)heros.html; "Public Welfare Foundation," http://www.publicwelfare.org/.

Amidon, Elias, and Elizabeth Roberts

(September 12, 1944– ; March 29, 1944–)
Editors, Educators

Elias Amidon and Elizabeth Roberts are environmental activists guided by a global environmental ethic. In their international efforts in support of the environment, women's health, and indigenous peoples, they helped found four schools for spiritually oriented environmental activism: The Institute for Deep Ecology in Occidental, California; the Spirit in Education Movement in Bangkok, Thailand; the Boulder Institute for Nature and the Human Spirit in Boulder, Colorado; and the Alliance for Sustainable Forests and Communities. Amidon and Roberts have also edited three anthologies of prayers that revere planet earth, *Prayers for a Thousand Years*, *Life Prayers Celebrating the Human Journey*, and *Earth Prayers*, which collectively have reached more than a million readers.

Elias Velonis Amidon was born in Denver, Colorado, on September 12, 1944. His summers in upstate New York's Columbia County were formative. There, in the early 1960s, he witnessed what he later described as "the suburbanization of a beloved landscape." After graduating from Antioch College in 1967 with a degree in literature, Amidon homesteaded on an island in British Columbia, living off the land as self-sufficiently as possible. But his island was not far enough off the grid for him, and he began a ten-year spiritual journey, studying with Sufis in Europe and India. This apprenticeship deepened both his skills in experiential education and his understanding of the spiritual roots of the environmental crisis. Convinced that one of the most crucial "front lines" of human-earth relations lies in the sustainable and human-scale design of the built environment, in 1978 Amidon founded Heartwood School in Massachusetts. The school was dedicated to teaching energy-efficient home design and construction. At Heartwood, Elias met Elizabeth Roberts.

Elizabeth Roberts was born in St. Louis on March 29, 1944. She began as an academically gifted student who went on to excel at Marquette University, earning a B.A. in philosophy and theology in 1966 and a Master's degree in philosophy in 1969. During the early 1960s she worked with Dr. Martin Luther King Jr. on civil rights marches and voter registration drives in Alabama. At the age of 25 she left Marquette to coordinate the White House Conference on Children and Youth, then went on to National Public Radio to develop the program *All Things Considered*, which, incidentally, she named. For the next five years—starting in 1974—she worked as a special assistant to John D. Rockefeller III, designing and implementing international programs for women in development, as well as leading national studies in the United States on population and sexuality. Roberts also served as a consultant and board member for a broad range of institutions, including the National Institute for Mental Health, the Public Broadcasting Service, and the U.S. Forest Service.

In 1984, when she turned 40, Roberts took a step back from her professional successes and began a two-year retreat. She spent time in nature, with her family and alone, and emerged dedicated to deep ecology (a movement that maintains that extended periods with nature are fundamental to experiencing balance and clarity, vision and morality), engaged Buddhism, environmental action, and prayer.

Roberts and Amidon began their first professional collaboration in 1986. It was a five-year consultancy with the Environmental Research Laboratory at the University of Arizona, researching ecologically sustainable and human-scale design patterns for desert cities. This work took them several times to Saudi Arabia, where they consulted on long-term ecological planning for the capital city of Riyadh. They were awarded a major National Endowment for the Arts grant to write a book on the design of desert cities, *The Soul of the Oasis*. For spiritual nourishment during this project (and to avoid sitting in front of their respective computers all day every day), Amidon and Roberts started taking groups into the desert on wilderness rites of passage, in which people from ages 18 to 80 participate in three- or four-day periods of solitude and fasting to discover or renew their life's purpose and direction.

In 1992 Roberts and Amidon were asked by Buddhist activist and writer Joanna Macy to help in the launching of a national school for environmental activists and educators, the Institute for Deep Ecology. The Institute would bring together many of the nation's leading deep ecology activists and educators who would begin to articulate the principles of deep ecology in environmental curricula throughout the world.

Beginning in 1994 they began administering the Boulder Institute for Nature and the Human Spirit, a nonprofit organization in Boulder, Colorado, dedicated to strengthening interfaith and intercultural understanding, cooperation, and action. Among the Boulder Institute's several projects is the Spirit in Action program in Southeast Asia. One aspect of the Spirit in Action program is the Interfaith Solidarity Walks that Roberts and Amidon lead in support of the indigenous peoples of northern Thailand who are threatened with relocation, racism, and cultural assimilation. The Solidarity Walks—part pilgrimage, part seminar, and part direct political action—bear witness to the wisdom of native cultures and the struggles they face for survival. Other Spirit in Action projects include training for indigenous leaders in bioregional mapping of ancestral lands, environmental education programs for activist Thai Buddhist monks and community leaders in northern Burma, and a microgrant program that supports projects in community self-reliance and sustainability. Exemplary of the spiritual activism that Amidon and Roberts espouse, Spirit in Action has helped facilitate the ordaining of trees in forests threatened by logging companies. Local beliefs hold that it is a grave sin for anyone to kill an ordained priest. So, a local shaman or Buddhist or Christian priest is called upon to ordain the oldest trees as priests. The trees are wrapped with a bright sash to make this designation visible to the peasants hired to fell a forest. Where this has happened, the forests remained standing.

Both Amidon and Roberts have long had an affinity with eastern mysticism and were attracted to the only Buddhist university in the Americas, Naropa University in Boulder, Colorado, where they became adjunct professors in the graduate program in environmental leadership. While at Naropa, they compiled their unique and poignant articulations of the earth ethic, *Earth Prayers*, *Life Prayers: Celebrating the Human Journey*, and *Prayers for a Thousand Years*. Used by Christians and Jews, Muslims and Buddhists, mainstream professionals and front-line activists, at weddings and workshops, festivals and funerals, demonstrations and dedications, these collections have become standard sourcebooks for prayers and expressions of interfaith spiritual values. In the introduction to *Earth Prayers*, they write: "Over and over the prayers in this book remind us of a universal marriage of matter and spirit. They call on us to rethink the dualism of our culture that separates the sacred and the secular, the natural and the supernatural, body and mind. They make it clear that we humans are not here simply as transients waiting for a ticket to somewhere else. The earth itself is Christos, is Buddha, is Allah, is Gaia."

Amidon and Roberts have continually renewed themselves and their commitment to nature through retreats to the deserts west of the Rocky Mountains and to the jungles of northern Thailand and Burma. In 1999 they left home and teaching positions in Colorado and began a open-ended pilgrimage as "itinerant activist-teachers" in an effort "to reduce the time spent administering programs and increase the time we spend in the field teaching and putting spirit into action." They have lectured internationally at numerous conferences and universities, as well as on television and radio programs.

Bibliography

"The Boulder Institute for Nature and the Human Spirit," http://www.earthprayers.com; "The Institute for Deep Ecology," http://www.deep-ecology.org; Roberts, Elizabeth J., and Elias Amidon, *Earth Prayers*, 1991; Roberts, Elizabeth J., and Elias Amidon, *Life Prayers: Celebrating the Human Journey*, 1996; Roberts, Elizabeth J., and Elias Amidon, *Prayers for a Thousand Years*, 1999.

Anderson, Adrienne

(February 10, 1952–)
Grassroots Public Health Organizer and Educator

Grassroots community organizer Adrienne Anderson has been a leader in toxic contamination struggles throughout the western United States since she became involved with the issue in 1983. Since 1992 she has also worked as an instructor at the University of Colorado–Boulder (CU), sharing her knowledge and research methods with students as they investigate contaminated site histories and how regulatory agencies respond to the problems.

Adrienne Anderson was born February 10, 1952, in Dallas, Texas. From a young age she was concerned about justice; as a child she protested segregationist politics by defending black children's rights to ride their bikes through her white neighborhood. Then as a student at Southern Methodist University (SMU), she challenged her sorority's race-based admissions policy. Anderson earned a B.A. from SMU in 1974 and worked for one year for the Texas Department of Public Welfare in Dallas. Although welfare casework was frustrating because Anderson realized that the system, for all its good intentions, tended to control and harass its clients, Anderson gained a respect for poor people's ability to organize and help one another.

Anderson began work in a doctoral program in sociology at the University of Oregon in 1975, where she concentrated on social policy, political economy, and field research methods. While studying there, she also worked for Lane County, Oregon, to design and implement an outreach program for the county's rural poor. Anderson took a leave of absence from her studies in 1979 to pursue her career in social change organizing full-time in Denver, Colorado. From 1977 to 1983 she served as executive director of the Mountain Plains Congress of Senior Organizations, which advocates for the rights of low-income elderly people living in the Rocky Mountain region.

By 1983 Anderson had focused her attention on energy and the environment and how these issues affected poor people. She became regional director of the national Citizen/Labor Energy Coalition and then organized and became director of the Colorado Citizen Action Network (CCAN). CCAN, which was affiliated with the national coalition Citizen Action (CA), was a statewide coalition of grassroots antitoxics groups and labor, farm, and senior citizen groups devoted to energy, environment, and health care issues. When CA increased its commitment to hazardous waste and public health issues in 1984, Anderson sent her staff on a canvass to garner support for stronger federal legislation to clean up the nation's Superfund sites—the most contaminated areas of the country that the government has committed to clean up. Canvassers returned with disturbing stories about birth defects and children's health problems in the Friendly Hills neighborhood of Denver. Anderson assisted the area in forming a local citizens' group affiliated with CCAN, the Friendly Hills Health Action Group, to investigate the neighborhood's problems.

The investigation revealed that the municipal water board, the county and state health departments, and the Environmental Protection Agency all were aware that military contractor Martin Marietta was violating the Clean Water Act. It was discharging scores of toxic chemicals, including the highly carcinogenic rocket fuel propellant hydrazine, into a waterway that ran through the site and into a neighboring Denver water supply treatment plant. The plant treated the water for bacteria but not for the toxic chemicals it contained and then piped it to Friendly Hills and other neighboring areas.

Anderson and Friendly Hills Health Action Group forced the water board to close the contaminated water treatment plant in 1985 and convinced the governor to order a criminal investigation of the cover-up conspiracy between Martin Marietta and the Denver Water Board.

In the midst of this campaign, Anderson became western director for the National Toxics Campaign (NTC). In addition to her work on Denver's interlocking toxic sites, such as Martin Marietta, the Rocky Mountain Arsenal, Lowry Landfill, the Rocky Flats nuclear weapons plants, and the Coors brewery,

Anderson also organized and assisted community toxics campaigns in other western states, including Wyoming, Oklahoma, and Alaska. She participated in a three-year campaign in Ponca City, Oklahoma, that eventually forced an evacuation of the entire south half of the town, where benzene poisoning from a Conoco refinery's contaminated groundwater swamped 400 neighboring homes. As part of the successful campaign for evacuation, Anderson helped form a local citizens' group, the Ponca City Toxic Concerned Citizens, and accompanied the group to Oklahoma City, 100 miles away, where they camped out on the grounds of the state capitol for months to protest the governor's failure to respond to the problem. Finally taking Conoco to court while continuing their public protests, the residents in 1990 won the largest private buyout of a contaminated community in the nation.

Over the years, working for CA, the NTC, and as a volunteer, Anderson has helped some 20 to 30 communities organize citizens' groups to respond to local environmental disasters. She currently concentrates her work in the Denver area, a city ringed by Superfund sites whose owners have traded toxic wastes between one site and another for several decades. A complicated, interlocking set of relationships among corporations, the corporate law firms that defend them, their public relations firms, and municipal and federal government agencies has made it difficult to address the problems. When Anderson, appointed by the mayor of Denver to the Metro Wastewater Reclamation District Board to represent the sewage plant's workers, uncovered documents about a plan to treat plutonium-contaminated groundwater from the Lowry Landfill Superfund Site east of Denver to the sewage plant, she alerted sewage workers, who protested the plan. Anderson's and her students' investigations of federal files revealed that the Lowry Landfill was contaminated with plutonium and had been used as a dump for years by the Rocky Flats Nuclear Weapons Plant. This resulted in threatening letters from the sewage plant's leadership. Anderson fought back, invoking the whistle-blowing protection statutes of several national environmental laws. For her work on this case, Anderson was given the Brown-Silkwood Health and Safety award for 1998 from the Oil, Chemical, and Atomic Workers Local 2-477.

Since 1992, Anderson has taught at the University of Colorado–Boulder, offering such courses as Environmental Ethics: Race, Class and Pollution Politics; Advanced Environmental Investigations; and The War Environment. She emphasizes research techniques and assigns her students to investigate local polluters and how the appropriate regulatory agencies have responded to them. Anderson invites regulatory officials to visit the class at the end of each semester; they are met by well-prepared students demanding explanations for the problems they have uncovered. Although Anderson's courses receive consistently high scores in student evaluations, and the University's large Environmental Studies program depends upon her to teach required courses for the program, her presence at CU has been officially protested by one of the targets of her investigations, the ASARCO mining company. (An Anderson investigation revealed that ASARCO had contaminated a Denver neighborhood, and the company was fined $38 million.) The public relations firm that has represented ASARCO, Shell, Martin Marietta, and other area polluters has also complained to the university about Anderson. Despite various threats by the CU administration to cancel Anderson's contract, students have successfully rallied to assure that she continues to teach.

Anderson lives in Denver with daughters Erin and Sarah.

BIBLIOGRAPHY

Cowan, Jessica, *Good Works: Jobs that Make a Difference*, 1991; Obmascik, Mark, "Listen! Money's Talking," *Denver Post*, 1996; White, Nadia, "Environmental Studies Made Demands at Campus Rally," *Boulder Daily Camera*, 1998.

Anderson, Ray

(July 28, 1934–)
Entrepreneur

Ray Anderson, chairman of the board and chief executive officer of Interface, Inc., the largest commercial carpet tile and interior fabrics manufacturer in the world, challenged his employees in 1994 to transform their company into a model of environmental responsibility and sustainability. Interface, Inc., is now a widely recognized model for what an industry can accomplish if its priorities are to work toward sustainability.

Raymond Christy Anderson was born on July 28, 1934, in the small town of West Point, Georgia, the youngest of three boys. His father had been forced to quit school in the eighth grade to help support his six siblings and had resented this during the remainder of his life, so he and Anderson's mother, a teacher, provided each of their sons with all the educational opportunities they could. Anderson believes that playing football through secondary school and college gave him the "never say die" attitude that would eventually serve him as he worked in the competitive world of business. Anderson graduated from Georgia Tech in 1956 with high honors and a degree in industrial engineering.

After a couple of years in unsatisfactory jobs immediately after graduating, Anderson joined Callaway Mills Co., a textile manufacturer, where he stayed for 17 years. That company was sold in 1968 to Deering Milliken, and Anderson assumed the directorship for the development of its floor-covering branch. He was responsible for importing carpet tile technology from Europe, and Milliken became the leading carpet tile company in the United States. As a floor-covering expert, Anderson became enamored with these European carpet tiles, 18-inch by 18-inch squares of free-lay carpet, which were practical for office floor coverings because they allowed easy access to subfloor electrical wiring. Anderson recognized the opportunity to found a new company specializing in carpet tiles, and he left Milliken in 1973. He recruited 20 investors to raise $1,250,000 in equity for his new company and joined an English supplier of carpet tiles, Carpets International, to found Carpets International–Georgia in April 1973.

Renamed Interface Flooring Systems, Inc., in 1983, Anderson's company acquired several competitors and related businesses during the mid-1980s and by 1988 had become the world's largest global producer of carpet tiles. Listed in 1988 as a Fortune 500 company, Interface enjoyed spectacular success, from a business point of view. Anderson also believed that the company was on the forefront of the environmental movement; in 1984 it had responded to the problem of indoor air quality–related illness and had developed an antimicrobial agent called Intersept to incorporate into its carpets. This reduced the so-called sick building syndrome that affected many office buildings.

By 1994, however, thanks to a growing popular interest in the environment, clients were frequently asking Interface about its contributions to the environment. Interface Research Corporation—the branch of the company devoted to research and development—decided to organize a task force to formulate the company's official environmental position. Anderson was asked to give the keynote speech, and he was at a total loss for words. Serendipitously, someone sent him a copy of PAUL HAWKEN's *The Ecology of Commerce* while he was trying to formulate his speech. In this book, Hawken describes a long series of ecological disasters over recent history, all caused by human overuse of nature's fragile and limited resources. Hawken accuses business and industry together of being the perpetrators of most of the destruction, but he believes that they are also the forces that can most effectively turn it around. Hawken's ideas resonated with Anderson, whose company was completely dependent upon nonrenewable petroleum-based products. Anderson was thus inspired to issue a dramatic challenge to Interface employees to make their company one that would exemplify how industry could help solve the world's environmental crisis.

Interface employees were surprised at Anderson's challenge, but they did accept it and chose 2000 as the year by which Interface would achieve sustainability. (Anderson has since predicted that it would take many more years.) The effort's first focus was on eliminating waste at its 29 factories. At each factory, employee suggestions for eliminating waste were solicited. Within five years, about half of its factories' waste was eliminated, and through this, $114 million was saved.

Anderson assembled what he calls his Dream Team, 11 sustainability experts including HUNTER AND AMORY LOVINS, Paul Hawken, DAVID BROWER, BILL MCDONOUGH, and others, who consult with Interface and have helped map its path toward sustainability. With their help, Interface has formulated six goals in addition to its original waste-reduction effort. They are as follows: to work toward having its factories emit only benign air and water emissions; to consume less energy and shift to alternative energy sources; to use organic, renewable, and closed-loop recycled raw materials to make the carpet tiles; to transport employees, information, and products more efficiently; to educate its suppliers, customers, competitors, and communities about the importance of seeking sustainability; and to "redesign commerce," which includes providing flooring services rather than just selling carpet tiles, allowing Interface to selectively replace tiles as they wear out, and recycle them when they do.

In addition to leading Interface on a path to sustainability, Anderson has convinced his alma mater, Georgia Tech, to commit itself in its mission statement to working toward a sustainable society. He was appointed to the President's Council on Sustainable Development in 1996 and served as cochair with Jonathon Lash.

Anderson resides in Atlanta, Georgia, with his wife, Pat. Anderson has two grown daughters, Mary Anne and Harriet, and four grandchildren.

Bibliography

Anderson, Ray, *Mid-Course Correction*, 1998; "Interface," www.ifsia.com; Interface Research Corporation, *Interface Sustainability Report*, 1998.

Andrus, Cecil

(August 25, 1931–)
United States Secretary of the Interior, Governor of Idaho

As secretary of the interior from 1977 to 1981, Cecil Andrus worked for the protection of the environment by attempting to eliminate the control of economic interests over the public domain and by concentrating decision-making power in the hands of a small group of staff members to create a unified conservation mission in a department that until that point had had none. He is a believer in the idea that environmental protection can and must coexist with economic development. "Protection," he once told *Newsweek*, "is no longer a pious sentiment. It is an element of survival of this race."

Cecil Dale Andrus was born on August 25, 1931, to Hal Stephen and Dorothy John Andrus in Hood River, Oregon. His father was a sawmill operator, and Andrus's own first jobs were in the forests and lumber mills of northern Oregon. He developed an early appreciation for the wilderness of the Pacific Northwest, hiking, camping, hunting, and fishing throughout his childhood. His memories of some of the region's wild rivers before and after the erosive effects of indiscriminate logging practices contribute to his conservationist sensibilities.

Andrus attended Oregon State University at Corvallis for a year in 1948 and 1949. He did not graduate. He served in the United States Navy in the Korean War from 1951 to 1955 and was discharged after his tour of duty in the Pacific theater of operations with the rank of aviation electronic technician second class. He began work as a lumberjack for the Tru-Cut Lumber Corporation in Orfino, Idaho, in 1955, eventually working his way up to the position of production manager.

His political career began in 1961, when he was elected to the Idaho state senate from the Clearwater County district. At just 30 years old, he was the youngest senator in the history of Idaho.

Cecil Andrus (Bettmann/Corbis)

He was elected to a second term, and he served until 1966, working during his tenure for legislation in the areas of education, conservation, agriculture, social services, and business. He also worked as assistant manager of the Workman's Compensation Exchange from 1963 to 1966. Andrus made a bid for governor in 1966 but was defeated. Andrus served in the state senate again from 1968 to 1970 and became an agent with the Idaho offices of the Paul Revere Life Insurance Company. By 1970, he had been promoted to state general manager.

Andrus ran for governor once again in 1970. One of the major issues in his campaign against the incumbent, Dan Samuelson, was the American Smelting and Refining Company's proposal to operate an open-pit molybdenum mine at the base of Castle Peak in Idaho's White Cloud Mountains in the Challis National Forest. Governor Samuelson supported the project, citing its benefits to Idaho's agricultural economy. Andrus, however, made it clear to the people of Idaho that he felt that temporary economic gain was not reason enough to destroy their irreplaceable natural resources. Andrus's position was popular among environmentalists but not well supported in the areas around Challis that would have reaped the economic benefits from the mine. Andrus won the election by a narrow margin and became governor of Idaho in 1971.

As governor, Andrus increased state aid to public schools, added $1,000,000 a year to state revenues through a revision of the state's income tax law, and reduced the number of state agencies from 268 to 19. He also was a vigorous supporter of conservation legislation, endorsing the 1972 federal legislation that established the Sawtooth National Recreation Area and opposing the construction of two dams in Hells Canyon on the Snake River. He did, however, support the construction of a dam on the Teton River that was considered by both engineers and environmentalists a dangerous dam site. They were proved correct when the dam collapsed in June 1976. Andrus's seemingly contradictory stands on these issues can be explained by his philosophy that conservation does not negate the possibility for development, nor does development negate the responsibility to conserve. He described his basic philosophy to *Reader's Digest* in this way: "If I am faced with a decision of development with adequate safeguards for the environment, I'll come down on the side of development. If

I am faced with development without adequate safeguards, I'll come down on the side of the environment."

Andrus was reelected in 1974. During his second term, the Idaho Land Use Planning Act of 1975 was passed, providing for local land-use decision making with the help and support of state agencies. In December 1976, Andrus was nominated by President-elect JIMMY CARTER as secretary of the interior. Andrus brought with him a much deeper concern for conservation and the environment than had previously been found in the Department of the Interior. He saw his job as being to eliminate the department's "three Rs" (rape, ruin, and run). He totally rewrote the organizational chart, taking decision-making capability away from the heads of the 16 separate, specialized bureaus, services, and administrations that make up the Interior Department and concentrating that power at the top of the hierarchy, with a staff that he himself appointed. In doing this he eliminated the disunity and lack of communication that had been the standard mode of operation and created an agency with a unified conservation philosophy and the decision-making structure to act on that philosophy.

During his four-year term, Andrus delighted environmentalists with his strong support of conservation legislation that curbed offshore drilling and strip mining and expanded natural parks. The accomplishment of which he was most proud was the Alaska Lands Conservation Act of 1980, which protected 100 million acres from industrial exploitation and created the largest wilderness area in U.S. history. In accordance with Andrus's views on the appropriate marriage between conservation and development, the act opened another 254 million Alaskan acres to development as well. As secretary, Andrus also concerned himself with breaking the ties of economic interest to decision making in the public domain, stating that public lands should be managed in the interest of all of the people, simply because they belong to all of the people. "The initials B.L.M. [Bureau of Land Management]," he quipped at one point, "no longer stand for the Bureau of Livestock and Mining."

In 1981, after Carter lost the presidential election for a second term and the newly elected Ronald Reagan appointed the decidedly promining and protimber industry James Watt to the post, Andrus joined three former Environmental Protection Agency officials in creating a New Jersey–based corporation that tests the toxicity of industrial and chemical wastes to assist companies in complying with federal and state environmental regulations. In Idaho, he also worked for the federal protection of a part of the Snake River that is a nesting ground for one of the most dense populations of raptors in the world, an endeavor for which he received the Audubon Medal in 1985.

Andrus began to devote himself full-time to environmental issues through the Andrus Center for Public Policy at Boise State University in 1995. His priorities at the Andrus Center have been on the Snake River Birds of Prey Area, Wild and Scenic Rivers, and nuclear waste disposal. He has been married to Carol Mae May since 1949. They have three daughters, Tana Lee, Tracy Sue, and Kelly Kay.

BIBLIOGRAPHY

Andrus, Cecil, *Cecil Andrus, Politics Western Style*, 1998; Boeth, Richard, William J. Cook, and John Walcott, "Environment; Interior Redesign," *Newsweek*, 1977; Carter, Luther J., "Interior Department: Andrus Promises 'Sweeping Changes,'" *Science*, 1977; Maxwell, Jessica, "Q and A Interview," *Audubon*, 1995; Miller, James Nathan, "Secretary Andrus Makes His Stand," *Reader's Digest*, 1977; Moritz, Charles, ed., *Current Biography Yearbook*, 1977; Seligman, Jean, Mark Kirchmeier, and John Accola, "Andrus: A Booster for the Way of the West," *Newsweek*, 1983.

Anthony, Carl

(February 8, 1939–)
Architect, Executive Director of Urban Habitat

Carl Anthony is the executive director of Urban Habitat, a San Francisco–based environmental justice organization that works toward a socially just and environmentally sustainable economy and builds multicultural urban, environmental leadership. Urban Habitat promotes the restoration and revitalization of urban environments, to provide an alternative to suburban sprawl and the social and environmental destruction that it entails.

Carl Anthony was born on February 8, 1939, in Philadelphia. His working-class parents sent him to a racially integrated elementary school—a rarity at that time—in hopes of providing him with the best possible education. After receiving a graduate degree in architecture from Columbia University in 1969, Anthony journeyed to Africa, where he spent more than a year traveling the continent in a Volkswagen van. He became especially fascinated with how well African traditional cultures had adapted, over generations, to climate and other aspects of the environment. This trip inspired Anthony to dedicate himself to ecological architecture.

Upon his return to the United States in 1971, Anthony moved to the San Francisco Bay Area and taught architecture at the University of California, Berkeley (UCB) College of Environmental Design. He continued this until 1978, at which point he opened his own architecture studio and began designing affordable housing in Berkeley and Oakland. Anthony has long felt that the lifestyle and housing option that middle-class white Americans have been choosing since the 1950s—living in the suburbs and becoming increasingly dependent upon automobiles to travel to stores and their workplaces—has had devastating effects on society at large. Low-income people of color, who have remained in urban centers as wealthier whites have poured out, have borne a disproportionate share of the costs of this suburbanization. They have suffered increased air and noise pollution as new highways to the suburbs have bisected their neighborhoods; lack of interest in their inner-city communities as private investment and public infrastructure investment have flowed to the new suburbs; and declining job opportunities and access to jobs as employment centers have moved to the suburban rim. Institutional racism in zoning and land use, such as racial covenants that excluded African American homeowners in suburban developments and redlining within communities of color, have combined to confine many of these same communities to less desirable and older neighborhoods adjacent to industrial and manufacturing facilities. An example of this is the community surrounding the Chevron industrial facility in Richmond, California. The plant was built first, but the area grew into a residential neighborhood when African American families, who were unable to buy homes in the more desirable areas nearby, settled near it.

In addition to experiencing increased exposure to toxics from nearby manufacturing plants, the neighborhoods of low-income people of color are often more likely to be locations for other undesirable land uses, such as solid waste recycling, truck transfer stations, or medical waste incinerators. The inordinate environmental burden that many communities of color face in both urban and rural settings gave rise to the national movement for environmental justice.

Anthony has been working on the urban front of the environmental justice movement throughout his career. As a consultant to the Berkeley Redevelopment Agency in the early 1980s, he was able to resolve a conflict between interests fighting for affordable housing, historic preservation, and economic development, by designing plans that integrated all three. In 1989, he was a protagonist in the defeat of a proposal to build a huge shopping development on the Berkeley waterfront. Although the proposal was supported by the area's African American community for the jobs it would offer, Anthony worked against it because he knew the jobs would be only temporary, and he worried about its effects on Berkeley's already declining downtown area. While serving as president of the Berkeley Planning Commission from 1990 to 1992, Anthony worked on a successful downtown revitalization effort that now serves as a model for other urban areas. Most recently, Anthony has worked with commissions to plan for the closing of Bay Area military bases. For the Alameda Naval Air Station, Anthony has successfully advocated such environmentally sustainable alternatives as recycling building materials from the structures that are torn down, building housing in a way that preserves open space along the water front, and encouraging environmentally friendly industry (an electric car factory was built). Anthony was appointed to the Presidio Council in 1992 to advise the National Park Service about the future of this converted Army base.

Anthony began working with the San Francisco–based Earth Island Institute (EII) in 1989. At EII, which nurtures projects that work on specific environmental issues and then helps them spin off into independent organizations, Anthony headed the Urban Habitat Program, which promotes the cultural and economic restoration of inner cities through its conferences, publications, educational programs, and advocacy. It networks environmental activists of color; by 1997 it had assisted more than 100 organizations working on environmental justice issues, including health, food security, recycling, energy, transportation, arts and culture, education, immigration and population, and parks and open space. Urban Habitat spun off from EII in 1997 and has since become a project of the Tides Center.

Anthony believes that because environmental problems disproportionately affect people of color, people of color should take the lead in solving them. To these ends, Urban Habitat has worked with the U.S. Forest Service—an agency currently under court order to diversify its staff—to set up an environmental education program in Oakland's public school, and to graduate a set number of people of color per year from the College of Natural Resources at UCB. Other projects include the Transportation Equity Project, which redirects regional transportation priorities away from suburban highways and back to urban transit; the Community Revitalization and Land Restoration Project to reclaim "brownfields" (vacant, blighted, or contaminated land); the Leadership Institute for Sustainable Communities, which provides training for activists; and the Hunters Point Environmental Health Project, which works with community residents on issues of environmental health and justice.

Anthony coedits the quarterly journal *Race, Poverty, and the Environment,* which he cofounded with Luke Cole of the California Rural Legal Assistance Foundation at the University of Oregon's 1990 Public Interest Law Conference. In 1993, he was appointed to the National Environmental Justice Advisory Council by Environmental Protection Agency administrator Carol Browner, and in 1996 he became a Fellow at Harvard University's John F. Kennedy School of Government. Anthony resides in the San Francisco Bay Area.

Bibliography

Gilliam, Harold, "Carl Anthony," *San Francisco Chronicle,* 1996; Inman, Bradley, "The Battle Against Environmental Racism," *San Francisco Examiner,* 1993; Kay, Jane, "Community Hero at Helm of Change," *San Francisco Examiner,* 1995; "Urban Habitat Program," http://www.igc.apc.org/uhp/

Audubon, John James

(April 26, 1785–January 27, 1851)
Artist, Ornithologist

John James Audubon is easily the best known and best remembered among nineteenth-century naturalists and ornithologists. The artwork of this skilled painter of wildlife first introduced the images of many bird species to the widespread public. Audubon's paintings and simple manner of writing made birding and ornithology accessible to the nonscientist, thereby contributing to their popularity as leisure activities. Audubon's name continues to pervade modern conservationist activity; Audubon societies exist worldwide and have become synonymous with the protection of wild birds.

Jean Jacques Fougère Audubon was born April 26, 1785, on his father's plantation at Les Cayes, Santo Domingo (now Haiti). He was the illegitimate son of Jean Audubon, a sea merchant, and a Creole woman of Santo Domingo, known only as

Jean Jacques (John James) Audubon (courtesy of Anne Ronan Picture Library)

Mlle. Rabin, who probably died within a year after his birth. Audubon, who was called Fougère, or sometimes Jean Rabin, and his younger half-sister Muguet, the daughter of another Creole woman, went to Nantes, France, with their father in 1789. They were warmly received by Jean Audubon's wife, Anne Moynet Audubon, who took charge of their upbringing as her husband became occupied with the brewing French Revolution.

Audubon received the standard bourgeois education; he was instructed in mathematics, geography, music, and fencing, though his lessons were sometimes neglected by his indulgent stepmother. Years later, Audubon regretted not having been drilled in writing in French. He did, however, become influenced by the revival of interest in nature that Rousseau, Buffon, and Lamarck had made popular; by the age of 15, he had already begun a collection of his original drawings of French birds. Recognizing his son's lack of discipline, Jean Audubon put him in military school for a year, but the experience had little effect. Having always encouraged his son's taste in natural history and drawing, Jean Audubon arranged for him to study drawing in Paris under the great French artist, David.

In the autumn of 1803, Audubon left France for America. Early in 1804, he arrived at Mill Grove, an estate near Philadelphia that his father had purchased in 1789. Living the life of a country gentleman, free of financial concerns, Audubon became an enthusiastic observer of nature. While the bird protection societies across the United States that carry his name have promoted Audubon as a passionate protector of wildlife, Audubon in reality enjoyed hunting with dog and gun and was an avid sportsman at this stage of his life, killing for amusement as well as for food. Nevertheless, Audubon did have a spirit of scientific inquiry, and it was during this period that Audubon conducted the first banding experiment on wild birds in America. He discovered a nest of pewees in a cave and fastened light silver threads to the legs of some of the baby pewees. The next spring, Audubon found that two of them had returned to the region and were nesting a little way up the creek from their place of birth.

The next few years saw Audubon return to France for a year before returning to the United States, selling Mill Grove and moving to Louisville, Kentucky, where he opened a general store with Ferdinand Rozier, the son of one of his father's business associates. The store suffered considerably because of the Embargo Act, which prohibited the importation of some goods the store would have sold, but all the same, in 1808 he went to Philadelphia to marry Lucy Bakewell, to whom he had been engaged since 1804, and brought her back to Louisville.

At the beginning of the nineteenth century, Kentucky was still almost a wilderness, which strengthened Audubon's passion for natural history. Out of touch with other ornithologists, Audubon worked as an artist and a lover of nature more than as a scientist as he continued to paint birds. Indeed, a seeming lack of interest in science pervades much of his bird paintings from this period. Much criticism has subsequently been directed at Audubon's carelessness in his paintings that has hurt his reputation as a naturalist. The criticism is generally the same, from publishers, ornithologists, and other scientists, who claim that Audubon's birds lack the veracity that science demands. Even his most loyal supporters, such as Elliott Coues, concede that many of his paintings show birds that are posed in anatomically impossible stances.

Audubon's consuming interest in painting birds resulted in his neglect of his business in Louisville and its eventual failure. As the town and business competition grew, Audubon and Rozier moved their business 125 miles down the Ohio River to Henderson, Kentucky, in 1810. There, the same pattern emerged: As Rozier tended to the store, Audubon roamed the country in search of rare birds; his fishing and hunting were often the only means of subsistence for both partners. Ultimately, the partnership was doomed to collapse, which it did, in 1818, though the two remained friends.

Subsequent business ventures also failed, and in 1819, Audubon was jailed for debt. He was released on the plea of bankruptcy with only the clothes he wore, his gun, and his original drawings. This disaster ended his business career, and he spent the winter doing crayon portraits before moving his family to Cincinnati, where he became a taxidermist in the new Western Museum. It was around this time that the idea of publishing his bird drawings came to Audubon. In October 1820, Audubon traveled down the Ohio and Mississippi Rivers, looking for birds to draw and paint. After some time in New Orleans, where Audubon worked as a tutor and a drawing teacher, Mrs. Audubon took a job as a governess and assumed the financial responsibilities of looking after the hungry family; for 12 years, she was the primary wage earner for the household.

Unable to publish his works in the United States, Audubon traveled to England, on funds saved from his wife's wages, where *Birds of America* was published in 1827. He was especially well received in Scotland, where he was elected to the Royal Society of Scotland in March 1827. In 1830, he and his wife bought a home in Edinburgh, where he began to write the text portion of *Birds in America*. The text was published separately in 1838 as *Ornithological Biography*.

Returning to the United States, Audubon purchased a large estate on the Hudson River, called Minnies' Land; this estate in New York is known today as Audubon Park. Audubon returned to the United States to work on a miniature edition of *Birds in America* but almost immediately began collaborating with the naturalist John Bachman on their three-volume *Viviparous Quadrupeds of North America.* Their friendship was cemented by the marriages of two of Audubon's sons to two of Bachman's daughters.

In his final years, Audubon continued to work on color plates for his work with Bachman and tutored several young ornithologists. Audubon suffered a debilitating stroke in early January 1851, leaving his son, John W. Audubon, to finish his work on *Viviparous Quadrupeds of North America.* Audubon died before the month was out, on January 27, at the age of 65. In 1896, the first Audubon Society was formed by a group of conservation-minded bird-watchers, headed by the ornithologist William Brewster, in Massachusetts.

Bibliography

Audubon, M. R., *Audubon and His Journals, with Notes by Elliott Coues*, 2 vols., 1897; Foshay, Ella M., *John James Audubon*, 1997; Herrick, Francis Hobart, *Audubon the Naturalist: A History of His Life and Time*, 2 vols., 1968.

Austin, Mary

(September 9, 1868–August 13, 1934)
Nature Writer

Mary Austin authored fiction, autobiography, and nature-inspired writings, addressing such issues as ecology, mysticism, spirituality, bioregionalism, and feminism. She is known for her commitment to and her portrayal of the land and people of the American Southwest. Her books laid the foundation for science and nature writing as it is practiced today and helped to influence the deep ecology movement.

Born on September 9, 1868, in Carlinville, Illinois, Mary Hunter Austin was the third of four children born to George and Susanna Savilla Hunter. Her relationship with her mother was a strained one, as were her relationships with her other family members (they would continue to be so throughout her life), except for those with her father and her younger sister, Jennie. Her father was a successful lawyer, and he encouraged Austin's literary interests. He had fought on the Union side of the Civil War. In 1878, when Austin was ten years old, he died of a malarial illness that he had contracted in his years as a soldier. Two months later, Austin's younger sister died of diphtheria. Of Jennie, Austin would later write, "She was the only one who ever unselfishly loved me."

In 1884, Austin enrolled in Blackburn College in her hometown, where she was editor of the college journal and was elected class poet. Mathematics and science, however, stimulated her imagination as well, and she sought to include both the sciences and the humanities in her studies. Years later, in 1922, in an essay entitled "Science for the Unscientific," Austin would argue that gifted writers should "immerse themselves in the data of science to the point of saturation." In this, she was a

Mary Austin (Ansel Adams Publishing Rights Trust/Corbis)

precursor to such scientist-writers as ALDO LEOPOLD, RACHEL CARSON, and Ann Zwinger.

Shortly after graduating from Blackburn in 1888, Austin moved with her family to the San Joaquin Valley in California, where her brother Jim wanted to homestead a parcel of land. The train ride west left a deep impression on Austin. She felt drawn to the vast space of the Mojave Desert. The arid lands of the western landscape were to have a profound impact on Austin's life and on her career as a writer. The Hunter family lived in a cabin consisting of a single room with bunk beds and the bare minimum of essential furnishings. Austin spent much of her time outdoors, studying unfamiliar native animals and plants. She slept little, often sitting for hours in the moonlight, watching the nighttime goings on of the San Joaquin Valley wilderness. In 1889, she had an essay published in her alma mater's magazine, *The Blackburnian.* Entitled "One Hundred Miles on Horseback," it is an account of her trip west with her family and contains vivid descriptive passages of a California that no longer exists.

The Hunter family, unfortunately, arrived in the midst of one of California's not uncommon periods of drought and found itself in a constant struggle to survive on the land. Austin took a teaching position away from her family with the Kern County School District in order to provide some income for herself and her struggling kin. During this time, she made the acquaintance of Gen. Edward Beale, owner of the vast Tejon Ranch and early California pioneer. He was an inexhaustible source of information about California and the West. Beale shared stories of the region, of its Indians and early settlers, that provided a substantial amount of material that later appeared in Austin's first books.

In the summer of 1890, after a monetary dispute with her mother that led to Austin's ostracism from her family, she met Stafford Wallace Austin, who was attempting to develop a 20-acre fruit farm. They married a year later, in 1891. When the farm failed, the couple left the San Joaquin Valley, and together they lived in various towns of the Owens Valley in California. Austin taught school, wrote, and observed nature, while Wallace failed at one agricultural endeavor after another and generally failed to provide for his new family. Austin would later refer to this time as "the desert years." This period provided much in the way of literary inspiration, but it was also a time of loneliness and frustration for Austin. Her marriage was deteriorating, and in 1892, Austin gave birth to a child, Ruth, who was mentally disabled. Austin had her daughter institutionalized until her death in 1918 at the age of 26. In 1903, Austin published *The Land of Little Rain.* Her best-known work, it is a collection of essays addressing her experiences with nature, the people of the southwest, and their religion.

In 1906, having visited Carmel two years earlier while researching her novel *Isidro,* Austin sold the house that she and Wallace had lived in since 1900 and bought land in Carmel, joining an artist colony centered around the poet George Sterling. Wallace did not follow her to Carmel. He charged her with desertion. Their divorce was finalized in 1914. While in Carmel, Austin completed two more novels, *Santa Lucia* (1908) and *Outland* (1910), and wrote stories that would be collected in *Lost Borders* (1909). Austin was diagnosed with breast cancer in 1907 and was given only nine months to live. She went to Italy, where she studied prayer and mysticism. While in Rome the pains she had been experiencing suddenly disappeared. She stayed in Europe, spending her time primarily in England and Italy for the next two years, returning to the United States in 1910.

Between 1911 and 1923, Austin divided her time between Carmel and New York. She continued writing, publishing many books during this time, including a novel *A Woman of Genius* (1912), which is considered by many to be her best fiction. In 1924, she moved to Santa Fe, New Mexico, where she had been named an associate in Native American literature at the School of Native American Research. Here, she published *The Land of Journey's End* in 1924 in a return to the subject matter (the landscape of the Southwest) and style of her earliest nature writings. In this book, she presents her perspective that "not the law but the land sets the limit." This has been interpreted as a forerunner of the modern day "deep ecology" philosophy, which attempts to create a nonanthropocentric vision of reality and nature. Austin believed, however, that humans do have a place in nature and that this place was to be created through the creation of working relationships and connections with the land.

While in Santa Fe, Austin also became heavily involved in the water rights battle over the Colorado River. She was appointed a delegate to the Second Colorado River Conference in 1927, where she argued against building a dam in Boulder Canyon (a dam was eventually built here—Hoover Dam was completed in 1936). She objected to the fact that the water from this dam would go to California, a state already beginning to sprawl, rather than to Arizona and New Mexico, states whose legitimate claims to the water would be ignored and lost.

In her autobiography, *Earth Horizon* (1932), Austin wrote that she "wanted to write books that you could walk around in." She considered herself a "naturist," not a naturalist. Her definition of natural history was that it concerns itself with facts and rationalization, two concerns that she most definitely did not have. In her books, the land is always something more than a series of physical, sensual, or intellectual features. Rather, all of its physical aspects (plants, animals, humans, climate, geology) are manifestations of what Austin considers to be "spirit." In her books she addressed issues such as feminism, ecology, and bioregionalism. Mary Austin died in her sleep, on August 13, 1934, in Santa Fe, New Mexico, after a heart attack the day before.

Bibliography

Fink, Augusta, *I—Mary, A Biography Of Mary Austin*, 1983; O'Grady, John P., "Mary Hunter Austin," *American Nature Writers*, John Elder, ed., 1996; Pearce, T. M., *Mary Hunter Austin*, 1965.

Ayres, Richard

(February 2, 1942–)
Attorney, Cofounder of the Natural Resources Defense Council

Cofounder of the Natural Resources Defense Council (NRDC), Richard Ayres is one of the nation's leading experts on the Clean Air Act. Ayres has shaped the nation's clean air program and laws since the passage of the Clean Air Act of 1970 and has litigated many of the key Clean Air Act cases. He has influenced many of the most important Environmental Protection Agency (EPA) rules interpreting the Clean Air Act and has played a leading role in congressional consideration of amendments to the Clean Air Act in 1977, 1980, and 1990.

Richard Edward Ayres was born on February 2, 1942, in Salem, New Jersey. In 1949, his family moved to Beaverton, Oregon, then a quiet Pacific Northwest town outside of Portland. The family spent weekends boating and fishing. Ayres credits his early experience of the beauty of nature in the Pacific Northwest as a formative influence in his later career, along with his father's interest in politics and public policy, sense of stewardship for the natural world, and his personal integrity.

Ayres graduated with honors in 1964 from Princeton University's undergraduate program in the Woodrow Wilson School of Public and International Affairs. In 1969 he received his LL.B. from Yale Law School, together with an advanced degree in political science from Yale University. He was an editor of the *Yale Law Journal* and received the Perez Prize, given for the best student-authored article of the year, for an empirical investigation of police interrogations, "Interrogations in New Haven: The Impact of *Miranda*" (1967).

Ayres cofounded the NRDC in 1970. He was instrumental in making it one of the nation's most influential environmental organizations, with a membership today of 475,000. NRDC seeks to safeguard the health of humans and the natural world. With a staff of lawyers, scientists, and others, it attempts to persuade governments, business, and other institutions to adopt more environmentally friendly policies. NRDC is credited with having major impacts on national and international policies in many areas, including air and water pollution, toxic chemicals, energy, transportation and urban development, protection of old-growth forests and other terrestrial and marine habitat, limitation of nuclear proliferation, and global warming. At NRDC, where he worked until 1991, Ayres became one of the most influential voices shaping the nation's clean air policies in all three branches of the federal government.

Ayres has handled nearly three dozen cases in the federal courts, including the Supreme Court of the United States, involving the interpretation and enforcement of the Clean Air Act and the National Environmental Policy Act. Ayres achieved the largest single reduction in pollution in American history in litigation with the Tennessee Valley Authority (TVA), then the nation's largest emitter of sulfur oxides. In settlements in seven federal district courts located in three different states, TVA agreed to cut sulfur oxides emissions by over one million tons per year, at the time approximately 5 percent of the total national emissions of sulfur oxides. Ayres also argued *Train v. NRDC* (1975) and *Vermont Yankee Nuclear Power Corporation v. NRDC* (1977) before the Supreme Court.

Ayres has participated in most of the important EPA rule-making proceedings under the Clean Air Act since 1971. Among the more notable are development of all 50 of the original State Implementation Plans, revisions to the National Ambient Air Quality Standards in 1972 and 1980, regulation of the use of tall smokestacks by large electric power generating plants, emission standards for new coal-fired electric generating plants, development of standards for hazardous air pollutants for several industries, standards for "reformulated" gasoline, adoption of emission trading guidance for states, standards for sulfur content of gasoline, and development of open Market Emission Trading guidance.

During congressional consideration of amendments to the Clean Air Act in 1977, 1980, and 1990, Ayres led the National Clean Air Coalition, which included the major environmental and public health organizations and labor unions, churches, and civic organizations. The Coalition successfully sought important additions to the national clean air program enacted by Congress in the 1970 Clean Air Amendments.

In 1977, the Coalition proposed what became the Prevention of Significant Deterioration program, designed to manage emissions growth in the interests of preserving high quality air and maximizing the potential for economic growth. As a means of achieving these goals, the Coalition successfully advocated requirements that new emissions units be required to install state-of-the-art pollution control technology.

In 1990, the Coalition successfully urged President Bush and the Congress to adopt programs to control acid rain, reduce emissions of toxic chemicals, and cut emissions from new motor vehicles. The acid rain control program requires major reductions in sulfur oxides emissions from electric power plants. It includes the most ambitious and successful emission trading program ever adopted, which has cut costs to about one-fourth of what was predicted when the legislation was being considered. The hazardous emissions control program replaced a previously ineffective section of the Clean Air Act with a list of 190 toxic chemicals and instructions for EPA on how to control them. Motor vehicle standards enacted in 1990 have cut allowable emissions from new cars by 75 percent.

In 1991, Ayres left NRDC for private practice, becoming a partner in the

Washington office of O'Melveny and Myers, where he headed its environmental department. He is currently a partner in the environmental practice of Howrey Simon Arnold and White of Washington, D.C., where he represents clients before EPA, the Federal Energy Regulatory Commission, and the federal courts and has acted as a dispute resolution facilitator. He also specializes in advising senior-level management on environmental issues and finding creative solutions to environmental disputes. His clients include companies in automobile, diesel engine, transportation, electric power, oil, natural gas, chemical, pulp and paper, and pollution control industries, as well as state governments. In 1998, Ayres participated in settling the largest mobile source air enforcement case in history, brought by the federal government against the manufacturers of diesel engines.

Ayres has served on a number of blue ribbon panels dealing with the nation's clean air policy. He was appointed by President Carter to the National Commission on Air Quality in 1978; was a member of the Carnegie Commission on Science, Technology, and Regulatory Decision Making from 1991 to 1994; and currently serves on the Environmental Protection Agency's Clean Air Act Advisory Committee and the Executive Committee of the District of Columbia Mayor's Environmental Council. In 1988, Ayres was recognized by the Yale Law School Association of Washington for his outstanding service to the public interest. In 1989, he was honored by the Yale Law School Environmental Law Association for his role in creating the public interest law movement.

Ayres also serves on several boards of directors of environmental and energy-oriented organizations. He chairs the policy committee of the board of trustees of the Natural Resources Defense Council and serves as vice chair of the board of the Keystone Center. He serves on the Vermont Law School Environmental Law Center advisory committee and is president of the Breakthrough Technologies Institute, a member of the board of the Global Forest Foundation, and a member of the advisory board of the Social Venture Capital Foundation.

Ayres lives in Washington, D.C., with his wife, Merribel, and children, Alice and Richard.

BIBLIOGRAPHY

Ayres, Richard, "The Clean Air Act: Performance and Prospects," American Bar Association *Natural Resources and Environment*, 1998; Ayres, Richard, "Developing a Market in Emission Credits Incrementally: An 'Open Market' Paradigm for Market-Based Pollution Control," *BNA Environment Reporter*, 1997; Ayres, Richard, and Richard Parker, "The Proposed WEPCO Rule: Making the Problem Fit the Solution," *22 ELR 10201*,1992.

B

Babbitt, Bruce

(June 27, 1938–)
Secretary of the Department of the Interior, Governor of Arizona

Bruce Babbitt has been secretary of the Department of the Interior since 1993. He brought a strong conservation ethic to this position, and he has encouraged cooperation between environmental and commercial interests. He founded the National Biological Service.

Bruce Edward Babbitt, the second of six children, was born on June 27, 1938, to Paul J. and Francis Babbitt. He grew up in Flagstaff, Arizona, where his family had originally settled in the 1880s and had ranched and operated trading posts. From his early childhood on, Babbitt was exposed to the type of outdoor activities that go along with a ranching heritage. He was an avid hiker and horseback rider. He attended the University of Notre Dame, where he studied geology, graduating with honors in 1960 with a B.S. He continued his education at the University of Newcastle-upon-Tyne in England, receiving an M.S. in geophysics in 1963.

After completing his M.S. degree, he reevaluated the direction of his career and decided to pursue politics rather than geology. To this end he enrolled in law school at Harvard University, graduating in 1965. During his time at law school he became active in the civil rights movement. He joined marches in Selma, Alabama, and upon graduating from Harvard he worked in the federal antipoverty program as a civil rights lawyer. From 1965 to 1967 he served as special assistant to the director of Volunteer Service to America (VISTA), after which he left government service and returned to Arizona, where he joined a private law firm in Phoenix.

In 1974, Babbitt was elected attorney general of Arizona. In this capacity he fought for consumer protection, cracking down on land sale frauds, price-fixing, and insurance irregularities. In 1978, after the sitting governor resigned to become an ambassador, Babbitt, being next in line, became the governor of Arizona. He served the remainder of the term and was elected to two more full terms in office. As governor, Babbitt's philosophy was that government should be streamlined and show fiscal restraint. He also

Bruce Babbitt (courtesy of Anne Ronan Picture Library)

believed that government should act as a protector of the environment and civil rights. He pushed for environmental controls and water management, and he supported education and child welfare programs. He did not seek a third full term. For a short time in 1988 Babbitt was a presidential candidate, before returning to his law practice in Phoenix.

In 1993, Babbitt was appointed secretary of the Department of the Interior by Pres. Bill Clinton. The holder of this position is responsible for managing the federal government's land holdings and natural resources. One of Babbitt's first actions as secretary was to lobby for the creation of a new governmental agency: the National Biological Service (NBS). In Babbitt's original vision, this agency would be based on the model of the U.S. Geological Survey and would be responsible for making inventories of the nation's biological resources. The basic premise behind the idea was that what was not known to exist could not be protected. The NBS was to provide a scientific foundation to support the management and conservation of the ecosystems of the United States. However, Congress was reluctant to appropriate funds for an entirely new government agency. The version of the NBS that we have today is watered down, with scientists already employed by government agencies such as the National Parks Service being assigned NBS duties on top of those they were already performing.

Babbitt's views on conservation distinguish him from past secretaries of the interior. His predecessor, Manuel Lujan, referred to Bureau of Land Management land as "a place with lots of grass for cows." Although some might argue that technically Lujan was not far wrong, Babbitt has employed a drastically different perspective in managing the government's resources. He is a believer in the concept of "public use." He has based his management practices on the idea that lands must be shared by accommodating environmental, recreational, and commercial interests. He is also a strong believer in coalition building and cooperation, as can be seen in his efforts in the Everglades National Park in Florida, where, in 1995, he and Environmental Protection Agency administrator CAROL BROWNER worked together to negotiate a settlement that would allow for the restoration of that environmentally degraded park.

That agreement called for a $700 million cleanup, half of which was to be paid for by the sugarcane farmers, who were also required to cut back on dumping phosphorus into the watersheds that feed the park. The other half was to be paid for by the taxpayers of Florida. According to Babbitt, this type of settlement is ideal because it leads to the restoration of an entire ecosystem, not just scattered pieces, and because it puts an end to the expensive, time-consuming legal warfare that leads to "environmental train wrecks." Neither the farmers nor the environmentalists were entirely happy with the deal. The farmers felt that they were forced to shoulder an excessive share of the financial burden, and some environmentalists thought that Babbitt had sold out to agricultural interests and had "sounded the death knell for the Everglades." Observers say that any multiple-use resource manager who draws such heavy criticism from two diametrically opposed interest groups must surely be doing his job.

Babbitt is married to Harriet Coons. They have two sons, Christopher and

Thomas Jeffery. Babbitt has published two books on the American Southwest: *Color and Light: The Southwest Canvases of Louis Akin*, 1973, and *Grand Canyon: An Anthology*, 1978.

Bibliography

"U.S. Department of the Interior," http://www.doi.gov; Williams, Ted, "On The Fire Line for Conservation," *Outdoor Life*, 1996; Wuerthner, George, "The Science Stalemate," *National Parks*, 1995.

Bahouth, Peter

(August 26, 1953–)
Executive Director of Turner Foundation, Former Executive Director of Greenpeace U.S.A.

Known for his refusal to compromise, Peter Bahouth has proven to be a strong and capable leader during his career, which includes heading two of the largest environmental organizations in the country. He served as executive director of Greenpeace U.S.A., which evolved from a scrappy, disorganized advocacy group into a powerful environmental organization with 2.2 million members during his leadership. While at Greenpeace, he planned direct action campaigns, marketed the organization to the nation's youth through television ads and a record album, and strove to maintain Greenpeace's integrity in an era of increasing compromises between environmental concerns and corporate interests. Bahouth now heads the Turner Foundation, where he helps to channel millions of dollars into environmental causes.

Peter Bahouth was born on August 26, 1953, in Syracuse, New York, to Frank and Anne Marie (Pietrafesa) Bahouth. He earned a B.A. in history from the University of Rochester in 1975 and a degree in law from the Northeast School of Law in 1978; he established a law practice in Boston that same year.

In 1979 he began volunteering his legal services at Greenpeace, an environmental organization known for its confrontational direct action campaigns, such as voyaging to sea to interfere with whaling, nuclear testing, and the killing of seal pups. He became increasingly involved with the organization; he was elected to the board of directors in 1982 and then two years later was named national chairman. In 1985, Greenpeace International sent its flagship, the *Rainbow Warrior*, on a mission to protest French nuclear testing in the South Pacific. While making preparations in a New Zealand port, the ship was sunk by French secret service agents, killing a photographer on board and resulting in an international scandal. The incident prompted an outpouring of sympathy for Greenpeace in the form of new members and contributions. In the aftermath, Bahouth negotiated for compensatory damages; France agreed to pay Greenpeace $8 million.

In 1987 Bahouth coordinated the production of a Greenpeace-sponsored album called "Rainbow Warriors." Among the performers who donated songs were U2, R.E.M., Sting, Peter Gabriel, the Talking

Heads, Dire Straits, and many other rock musicians, all of whom gave Greenpeace instant credibility to millions of young people. At this point Bahouth had given up his law practice and had begun working to consolidate Greenpeace's seven regional chapters into Greenpeace U.S.A., of which he became executive director in 1988. He was responsible for planning campaigns, managing personnel, and handling the budget and the media. He participated in many of the direct action campaigns himself, including chaining himself to railroad tracks near the DuPont Company's headquarters to protest shipments of toxic pollutants.

Bahouth also continued helping Greenpeace recruit the nation's youth. Capitalizing on its radical reputation, Greenpeace marketed its environmental activism through a series of "World Alert" commercials on the cable music channel VH-1, featuring celebrities and the quick cuts and hypnotizing graphics used in the channel's music videos. The response from the ads, which ran in 1989 and 1990, was highly enthusiastic, with up to 7,000 phone calls a month coming in from interested viewers, and Greenpeace U.S.A.'s membership reached an all-time high of 2.2 million.

With the influx of money and status, Bahouth had to struggle to resist the pressure to join with the more mainstream environmental groups in forging alliances with what he called the "regulatory-industrial-negotiating complex." When invited to join the Group of Ten, a coalition of mainstream environmental organizations, Bahouth balked, saying he wouldn't join unless LOIS GIBBS (a grassroots hazardous wastes activist) was invited too. As Greenpeace grew, many felt it was losing its vision, and in 1991 Bahouth left the organization and spent a year campaigning in Montana to save over four million acres of publicly owned wilderness from development and logging.

In the spring of 1993, Bahouth became executive director of the Turner Foundation, a private conservation organization founded by billionaire media mogul TED TURNER and his family in 1990. The Turner Foundation, which is endowed with earnings from Turner's Cable News Network and its acquisition by Time Warner, directs funds toward local, national, and international groups that work to protect water quality and wildlife habitat, control world population through family planning, reduce consumption, and conserve natural resources. In 1996, the Turner Foundation gave away $8.9 million in grants. In 1997 this amount doubled to $18 million, and then nearly tripled in 1998 to $25 million. From 1990 to 1997 it gave away $26.7 million in about 1,100 grants. The foundation generally favors grassroots organizations and watchdog groups over the larger national organizations and instructs applicants to limit their grant requests to three pages so as not to waste paper.

In addition to its work at the international level, the Turner Foundation has directed efforts toward local environmental issues in its home state of Georgia. In August 1997, Bahouth and the board of directors at the Turner Foundation convened a working group of local and national environmental advocacy groups to devise strategies for fighting urban sprawl in Atlanta, which has shown the fastest outward growth of any human settlement in history and impacts everything that the foundation fights for, from air and water quality to helping inner city youth. Other locally directed efforts have focused on energy use. In mid-1997, the Turner Foundation made an offer to the city of Atlanta to finance a $1.5 million energy conservation program that could have saved the city as much as 30 percent of its energy bills. In a decision that garnered much criticism for city officials, the offer—which could have saved taxpayers at least $3 million a year—was rejected. Bahouth, while remaining open to discussing further ideas with the city, was disappointed at the missed opportunity to improve the quality of life in Atlanta, where he makes his home.

Bibliography

Dowie, Mark, *Losing Ground: American Environmentalism at the Close of the Twentieth Century*, 1995; Goldberg, David, "Turner Foundation Funds Efforts to Study, Curb Urban Sprawl," *Atlanta Constitution*, 1997; "Greenpeace: An Antidote to Corporate Environmentalism," *Multinational Monitor*, 1990; Kowet, Don, "A Natural Made-for-TV Movement," *Insight on the News*, 1990; Saporta, Maria, "Atlanta Rejects Turner Funds for Energy Conservation Plan," *Atlanta Constitution*, 1998.

Ball, Betty, and Gary Ball

(December 26, 1941– ; December 15, 1948–)
Community Organizer, Environmental Activist; Environmental Activist and Analyst

Betty and Gary Ball cofounded the Mendocino Environmental Center (MEC) in Ukiah, California, and codirected it as it became the hub for many of the activists working on a range of environmental issues in northern California, including redwood forest clear-cutting.

Betty Ball was born Elizabeth Louise Johnston on December 26, 1941, in Milwaukee. Ball's father, a YMCA director, was transferred to Scottsbluff, Nebraska, to establish a new YMCA when Betty was six years old and again to do the same in Lubbock, Texas, when she was ten. Ball followed in her father's footsteps by graduating from the YMCA's George Williams College in Chicago in 1967 with a B.A. in group work and psychology. Her first job out of college was as program coordinator for a branch of Jane Addams's Hull House, where she organized recreational programs for disadvantaged children.

In the summer of 1969 she moved back to Boulder, Colorado, where she had spent many childhood summers visiting her grandparents. Ball worked at a home for developmentally disabled adults in Boulder and for the Colorado Department of Social Services in Nucla, in western Colorado. She briefly attended the Jane Addams School of Social Work in Chicago in 1971 but became disenchanted with the program after a professor of a class about the history of social welfare reform refused to allow his class to discuss the U.S. bombing of Cambodia the day after it happened.

Gary Ball was born on December 15, 1948, in Denver, Colorado. At the age of

Betty and Gary Ball (courtesy of Anne Ronan Picture Library)

two, he was stricken with polio. Isolated from other children by his disability, Ball found a refuge in music and his intellect. He attended the Boettcher school for disabled children and despite its lack of a precollege curriculum, went on to graduate from the University of Colorado in 1970 with a degree in psychology.

The Balls met in 1970, and together they worked with various community organizations in Boulder, Colorado. They helped run the Boulder Communication Center, an alternative social service center for the town's numerous hippies and transients. A local ecumenical religious center donated space for the Center, and the transients had a place to hang out, leave their backpacks, get messages and mail, receive counseling, find temporary work, catch a free nightly bus to camp in the mountains, eat for free, and so on. Gary Ball's paid job during most of the 1970s was with a team that conducted early studies on biofeedback, a therapeutic technique in which a patient is given information about his or her physiological functions that he or she is not able to perceive otherwise, with the object of trying to gain control over them. Biofeedback is done with modified polygraph equipment; Gary Ball became a computer data analyst during his work on the studies. After these projects were finished, he went on to help analyze data for the National Oceanic and Atmospheric Administration from studies having to do with forecasting hail storms. He left once he discovered that one ulterior motive of this study was to learn if it would be possible to use severe storms as weapons. Betty Ball worked for the National Conference of Christians and Jews from 1973 until 1977. The Balls moved to the mountain town of Nederland, Colorado, where she became town clerk. She also worked with Ideas, a program based on the "Foxfire" concept, which engaged various groups of children and adults from different cultural groups in collecting and publishing the oral histories of elders in their communities.

In 1984, the couple moved together to California. They spent their first year there as nomads, selling environmental T-shirts at different fairs every weekend. In each destination they learned as much as possible about the area's environmental problems and grassroots environmental organizations. They were invited to work in Ukiah, California, at a unique store that sold comic books, science fiction literature, and educational materials and provided desktop publishing services as well. There they became acquainted with the local activist community and learned about the serious environmental problems facing the area. Clear-cutting of the old-growth redwood forests was devastating ecosystems and waterways. The salmon were no longer able to spawn, which ruined the fishing industry. Siltation ruined riverine habitat for many other aquatic species as well. Gary Ball became involved in the local chapter of the Sierra Club and served as its chair for a term and a half.

A year after moving to Ukiah, the Balls were approached by a local environmentalist who had recently inherited a storefront building in downtown Ukiah, across the street from the courthouse, which he wanted to establish as an environmental center. The Balls offered to help create and staff the center, and owner John McCowen readily agreed. The Mendocino Environmental Center (MEC) opened in March 1987, and it soon became a center for the region's grassroots environmentalist and social justice efforts. Previously, groups were working independently from activists' homes; MEC provided space and resources and allowed the groups to collaborate and avoid duplication of efforts. Such organizations as the Redwood Coast Watershed Alliance, Earth First!, and the Sierra Club came to use MEC. What had been a disparate group of environmentalists eventually became one of the best-known and most effective regional environmental movements in the country. Well-known environmentalists like JUDI BARI worked out of the center, organizing the mass protests of the annual Redwood Summers.

When the Wise Use Movement emerged with its assault on environmentalism, Gary Ball authored an extensive article about it in the summer/fall 1992 MEC newsletter. "World War III: Meet the Wise Use Movement" was an exposé of the Wise Use Movement's agenda to disable environmental protection laws and regulations and chill environmental activism. The article identified major Wise Use activists and the movement's main organizations and their financiers and summarized the texts that provide the Wise Use Movement with its ideology. The article is still useful for activists and students of environmental movements and history; reprints are available from MEC.

Once MEC became enough of an institution to sustain itself, the Balls returned to Colorado. Betty Ball currently works at the Rocky Mountain Peace and Justice Center in Boulder, coordinating the nonviolence education collective that seeks to deescalate tensions that could lead to continued violent conflicts between University of Colorado students, community residents and merchants, and local police; she also coordinates the campaign to work collaboratively with other groups in the area to reduce violence against women. Both Balls work with the Center's Environmental Collective, Boulder Environmental Activists' Resource (BEAR), which seeks to draw the connections between the devastation of the earth, human rights violations, and the structural violence and oppression in our society and to provide support for environmental efforts being undertaken by other groups and individuals. BEAR also serves as the clearinghouse for RAID (Ridding Activism of Intimidation and Disruption), a national group that works to reduce violent attacks on environmental activists and disruption of their work by harassment. BEAR works to increase awareness of these attacks and to elicit support for the activists experiencing attacks and harassment. Betty Ball says that her ability to enthusiastically continue her environmental work despite the odds against winning every struggle is due to her understanding that "tangible results may not be visible for years, so it's important to do the work because it is the right thing to do, and not to be attached to an immediate or specific outcome."

Gary Ball currently works for Instead.com, a newly emerging Internet company that offers earth-friendly products that are produced in non-environmentally damaging ways, with fair and nonabusive labor practices.

Bibliography

Ball, Betty, "Oh What a Tangled Web We Weave, or Stop Me before I Email Again," *Mendocino Environmental Center Newsletter*, 1996; Ball, Gary, "World War III: Meet the Wise Use Movement," *Mendocino Environmental Center Newsletter*, 1992; "Mendocino Environmental Center," http://www.pacific.net/~mec/; "Rocky Mountain Peace and Justice Center," http://www.rmpjc.org.

Bari, Judi

(November 7, 1949–March 2, 1997)
Revolutionary, Labor and Earth First! Organizer

Earth First! activist Judi Bari organized nonviolent mass actions to stop what she and her allies called "the corporate liquidation of Northern California's" redwood forest. Identifying the enemy as the corporate interests that sought to cut the area's forests as quickly as possible without regard for the long-term interests of workers or environmental concerns, Bari used her experience as a labor organizer to help workers and environmentalists find common cause. Bari spent the last seven years of her life severely disabled and in pain as a result of a car-bombing attempt on her life. She dedicated this time to continuing the struggle against clear-cutting as well as to suing the Federal Bureau of Investigation (FBI) for illegally arresting her and fellow victim Darryl

Judi Bari (courtesy of Anne Ronan Picture Library)

Cherney and for improperly investigating the bombing.

Judith Beatrice Bari was born on November 7, 1949, in Baltimore, Maryland. Bari was an enthusiastic activist and organizer from adolescence, participating in student government and the "school spirit" club. Politics caught up with Bari while she was studying at the University of Maryland. She transferred her organizing abilities to the anti–Vietnam War struggle, coordinating mass actions to close major highways to protest the war. The war, as well as her growing interest in the labor union movement, drew Bari away from her studies and toward a life dedicated to activism.

Bari dropped out of college as a senior in 1972, in order to join the labor union movement. She got a job as a checker at a local Grand Union grocery store and was soon elected vice president of her local. When the demand that she put forward for a $1/hour wage raise was supported by her union but denied by store officials, she led 12,000 workers on strike. Fired for her union activity several times, Bari was transferred to the same company's Korvette stores and assigned to the cosmetics counter, a real oxymoron, as anyone who knew her would attest. She was fired after she was not able to assist cosmetics customers in an appropriate manner. Bari then began working for the U.S. Postal Service (USPS) at a Bulk Mail Center and became active in its union. Her cause there was mandatory overtime, which exhausted workers and left them vulnerable to injury while working with dangerous mail-sorting machinery. She helped a researcher for Jack Anderson's investigative column sneak into the plant to witness the problem, and she edited and surreptitiously distributed an underground publication, *Postal Strife*, a take-off on the official USPS publication *Postal Life*. *Postal Strife* rallied workers to resist mandatory overtime and sported a cigarette-smoking buzzard instead of the USPS golden eagle.

In 1979, Bari married union organizer Mike Sweeney and moved with him to northern California. She was active in the antinuclear, prochoice, and Central America solidarity movements of the 1980s; she had two daughters. When her marriage dissolved in 1988, she began to work as a carpenter to support herself. Soon she made the connection between the magnificent redwood lumber her clients demanded and the clear-cuts that were denuding the mountains throughout the area. Ever bold, when Bari was invited to the posh housewarming party of one wealthy client, her gift was a photograph of the clear-cut that his house had been built from.

Attracted by the no-compromise position, impetuous antics, and creative music of Earth First!, Bari became involved in that burgeoning environmentalist movement of northern California in 1988. Clear-cutting had been accelerated as such large corporations as Louisiana Pacific, Georgia Pacific, and Pacific Lumber were cutting the last remaining ancient rain forests of California, some on private lands and others in national forests. Earth First! and other local environmental groups responded with a variety of tactics, including direct action protests, preventing workers from entering, or blocking roads so that timber trucks could not leave sites. Throughout these potentially incendiary conflicts,

Bari and other environmental leaders insisted on a policy of nonviolent resistance and in 1990 signed a moratorium on tree-spiking (driving spikes into trees to discourage loggers from cutting them) and other acts of "ecotage."

Bari always encouraged fellow environmentalists to blame corporate management of logging companies rather than the workers. She was one of the first to make the connection between the safety and stability of lumber industry workers' jobs and the concerns of environmentalists. In 1989, when she and activists Anna Marie Stenberg and Darryl Cherney learned of a serious polychlorinated biphenyl (PCB) spill at the Georgia Pacific mill, they reported the incident to the Occupational Safety and Health Administration. The workers were never compensated for their exposure to the chemicals. When Louisiana Pacific laid off 1,000 workers in six mills throughout northern California, forced its remaining workers to work overtime despite the dangers of operating heavy machinery when overtired, and then opened up a new mill in Baja California, Mexico, it became even more evident that the real rift was between management and workers, not environmentalists and workers.

Still, Bari and the environmentalists were subject to the powerful public relations departments of timber corporations, which whipped up hatred for environmentalists by blaming them for the dwindling timber supply. There were a few violent face-to-face incidents in which environmentalists were hurt by loggers during protests, and in 1989, after leaving one event, Bari's car was intentionally rammed from the rear and forced off the road by one of the logging trucks that had been blockaded during the protest. In the spring of 1990, Bari began to receive regular death threats, and false Earth First! press releases were being distributed to local newspapers. At this time, Bari and Darryl Cherney were organizing "Redwood Summer," a series of mass protests through the region that would attract environmentalists and attention from around the country. On May 24, 1990, as Bari and Cherney were driving from Oakland, California, to Santa Cruz to recruit Redwood Summer participants, their car blew up. A bomb under Bari's seat, which was triggered to explode when the car began moving, almost killed Bari and injured Cherney. As soon as the FBI and the Oakland Police Department arrived at the scene, and in defiance of all the physical evidence, they falsely concluded that Bari and Cherney had planted the bomb themselves and blown themselves up and arrested them. Bari was hospitalized for six weeks and underwent a long and painful rehabilitation before she could walk again.

During the next seven years, despite severe pain and disability, Bari remained a key organizer in the Earth First! movement, investigated the bombing, and worked diligently and meticulously on a law suit against the FBI and the Oakland Police Department. Her research led her to think that Earth First! was being infiltrated by the FBI just as the Black Panthers and the American Indian Movement had been during the 1960s and 1970s. Bari's work led her to documents proving that the FBI had illegally infiltrated and actually instigated the power-line downing attempt in Arizona that ended with the arrests of Arizona Earth First! activists, including Dave Foreman, in 1989.

Bari discovered a lump in her breast in July 1996 and learned that she had breast cancer that had already metastasized to her liver in September of that year. She died on March 2, 1997. The Redwood Summer Justice Project in Santa Rosa, California, coordinates the ongoing legal struggle about the bombing, and the fight against continued forest depletion in the region continues in the hands of Bari's environmentalist colleagues in Mendocino and Humboldt Counties.

BIBLIOGRAPHY

Bari, Judi, *Revolutionary Ecology: Biocentrism and Deep Ecology*, 1998; Bari, Judi, *Timber Wars*, 1994; Breton, Mary Joy, *Women Pioneers for the Environment*, 1998; "Redwood Summer Justice Project's Official Judi Bari Home Page," http://www.monitor.net/~bari; Wilson, Nicholas, "Judi Bari Dies but Her Spirit Lives on," http://www.pacific.net/~mec/NEWSL/ISS23/23.01bari.html; Zakin, Susan, *Coyotes and Town Dogs: Earth First! and the Environmental Movement*, 1993.

Bartlett, Albert

(March 21, 1923–)
Nuclear Physicist, Educator

Since 1969, nuclear physicist Albert Bartlett has dedicated himself to educating the public about exponential growth of the human population. In his stump speech "Arithmetic, Population, and Energy," which by 1999 he had presented more than 1,300 times, Bartlett uses simple arithmetic to demonstrate how quickly even a so-called moderate rate of population growth will lead to overpopulation. He also deconstructs such oxymoronic catchphrases as "sustainable growth" and explicates the economic, social, and environmental consequences of population growth.

Albert Allen Bartlett was born on March 21, 1923, in Shanghai, China, where his father was principal of the Shanghai American School, a private school for American children. He came to the United States at three months of age and grew up in Ohio. Bartlett began college at Ohio's Otterbein College. After his freshman year, he took time off to wash dishes and cook on two Great Lakes freighters. When he decided to return to his studies, he applied to Colgate University in Hamilton, New York. He earned a B.A. in physics there, graduating summa cum laude in 1944. As Bartlett recounts, "I then hitchhiked, drove trucks, and hopped freights to get to a job at P.O. Box 1663 in Santa Fe where they were hiring physicists for war work. This turned out to be Los Alamos Scientific Laboratory where I worked for the rest of the war." Bartlett spent some time also on the Bikini Islands, taking high-speed photos of atomic bomb test explosions. After the war, Bartlett moved on to Harvard University for his M.A. and Ph.D. in physics (1948 and 1950, respectively). Bartlett married his wife, Eleanor, in 1946, and they have four daughters, Carol, Jane, Lois, and Nancy.

Bartlett was recruited by the Physics Department of the University of Colorado–Boulder (CU) in 1950, where he developed CU's first teaching laboratories for modern physics and worked in the rapidly expanding field of nuclear physics. While at CU, Bartlett was awarded several National Science Foundation grants for his work on radioactive isotopes and the beta-ray spectroscopy. He also received many teaching prizes from the University of Colorado, including the 1972 Thomas Jefferson award for outstanding service to the educational community, the 1974 Robert L. Stearns Award for outstanding achievement, and a University Medal from the University's Board of Regents in 1978. He served as chair of the four-campus Faculty Council of the University of Colorado for two years (1969–1971) and in 1978 was elected national president of the American Association of Physics Teachers. Despite his retirement from full-time teaching in 1988, Bartlett remains active on faculty committees at the university.

After nearly a decade of living in Boulder, which was then a small town nestled at the foot of the Front Range of the Rocky Mountains, Bartlett and other Boulder residents began to worry that Boulder would suffer the same ugly sprawl as a growing number of U.S. communities. In 1959, Bartlett helped found the local environmental group People's League for Action Now (PLAN)–Boulder County, which was to become an important protagonist in the establishment of Boulder's vanguard Open Space program. A 1959 City Charter Amendment proposed by PLAN–Boulder County activists limited development in the foothills west of Boulder. Bartlett served on the City of Boulder's Parks and Recreation Advisory Board for five years in the mid-1960s, chairing it in 1967, the year of a city sales tax measure that funded the purchase of open space lands. As of 1999, the Boulder Open Space Program had allocated over $100 million for 26,000 acres of mountain and plains parks. Bartlett and PLAN–Boulder County members also inspired Boulder's unique system of bikeways, which have made Boulder one of the most bike-friendly cities in the nation.

In 1969, Bartlett developed a speech on the hazards of growth that he would give, by 1999, over 1,300 times, to an estimated 130,000 interested Americans, including members of Congress and their staff and regulatory, environmental, professional, and educational groups. The speech, which the University of Colorado also sells on videotape, shows the actual numerical increases that result from various growth rates and the effects of growth on a community's resources. For example, the world's 1999 growth rate was 1.3 percent. Its population, estimated in 1999 to be six billion, was increasing at the rate of 80 million/year, and at this rate will double in 51 years. Bartlett deftly exposes the blithe claims of optimists such as economist Julian Simon, who wrote, "We have in our hands now . . . the technology to feed, clothe and supply energy to an ever-growing population for the next seven billion years." After Simon admitted that he had meant seven million—not billion—years, Bartlett did the arithmetic. He found that if Simon assumed that the population could grow at the modest rate of 1 percent per year for seven million years, the size of the population would be 2.3 x 10 to the 30,410th power.

Bartlett's message incorporates the quantitative analysis of Thomas Malthus, who in the eighteenth century warned that population had the potential to grow much more rapidly than food production could increase and that famine and starvation would inevitably curb population growth. This may still turn out to be the case, although it has not happened on a global scale as quickly as Malthus predicted. Bartlett reminds his audience of Malthus's premise that a population was doomed if it exceeded its environment's ability to provide for its human inhabitants. Modern environmental thinkers often warn people to heed an area's "carrying capacity."

Bartlett avoids the frequent conclusion that the problem is most serious in underdeveloped countries where population is growing at the fastest rates. He offers the observation that although the United States is growing more slowly than many other countries (at the rate of 1 percent per year), it has the most serious population problem because of the quantity of resources consumed here. An average American uses 30 times more resources than a person from an underdeveloped country. Bartlett uses the following equation: I = P A T, where *I* is the impact on an area's environment of any group of people, *P* is the size of the population, *A* is the per capita affluence, and *T* is the environmental damage caused by the technologies used there.

The damage done by continuous population growth is wide-ranging. It includes social disorganization, increased economic gaps between the poor and the wealthy, and the following environmental problems: the ozone hole, global climate change, the drop in food grain production per capita, the decline in oceanic fish catch and possible collapse of world fish stocks, and the availability of fresh, potable water. The solution? Most important: STOP population growth worldwide and reduce consumption of nonrenewable resources. Alternative, nonpolluting technologies, development of renewable fuels and other resources, and work on social injustice are all important, but gains in those areas can be canceled out by continued population growth. Bartlett insists that Americans concerned about population growth should start their work in the United States but that U.S. foreign assistance should be contingent on recipient countries having effective family planning policies.

Bartlett is not optimistic that humans will be able to curb growth ourselves, but he warns in a 1993 Boulder *Daily Camera* article, "Nature will take care of the problem if we don't first, and there's not much evidence that we can." Still, he perseveres with his message. "Wouldn't you like to believe there's intelligent life on earth? We just have to keep trying to educate people."

Bibliography

Bartlett, Albert, "Arithmetic, Population, and Energy," 1994 (available from CU's Information Technology Services at 303-492-1857); Bartlett, Albert, "Is There a Population Problem?" *Wild Earth*, 1997; Bartlett, Albert, "Reflections on Sustainability," *Population & Environment*, 1994; Cosper, Doug, "Talk to Focus on Overpopulation," *Boulder Daily Camera*, 1993.

Bartram, John, and William Bartram

(May 23, 1699–September 22, 1777; April 9, 1739–July 22, 1823)
Botanist; Botanist, Artist

The father-son botanical team of John and William Bartram was among the first to collect and describe indigenous American plants. Their work raised awareness in America and abroad of the rich flora that blanketed the North American continent and inspired naturalists, poets, and artists of their time and in the centuries to come.

John Bartram was born on May 23, 1699, in Marple, Pennsylvania, to William Bartram and Elizabeth Hunt Bartram. His mother died when he was two, and when his father and stepmother moved to North Carolina, he stayed in Pennsylvania with relatives. His only formal education was a few years' study at a local country elementary school. He inherited a farm from an uncle and did well as a farmer. He had two children with his first wife, who died in 1725, and with his second wife, Ann Mendinghall, Bartram had seven more. During the 1720s, Bartram bought a 100-acre farm along the Schuylkill River near the town of Kingsessing, on which he experimented with methods for increasing agricultural productivity, including crop rotation and the use of fertilizer. Part of the plot became Bartram's famous botanical garden, which survives today as part of the Philadelphia park system.

Bartram developed an interest in native flora early in life. His admiration of it was documented by Hector St. John de Crèvecoeur in his 1782 book, *Letters from an American Farmer.* Bartram was said to have sat down for a rest from plowing, plucked a daisy, looked at it carefully, and reflected, "What a shame, said my mind, that thee shouldst have employed so many years in tilling the earth and destroying so many flowers and plants without being acquainted with their structures and uses!" While this "quotation" was most likely imagined by de Crèvecoeur, it describes a man who, despite his lack of formal education, was constantly seeking information from the natural world.

Bartram's career as a botanist began about 1733, when he began to correspond with London merchant and botanist Peter Collinson. Collinson and other correspondents, with whom Bartram came into contact through Collinson, recognized Bartram's gift and sent him the most important botanical texts of that time. Bartram hired a Latin tutor in order to understand these books and to be able to correspond with their authors and other botanists. Collinson offered to hire Bartram to collect American botanical specimens and ship them to him. He was the first of many "subscribers" who funded Bartram's research. Bartram scoured the Pennsylvania countryside for unusual plants and made long excursions, the first with a group of scientists to Lake Ontario in 1743. When his son William grew old enough, Bartram included him on his expeditions. Over John Bartram's lifetime, they traveled as far south as Georgia, north through New England, and west to Ontario.

William Bartram was born to John Bartram and Ann Mendinghall Bartram on April 9, 1739, on their farm in Kingsessing, Pennsylvania. He was given a more complete education than his father had

William Bartram (Archive Photos)

had; he attended the Academy of Philadelphia (which later became the University of Pennsylvania), graduating in 1756, and apprenticed with a Philadelphia merchant for four years after that. Both his schooling and his apprenticeship were broken by frequent trips with his father. Beginning in 1755, with an expedition to the Catskills, his father included him on his botanical expeditions, and young William avidly sketched the plants and animals they found. In 1760, they journeyed to the Carolinas, in 1761 to the forks of the Ohio River, and in 1765, to Florida. William Bartram made his own four-year expedition starting in 1773 to Florida, Georgia, and the western Carolinas.

Both William and John Bartram kept detailed notes during their expeditions, and these notes formed the body of the books, letters, and reports for which they became famous. John Bartram's prose was dense and difficult to penetrate, since his abbreviated education did not include conventional spelling of the time, nor accepted sentence structure. He defended his style by claiming that he preferred to write not by the rules of grammar but rather by "nature." The only book he published was *Observations on the Inhabitants, Climate, Soil, Rivers, Productions, Animals, and Other Matters . . . From Pennsilvania to Onondago, Oswego and the Lake Ontario in Canada,* which described his first expedition in 1743 and came out in 1751. Otherwise, he wrote reports and long letters, primarily for his patrons, including Collinson and such other seventeenth-century luminaries as American naturalist Mark Catesby, American physician William Byrd II, American inventor and politician Benjamin Franklin, Dutch physician John Frederic Gronovius, Finnish-Swedish naturalist Peter Kalm, Swedish taxonomist Carolus Linnaeus, and many others. Other professional and amateur botanists eagerly read reprints of his papers, since they knew that he was, as Linnaeus called him, the "greatest contemporary 'natural botanist' in the world." His work contained not only elaborate descriptions of climate, landscape, soil conditions, and flora and fauna, but also observations about the indigenous people he encountered: particularly which plants they used for food and supplies.

William Bartram's writings were disseminated to a broader readership and were illustrated with exacting drawings of flora and fauna. Historians now recognize him as the most important natural history artist before JOHN JAMES AUDUBON. William Bartram's journal from his four-year expedition, entitled *Travels through North and South Carolina, Georgia, East and West Florida, the Cherokee Country, the Extensive Territories of the Muscogulges, or Creek Confederacy, and the Country of the Choctaws*, published in 1791, identifies 215 native birds of the Southeast, the most complete list of the day. He described the complexity of ecosystems, marveling at the perfect bodies of even the tiniest insects and at how their great numbers served to nourish fish and other predators on up the food chain. In addition to this book's usefulness for naturalists, it inspired the growing Romantic literary movement, which celebrated wilderness as an expression of the beneficence, rationality, and perfect organization of God's creation. William Bartram wrote of his awe at the spectacular wild landscapes of Appalachia and his feeling of solitude within such places. He was at once exhilarated and terror-struck while alone in the wilderness. Poets Samuel Coleridge and William Wordsworth were said to have been especially inspired by Bartram's work.

Both of the Bartrams made their base at the Bartram botanical garden in Schuylkill. Thanks to generous patronage from King George III, who named John Bartram the king's botanist, and wealthy English physician John Fothergill, who was a correspondent of John Bartram and sponsor of William Bartram, the Bartrams were able to dedicate themselves full-time to botany. Although they sent many specimens to their patrons abroad, they kept many for their own botanical garden, which William and his brother John converted into a nursery after their father's death.

John Bartram died on September 22, 1777, in Kingsessing, Pennsylvania. William Bartram returned to Kingsessing from his last expedition shortly after his father's death, in January 1778. From that point on, William Bartram remained at Kingsessing. He prepared his travel notes for publication; *Travels . . .* saw several editions in English and was translated into Dutch, German, and French before 1800. William Bartram spent the remainder of his life writing and drawing and receiving famous visitors to Kingsessing. He died on July 22, 1823, while walking in his garden.

BIBLIOGRAPHY

Nash, Roderick, *Wilderness and the American Mind*, 1967; Scheick, William J., "John Bartram," in *Dictionary of Literary Biography, American Colonial Writers*, Emory Elliott, ed., 1984; Terrie, Philip G., "William Bartram," in *American Nature Writers*, John Elder, ed., 1996.

Bates, Marston

(July 23, 1906–April 3, 1974)
Zoologist, Writer

One of the foremost zoologists in the United States, Marston Bates insisted that scientists should address a larger audience in their writings. Following his own decree, Bates wrote a series of popular volumes that made the problems of environmental science accessible to a wide audience. Though criticized by his peers for his offering little to students of ecology, Bates found a broader audience; his engaging writing style helped spread awareness and understanding of global ecological problems.

The only son of Glenn, a farmer and horticulturist in Florida, and Amy Mabel (Button) Bates, Marston Bates was born in Grand Rapids, Michigan, on July 23, 1906. Bates attended public schools in Fort Lauderdale, Florida, and then majored in biology at the University of Florida, Gainesville, receiving his B.S. degree in 1927. Before continuing his studies, Bates spent three years, from 1928 to 1931, working for Servicio Técnico de Cooperación Agrícola of the United Fruit Company in Honduras and Guatemala. He started as a research assistant in entomology, working his way up to director. Bates did his graduate studies at Harvard University, receiving his A.M. degree in 1933 and his Ph.D. a year later. His dissertation was entitled "The Butterflies of Cuba."

Bates joined the staff of the Harvard Museum of Comparative Zoology in 1935 but spent most of his two years there on leave with the Rockefeller Foundation, looking into the biology of mosquitoes in Albania. He resigned from the Museum of Comparative Zoology in 1937 and continued his work in Albania until 1939. Bates also studied malaria in Egypt and then, at the onset of World War II, studied yellow fever in Colombia. He remained employed by the Rockefeller Foundation until 1952, except for one year of postdoctoral study at Johns Hopkins University (1948–1949). During this year, Bates wrote *The Natural History of Mosquitoes*, which was widely recognized as a masterpiece and of immense value, especially to scientists but also to the lay public. Understanding the entomology of mosquitoes was essential before control of mosquito-borne diseases could be possible. Bates left the Rockefeller Foundation in 1952 to join the zoology faculty of the University of Michigan.

Bates saw little value in not making scientific findings widely accessible. Writing for a small number of scientists would not help solve the many environmental problems the world faced and that needed a massive popular response to solve. As a faculty member at the University of Michigan, Bates wrote 12 books aimed for a popular audience. The topics of his books ranged from a study of populations and problems in demography in *The Prevalence of People* (1955), to his own observations about the problems of coexistence of several species in *A Jungle in the House: Essays in Natural and Unnatural History* (1970). This last book describes the greenhouse attached to Bates's home in Ann Arbor, Michigan, which he stocked with tropical plants, birds, and animals. His observations about problems of coexistence in his "jungle" led to discussions about the larger environment in which humans, plants, and animals coexist. Opinions of Bates's books were mixed. Scientists often criticized his work, citing that he was too prone to accept the conclusions of authorities in fields where he himself had not specialized. Moreover, his books generally had little to offer professional scientists or their students. Nonetheless, they were praised outside of academia; for many lay readers they were a first, compelling introduction to the natural world.

During his twenty years as a member of the zoology faculty at the University Michigan, Bates continued to travel extensively. In 1954, he was Timothy Hopkins Lecturer at Stanford University, and between 1956 and 1957, the director of research at the University of Puerto Rico. He also spent two years (1956–1958) as chairman of the National Science Foundation's division of comparative biology and medical sciences, three years (1955–1958) on the advisory board of the Guggenheim Foundation, and seven years (1955–1962) as a trustee at the Cranbrook Institute of Science. In 1967, Bates was the recipient of the Daly Medal of the American Geographical Society. The following year, the University of Michigan gave him its Distinguished Faculty Achievement Award.

Bates married Nancy Bell Fairchild, the daughter of botanist David Fairchild, in 1939. She went with him to Colombia and shared their experiences in a *National Geographic Magazine* article, "Keeping House for a Biologist in Colombia," in 1948. Together they had four children, one son and three daughters. Bates died April 3, 1974, in Ann Arbor, Michigan.

BIBLIOGRAPHY

Bates, Marston, *The Forest and the Sea: A Look at the Economy of Nature and Ecology of Man*, 1960; Bates, Marston, *Gluttons and Libertines*, 1968; Bates, Marston, *The Nature of Natural History*, 1950; Bates, Nancy Bell Fairchild, *East of the Andes and West of Nowhere*, 1948.

Bauer, Catherine

(May 11, 1905–November 22, 1964)
Urban Planner, Houser

Catherine Bauer was a leading voice in the development of the fields of housing and urban planning. She wrote many influential articles and a 1934 book, *Modern Housing*, which laid out many of the principles by which federal housing programs were developed. During the 1930s and 1940s she served in executive positions in labor organizations and in federal housing programs. She lobbied successfully for the passage of the United States Housing Act of 1937, an important milestone in the history of the development of national housing policy, leading eventually to the establishment of the Department of Housing and Urban Development (HUD). As administrator, policy advocate, and academic, Bauer successfully championed a new, integrated approach to urban planning that was attentive to the whole environment, including poverty, aesthetics, land use, transportation, and open space issues.

Catherine Krouse Bauer was born in Elizabeth, New Jersey, on May 11, 1905. The oldest of three children, Bauer was raised in an affluent family. Her father, Jacob, was a well-regarded highway engineer; her mother, Alberta, a self-taught biologist and botanist who passed her love of nature on to her children. Bauer graduated at the top of her high school class in 1922 and went to Vassar College. She studied at the Cornell University College of Architecture for one year in 1924 and eventually graduated from Vassar in 1926. Bauer spent the year following graduation in Europe, where her interest in architecture grew into an expertise on developments in European modernism and urban planning. On her return to New York she took a job at the prestigious publishing house Harcourt, Brace, where she first met the architectural critic and social philosopher LEWIS MUMFORD. Mumford and Bauer had a long affair, which played a vital role in shaping Bauer's ideas about urban and regional planning. Both Mumford and Bauer were involved in the Regional Planning Association of America (RPAA), an influential group that advocated for community-based planning and "garden cities," modeled after the ideas of British theorist Ebenezer Howard.

In 1933 Bauer took a job in Philadelphia with the Labor Housing Conference (LHC), a lobbying arm of the Pennsylvania Federation of Labor. Here Bauer's role as "houser" became increasingly political. The term *houser* was used in this period to describe those committed to improving housing for the urban poor. Bauer was involved in the effort to pass comprehensive housing legislation to address the housing crisis facing poor Americans. She fought for federal support of local initiatives that would provide decent living space, with light and open space, parks, and integration of land and water use planning. She eventually moved to Washington, D.C., where she lived until 1939. While in Washington she developed friendships with a number of influential planners and environmentalists, including Ernest Bohn, president of the National Association of Housing Officials, and ROBERT MARSHALL, founder of the Wilderness Society.

Growing weary of the demands of Washington politics, Bauer took a position at the University of California in Berkeley in 1939. She taught classes in housing policy and became involved in California regional planning. In 1941 she helped the California Housing Association expand into the California Housing and Planning Association, a group she served as vice president for three years. In 1942 she married leading Bay Area architect William Wurster; the couple had one daughter, Sarah Louise, born in 1945. In 1943 Bauer moved with her husband to Cambridge, Massachusetts, where he worked on a graduate degree and she continued her advocacy work. In 1950 they returned to California, where Bauer lived for the rest of her life. She taught in the University of California's Department of City and Regional Planning, winning a number of grants to study the effects of California's rapid growth, including a 1956 grant to study the effects of urban development on California's wild places. In 1960 she founded, along with William Matson Roth and Alfred Heller, California Tomorrow, a group dedicated to including conservation in state planning. In 1962 she and Clark Kerr staged a conference on "The Metropolitan Future," which brought together a wide range of experts, including business leaders, planners, social scientists, and politicians.

Bauer died on November 22, 1964, of exposure, after falling on California's Mount Tamalpais.

Bibliography

Frieden, Bernard, and William Nash, *Shaping an Urban Future: Essays in Memory of Catherine Bauer Wurster*, 1969; Newbrun, Eva, and H. Peter Oberlander, *Houser: The Life and Work of Catherine Bauer*, 1999; Sussman, Carl, *Planning the Fourth Migration: The Neglected Vision of the Regional Planning Association of America*, 1976.

Bean, Michael

(July 3, 1949–)
Wildlife Lawyer

Michael J. Bean has been a leader in wildlife law and endangered species protection since 1977, when he joined the staff at the Environmental Defense Fund (EDF; now known as Environmental Defense, ED) as chairman of its Wildlife Program.

Michael Bean was born on July 3, 1949, in Fort Madison, Iowa, a small town of 12,000 along the Mississippi River where he first became interested in the natural world. Bean spent as much time as possible exploring the Mississippi riverbanks and an intermittent stream in his backyard. As a child, Bean caught and raised snakes and collected insects. His grandmother shared his interest in insects and helped him amass his bug collection. Bean decided to pursue a degree in entomology when he entered Iowa State University in 1966 but reevaluated his decision after realizing that this field concentrated on controlling insects in the state's croplands. His decision was solidified in 1967 when Bean discovered a robin suffering from what he suspected were the effects of dichlordiphenyltrichlor (DDT) poisoning near the campus. Coincidentally, scientists in Stoney Brook, New York, at just that time were forming the EDF to combat the use of DDT in the United States. EDF's efforts to ban DDT played a significant role in Bean's eventual acceptance of a position at EDF. Certain that he did not want to major in entomology, Bean transferred to the University of Iowa in 1968 and studied political science. Having earned a Phi Beta Kappa key and graduated summa cum laude from the university in 1970, Bean entered Yale Law School later that year.

Bean received his J.D. in 1973 and began work in antitrust law with a well-respected Washington, D.C., firm, Covington & Burling. Finding the work unsatisfying, Bean considered combining his interests in the natural world with his legal education. While working for Covington & Burling, Bean wrote an article about the "impending extinction of the desert-dwelling pupfish of Devil's Hole National Monument" that was published in the *Washington Post* and a review of a book on endangered species in *Natural History Magazine* in 1976.

These article proved to be well-timed, because to celebrate the bicentennial in 1976, the President's Council on Environmental Quality commissioned a book on wildlife conservation law in the United States. The Environmental Law Institute was selected to write the book, but it determined that no one on its staff was qualified. Bean later told an EDF reporter that "in 1976, there weren't any experts in wildlife law; indeed, the very term 'wildlife law' was a novelty. There were environmental lawyers then, including some who had handled cases pertaining to wildlife, but no one who had developed that expertise as a specialty. The Institute concluded that there weren't any qualified candidates, so they hired me instead." Nine months after he started the book, he had finished the first edition of *The Evolution of National Wildlife Law*. As Bean says, "It was instantly recognized as the leading book on the subject (because there weren't any others), and it suddenly opened a lot of doors for me." The Environmental Defense Fund offered Bean a position, which he accepted and has never regretted. Not only was EDF working on endangered species recovery, the group had succeeded in ensuring a ban on DDT in the United States—two issues about which Bean cared deeply.

Over the past 22 years at EDF, Bean's work has focused extensively on endangered species conservation. Spending time lobbying and litigating, Bean has made headway into endangered species recovery. Owing in part to his work on sea turtle protection, regulations are in place that protect six threatened species. Bean also worked to protect wildlife refuges from oil and gas exploration under Secretary of the Interior James Watt. Watt's proposal would have opened up to oil and gas leasing nearly all the wildlife refuges that were created for endangered species protection and almost all the refuges east of the Rocky Mountains. From 1977 to 1987, Bean influenced the initial implementation of the Convention on International Trade in Endangered Species (CITES) that was signed in 1972 and enforced in 1973.

Recently, Bean's work has centered around protecting rare species on private land. In describing his work, Bean recalls ALDO LEOPOLD's words, "the only progress that counts is that on the actual landscape." That is, the number of law suits won is ultimately less important than what actually happens to the species themselves.

As old methods of protection are failing, Bean recognizes the need to create new methods of species conservation that will work in practice on private lands. One example is a "safe harbor" agreement that serves to reward species conservation. Bean put this new idea into practice in the sandhills area of North Carolina, where the endangered red-cockaded woodpecker now has a better chance at survival than it had before.

Along with his on-the-ground success, Bean is a prolific author and speaker. He has written over 75 articles on endangered and rare species issues and has been a panelist in numerous wildlife events and an adviser to wildlife conservation organizations. His book, *The Evolution of National Wildlife Law*, published in 1977, has been twice revised, most recently as *The Evolution of National Wildlife Law: Revised and Expanded Edition* (1997) with Melanie Rowland. He is still considered the leading source of information for wildlife law.

Bean has received several awards, including Pew Charitable Trusts Conservation and the Environment Scholar (1990), the Society for Conservation Biology's Distinguished Achievement Award for the legal defense of endangered species (1988), and the Desert Tortoise Preserve Committee's "Golden Tortoise" Award for securing legal protection for the desert tortoise (1990).

Although Bean no longer collects insects, he spends much of his free time in the field, identifying and observing dragonflies and butterflies. He lives in the Washington, D.C., area, with his wife, Sandy, and two daughters, Amanda and Emily.

BIBLIOGRAPHY

Bean, Michael F., "The Endangered Species Act and Private Land: Four Lessons Learned from the Past Quarter Century," *Environmental Law Reporter: News and Analysis*, 1998; Bean, Michael F., "Endangered Species, Endangered Act?" *Environment*, 1999; Bean, Michael F., with Melanie Rowland, *The Evolution of National Wildlife Law: Revised and Expanded Edition*, 1997; Bean, Michael F., "The Private Land Problem," *Conservation Biology*, 1997; Bean, Michael F., "A Tool Kit for Conservation Issues, a Review of Private Property and the Endangered Species Act," *Bioscience*, Josen F. Shogren, ed., 1999; Geniesse, Jane, "Michael J. Bean," EDF-Letter, 1986, http://www.edf.org.

Beattie, Mollie

(April 27, 1947–June 27, 1996)
Director of the U.S. Fish and Wildlife Service, Forester

As director of the U.S. Fish and Wildlife Service (USFWS) from 1993 to 1996, Mollie Beattie presided over the agency in a time of shrinking budgets and an often hostile Republican Congress. The first woman and nonhunter ever to hold the position, Beattie successfully defended the Endangered Species Act, oversaw the reintroduction of wolves into the northern Rockies, and developed successful public/private partnerships to protect wildlife and wilderness. Beattie pushed the agency to put the issue of endangered species into a larger context and moved policy from a crisis-oriented, species-centered approach to a more forward-thinking, ecosystems approach to preservation.

Mollie Hanna Beattie was born on April 27, 1947, in Glen Cove, New York. She was raised in Connecticut and later attributed her love of nature to her grandmother, Harriet Hanna, who lived on a farm in upstate New York. Hanna was a self-taught botanist who knew the scientific names of all the local plants and kept a number of wild animals, including raccoons, a crow, and a deer with an artificial hip. Beattie said that she learned an important lesson from her grandmother: "If it moves, feed it." After graduating from Marymount College with a B.A. in philosophy in 1968, Beattie worked for several years as a journalist.

In 1973 she was offered a job writing for *Country Journal*. In order to gain the outdoor skills she needed for the job, Beattie enrolled in an Outward Bound course in the Colorado Rockies. She had little wilderness experience and arrived with luggage that included electric curlers, a hair dryer, clean sheets, and stiff new hiking boots with their tags still on. At the end of the course, after first swearing the brutal final hike would be her last for life, Beattie signed on to be an instructor, and she began a new career in wilderness pursuits. In 1979 she earned a master's degree in forestry from the University of Vermont. During her studies, Beattie chose wildlife management pioneer Aldo Leopold as a role model for her approach to wildlife conservation. Leopold saw the forest not as timber but as an ecosystem, considering it, in Beattie's words, as "all that it produces and all the things that are seen and heard there." From 1980 to 1982 she worked for the University of Vermont Extension Service, teaching forestry and wildlife management to private landowners and coordinating federal, state, and private land management efforts. From 1983 to 1985 Beattie was the program director and lands manager for the Windham Foundation in Grafton, Vermont, where she gained more experience directing public/private cooperative conservation efforts. While living in Grafton she and her husband, builder Richard Schwolsky, built a solar house, a mile away from the nearest utility pole. (Schwolsky once installed solar panels on the White House, during Jimmy Carter's presidency, though Ronald Reagan ordered their removal.)

In 1985 Beattie was appointed by Madeleine Kunin, Vermont's first female governor, as commissioner of the Department of Forests, Parks, and Recreation for the state of Vermont. This position prepared her for the USFWS in a number of ways, one of them being the fact that she was the first woman to hold the post. During her tenure Beattie addressed a number of pressing environmental issues, including clear-cutting, resort development, overuse of pesticides, and air and water quality concerns. Beattie pushed for local control over development decisions and initiated a task force that eventually became the federally funded Northern Forest Council, an alliance of conservation efforts in Maine, New Hampshire, New York, and Vermont. She also oversaw the acquisition of several thousand acres of land for state parks and forests. From 1989 to 1990, she continued her work for Vermont as deputy secretary for the Agency of Natural Resources. In 1991 she earned an M.P.A. from Harvard University's Kennedy School of Government.

In 1993 she became President Clinton's USFWS director. Much of the confirmation hearing addressed her view of the Endangered Species Act, and she signaled in the hearing her commitment to working with private landholders to protect species and habitat before species become officially listed as endangered. Too often, she said, the act is used as an emergency measure when it should be seen as an ethical principle to protect whole ecosystems as wildlife habitat. She followed through on this principle throughout her years at the USFWS, developing regulations to encourage landholders to contribute to the preservation of species as a whole, while allowing flexibility for farmers and ranchers to kill single, troublesome members of those species. Beattie came under fire for allowing fishing and hunting in about half the National Wildlife Refuges, which she permitted as long as scientists determined these uses were compatible with wildlife preservation goals. Beattie oversaw the return of wolves to Yellowstone National Park and was effective in preserving wilderness in Alaska, where she negotiated the competing interests of Alaska natives, sportsmen, industry, and habitat preservation. In addition, Beattie presided over the addition of 15 national wildlife refuges to the existing system.

Mollie Beattie died in Townshend, Vermont, on June 27, 1996, from brain cancer. Since her death, portions of the Arctic National Wildlife Refuge in Alaska and 76 acres of forest and bog near Island Pond, Vermont, have been named in her honor.

Bibliography

Dicke, William, "Mollie Beattie," *New York Times*, 1996; Gup, Ted, "Beattie's Battle," *Audubon*, 1994; Walsh, Barry Walden, "The Ecosystem Thinking of Mollie Hanna Beattie," *American Forests*, 1994.

Beebe, C. William

(July 29, 1877–June 4, 1962)
Marine Biologist, Ornithologist, Nature Writer

William Beebe, director of tropical research with the New York Zoological Society for nearly 63 years, was a professional field biologist who started his career studying jungle birds and later became famous for his deep sea dives to study ocean life. In addition, he was a skillful writer, and with a cordial and entertaining style he wrote popular accounts of his fieldwork; in the process becoming the best known American nature writer of his time. His writings demonstrate his strong interest in conservation; he wrote of his distress at the devastation already occurring in the tropical forests, predicted a time when human destruction would wipe out whole populations of birds, and spoke of nature's most terrible enemy—humans. His vivid presentation made his message especially compelling, and he is credited with popularizing the study of zoology and giving many people a new concern for threatened species and habitats.

Charles William Beebe was born on July 29, 1877, in Brooklyn, New York, the only child of Charles and Henrietta Marie (Younglove) Beebe. He was raised in East Orange, New Jersey. As a boy he enjoyed reading the scientific adventure stories of Jules Verne and others, and he cultivated a taste for excitement and exploration. In

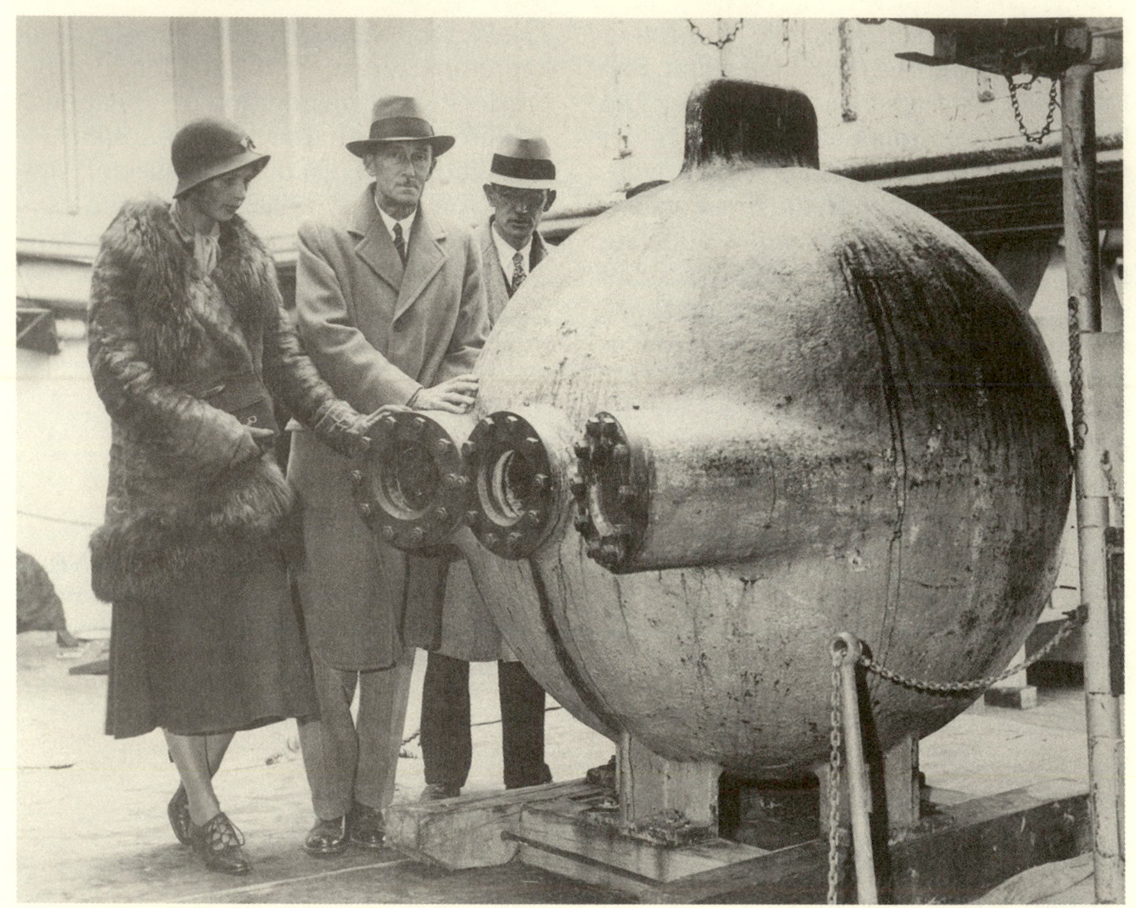

Gloria Hollister, William Beebe, and John Teevan are seen here with the bathysphere in which Beebe and Otis Barton made a famous descent in August 1934. Here the team arrives in New York from Bermuda on November 23, 1934. (Bettmann/Corbis)

1891 he entered East Orange High School, where he would achieve highest marks in geology, botany, physiology, and zoology. He attended Columbia University for three years as a zoology student, and although he subsequently claimed a B.S., he never actually completed the classes required for a degree. One of his teachers, Henry Fairfield Osborn, encouraged and inspired Beebe's curiosity and helped Beebe get a job as assistant curator of birds at the New York Zoological Park (later known as the Bronx Zoo) in 1899. Three years later he became full curator, but he was becoming more interested in field studies and began taking expeditions to study birds. In 1902 he married Mary Blair Rice, and the two of them traveled to Mexico and collaborated on a book for the general public called *Two Bird Lovers in Mexico* (1905). They later wrote another book together, *Our Search for a Wilderness* (1910), about the natural history of the jungles of Venezuela and British Guiana (now Guyana). Mary eventually divorced him in 1913 and made a name for herself writing novels and travel books.

In order to continue his studies in the tropics, Beebe established the New York

Zoological Society's department of tropical research in British Guiana in the mid-1910s. The results of his first year of fieldwork in ornithology and entomology there yielded a technical work titled *Tropical Wild Life in British Guiana* (1917), written in collaboration with his chief assistants, C. I. Hartley and P. G. Howes.

In 1917 and 1918, Beebe served as an aviator in World War I, an experience that left him agitated and unhappy. He returned to the jungles of British Guiana for healing and solace and compiled a series of essays originally published in the *Atlantic Monthly* and later collected in *Jungle Peace* (1918). Another work published around this time was the result of years of studying pheasants in southeast Asia and other parts of the world. The beautifully illustrated four-volume series was later abridged to a one-volume version, *Pheasants: Their Lives and Homes* (1926), which was more accessible to the general public.

Beebe's works endure largely because of their appeal to a large audience. Although he was a detail-oriented researcher, he was not so taken with science as to be obsessed and instead brightened his observations with humor and enjoyment of life. He conveyed a contagious excitement over his discoveries and observations of the natural world and offered readers a stimulating introduction to science in general and the tropical jungles in particular. His concern for the future of the jungles is also contagious, and often his books reveal these conservation ideals. Even though it was early in the twentieth century, he wondered if there was any place on earth left untrampled by people and lamented the tragic succession of chopping, overcultivation, and overgrazing in the tropics where he worked. Proclaiming humans to be the worst enemy of the natural world, he pointed out that everywhere they have gone, they have worked havoc on the wild plants and animals. Beebe's writings describe the interrelationships that bind different organisms together, creating a web of life that is in constant danger of being upset by environmental degradation—a concept that he articulated far earlier than the rest of the scientific community. He frequently details how particular organisms interact with their larger habitat, providing many of his readers with their first taste of the principles of ecology. Among his many admirers was RACHEL CARSON, who acknowledged Beebe as an inspiration.

In 1927 Beebe was married to the novelist Elswyth Thane and the following year set up a six-acre tropical research laboratory on Nonsuch, an island in Bermuda. His career as a marine biologist had begun three years earlier on a trip to the Galapagos Islands where he became interested in collecting ocean specimens. From then on he shifted his focus from birds to undersea life and began a series of deep-sea exploratory expeditions. Several of his close studies of Bermuda sea life were documented in *Nonsuch: Land of Water* (1932), a book that demonstrated the links between land and sea and the interrelationships of life on the shore. His studies led him deeper and deeper underwater as well, and he became famous for a dive he made with Otis Barton on August 15, 1934, in a bathysphere. Barton and Beebe had designed this contraption—a spherical chamber made of a single casting of finest grade open-hearth steel, just large enough for two men to fit inside. Oxygen tanks were fixed to the sides, along with trays of powdered chemicals for absorbing carbon dioxide and water vapor. Curled into the bathysphere, Barton and Beebe descended to 3,028 feet, deeper than any human had ever dived beneath the ocean's surface.

Beebe chronicled the experience in *Half Mile Down* (1934) and continued oceanographic studies through the 1930s in the Pacific and Central America. By the early 1940s, he was ready to return to his studies of tropical land, and he began a series of expeditions to the Venezuelan Andes. In 1949 he established another New York Zoological Society research station, this time at Simla, Trinidad, where he worked until his death. *High Jungle*, the last of his full-length books, was published that same year. In the latter part of 1955, Beebe took his last major expedition, a 144-day trip that included visits to Naples, India, and Singapore. After three years of slowly failing health, William Beebe died of pneumonia on June 4, 1962, at his Simla station.

Bibliography

Beebe, Mary Blair, and C. William Beebe, *Our Search for a Wilderness*, 1910; Beebe, William, *High Jungle*, 1949; Pollard, Jean Ann, "Beebe Takes the Bathysphere," *Sea Frontiers*, 1994; Tracy, Henry Chester, *American Naturists*, 1930; Welker, Robert H., *Natural Man: The Life of William Beebe*, 1975.

Begley, Ed, Jr.

(September 16, 1949–)
Actor

Ed Begley Jr. began his acting career in 1967 at the age of 17 on an episode of the television program *My Three Sons*. Now a Hollywood veteran of more than 30 years, Begley has appeared in numerous television and movie roles ranging from *The Simpsons* to *Star Trek Voyager* to *The Accidental Tourist* to *She-Devil*. Best known as an actor for his portrayal of Dr. Victor Ehrlich on the television series *St. Elsewhere*, Begley is also highly respected as a passionate and dedicated environmentalist. He gives about 90 percent of his time to environmental causes, serving on the boards of several environmental organizations. He is also well known for putting his principles into practice with an ecologically sound lifestyle that includes alternative transportation, a solar home, a vegetarian diet, and intelligently informed activism.

Edward James Begley Jr. was born on September 16, 1949, in Hollywood, California. His father, Ed Begley Sr., worked as a factory laborer in Hartford, Connecticut, before finding later success as a radio personality and actor. In 1962, at the age of 62, Ed Begley Sr. won an Oscar for best supporting actor for his role in *Sweet Bird of Youth*. Following in his father's creative footsteps, Ed Begley Jr. decided at the age of five that he wanted to be an actor and began auditioning. His first role was as a friend of Chip's on an episode of the famous 1960s situation comedy *My Three Sons*.

Begley grew up attending private Catholic and military schools until his senior year. Then he attended Van Nuys High School, the alma mater of several well-known Hollywood actors, including Marilyn Monroe, Robert Redford, and Natalie Wood. After high school, Begley attended Los Angeles Valley College, where he became friends with actor Michael Richards. The chemistry between the two was magnetic, and soon they were doing improvisations on stage at the Troubadour nightclub in West Hollywood. After they went their separate ways, Begley decided to pursue work as a cameraman, which he did for several years until moving to Boulder, Colorado, in 1971. In Boulder, where he lived for only six months, Begley had his first solo nightclub performance as a stand-up comedian. His comedy routine landed Begley work opening for singers and musicians from Don McLean to Barry Manilow at venues from New York City to Kansas City.

Returning to Hollywood, Begley began acting again with a part in *Stay Hungry*, a film starring Jeff Bridges and Sally Field. His career included many cameos in movies such as *This Is Spinal Tap*, *Streets of Fire*, *Eating Raoul*, *Cat People*, *The Accidental Tourist*, and *Protocol*. Begley has also guest starred in many television shows, including *Happy Days; Mary Hartman, Mary Hartman; MASH; Roseanne; The Simpsons; The Drew Carey Show; Star Trek Voyager;* and *Providence*. His most recent large-scale success was as Dr. Victor Ehrlich on the television series *St. Elsewhere*, a role for which Begley was nominated twice for an Emmy.

Begley's personal life in his teens and twenties was often filled with turmoil, and Begley admits to having an addictive, obsessive personality. He has struggled with alcoholism and gambling and has even been known to clean obsessively. Begley told Mark Morrison in *Rolling Stone* that in 1976, he was drinking a quart of vodka a day. Begley married Ingrid Taylor in 1977, and they had their first child, Amanda. The following year, at the age of 29, Begley stopped drinking alcohol for good. Married for 13 years, Begley and Ingrid had their second child, Nick, before divorcing in 1989.

Begley states on his official website (http://www.edbegleyjr.com) that his "passion for preserving this beautiful planet" began when he was a child in the late 1950s living with his family in rural Long Island, New York: "Anyone camping in those wild and beautiful areas couldn't help but grow up with a profound respect for natural systems." Begley also says that in 1969, when he saw the first photographs of Earth from the moon, he was deeply moved. "It had a profound effect upon me because I saw that, for better or worse, we have nowhere else to go. . . . And I don't think it's any coincidence that not too long after, we had the first Earth Day." According to Begley, that was the beginning of his activism on behalf of the environment; he became a vegetarian, started recycling and composting, and bought his first electric car.

Begley receives praise for his ecologically sound lifestyle and activism from all ranks of the Hollywood and environmental communities. "Ed is a very political guy in the best sense of the word," says Lucy Blake of the California League of Conservation Voters. "He understands the relationship between political leadership and environmental quality, and he's been very powerful in getting that message across to the public." Sandra Jerabek, executive director of Californians Against Waste, says, "No doubt about it, Ed's a fully committed environmentalist. . . . He drives an electric car, he rides his bike, he takes the bus and the train. When he renovated his garage, he found a way to recycle the broken concrete they ripped out. I mean, how many people take their commitment that far?"

Begley himself is unabashedly enthusiastic when he talks about what he has done, and what he encourages others to do as well, to make life more ecologically sustainable. His official website is a virtual environmental education center where readers can log on and find out almost anything they want to know about how to live more lightly on the planet. Begley talks about everything from biodegradable soap and recycled toilet paper, to compact fluorescent light bulbs and energy-saving thermostats, to how he retrofitted his existing house to be off the power grid, using thermal collectors for hot water and solar energy for electricity. Begley's most ardent advice for anyone interested in doing something that will help to protect the natural environment is to avoid automobiles and use alternative transportation as much as possible. He is well known for riding his bike, taking the bus, and driving his electric car, a GM EV-1. Even his exercise bike has a generator that returns power to the same battery array that stores the solar energy for the house.

Begley serves on the boards of the Walden Woods Project, the Environmental Media Association, Earth Communications Office, and Environmental Research Foundation. He is often asked to speak at events, including the 2000 Earth Day celebration in Washington, D.C. Currently, Begley lives in Los Angeles with his girlfriend, Rachelle Carson, and their baby, Hayden Carson Begley.

BIBLIOGRAPHY

"Ed's World—The Official Website for Ed Begley Jr.," http://www.edbegleyjr.com; Morrison, Mark, "The Trivial Pursuits of Ed Begley Jr.," *Rolling Stone*, 1984; "MSN Chat," http://chat.msn.com/msnlive/features/ebegleyjr2-42800.asp; Stark, John, "No High-Pollutin' Actor, Ed Begley Jr. Successfully Recycles His Love Life and Career," *People Magazine*, 1990.

Bennett, Hugh Hammond

(April 15, 1881–July 7, 1960)
Soil Scientist, Director of U.S. Soil Conservation Service

Soil scientist Hugh Hammond Bennett is known as the father of soil conservation. He discovered the connection between soil erosion and loss of agricultural productivity in 1903 and worked for 50 years through the U.S. Department of Agriculture to educate farmers on how to prevent it. Bennett was the first director of the Soil Erosion Service, which later became the Soil Conservation Service.

Hugh Hammond Bennett (Library of Congress)

Hugh Hammond Bennett was born on a farm near the town of Wadesboro, North Carolina, on April 15, 1881. His family cultivated cotton on poor, tree-stripped land; Bennett was later to realize that his family had broken the soil conservation rules that would have conserved its productivity. Despite his family's financial hardship, Bennett studied chemistry at the University of North Carolina. To pay for his studies, he took time off to work clearing trees from future cotton fields.

After graduating with a B.S. in 1903, Bennett joined the Bureau of Soils of the U.S. Department of Agriculture. His first assignment was to learn why crop yield was so poor on the farms of Louisa County, Virginia. He compared the rich loam in virgin hardwood forests to the hardpan of the sloping cotton fields that bordered them. Bennett discovered that rain falling on cultivated hillsides gradually washed away layer after layer of topsoil—he called this process "sheet erosion." When three years later Bennett was promoted to soil scientist at the Bureau of Soils and took charge of soil surveys throughout the eastern half of the country, he began to compile a list of farming practices that led to loss of topsoil and, hence, productivity. For each problem he developed an alternative practice. Cultivation of steep land in vertical rows was common; Bennett recommended plowing along the contours of the land and terracing so that rainwater draining downhill would be caught in trenches along each terrace. Farmers often left the soil surface bare after the harvest, which facilitated erosion by rain or wind; Bennett instructed farmers to "stubble-mulch," or leave the roots and part of the plants in the ground over the winter, in order to hold the soil more effectively.

Bennett decided that soil conservation was the nation's major agricultural issue. During the almost 20 years he was in charge of soil surveys, he became an effective, charismatic educator about soil erosion. His demonstrations included one in which he poured a glass of water (representing rain) onto a towel (vegetation) on a table (soil). Of course the towel absorbed the precipitation. But then when he removed it and the water was poured right onto the table, a mini-flood ensued. He used drama again when lobbying Congress in 1928 for a national erosion control program. The dust bowl was in its peak, and Bennett timed his testimony to coincide with the arrival over Washington, D.C., of a huge dark cloud of wind-eroded topsoil that was blowing east from New Mexico. Congress readily passed the 1929 Buchanan Amendment to the Agricultural Appropriations Bill that established a fund to study and control erosion. This program became the Soil Erosion Service, later renamed the Soil Conservation Service, which Bennett directed for the duration of his career.

Under Pres. Franklin D. Roosevelt, who recognized the importance of conservation and promoted it during his four terms in office, Bennett's Soil Conservation Service reached thousands of farmers. In its first year of full operation, 1935, 147 erosion control demonstration projects were set up, each 25,000 to 30,000 acres in size. Fifty thousand farmers were trained in erosion control techniques every year. They were taught such techniques as terracing and contour plowing, crop rotation, fertilization, soil strengthening through planting of grasses and legumes, planting trees as windbreaks, and strip cropping. Farmers found the program so worthwhile—it was said to improve farmers' incomes by up to 20 percent—that in 1937 states began to organize "soil conservation districts." The federal government was proud of these because they were so economical; all the federal government had to provide was technical assistance. There are now over 3,000 soil conservation districts nationwide.

Bennett's influence extended even beyond the borders of the United States. During the 1920s he studied soils on sugar and rubber plantations in Cuba and South America. Eighty-eight countries sent over 1,100 technicians to study Bennett's methods in the United States, establishing similar erosion control programs in their own countries.

Bennett directed the Soil Conservation Service until 1951, when age forced him into mandatory retirement. A *New York Times* editorial celebrating his 50 years at the U.S. Department of Agriculture claimed that "this 'father of soil conservation' stands among the nation's most useful citizens."

Bennett died of cancer in Burlington, North Carolina, on July 7, 1960. He left his second wife, Betty Brown Bennett, to whom he had been married since 1921, and two children, Sarah and Hugh.

Bibliography

Bennett, Hugh Hammond, and William R. Chapline, "Soil Erosion, a National Menace," USDA Circular No. 33, 1928; Brink, Wellington, *Big Hugh: The Father of Conservation*, 1951; "Hugh Hammond Bennett Dead; 'Father of Soil Conservation,' 79," *New York Times*, 1960; Petulla, Joseph M., *American Environmental History*, 1977.

Berg, Peter

(October 1, 1937–)

Bioregionalism Philosopher, Founder and Director of Planet Drum Foundation

One of the leading advocates of bioregionalism, Peter Berg is the founder and director of Planet Drum Foundation, the contributing and managing editor of *Raise the Stakes: The Planet Drum Review*, a noted ecologist, and a popular public speaker on several continents. He is widely acknowledged as an originator of the use of the terms *bioregion* and *reinhabitation* to describe land areas in terms of their interdependent plant, animal, and human life. Berg believes that the relationships between humans and the rest of nature point to the importance of supporting cultural diversity as a component of biodiversity.

Peter Stephen Berg was born October 1, 1937, in Jamaica, Long Island, New York. When he was six, his family moved to Florida. At the University of Florida in Gainesville, Berg discovered beat poetry and was introduced to the emerging revolution that it expressed. Berg joined an underground minority at the overwhelmingly conservative institution and became involved in the civil rights movement. Leaving the University of Florida while still a teenager, Berg hitchhiked across the United States, at which time he first visited San Francisco. In 1964, he settled in the city, where he joined the San Francisco Mime Troupe. He later helped create the Diggers in San Francisco, who served free food at the Be-In and began, as he puts it, "ecologizing the left."

As the revolution died down at Haight-Ashbury, Berg and a caravan of former Diggers set out on a cross-country tour in the summer of 1971 to determine what common threads existed in the nation's land-based communities. By winter, he had reached Nova Scotia and visited the expatriate American poet, Allen Van Newkirk, who also studied the connections between society and ecology. Interested in the research, classification, and preservation of the natural features within a given geographic area, Van Newkirk—along with Berg, Raymond Dasmann, and others—began promoting the idea of the bioregion. Berg and Van Newkirk both felt that the environmental movement was incapable of dealing with the underlying problems that industrial society posed for the biosphere. Rather than cleaning up after disasters, both felt the disasters needed to be prevented. Whereas Van Newkirk had explored the possibility of the bioregion as an arena for wildlife conservation, Berg proposed the inclusion of humans into the bioregion as an active—not dominant—species in that habitat. In essence, this was an exercise in reinhabitation; humans had to learn how to live in nature, not with dominion over it. As Berg states in his essay, "Beating the Drum with Gary," the only way to succeed at preventing further environmental disasters "was to restructure the way people satisfied basic material needs and related to the natural systems upon which their own survival ultimately depended." Pushing ecological concerns to the center of society was the only tangible approach that might successfully broach this problem. As a movement, bioregionalism was born.

Berg and others took these new ideas to the 1972 United Nations (UN) Conference on the Environment in Stockholm. In the company of thousands of activists and demonstrators from all over the world, Berg discovered that ecology was not just a North Atlantic cause. Berg mixed with groups of Japanese mercury-poisoning victims, Eritrean rebels, Laplanders from the Arctic Circle, Native Americans, and countless others, who made up what Berg called "the planetariat." For most of "the planetariat," no real answers to their issues emerged from the official gathering. Instead, their experience at the conference left them with increased frustration about the inability of any established institution to deal with planetary problems.

Returning to the United States with these frustrations, Berg was determined to find a method for constructing a forum for human and ecological sustainability in the biosphere. His focus naturally shifted from the global to the local or regional and resulted in the founding of the Planet Drum Foundation in San Francisco in 1973.

Planet Drum's mission is to determine the cultural and ecological dimensions of a human-scale geographical region. Given the relative failure of the 1972 UN conference, Berg became convinced that breaking down the world into separate biotic provinces or bioregions would help find plausible routes toward sustainable living for the earth as a whole.

In 1979, Berg introduced the Planet Drum Foundation review, *Raise the Stakes*. A radical review that argues that environmentalism is not demanding enough from the corporate government, Berg suggests that bioregionalism is postenvironmentalist in that it pushes the limits of the environmental movement; that it "raises the stakes." Modern environmentalism does not deal sufficiently with the currently important issues of ecosystem restoration and urban sustainability. Bioregionalism proposes a whole new philosophy necessary if these goals are to be reached. *Raise the Stakes* helped popularize the notion that health, food, and culture are all bioregional issues, profoundly affected by the place in which they are situated.

Unlike many environmentalists and ecologists, Berg looks to the future with a certain degree of optimism. He believes that the localization of politics will eventually take a bioregional turn. While there is concern that globalization appears to be a dominant force that even threatens the nation state, Berg insists that localization, the forwarding of ethnic autonomy and home rule, for example, is playing an equally influential role in this movement away from the nation state and toward regional ecology.

One of Berg's current projects is the creation of a wilderness corridor on the coast of Ecuador. Already, the group with which Berg is working has reintroduced 12 native plant species in an attempt to bring the wilderness back to the city and to demonstrate the connections between the two. While the project is ecologically motivated, its pragmatic motivations demonstrate the link between culture and ecology; the revegetation of the hills will prevent destructive mudslides during the next El Niño cycle and also bring nature-loving visitors to the region.

In 1998, Berg was awarded the Gerbode Professional Development Program Fellowship for outstanding nonprofit organization executives. He lives and works in Shasta Bioregion in northern California. Berg and Judy Goldhaft have two children, Aaron and Ocean, and one granddaughter, Florence Amelia.

Bibliography

Berg, Peter, "Beating the Drum with Gary," in *Gary Snyder: Dimensions of a Life*, John Halper, ed., 1991; Berg, Peter, *Discovering Your Life-Place: A First Bioregional Workbook*, 1995; Berg, Peter, *Figures of Regulation: Guides for Re-Balancing Society with the Biosphere*, 1981; Berg, Peter, *A Green City Program for the San Francisco Bay Area and Beyond*, 1990; Berg, Peter, ed., *Reinhabiting a Separate Country: A Bioregional Anthology of Northern California*, 1978.

Berle, Peter

(December 8, 1937–)
Attorney, Chief Executive Officer of the National Audubon Society,
Radio Show Director and Host

Peter Berle has been active as an environmental lawyer, legislator, administrator, and educator. As a lawyer he has won many important environmental cases of national significance. As legislator and administrator, he helped to write and enact much of New York State's early environmental legislation. He was chief executive officer of the National Audubon Society from 1985 to 1995. And today, he hosts the "Environment Show," which airs on National Public Radio and on ABC radio stations.

Peter Adolf Augustus Berle was born on December 8, 1937, in New York City. His father, Adolf A. Berle Jr., was a lawyer and a professor of law, and his mother, Beatrice Bishop Berle, a medical doctor. Growing up, Berle spent his weekends and summers on the family farm in Massachusetts. The farm, which is still operated by the Berle family, occupies land adjacent to the Appalachian Trail, and Berle and his father would go on backpacking trips along this trail together, hiking all the way up the trail to Maine.

Berle attended Harvard University, where he earned a B.A. degree in 1958 and an LL. B. degree in 1964. In the time between these degrees, from 1959 to 1961, he served as an Air Force officer and parachutist in Southeast Asia. Upon graduating from Harvard Law School, Berle accepted a position with the large New York law firm, Paul, Weiss, Rifkind, Wharton and Garrison. One of his first assignments was to the legal team representing an environmental group that challenged the Federal Power Commission. The issue was the Commission's issuance of a permit to the Consolidated Edison Company to construct a pump storage plant on Storm King Mountain on the scenic Hudson River. The federal court invalidated the permit, accepting the legal team's argument that the Federal Power Commission could not make a decision simply by calling "balls and strikes" at a hearing when the public interest was not represented. The court held that it was the responsibility of the Power Commission to see that a full record was developed. This was a landmark case in environmental law. It provided direction and set precedent for government regulatory agencies. It set the forces in motion that would eventually result in the passage of such national legislation as the National Environmental Policy Act (NEPA). Berle was also heavily involved in preparing and presenting witnesses for the congressional hearings that would lead to the Alaska Native Lands Claims Settlement Act of 1971. He spent much of 1967 and 1968 commuting to Alaska to assist in organizing the effort. The act dealt with long unsettled claims of native Alaskans. It provided $962 million and 40 million acres of land and established 12 native regional corporations to manage the acquired resources.

Berle left Paul, Weiss, Rifkind, Wharton and Garrison in 1971 to start his own firm, Berle, Butzel and Kass. He continued practicing law until 1976, concentrating, whenever possible, on environmental issues. One significant case he was involved with during this time centered on the United States Postal Service, which, being only a quasi-government agency, claimed not to be subject to NEPA. The Postal Service was making plans to build a facility that was to run diesel trucks continuously in a densely populated area of Manhattan. The court found that the Postal Service was responsible for adhering to the requirements of NEPA, and the shipping facility was not constructed. Representing Long Island homeowners in a class action suit, Berle successfully sued Union Carbide over groundwater contamination caused by Aldicarb, a pesticide the company manufactured.

In 1968, Berle was elected to the New York state legislature from the east side of Manhattan. He served for three terms, until 1974. While in the legislature he became the ranking member of the Environmental Conservation Committee and managed the floor fight to pass the Adirondack Park Act of 1971, which established a zoning, land use agency to oversee management of Adirondack Park, the largest state park in the lower 48 states. Berle also introduced the first New York legislation to require projects to be subject to environmental impact assessment.

In 1976, Berle became commissioner of the New York Department of Environmental Conservation, one of the most comprehensive state environmental agencies in the nation. As commissioner, Berle was responsible for bringing the state of New York into compliance with many of the laws he had proposed while in the legislature. He was also a key player in the infamous Love Canal toxic waste crisis in 1979. He and his agency were the first public authorities to recognize the impact of the contamination and take action. He and the local regional administrator of the federal Environmental Protection Agency convinced President Carter to designate Love Canal a disaster area. This marked the first time a crisis created by the acts of humans was named a disaster area under federal law, a designation heretofore reserved for natural disasters such as floods and earthquakes. As commissioner, Berle was also able to free up state funding for land acquisition. On behalf of the state, he purchased the last 11 high peaks in the Adirondack Park that were in private ownership, as well as substantial lake and forest properties.

Berle returned to private law practice for six years in 1979, before becoming president and chief executive officer of the National Audubon Society in 1985. One of his main goals in this position was to combine the Audubon Society's traditional interests with an increased degree of activism. He wanted people to see birds as indicators of environmental health and thus expanded the scope of the Society and its members to include such issues as toxic waste and pesticides. Berle orchestrated an increase in grassroots activism through the 500 local chapters of the Audubon Society and continued to emphasize the importance of education. He increased the Audubon Society's presence in Washington, D.C., and brought about the creation of the Audubon House, the organization's new headquarters in New York City. When the renovation of this historic building was completed in 1992, the structure was heralded as one of the most environmentally benign and energy-efficient structures in the United States. Under Berle's leadership the National Audubon Society's budget increased from $23 million to $46 million per year. By the time his 10-year stint as president was completed in 1995, Audubon had been involved in many high-profile environmental battles, including the fight to protect the Arctic Wildlife Refuge from oil development in 1987.

After leaving the Audubon Society in 1995, Berle considered returning to his environmental law practice but instead decided to pursue a path that he hoped would lead to a greater public understanding of environmental issues. He became director and host of the "Environment Show," which airs on National Public Radio and ABC stations nationwide and abroad on Voice of America and Armed Forces Radio. The show addresses a diverse array of environmental issues from genetically modified foods to urban sprawl. The show's mission is to educate and to help stimulate thought on the idea that everything is interconnected. It includes short segments that reflect on the importance of place, debates between opposing viewpoints, and a highlight of local activism. The show also includes Berle's interviews with prominent activists, politicians, and scientists about pressing environmental issues and potential solutions. Berle's show tackles moral and ethical issues, addresses the importance of urban spaces, portrays humans as a part of a biologically diverse system, and attempts to deliver the message that everyone can make a difference. Berle is also currently a presidential appointee on the Joint Public Advisory Committee on Environmental Cooperation (JPAC), which was created under the North American Free Trade Agreement (NAFTA) environmental side agreement, negotiated by President Clinton. He also serves as a board member of the New York Independent System Operator (NYISO), an entity set up to manage the sale and distribution of electricity in New York's deregulated market. Over the years, Berle has taught as a lecturer and adjunct professor at various academic institutions, including Syracuse University, the Century Foundation, and New York State University. His wife since 1960, Lila Berle, runs the largest herd of sheep in Massachusetts on the Berle family farm. Berle and his wife have four grown children.

Bibliography

Begley, Sharon, "Audubon's Empty Nest," *Newsweek*, 1991; Berle, Peter, "Building for a Sustainable Future," *Audubon*, 1995; "WAMC Public Radio—Meet Peter Berle," http://www.wamc.org/peter.html.

Berry, Wendell

(August 5, 1934–)
Poet, Farmer

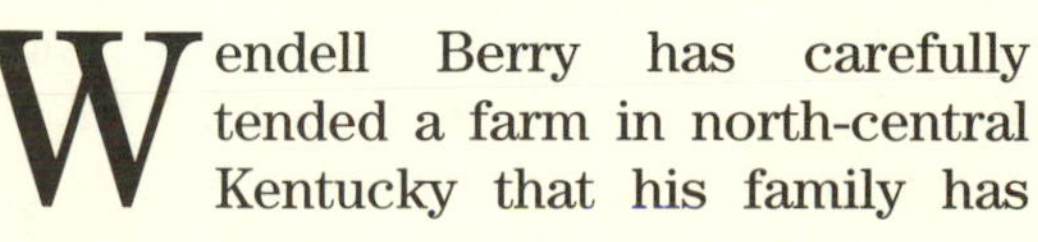

Wendell Berry has carefully tended a farm in north-central Kentucky that his family has tilled since 1803, while simultaneously conducting a formidable literary and academic career. His many novels, short sto-

Wendell Berry *(Louisville Courier-Journal)*

ries, poetry, and essays on rural life, land use, conservation, and forestry have encouraged and inspired the modern environmental movement in the United States. Berry's principal preoccupation as a writer derives from his work as a farmer: transforming the man *versus* nature paradigm into the more appropriate man *with* nature.

Wendell Erdman Berry was born on August 5, 1934, in Henry County, Kentucky, and grew up a studious and hardworking child on land that had provided for over a hundred years of Berrys before him. In Port Royal, Kentucky, along the Kentucky River, at a riverside cabin on stilts built by his great-uncle, young Wendell Berry discovered his peace with nature and his facility for contemplation and writing. There he learned to use the hand tools of his life work: vice grips, hatchets, pens, and notebooks. At 18 years of age his formal education took him to the University of Kentucky at Lexington, where he attained a B.A. in English in 1956, staying on to complete an M.A. in 1957. Upon graduation Wendell Berry married Tanya Amyx and took up residence at the "long-legged house" by the river.

At this point in his life Berry committed himself deeply not only to what would be a long and happy marriage to his wife, but also to the land that had nurtured his forebears. He accepted a teaching position nearby at Georgetown College and laid down the groundwork for his first novel. However, in 1958 Berry was awarded a Wallace Stegner Writing Fellowship and moved to California to study creative writing at Stanford University. In 1961, a Guggenheim Fellowship sent him to Europe for a year, and the next year he joined the English Department at New York University. But, as he wrote in his autobiographical essay, "The Long-Legged House," "for reasons that could perhaps be explained, I never lost affection for this place, as American writers have almost traditionally lost affection for their rural birthplaces. I have loved this country from the beginning." He decided to move back to Kentucky after the release of his first book of poems, *The Broken Ground,* in 1964. He took a position teaching English at his alma mater, settling first in Lexington and slowly retreating back to his farm in Port Royal and the long-legged house.

Berry's fiction writing found fertile ground in Port Royal. Beginning with his first novel, *Nathan Coulter* (1960), Berry introduces characters who inhabit Port William, a thinly disguised Port Royal. Over the course of four novels, Nathan

Coulter, Jack Beechum, Mat Feltner, and their progeny moved over the same land that Berry still cultivated.

Wendell Berry's writing is largely autobiographical. He uses family history to chart the evolution of characters who are wedded to the Port Royal setting. By describing his own life, through a humble farmer's-eye view, his fiction gains the detail necessary to be authentic. In his essays the sincerity of the first person voice aids his arguments. Berry's first five books of essays are compiled in *Recollected Essays: 1965–1980* (1981). He diagnoses the havoc wreaked by white man's mistreatment of the land and white society's lack of discipline in politics, production, and consumption, and he prescribes approaches to the necessary healing: going local (living responsibly in one small part of the planet), getting personal (reforming lifestyle; learning to treat our bodies as we should treat the earth), and getting away from a bottom-line-driven, industrial capitalist culture. His polemic book *The Unsettling of America* (1977) depicts modern agriculture as a practice that is as predatory to communities as the industrial economics behind it.

Berry asks broad questions in the essay "Preserving Wildness" in *Home Economics:*

> What is the proper amount of power for a human to use? What are the proper limits of human enterprise? . . . Such questions may seem inordinately difficult, but that is because we have gone too long without asking them. One of the fundamental assumptions of industrial economics has been that such questions are outmoded and that we need never to ask them again. The failure of that assumption now requires us to reconsider the claims of wildness and to renew our understanding of the old ideas of propriety and harmony.

Wendell Berry's poetry could also be called autobiographical, even as its focus strays to nonhuman elements—the sparrows, the soil, the river, the locusts, the lilies, the minute particulars of his own locale. His method is simple. He carries a small notebook in his workshirt pocket for jotting down his epiphanies and observations. Berry has always been taken by HENRY DAVID THOREAU's notion of the "hypaethral," or roofless book, written out of doors under the open sky. The circle of life is the chief concern of his early books of verse, *The Broken Ground* (1964), *Openings* (1968), and *Farming: A Handbook* (1970).

In a later collection of essays that recounts his own ecological creed, *The Gift of Good Land: Further Essays on Culture and Agriculture* (1981), Wendell Berry joins RALPH WALDO EMERSON in his insistence on a religious understanding of ecology and repeats Emerson's notion of the Chain of Being. His message is that the dead coinhabit the land of the living, that all things—dead and alive—are interconnected. In "The Man Born to Farming" in *Farming: A Handbook*, he wrote:

> The grower of trees, the gardener, the man born to farming,
> Whose hands reach into the ground and sprout,
> To him the soil is a divine drug. He enters into death
> Yearly, and comes back rejoicing. He has seen the light lie down
> In the dung heap, and rise again in the corn.

Religious implications aside, the poetry of Wendell Berry is essentially bucolic. Some critics complain that Berry writes too much in the shadow of Thoreau and that his work represents pastoral works of withdrawal rather than the engagement to which he seems to aspire. But Berry is a writer with an ear not so much for what critics say as for what his experience on the land has told him. His unique nonindustrial means of farming and creative production (he still writes with a pencil and paper tucked in his workshirt pocket) bring an integrity to his voice and to the second half of twentieth-century American literature.

Few American writers have delivered such ecologically minded convictions in as many varied forms as Berry. His contribution of fiction, nonfiction, and poetry continues to grow each year. He has been awarded the Lannan Foundation Award for nonfiction (1989), the Jean Stein Award from the American Academy of Arts and Letters (1977), the Vachel Lindsay Prize (1962), and numerous honorary doctorates and accolades. He still lives at Lanes Landing Farm in Port Royal, Kentucky, with his wife, Tanya. They have two children, Mary Dee and Pryor Clifford.

Bibliography

Berry, Wendell, *Collected Poems: 1957–1982*, 1985; Berry, Wendell, *A Continuous Harmony: Essays Cultural and Agricultural*, 1981; Berry, Wendell, *The Hidden Wound*, 1970; Berry, Wendell, *The Memory of Old Jack*, 1974; Berry, Wendell, *What Are People For?* 1990; Berry, Wendell, *The Wild Birds: Six Stories of the Port William Membership*, 1986; Nibellink, Herman, "Wendell Berry," *American Nature Writers*, John Elder, ed., 1996.

Bien, Amos

(February 12, 1951–)
Biologist, Ecotourism Pioneer, Entrepreneur

Tropical biologist Amos Bien is a pioneer of ecotourism, establishing Rara Avis, Costa Rica's first rain forest reserve devoted to conservation, sustainable development, and education, in 1983. During the 1990s, Bien helped found and has served as president of the Costa Rican Natural Reserve Network, an association of private reserve owners that promotes conservation on privately held land of ecological value.

Bien was born on February 12, 1951, in New York City and grew up in Lynbrook, New York. He spent formative years studying nature and learning about civic responsibility with the Boy Scouts. While attending the University of Chicago, Bien and three friends helped found the country's first recycling program, which is still in operation and is now the largest in the region. The same group of friends also used the equipment in the university's chemistry lab to test Chicago-made clothing detergents for their phosphorous content. Once they found that the detergents contained far more phosphorous than the labels claimed, they called the press. Soon the City of Chicago was demanding that the companies change their recipes. After graduating from the University of Chicago in 1973 with a degree in biology, Bien went on to State University of New York–Stonybrook for master's and doctoral work in ecology, specializing in sea slugs and tropical forest structure.

Bien moved to Costa Rica in 1979 to carry out doctoral dissertation research at La Selva Biological Station. Within six months, he was invited to coordinate the Station. While there, he began to hear frightening predictions about the disappearance of rain forests. He was pained by the irony that he and his colleagues were studying the forest's secrets just as massive deforestation was coming to claim them. So he began to visit La Selva's neighbors, the men armed with chainsaws and machetes, to find out why they were cutting down trees. He learned that far from being ill-intentioned forest-haters, they were just trying to make a living the best way they knew how.

Bien decided to see if he could beat that. He made some calculations and found that cutting trees, leaving them to rot on the ground, and setting cows out to pasture was one of the most inefficient ways of making money off of the land; annual profit was rarely more than $20 per hectare. Harvesting and selling one tree from one hectare of land per year was more profitable than grazing cattle on that hectare, in fact. And even more profitable than cutting the trees at all, Bien imagined, was leaving them there and inviting tourists to come admire the rainforest.

In 1983, Bien put his ideas into practice and founded his mountainous rain forest reserve, Rara Avis. The first tourist accommodations at Rara Avis were in an old building that had been a penal colony guardhouse. The bunk-style accommodations there were rustic but comfortable for the curious tourists who braved the four-hour ride by horse or tractor through the foothills to Rara Avis. Through the 1980s, Bien added what was needed to make Rara Avis the model ecotourism development it has become. Nature trails criss-cross the forest, and for cooling down after the day's hike there is a swimming hole at the foot of a spectacular waterfall. Bien offers a variety of accommodations ranging from a luxurious lodge, to a treehouse, to riverside cabins. In his on-going experiments with making a living from the rain forest without destroying it, Bien and his associates have established a butterfly research and production project and a nursery for endangered rain forest trees and orchids. Bien is now internationally recognized as a pioneer in the world's growing ecotourism industry.

In addition to managing his successful rain forest resort, Bien serves as president of the Costa Rican Natural Reserve Network, an association of owners of private nature reserves. As government and international resources for conservation shrink, Bien and his colleagues in the Natural Reserve Network are showing that private landholders can be important collaborators in national conservation strategies, because they conserve habitat and watersheds that the government alone could not afford to protect. The private reserve owners can also contribute to the national economy by allowing bioprospectors to survey their land for pharmaceuticals or other substances of value to industry. Bien and his colleagues lobby for economic credits for landholders who commit to conserving their land in a natural state and for more input into government conservation policies.

Bien has been honored with the Costa Rica Tourism Professionals Association's 1996 Amigo de la Naturaleza award, given annually to a prominent "Friend of Nature." He has taught conservation and ecotourism at Costa Rican universities, makes frequent appearances on international television and radio programs devoted to nature and conservation, and is an active participant in the public debate on conservation in the Costa Rican media. Bien is married to Damaris Reyes and has three children.

Bibliography

Auerbach, Jonathon, "Journey Deep into a Rainforest," *Christian Science Monitor*, 1991; Honey, Martha, *Ecotourism and Sustainable Development: Who Owns Paradise*, 1999; "Rara Avis Rainforest Lodge," http://www.rara-avis.com; Tripoli, Steve, "Tramway in the Treetops, Bird's Eye View," *Christian Science Monitor*, 1989.

Bierstadt, Albert

(January 7, 1830–February 18, 1902)
Landscape Painter

Landscape painter Albert Bierstadt produced some of the first images of the vast, breathtaking topography of the American West. His paintings resonated with Americans of the nineteenth century because they called attention to the beauty of the briskly disappearing wilderness of the American West and because they reinforced a sense of national identity.

Albert Bierstadt was born on January 7, 1830, in Solingen, Germany, near Düsseldorf on the Rhine River. He was the youngest of six children born to Henry and Christina Bierstadt. The Bierstadt family emigrated to the United States in 1832, settling in New Bedford, Massachusetts. Little is known about Bierstadt's childhood. He attended the local public schools. His formal education, though, never extended past this rudimentary level. His artistic career began in May 1850 when he published a flyer offering to teach "Monochromatic Painting at Liberty Hall," 24 hours of instruction for just three dollars, promising "every picture the scholars make worthy of a frame."

Bierstadt traveled to Düsseldorf in 1853 to study art. He never formally enrolled at the academy in Düsseldorf, but instead gained knowledge of color and compositional techniques through his friendships with artists such as Emanuel Lenze, Worthington Whittridge, and Sanford S. Gifford, who were formal students at the Düsseldorf Academy. It was here that Bierstadt began to master techniques that would make his later landscape paintings so powerful. He remained in Europe for four years, traveling the countryside and sending home his Düsseldorf-influenced landscapes with their volatile skies, sharply contrasted lighting, and minute brush strokes. When he returned home in 1857, he unveiled several new paintings, all of which were praised by the New Bedford newspaper. Significant among his paintings of this time period are his *Westphalian Landscape* (1854), *Gosnold at Cuttyhunk* (1858), and *Lake Lucern* (1858), which he submitted to the National Academy of Design exhibition in New York.

In 1859, Bierstadt joined Col. Frederick Lander's South Pass Wagon Road Expedition in its search for an overland route from St. Louis, Missouri, to the Pacific. He left the expedition to explore the Wind River country of the Rocky Mountains, where he found scenes that lent themselves perfectly to his ambitious and dramatic style of painting. The sketches he made on this trip would supply him with subjects for the rest of his life. In one letter, he wrote "such beautiful cloud formations, such fine effects of light and shade, and play of cloud shadows across the hills, such golden sunsets, I have never before seen. Our own country has the best material for artists in the world." When Bierstadt returned from his journey, he settled in New York City and began painting huge landscapes (up to 9 feet by 15 feet), tremendous in scale yet containing the minutest of details. His paintings provided many Americans with their first view of the West. Bierstadt's reputation soared, and collectors enthusiastically purchased his paintings. In 1860, he was elected to the National Academy of Design. Of one of his most famous works, *The Rocky Mountains, Lander's Peak* (1863), critic and historian Henry T. Ackerman wrote, and is quoted by Tom Robotham in his book about Bierstadt, "No more genuine and grand American work has been produced in landscape art."

After the outbreak of the Civil War in 1861, Bierstadt was granted permission to visit the country around Washington, D.C., in search of material. *Guerrilla Warfare* (1876) and *The Ambush* (1876) were likely products of this excursion. Another Civil War painting of his, *Bombardment of Fort Sumter* (1863?), was purely a product of his imagination. At the time the painting was done, there is no way Bierstadt could have accessed Charleston. He likely created the painting based on a series of eyewitness accounts of the bombardment, which were published in the New York newspapers.

Bierstadt once again traveled west in the spring of 1863, this time in the company of Fitz Hugh Ludlow. Ludlow, a journalist, published an account of their journey, entitled *The Heart of the Continent* (1870). They passed through Kansas and Nebraska, visited Denver and Salt Lake City (where they met Brigham Young), and made their way to California. Bierstadt sketched for six weeks in the areas in and around Yosemite Valley. With these sketches he would produce, among others, *Yosemite* (1863), *Looking Down Yosemite Valley* (1865), and *The Yosemite Valley* (1868). After venturing into Oregon, the pair returned to New York via San Francisco and the Isthmus of Panama in late 1863.

The period of 1864 to 1873 brought Bierstadt growing critical acclaim as well as growing financial success. His paintings were some of the highest priced of nineteenth-century America. *The Rocky Mountains, Lander's Peak* sold for $25,000, and *Storm in the Rocky Mountains, Mount Rosalie* sold for $35,000. In 1866, he married Rosalie Osborne, the ex-wife of his friend Fitz Hugh Ludlow, and began plans for a large estate on the Hudson River. The house was completed in the next year and was, according to the *Home Journal*, one of the most "commanding and noticeable" on the Hudson. It would burn to its shell in 1882. In 1867, Bierstadt and his new wife sailed for Europe. Over the course of the next two years they visited London, Düsseldorf, Paris, Rome, Munich, Dresden, Berlin, Vienna, Switzerland, and Spain. They returned to their home on the Hudson in 1869.

In 1871, Bierstadt made his third trip west. This time he took his wife along and rode the newly completed transcontinental railroad. They stayed on the Pacific coast until 1873, sketching in the High Sierras and in Yosemite and occupying a studio in San Francisco for a time. During this period, he produced at least four paintings of coastal seal rocks, including *Seal Rocks, San Francisco* (1872) and *Seal Rocks, Farallon Islands* (1873). After two years on the Pacific Coast, the Bierstadts returned to the East. At this point in his career, his popularity began to wane. It became difficult for him to sell his paintings, for the tastes of critics and collectors had changed. They were no longer enamored with Bierstadt's sweeping and sublime interpretations of the American West and were turning their attention more toward the French impressionist painting style that was coming into vogue. In 1889, when Bierstadt submitted *The Last Buffalo* (1888) to the American Selection Committee for the Paris Exposition, it was rejected. It was, according to one committee member, "too big and not representative of his style." His later years would, in fact, be marked by a departure from his earlier style. He turned toward a greater diversity of subjects, including paintings tranquil and less dramatic in scope. In 1893, his wife died after a long battle with consumption. Bierstadt remarried in 1894, wedding a widow, Mrs. David Stewart.

Bierstadt's last years were busy ones. He became involved in a project to establish a National Academy of Art in Washington, D.C., continued painting, and applied for a series of patents concerned with improvements to railway cars. He also spent time promoting a gun that had been invented by Henry Schulhof and became interested in the gas and electrical projects of Charles Sanders Pierce. Bierstadt died on the morning of February 18, 1902.

BIBLIOGRAPHY

Hendricks, Gordon, *Albert Bierstadt, Painter of the American West*, 1974; Lewison, Florence, "The Uniqueness of Albert Bierstadt," *American Artist*, 1964; Robotham, Tom, *Albert Bierstadt*, 1993.

Bingham, Eula

(July 9, 1929–)
Director of the Occupational Safety and Health Administration, Toxicologist

The first woman director of the Occupational Safety and Health Administration (OSHA), Eula Bingham led the agency to focus on health concerns raised by worker exposure to chemical toxins. Trained as a toxicologist, Bingham pushed for recognition of the cancer-causing properties of a broad range of chemicals, making it easier to enforce stricter exposure regulations. In 1977, when Bingham was appointed assistant secretary of labor for occupational safety and health, she inherited an agency with a reputation for arbitrary rules and unnecessary paperwork. During her three years as OSHA director, she worked to ease the regulatory burden on industry and improve the agency's image, while ensuring that OSHA stayed true to its role as advocate for worker health and safety. As professor in the Department of Environmental Health at the University of Cincinnati, Bingham has conducted scientific studies on numerous environmental toxins, labored for government accountability for protecting citizens from the effects of chemical agents, and connected community activists with scientific investigators to work for safer environments.

Eula Mae Bingham was born in Covington, Kentucky, on July 9, 1929. Bingham was an only child, her father a railroad worker who became a farmer after losing his job during the Great Depression. Bingham lived in Covington until she went to Eastern Kentucky University in Richmond. She graduated in 1951 with a B.S. in chemistry and worked for one year at the Hilton Davis Chemical Company in Cincinnati, Ohio. Interested in pursuing a career as a research scientist, Bingham studied zoology at the University of Cincinnati, receiving an M.S. in 1954 and a Ph.D. in 1958. She was hired by the University of Cincinnati's College of Medicine, where she began conducting research into the carcinogenic properties of a variety of agents and eventually was recognized as one of the leading national experts on environmental health.

In the early 1970s, Bingham served on a series of national committees investigating and recommending policy on environmental hazards. During 1974–1975 she served as chair of the Department of Labor's Standards Advisory Committee on Coke Oven Emissions, where she came to the attention of Ray Marshall, who was to become Pres. JIMMY CARTER's secretary of labor. Bingham was nominated as assistant secretary of labor for occupational safety and health in 1977. The late 1970s were a time of high inflation and growing conservatism, and Bingham inherited an agency that had come under very public attack, as a symbol of the government overregulation perceived to be crippling the economy. In her nomination hearing she promised to rescind unnecessary and unproductive regulations and to concentrate OSHA's resources on addressing genuinely life-threatening workplace hazards. She also promised to turn the agency's attention more fully to health issues in addition to its traditional concern with safety.

Bingham was sympathetic to the concerns of labor unions and environmental activists and led OSHA to support a number of grassroots, local, and union-based environmental health initiatives, through the New Directions Grant Program. She also pushed for a more aggressive approach to setting permissible exposure limits for carcinogens. Prior to Bingham's term in office, OSHA set standards on a case-by-case basis, only after definitive scientific evidence proved the substance was harmful to humans. Bingham proposed to treat chemical agents in broad categories of known and suspected carcinogens and to set exposure limits at minimum feasible levels. This, it was hoped, would speed and simplify the regulatory process and protect the greatest number of workers. Bingham was not entirely successful in changing these regulations, because the Carter administration came under increasing pressure to limit the cost of environmental protection. A case in point is that of regulating worker exposure to cotton dust. Though the scientific evidence demonstrating a link between cotton dust and brown lung disease was irrefutable, Pres. Carter backed away from Bingham's proposed exposure limits because industry argued that meeting the new standard would be too costly and inflationary. It was widely reported that Bingham nearly resigned over this dispute, though she did eventually serve her full term in office. She was more successful in setting tough standards in other cases, including benzene, lead, and arsenic, establishing exposure limits at the lowest feasible levels. Bingham also achieved an impressive inspection and enforcement record, decreasing the inspection rate for minor violations while increasing inspections and penalties for the most serious violations.

After Ronald Reagan's election in 1980 ended her tenure at OSHA, Bingham helped spearhead organizations to press for environmental safety and health. These organizations included the Occupational Safety and Health Political Action Committee, which Bingham chaired in 1982, and the Regulatory Audit Project, a group of former environmental and consumer administrators who joined together in 1983 to educate the public about the costs of failing to enforce health and safety regulations. Bingham also chaired the Department of Veterans Affairs' Persian Gulf Expert Scientific Committee in 1996 and pushed for full scientific investigation into and government accountability for the effects of exposure to low levels of chemical weapons and other agents present during the Gulf war. Bingham has continued her work at the University of Cincinnati, which has included directing the Community Outreach and Education Project for the Center for Environmental Genetics.

Bingham has been elected to the National Academy of Sciences Institute of Medicine and the Ohio Women's Hall of Fame. She lives in Ohio.

Bibliography

"Community Outreach and Education Project," http://www.niehs.nih.gov/centers/coep/cincoep.htm; Noble, Charles, *Liberalism at Work: The Rise and Fall of OSHA*, 1984; Shenon, Philip, "Panel Disputes Studies on Gulf War Illness," *New York Times*, 1996.

Blackgoat, Roberta

(1917–)
Dineh (Navajo) Relocation Resistance Leader

Roberta Blackgoat has been a leader of her Dineh (Navajo) people's resistance movement to a government relocation order since the passage of the Navajo-Hopi Land Settlement Act in 1974. She and the 200 Dineh families who refuse to leave their ancestral lands believe that they were "planted" by the Creator on their land and charged with taking care of it and that to leave would be to abandon the wishes of the Creator.

Born in 1917 on Thin Rock Mesa, Arizona, where her family has lived for at least ten generations, Roberta Blackgoat had a traditional Dineh childhood during which she was taught how to survive in her people's dry, rugged homeland. The land her people claim as home is defined by four peaks spanning the Four Corners region of the Southwest: Mount Blanca in Colorado to the northeast, Mount Taylor in New Mexico to the southeast, the San Francisco Peaks in Arizona to the southwest, and Mount Hesperus in Colorado to the northwest. The Dineh people consider these mountains sacred, and the land between them their church. Their hogans, built in the western area of the quadrant, are an altar, according to Blackgoat.

Historians say the Dineh people migrated from northern Canada to what is now the southwestern United States in about 1400 A.D. They herded sheep and lived relatively peacefully with their Hopi neighbors for hundreds of years. Both peoples resisted Spanish and Anglo invaders of their lands, but the Dineh, because their shepherding culture required larger expanses of land, encountered more violence from the invaders. In 1864, 8,000 Dineh people, among them Blackgoat's grandparents, were herded by Kit Carson's soldiers on the "Long Walk"—a forced journey by foot from their homeland to Fort Sumner, New Mexico. It resulted in the death by exhaustion, famine, disease, beating, and bayoneting of about 4,000 Dineh people, before the government signed a treaty with Dineh representatives to establish a Navajo reservation straddling northeastern Arizona and northwestern New Mexico in 1868. Blackgoat's grandmother and her family returned to Thin Rock Mesa.

In 1882, boundaries for a Hopi reservation were drawn by the government, quickly and rather arbitrarily. Hundreds of Dineh people, scattered in family groups, were living on the land designated as the Hopi reservation, and a few Hopi people were living on Navajo reservation land. One Hopi village was totally left out of the reservation. Later additions to the Navajo reservation left the Hopi reservation totally encircled by the Navajo reservation. The Dineh and Hopi peoples lived with this ambiguity for almost a century. There were some territorial problems, but their coexistence was relatively peaceful. But in the mid-1960s, Peabody Coal announced its intention to mine the rich 100-square-mile coal deposit at Black Mesa, which was officially Hopi land but was inhabited principally by the Dineh. The Peabody negotiator, lawyer John Boyden, understood that Peabody could not sign leases unless land ownership was official and boundaries were clear, so he wrote legislation that divided the disputed parts of the Hopi reservation evenly between the Navajo and the Hopi. Sen. Barry Goldwater pushed this Navajo-Hopi Land Settlement Act through Congress in 1974. Ten thousand Dineh people were determined to be living on what were referred to as Hopi Partitioned Lands and were told to leave, the largest relocation effort in the United States since the World War II internment of Japanese Americans. Most of the people did leave, but a few hundred Dineh families defied the order. Within three years, one-third of those who had accepted relocation benefits and moved off the reservation had lost their new homes due to exploitation and inexperience with the Anglo economic system; unemployment, alcoholism, and suicide rates among relocated Dineh people are much higher than other Dineh populations.

Although most of her six children agreed to relocate, Blackgoat, who was widowed during the 1960s, has remained in her hogan on Hopi Partitioned Land. She lives 30 miles from the nearest paved road and has neither running water nor electricity. At first, Blackgoat did not take the relocation order seriously, but once the government erected a fence in 1977 to keep the Dineh people and their sheep out of the Hopi Partitioned Lands, she understood the act's threat to her livelihood. In 1979, Blackgoat chaired a council of 64 members of the Independent Dineh Nation at Big Mountain. They wrote a declaration of independence, claiming that the U.S. government and the Navajo Tribal Council (which had signed coal leases with Peabody) had "violated the sacred laws of the Dineh nation" and had allowed Mother Earth to be "raped by the exploitation of coal, uranium, oil, natural gas and helium." They declared that they "speak for the winged beings, the four-legged beings, and those who have gone before us and the coming generation. We seek no changes in our livelihood because this natural life is our only known survival and it's our sacred law."

Deadlines for relocation have come and gone over the decades, yet Blackgoat and the remaining 200 Dineh families on Hopi Partitioned Land have refused to leave. After the original relocation during the 1970s, there have been several renewed offers by the U.S. government of relocation benefits or accommodations to stay on the land. The most recent offer, authorized by Congress in 1996, allowed Dineh relocation resisters to stay on their land for 75 years if they reduced their herds of livestock, built no more dwellings, cut no more wood, and did not bury their dead on the land. Blackgoat and other resisters see these restrictions as unacceptable and have not signed on to them.

Blackgoat says that by remaining on the land she is following Dineh spiritual laws; the Creator charged the Dineh with the care for their given area of Mother Earth. She is deeply disturbed by the mining, which she believes is killing Mother Earth. Coal, she was told by her grandfather, is Mother Earth's liver, crucial to Mother Earth's functioning and survival. Cutting the liver out of Mother Earth without putting anything back inside or allowing her to heal is hurting her seriously. A direct effect is the air and water pollution caused by the mining activity, but Blackgoat also sees a relation between the mining and increased tornadoes, floods, earthquakes, loss of biodiversity, and other disasters.

Blackgoat and other grandmothers, including Pauline Whitesinger and Katherine Smith, have led the Dineh resistance movement. Being elderly women, they know that they are less likely targets of police violence than young men, so they are usually at the front of marches and protests. Blackgoat has spoken worldwide about the plight of the Dineh relocation resistors; she has become an international symbol of indigenous people struggling to maintain a traditional relationship with the Earth. Younger generations of Dineh resisters, led by such activists as Louise Benally who has accompanied Blackgoat on her speaking tours, have vowed to continue the resistance of their elders.

Bibliography

Benedek, Emily, *The Wind Won't Know Me: A History of the Navajo-Hopi Land Dispute*, 1992; Cockburn, Alexander, "Indian Rights: The Forced Relocation of Navajo families Is a Triumph of Greed," *Los Angeles Times*, 1997; Draper, Electa, "Forced Relocation Tears at Tribal Soul," *Denver Post*, 1999; Isay, David, and Harvey Wang, *Holding On*, 1996; as told to Johnson, Sandy, *The Book of Elders*, 1994; Kammer, Jerry, "Dividing the Sky," *The Arizona Republic*, 2000.

Bookchin, Murray

(January 14, 1921–)
Anarchist Philosopher, Writer

Anarchist philosopher Murray Bookchin founded the field of social ecology, a school of thought that holds that human destruction of the environment has its roots in social hierarchy in which elites dominate and exploit the great mass of humanity and that humanity's relationship with nature will improve only if such hierarchies are dissolved. He is credited for introducing ecology into the agenda of the radical political movements of the 1960s and, in later decades, for inspiring the emergence of Green parties throughout the United States.

Murray Bookchin was born on January 14, 1921, in New York City to Nathan and Rose Bookchin, Russian-Jewish immigrants who had been active in the revolutionary movement under tsarism. As a child he joined communist youth groups but became disillusioned with their authoritarian leaders and with the Stalinist betrayal of the workers' and peasants' revolution during the Spanish civil war in the late 1930s. He was expelled from the Young Communist League in 1939 for his "Trotskyist-anarchist deviations" and joined the American Trotskyists, but he left that group too, disappointed by its authoritarian leadership. He worked as a foundryman in northern New Jersey during the late 1930s and early 1940s, organizing unions for the Congress of Industrial Organizations (CIO). He served in the U.S. Army during World War II, and after the war, he became an autoworker, actively involved in the United Auto Workers (UAW), participating in the massive General Motors strike of 1946. When

Murray Bookchin (courtesy of Anne Ronan Picture Library)

the resolution of that labor conflict converted the UAW into what he considered a proponent of the status quo, Bookchin left the traditional U.S. labor movement and began working with a group of German immigrant Trotskyists who were developing a libertarian socialist movement in New York City. It was during this period, the early 1950s, that Bookchin's articles began appearing in that group's periodical, *Contemporary Issues.* Because Sen. Joseph McCarthy was persecuting the Left at this time, Bookchin used several pseudonyms to sign his articles: M. S. Shiloh, Lewis Herber, Robert Keller, and Harry Ludd.

Bookchin had always been concerned about improving the quality of life for his fellow human beings, but his first major work to explicitly address environmental quality was a 1952 article in *Contemporary Issues* entitled "The Problem of Chemicals in Food" (written under the name of Lewis Herber). It was inspired by congressional hearings on new chemical inputs that had been developed by the same postwar chemical industrial boom that created new pesticides, later decried by Rachel Carson in *Silent Spring.* Herber's article was published in book form in Germany in 1954 and was expanded into a full-length book in the United States in 1962 under the title *Our Synthetic Environment,* published six months before Carson's bestseller. Bookchin's scope in this volume was wider than food safety: he covered agriculture, various environmental carcinogens, and social health questions. In the 1950s Bookchin came out as an early opponent of nuclear power plants, because of the dangers of radioactive fallout were so great. He founded the Citizens Committee on Radiation Information in 1963, which successfully opposed the construction of the Ravenswood Reactor in New York City.

During the 1950s and 1960s, Bookchin developed the historical and philosophical framework for social ecology. According to Bookchin's historical analysis, human society originally was cooperative, with men and women, old and young, working together within the same community, without significantly harming the natural world. As males came to dominate females, the young, and other males, they also began to develop the idea of dominating nature. At their root, the environmental ills society faces today have emerged from society's domination and exploitation of human beings, as well as the natural environment, according to social ecology proponents.

Social ecology calls for the dissolution of the hierarchical and class institutions that allow people to dominate one another and nature. A society whose citizens do not allow themselves to be dominated and exploited, Bookchin believes, must be based on directly democratic and confederated local governments, a design he calls libertarian municipalism. People would organize on the most local level possible, for example, in their own neighborhoods or townships, and send delegates to free confederations to adjust differences in a democratic manner. This concept is based on the political institutions of the Athenian *polis* or city-state, in which citizens resolved differences and formulated policies in a manner later adopted in New England town meetings. Bookchin believed that technological advances would provide working people with ample free time to participate in such self-government.

Bookchin introduced social ecology to radical students during the social upheaval of the 1960s, in the hopes that their youth would give them the energy to hoist off the restraints of dominant hierarchies. His numerous speeches and publications during the 1960s resulted in the acceptance of an environmental agenda by the radical and progressive movements of that time. He was an active participant in the civil rights movement and in various groups for human freedom in New York City and taught in the late 1960s at the Alternative University in New York, a free university at the City University of New York.

Once a true environmental movement blossomed in the United States, following the blast of publicity on Earth Day, 1970, Bookchin wrote extensively, developing an ethic of ecology based on anarchist and libertarian political ideals and promoted a nonhierarchical, decentralized approach to decision making. His ideas from this period appear in several books, including *Toward an Ecological Society* and *The Ecology of Freedom*. In the 1970s, many radical Greens drew their ideas from his writings, speeches, and lectures, initially basing their politics and practice on Bookchin's articulation of social ecology.

During the 1980s Bookchin became concerned about certain tendencies adopted and publicized by some of the founders of Earth First!, a loosely organized environmentalist group known mainly for its acts of civil disobedience to stop logging, mining, and road building in undeveloped wilderness areas. These founders, including DAVE FOREMAN, often tied criticism of environmental abuses to calls for population reduction, which Bookchin found reactionary and potentially racist. In 1987, an article in *Earth First! Journal* by one Miss Ann Thropy, a.k.a. Christopher Hanes, an *Earth First! Journal* writer, welcomed the acquired immunodeficiency syndrome (AIDS) epidemic as a means of population reduction. In an interview, Foreman seemed to agree with this view, and advocated letting "nature taking its course" in Third World countries, where starvation was claiming the lives of thousands of people. Bookchin wrote an extensive response, criticizing the deep ecology movement—the philosophical basis for Earth First!—for its antihumanism. Bookchin published the article as "Social Ecology vs. 'Deep Ecology'" in his own newsletter, *Green Perspectives*, in 1987. He challenged the deep ecology position that humanity is a hindrance to nature; he insisted that people as such do fit into a natural evolution and that they can even be beneficial to the natural world.

Bookchin engaged Foreman in a debate between social ecology and deep ecology transcribed and published in 1991 as *Defending the Earth, A Dialogue between Murray Bookchin and Dave Foreman.*

In addition to his work as a teacher in New York during the 1960s, Bookchin taught at Goddard College during the 1970s and was professor of social theory in the School of Environmental Studies at Ramapo College in New Jersey from 1974 to 1983. Since 1974 he has lectured at the Institute for Social Ecology in Plainfield, Vermont. He resides in Burlington, Vermont, with his companion, social ecology writer Janet Biehl.

Bibliography

Biehl, Janet, ed., *The Murray Bookchin Reader,* 1997; DeLeon, David, ed., *Leaders from the 1960s: A Biographical Sourcebook of American Activism,* 1994; "Murray Bookchin," http://homepages.together.net/~jbiehl/Mbbiograph.htm; Plant, Christopher, and Judith Plant, *Turtle Talk: Voices for a Sustainable Future,* 1990.

Boulding, Kenneth

(January 18, 1910–March 19, 1993)
Economist

Kenneth Boulding was an economist known for his integration of insights from many social sciences, humanities fields, and ecology into his analyses of the "social system," the organization and inner workings of society. He was an early voice for a new ethic for life on "Spaceship Earth," declaring in 1965 that in order to solve the problems of a growing population, a rapidly depleting supply of fossil fuels, and increasingly polluted air and water, humanity should redefine the planet as a "closed-cycle" system that has no sewer and upon which "unrestrained conflict" would no longer be viable.

Kenneth Ewart Boulding was born on January 18, 1910, in Liverpool, England. He grew up in what he characterized later as a slum—"really rather an exciting neighborhood"—and at the age of nine he began to publish a neighborhood newspaper, which he typed with several carbon copies and distributed to his friends. He was the first in his family to continue his formal education beyond elementary school; his father was a plumber by trade and a lay preacher in the Methodist church, and his mother had been a maid before having children. Boulding attended high school on scholarship and received another scholarship to study chemistry at New College, Oxford. He shifted his emphasis to politics, philosophy, and economics after his first year and graduated with honors in 1931. Shortly after graduation his first paper was accepted for publication by John Maynard Keynes's *Economic Journal.* This paper, which challenged the concept of displacement costs—accepted without question at that time by nearly all working economists—was bold and controversial and established Boulding as an important economist with fresh ideas and a cogent writing style.

In 1932, Boulding received a two-year Commonwealth Fellowship to study at the University of Chicago and Harvard University, where he got a first-hand view of the Great Depression. The United States exhilarated Boulding, in part because his working-class background did not subject him to discrimination, as it had in England.

After returning to England in 1934 and teaching at the University of Edinburgh until 1937, Boulding returned to the United States, this time permanently. His first post was at Colgate University, where he stayed until 1941, and after that, he spent time at Princeton University, Fisk University, Iowa State College, University of Michigan, Stanford, University of the West Indies in Jamaica, International Christian University in Tokyo, Japan, and finally, University of Colorado–Boulder, where he was hired in 1967 and stayed for the remainder of his career.

At each post, Boulding absorbed whatever he could from his colleagues—economists and other social scientists—and early in his career he began to realize that economics as a social science must integrate observations from other social science perspectives to become a more accurate science. He felt that sociology, anthropology, political science, and psychology (and even the humanities fields of philosophy and theology) each approached the social system from a particular vantage point but that the social system could only be understood completely through an integration of all of these approaches. Later, Boulding realized that biology too shed light on social systems and he advocated that research on biological ecosystems be taken into consideration when trying to understand the social system. For example, he drew a parallel between the ecological niches of an ecosystem (physical locations or survival strategies for various organisms) and the niches of an economic system (strategies human beings use to earn a living). Boulding believed that social systems evolved just as ecosystems did, with niches opening and being filled by opportunistic species. "Human artifacts" such as the automobile, for example, competed with horses and reduced the number of horses in the social system. These ideas were articulated in his 1978 book *Ecodynamics.*

At the same time that he urged economics to adopt some insights about the social system from ecology, he criticized one of ecology's sacred tenets, that of Darwin's principle of survival of the fittest. In *Ecodynamics*, he recommended restating this principle as "survival of the fitting . . . fitting being what fits into a niche in an ecosystem."

Boulding's intellect spanned many fields and he rejected the boundaries that more orthodox scholars had erected between fields. His 1956 book *The Image: Knowledge and Life in Society* suggested that human behavior was not dictated as much by immediate stimulus, as economists and other social scientists influenced by behaviorism generally believed, but rather that people used their consciousness to form images of the world that influenced how they behaved. In 1965, he presented a paper at Washington State University entitled "Earth as a Spaceship" that warned that increased knowledge about space and how tiny planet earth was in comparison to other possible worlds, along with an ever-expanding human population, would require society to change its practices and develop "symbiotic relationships of a closed-cycle character with all the other elements and populations of the world of ecological systems." He hoped that a new image of the planet's plight would allow society to realize that "unrestrained conflict" was no longer viable and that "in a spaceship there are no sewers." Although he admitted that "once we begin to look at earth as a space ship, the appalling extent of our ignorance about it is almost frightening," Boulding often expressed the hope that human intelligence would help society evolve to become more ecologically harmonious and peaceful.

Boulding received several prestigious awards for his work in economics, including the 1949 John Bates Clark medal, given biannually to an economist under the age of 40 who shows exceptional promise. He served as president of the American Economics Association in 1969, and in 1973–1974 presided over both the American Academy for the Advancement of Science and the International Studies Association. He is known as a forefather of a diverse group of publications and institutions, including the *Journal of Conflict Resolution*, International Society for General Systems Theory, Association for the Study of the Grants Economy, and the fields of evolutionary economics and ecological economics. In addition to his more than 1,000 scholarly publications, including 40 books, Boulding published several volumes of rhymed poetry.

Kenneth Boulding died on March 19, 1993, in Boulder, Colorado, after a long struggle with cancer. He was survived by his wife, Elise Boulding, and their five children.

Bibliography

Boulding, Kenneth E., "My Life Philosophy," *American Economist*, 1985; "Kenneth Boulding," http://csf.colorado.edu/authors/Boulding.-Kenneth/; Latzko, David A., "Kenneth E. Boulding," *Proceedings of the American Philosophical Society*, 1995; Nasar, Sylvia, "Kenneth Boulding, an Economist, Philosopher, and Poet, Dies at 83," *New York Times*, 1993; Solo, Robert A., "Kenneth Ewart Boulding: 1910–1993. An Appreciation," *Journal of Economic Issues*, 1994.

Bramble, Barbara

(February 27, 1947–)
International Program Director and Vice-president of Strategic Programs Initiative at National Wildlife Federation

Barbara Bramble has played a leading role in ensuring that sustainable development be a central goal of environmentalists and developers domestically and abroad. Her efforts to link the goals of environmental protection with economic and social realities in domestic and international issues have been invaluable to the environmental movement.

Barbara Janet Bramble was born in Washington, D.C., on February 27, 1947, and spent much of her childhood in the nation's capital. Her parents, though from impoverished backgrounds, both worked their way through college to become economists and then joined Pres. FRANKLIN D. ROOSEVELT's New Deal administration in the 1930s. They took the family abroad to Latin America for several years at a time. The disparity between the standard of living of the rich and poor concerned Bramble and played a critical role in her decision to pursue a career as an environmental and social activist. Bramble's father, while a western pioneer at heart who built the family's home and loved urban subsistence gardening, was the U.S. economic counselor to embassies in Mexico, Venezuela, and the Dominican Republic. In Venezuela, at age nine, Bramble became aware of the oppressive reality of quality of life for the poor in developing countries. Later, as a teenager, Bramble lived in the Dominican Republic, a society yearning for democracy. In both countries, she observed deforestation, repression, miserable living conditions of the poor, and political revolution. At the same time that Bramble dealt with the reality of life in developing countries, she explored each country's unspoiled beauty with her family and learned first hand the value of natural environments. Her father, ahead of his time in the 1950s, suggested what is now called "ecotourism" as an option to resource extraction to provide jobs and increased income for indigenous people. The family once traveled to a jungle area of Venezuela where they stayed with an Indian tribe and delighted in the timeless landscape.

Bramble attended the University of the Americas in Mexico for her first year of college. She graduated with a B.A. in history from George Mason University in 1969. Inspired by colleagues involved in public interest law, Bramble attended George Washington University (GWU) Law School, and graduated with a J.D. in 1973. During Bramble's first fall in school, the law school expanded its environmental law program and provided the most extensive clinical program in the country at the time. Not only was the first Earth Day celebrated the spring before she began school, but GWU was in the middle of Washington, D.C., where the first environmental legislation was being passed. Bramble says environmental law fascinated her from the start, and she has never looked back.

After law school, Bramble worked as an environmental lawyer on domestic energy cases with a small firm, Wilson and Graham. While there, Bramble was introduced to the importance of linking economic viability of potential projects with environmental protection. Along with her partner, Ronald J. Wilson, she successfully stopped a gas pipeline from being built through the Arctic National Wildlife Refuge (ANWR) in Alaska and a hydro-electric "pumped storage" project in West Virginia. Bramble realized that touting the environmental value of ANWR and West Virginia's Canaan Valley would not have been sufficient to stop the projects. Instead, she brought in supply-and-demand economics to explain that the deals were not economically viable. Through litigation she held up the projects long enough to allow the financial backers to pull out of the projects. An important tactic in all of Bramble's work is to seek out the economic arguments for protecting the environment, which often prove the folly of imprudent development.

After working with Wilson and Graham, Bramble served as legal counsel to the Council on Environmental Quality (CEQ) from 1979 to 1981, in the Executive Office of the President, where she mediated environmental disputes among federal agencies and handled international legal matters for the Council. She also wrote the CEQ procedures directing federal agencies how to comply with National Environmental Policy Act regulations.

In 1982 Bramble moved to the National Wildlife Federation (NWF) to begin and direct an international program, a position she held for 16 years. NWF's International Program focuses on global environmental issues in which the U.S. government plays an influential role and in which NWF's large membership can impact government positions. Its goal is to encourage the U.S. government to make socially and environmentally responsible decisions, particularly to protect biological diversity. During Bramble's time as director, issues ranged from reducing the export of pesticides banned in the United States to supporting increased U.S. contributions for international family planning to reforming international trade rules. She cofounded the worldwide citizens campaign to reform the World Bank and other multilateral development banks and advanced the campaign's evolution from simplistic reforms within the World Bank (such as pushing for Bank-funded projects to have environmental impact assessments) to more complex goals (such as the involvement of local, and particularly minority, people in development decisions). Bramble's strategy was to explain the real-life damages to both people and ecosystems in developing countries to argue that large-scale government-supported development is not necessarily the best option. For example, she helped form alliances between U.S. environmental groups and Brazilian nongovernmental organizations (NGOs), the National Council of Rubbertappers and its founder, Chico Mendes, and several indigenous tribal leaders. Together, they provided evidence of the value of "extractive reserves" (preserved forests where local people may harvest nontimber resources, such as rubber and Brazil nuts) over government projects (such as dams, mines, and inappropriate agriculture) that would lead to the destruction of tropical forests. As a result, millions of acres of extractive reserves are conserving a significant fraction of the Amazon rain forest.

In the 1980s, when many developing countries faced enormous debt, Bramble was an architect of the concept of "debt for nature" swaps that are carried out by groups such as The Nature Conservancy and World Wildlife Fund. The successful concept has spread to dozens of countries; its benefits include debt relief and increased funding of sustainable development projects.

Bramble worked effectively with other NGOs as well as with local people, political leaders, and international development banks. As an elected member to the steering committee for the International NGO Forum at the 1992 Earth Summit in Rio, Bramble helped organize 3,200 NGO representatives—the largest conference of NGOs at that enormous event. The conference produced alternative NGO "treaties" that many called the most concrete outcome of the United Nations Conference on the Environment and Development (UNCED). Through the UNCED process, members of NGOs recognized that environmental democracy and finding economic solutions to development issues challenged all countries, rich and poor. In fact, Bramble and other NGO participants from industrialized countries took lessons from international development dilemmas and returned home to make sustainable development a focal point in Europe and North America. In the United States, NWF helped push through the formation of the President's Council on Sustainable Development. NWF now has programs in urban areas to advocate smart growth, restoring center city cores, greenbelts, and urban wildlife habitat as a strategy for conserving biological diversity. The experiences of the International Program are benefiting Americans at home.

After 16 years as the director of the International Program, Bramble became the vice president of NWF's Strategic Programs Initiative, an internal planning program for NWF. She will help guide NWF's organization in the twenty-first century by developing new programs and initiatives. Her position allows her to bring the concepts of sustainable development from the International Program into the U.S. domestic agenda.

Bramble lives in Washington, D.C.

BIBLIOGRAPHY

Bramble, Barbara, "Financial Resources for the Transition to Sustainable Development," in *The Way Forward: Beyond Agenda 21*, Felix Dodds, ed., 1997; Bramble, Barbara, "Non-Governmental Organizations and the Making of U.S. International Environmental Policy," in *The International Politics of the Environment*, Andrew Hurrell and Benedict Kingsbury, eds., 1992; Bramble, Barbara, "Swapping Debt for Nature?" *Hemisphere*, 1998.

Brand, Stewart

(December 14, 1938–)
Publisher, Editor, Technology Consultant

Stewart Brand, author, publisher, business consultant, and proponent of computers and communications technology, shaped the counterculture environmental movement in the late 1960s with the publication of his *Whole Earth Catalog: Access to Tools and Ideas* (1968 to 1971). This highly original catalog empowered the country with information—presenting books, tools, and advice on community, self-reliance, and ecological living. Following the catalog, Brand published the magazine *Co-Evolution Quarterly*, which later evolved into *Whole Earth Review* (now known as *Whole Earth*), continuing his mission to educate people about the power of ideas and the importance of putting the right tools in the right hands. Dismissing the commonly held concept that advancements in technology must be at odds with conservation, Brand believes that computer technology has the potential to create innovative and efficient solutions to many environmental problems.

Stewart Brand was born on December 14, 1938, in Rockford, Illinois, to Arthur Barnard and Julia (Morley) Brand. From early childhood Brand was driven by an insatiable curiosity and restlessness that would stimulate all his future endeavors. At the age of 16 he borrowed his parents' car so that he and his high school friends could leave Rockford and head to California to pan for gold. He cultivated quirkiness and began wearing a beret, to the chagrin of his father and to the amusement of his somewhat offbeat mother. Yet he possessed a conservative streak as well, and from 1954 to 1956 he went to high school at Phillips Exeter Academy, a boarding prep school. He then attended Stanford University and received his B.S. in biology there in 1960. He also studied design and photography at San Francisco Art Institute College and San Francisco State College (now University). On completing his studies, he served in the United States Army from 1960 to 1962, teaching basic infantry training and working as a photojournalist out of the Pentagon. During these years Brand began getting involved in the performance art scene in New York, and when his Army stint ended he moved to San Francisco and became a multimedia performance artist. He was the founding artist of the performance piece "America Needs Indians," in which he performed until 1966. In the meantime, he had joined up with novelist Ken Kesey's band of Merry Pranksters in 1964 and went along on the psychedelic bus journey chronicled in Tom Wolfe's *The Electric Kool-Aid Acid Test* (1968). In December 1966, he and Lois Jennings were married (the marriage ended in divorce in 1976).

Brand's interest in technology and its tools began infiltrating his work, and he was especially intrigued by the possible connections between ecology and technology. During his travels he had conceived and sold buttons that read "Why haven't we seen a photograph of the whole Earth yet?" Eventually NASA did succeed in producing a photograph of the world during the Apollo program in 1969, and Brand contends that this helped to create an awareness of the global environment and that it was no accident that the first Earth Day came one year later. His preoccupation with the earth, information, and using appropriate tools led to the creation of one of his most famous endeavors, *The Whole Earth Catalog*, which was first published in 1968. The catalog, which swiftly became a mainstream hit, was a wide-ranging guide to tools and books, providing unorthodox advice and philosophical commentary on returning to nature and being self-reliant. It promoted select high-tech items—such as computers, ham radios, and solar electric systems—as well as low-tech approaches—such as hand grain-grinders, adobe buildings, and organic gardening. The book became the bible of the counterculture and spawned a trend toward more earth-friendly thinking. It was updated continually and issued for three years, and in 1972 its final edition, *The Last Whole Earth Catalog*, won a National Book Award.

In 1972 Brand founded the nonprofit Point Foundation to run activities related to the catalog and other subsequent Whole Earth projects. During its first few years it gave away $1 million to various ecology groups. Following the success of *The Whole Earth Catalog*, Brand went on to write other titles concerning topics such as environmental restoration and communication technologies. Then in 1974 he started a new periodical titled *Co-Evolution Quarterly*, a practical magazine that encouraged social change and new ideas. It introduced concepts such as the Gaia hypothesis and watershed consciousness and continued where *The Last Whole Earth Catalog* left off by providing book and tool reviews and promoting access to information. Also in 1974, Brand published *Two Cybernetic Frontiers*, a book on new trends in computer science, and the first ever to use the term *personal computer.* Along with his writing and editing at the magazine, Brand served as special adviser to California governor Jerry Brown for two years, until 1979, and continued exploring communications technology and the use of the computer as a personal tool. In 1984 he helped found the WELL (Whole Earth 'Lectronic Link), an experimental computer networking system that became a pioneer of electronic discussion and now has over 11,000 active users worldwide.

In 1985 *CoEvolution Quarterly* changed its name to *Whole Earth Review* and in the years since has hosted discourse on such topics as cyberspace, community building, and ecosystem restoration. Over the next few years, Brand also worked as a business consultant for Royal Dutch/Shell Group, an oil giant that wanted Brand to help them strategize ways to stay up to date and competitive. Brand's technological knowledge and his ability to think outside of the box made this endeavor particularly successful, and in 1988 he helped found the Global Business Network (GBN), a management consulting firm that explores strategic development and global futures for large institutions. GBN's client list has since swelled to 90 multinational corporations, including Xerox, IBM, Monsanto, Disney/ABC, and Texaco.

Since GBN was started, most of Brand's time is devoted to his consulting work. His perspective on the global environment has evolved since the years of *The Whole Earth Catalog.* Immersed in the world of technology, Brand sees computer hackers as the new grassroots contingent. He senses that the environmental movement these days has an antitechnology bias, which he condemns, saying that computer and communications technology can help lighten the load on the environment and increase efficiency—through telecommuting, for example.

As the new millennium approached, Brand and a few of his friends came up with an invention for the world's largest and slowest mechanical clock, hoping to install it in the desert in eastern Nevada where it would stand for 10,000 years as a testament to the future. Brand's motive in the project is to reverse the current short attention span most people have developed in a fast-paced world. Even if the 80-foot-tall clock fails to work, it still sends a message to take the long view and recognize that decisions made today can have ramifications for future generations. He wrote a book about the project called *The Clock of the Long Now: Time and Responsibility* (1999), which explores the significance of a 10,000-year "now."

Brand and his second wife, Patricia, a software agent and director of a medical resource center, live in a renovated 83-year-old tugboat docked on the waterfront in Sausalito, California.

Bibliography

Betts, Kellyn S., "Stewart Brand: Whole Earth Vision for the 21st Century," *E*, 1996; Kleiner, Art, and Stewart Brand, eds., *Ten Years of CoEvolution Quarterly: News That Stayed News*, 1986; "Stewart Brand's Home Page," http://www.well.com/user/sbb/; Stipp, David, "Stewart Brand: The Electric Kool-Aid Management Consultant," *Fortune*, 1995; Sussman, Vic, "A Born-Again Whole Earth Catalog: Looking toward the Millennium, The Editors Show How New Technologies Can Perk Folks Up," *U.S. News & World Report*, 1994.

Brandborg, Stewart

(February 2, 1925–)
Conservation Activist, Executive Director of the Wilderness Society

Stewart Brandborg was a key player in assuring the passage of the Wilderness Act in 1964 and devoted the next 12 years, during which he served as executive director of the Wilderness Society (TWS) , to the expansion of the newly established National Wilderness Preservation System. Through Brandborg's efforts, millions of acres of wilderness were added to the National Wilderness Preservation System, and more than 100 million acres in Alaska were preserved in the National Park, National Forest, Wildlife Refuge, and Wilderness systems. Brandborg's specialty was in training concerned citizens to become environmental activists and participate in the designation process. Over the years, hundreds of grassroots leaders attended the training seminars that he designed and oversaw.

Stewart Monroe Brandborg was born in Lewiston, Idaho, on February 2, 1925, son of Edna Stevenson Brandborg and Guy Mathew Brandborg. His father, who spent his entire forty-year career with the U.S. Forest Service (USFS), was supervisor of the Bitterroot National Forest for 20 years. As a strong wilderness advocate, he passed down to his son a love of nature and commitment to responsible stewardship. It was through his father that Brandborg as a young man had known Gifford Pinchot, founder of the USFS, and Robert Marshall, pioneer wilderness advocate and founder of the Wilderness Society. Starting at the age of 17, Stewart Brandborg worked seasonally in various national forests throughout Montana, Idaho, and Oregon, on range and forest surveys, trail maintenance, and as a lookout fireman. By 1944, when he was 19 years old, he had risen through the ranks to train other lookout firemen in fire suppression. Brandborg graduated from the University of Montana in 1948 with a B.S. in wildlife technology; in 1951 he earned his M.S. in wildlife management from the University of Idaho.

As a research fellow for the Idaho Cooperative Wildlife Research Unit at the University of Idaho, Brandborg completed seven years of field research on the life history of the mountain goat and wrote the first (and thus far only) monograph about this species. He performed population, range, and management studies of other major big game species of the region for the Montana and Idaho Departments of Fish and Game from 1947 to 1954.

Brandborg left the Northern Rockies for Washington, D.C., in 1954, when he was appointed assistant conservation director of the National Wildlife Federation (NWF). With a background in public lands and wildlife ecology, his duties included planning and overseeing conservation education programs and editing NWF's weekly "Conservation Report," a digest of all the legislation on natural resource and conservation issues before the U.S. Congress. It was with NWF that he developed many useful skills; for example public interest advocacy—lobbying and testifying before Congress on behalf of wildlife, public lands, and a broad range of environmental issues—and mobilizing conservationists throughout the nation in support of these measures.

In 1956, TWS elected him to its council, where he worked closely with TWS director Howard Zanhiser on an issue of concern to both TWS and NWF. The Wilderness Bill, drafted by Zanhiser in 1956, was the first legislation to propose a system to designate federally owned wild lands as wilderness, giving them protection within a National Wilderness Preservation System. TWS offered Brandborg the position as director of special projects in 1960. He was the closest collaborator with Zanhiser in gaining passage of the Wilderness Bill. When Zanhiser died suddenly in May 1964, Brandborg was appointed to succeed him as executive director. He escorted the Wilderness Bill through its last crucial steps in Congress and final passage in September 1964.

The Wilderness Act initially provided for the protection of some nine million acres of wilderness. Conservationists set what was then the ambitious goal of adding at least 50 million more acres of wilderness to the system, and TWS, under Brandborg's leadership, worked hard to achieve this. One of the stipulations of the Wilderness Act was that, for each proposed new wilderness area, there would be a public hearing and review by the wilderness agencies and Congress. Drawing on his experience with Congress and with local and state environmental groups and leaders, Brandborg initiated training programs for citizen activists. These programs taught them to carry out field studies of the areas in question, work with the agencies charged with managing the areas, testify effectively, and carry out grassroots publicity campaigns to congressional designation of the areas. TWS brought selected state activist leaders to Washington for week-long seminars on how Congress and government agencies work, while providing assistance and advice for formation and funding of local groups. Between 1964 and 1974, 150 new areas in 40 states were reviewed for protection under the Wilderness Act. A total of more than 103 million acres have now been designated as wilderness in Alaska and in the lower 48 states, many of them as a result of the dedication of TWS-trained volunteers.

While with TWS, Brandborg made other notable contributions to preservation of wilderness and the nation's public lands as well. He led a legal fight against the Trans-Alaska Pipeline, which was found by the courts to violate the 1969 National Environmental Policy Act (NEPA). This landmark, precedent-setting case strengthened the environmental impact assessment procedures of NEPA. Brandborg also directed a two-year lobbying campaign that eventually resulted in the designation of more than 100 million acres of public lands in Alaska as wilderness, national parks, and wildlife refuges. He enlisted Congressman John P. Saylor of Pennsylvania (the original House of Representatives sponsor of the Wilderness Act) in sponsorship of the Wild and Scenic Rivers Act, drafted by JOHN AND FRANK CRAIGHEAD. This measure, with TWS's support, was enacted in 1968, establishing the National System of Wild and Scenic Rivers.

Brandborg worked on other environmental issues in addition to wilderness preservation. Prior to the first Earth Day in April 1970, he gave strong backing to Earth Day organizers, providing start-up funding for the celebration. From 1971 to 1975, he served as cochair of the Urban Environmental Conference, one of the nation's first groups to focus on the joint concerns of environmentalists, urban reform groups, and organized labor. Members of its board included senior staff of such national organizations as TWS, United Auto Workers, and the National Urban League. The issues it addressed in its congressional lobbying work included lead poisoning, occupational health, clean air and water, energy conservation, urban transit, and land use planning.

When Brandborg left TWS in 1976—he was dismissed by the governing council, some of whose members believed that he had devoted too much time to wilderness preservation in Alaska—the organization's membership had grown to 130,000, five times larger than when he had become executive director 12 years earlier. The organization's budget was $1.8 million per year, and there was a full-time staff of 43, plus 12 part-time regional organizers. Environmental movement historian MARK DOWIE called Brandborg a "committed warrior" and "the last true activist to lead" TWS.

Brandborg spent the remainder of the 1970s working for the government, first as special assistant to the assistant secretary of the interior and later as special assistant to the director of the National Park Service. In these posts he worked with parks personnel, teaching them how to work more effectively with citizen activists in building support for the National Park System. From 1982 to 1986, he was the national coordinator of the Regional Environment Leadership Conference Series, a collaboration of the ten largest national environmental organizations. This was a series of nine regional activist training seminars for environmentalists and representatives of labor, ethnic minorities', women's, and urban groups.

In 1986, Brandborg moved back to western Montana's Bitterroot Valley, where he quickly became involved in public land and other environmental issues in the Northern Rockies. He was founder and president of Friends of the Bitterroot from 1988 to 1990 and serves on the boards of Wilderness Watch and several state and regional environmental groups. Long active in land planning, he is currently president of Bitterrooters for Planning, working to develop a growth plan in his home valley. He lives near Darby, a small town in western Montana, with Anna Vee, his wife of more than 50 years. They have five grown children.

Bibliography

Dowie, Mark, *Losing Ground: American Environmentalism at the Close of the Twentieth Century*, 1995; Frome, Michael, "Stewart M. Brandborg: Wilderness Champion," *Wilderness Watch*, 1999; Stroud, Richard, ed., *National Leaders of American Conservation*, 1985.

Bresette, Walt

(July 4, 1947–February 21, 1999)
Native American Treaty Rights and Environmental Activist

Walt Bresette led grassroots Native American treaty rights and environmental movements in the Lake Superior region during the 1980s and 1990s. He defended the Lake Superior Chippewa people's treaty rights to harvest fish, game, and plants from public lands and fought successfully for a moratorium on mining in the state of Wisconsin. He was known for his skill in empowering activists and for unifying Native and nonnative people committed to environmental justice. One of his last efforts was to author and promote a visionary Seventh Generation or Common Property constitutional amendment proposal, which would force lawmakers to think seven generations into the future when making decisions about common property such as water, air, and public lands.

Walter Albert Bresette was born on July 4, 1947, in Reserve, Wisconsin, to Henry and Blanche Bresette. His father was a lumberjack and gardener; his mother raised seven sons. The family lived on the Red Cliff reservation, located on the southern shore of Lake Superior. Blanche Bresette's sister, Victoria Gokee, was one of the first female tribal chairs in the United States, and she recruited young Walt and his brothers and cousins to help research issues affecting the Red Cliff Band of the Lake Superior Chippewa. As a teenager, Bresette and his cousins held their first nonviolent environmental protest and succeeded in stopping loggers from illegally cutting a pine grove on Madeline Island, a sacred place for the Lake Superior Chippewa.

Bresette was drafted by the United States Army in 1964 and served four years as a noncombatant with an electronic intelligence unit, mostly in Japan. When he returned in 1968, he received a Bureau of Indian Affairs grant to study advertising art in Chicago. Bresette became a graphic artist, working freelance during the 1970s in Madison and Red Cliff. From 1978 to 1980, Bresette served as a public information officer of the Great Lakes Intertribal Council, a statewide organization that provides social and economic services to ten Indian tribes in Wisconsin. In an attempt to unify scattered Chippewa (who also call themselves by their own names: Anishinaabe or Ojibwe), Bresette organized an event to commemorate the 125th anniversary of the final treaty that the Chippewa had signed with the U.S. government. The three treaties that the Chippewas signed under pressure during the mid-nineteenth century ceded their rich territory in what is now Wisconsin, Michigan, and Minnesota to the U.S. government. The Chippewa people agreed to live on several reservations, including the six that dot northern Wisconsin. The 1979 treaty commemoration held in Red Cliff and Madeline Island was the first of its kind; it celebrated tribal history, language, and culture. Bresette and the Chippewa attendees hoped that it would mark the beginning of a reunification of the Chippewa.

In 1982, Bresette became news and public affairs director at the northwestern Wisconsin WOJB public radio station. As he was preparing for his nightly news broadcast on January 25, 1983, a story came over the wire that Lake Superior Chippewa Indians had just been granted "unlimited hunting and fishing rights" by a panel of three federal judges. The so-called Voigt decision, named after Lester Voigt, head of the Wisconsin Department of Natural Resources (WDNR), came after a nine-year legal battle by Chippewa spearfishers to reaffirm their treaty rights to hunt and fish on treaty-ceded territory. Bresette was outraged at the report; he knew that the spearfishers neither desired to nor would be able to fish to an "unlimited" extent. Soon, a virulent antispearfishing, anti-Indian movement emerged in northern Wisconsin.

Spearfishing for walleyed pike is traditionally done at night during the spring. When native spearfishers set their boats out in northern Wisconsin's small lakes after the Voigt decision, they were harassed by mobs of racist white sportsmen who sometimes physically endangered the Indians by rocking their boats, blinding them with spotlights, or threatening to shoot them. Bresette later recalled that "the only people fishing in Northern Wisconsin during that time were those willing to risk their lives."

In an attempt to protect spearfishers, Bresette appealed to a statewide network of environmentalists, civil rights and church activists, and concerned citizens from throughout Wisconsin to come to the boat landings in the spring as peaceful observers. This "witness" provided important support as the antispearfishing protests grew larger and the protesters more violent. The Native-nonnative collaboration that Bresette nurtured during the worst years of spearfishing protests has become a model for nonnatives doing solidarity work with Native peoples. It is described in *Walleye Warriors*, which Bresette coauthored with Green activist Rick Whaley.

Bresette himself became a protagonist in the definitions of treaty rights. As an artisan, Bresette earned a living by making and selling dream catchers containing bird feathers, some of them from migratory birds. This was illegal under the Migratory Bird Act. He and fellow artisan, Chippewa elder Esther Nahgahnub, were arrested in 1989. "Feathergate," as it came to be called, yielded a landmark decision on Chippewa harvesting rights. In 1991, a U.S. District Court ruled that Bresette and Nahgahnub were not causing severe impact on endangered migratory bird populations with their activity and that, furthermore, feather gathering on treaty-ceded land was allowed under Chippewa treaty rights.

After the spearfishing problems subsided, Bresette turned his attention to other serious environmental problems in northern Wisconsin. In 1990, he cofounded Anishinaabe Niijii, an Ojibwe mining watchdog group. Throughout the 1990s, Bresette led and participated in grassroots protests against northern Wisconsin's copper mines, adding his signature charisma and humor to them. One story often retold is of a protest at the Ladysmith open pit copper mine site in 1992, during which he scaled a fence, jumped into the site holding a war club that had once belonged to Sauk and Fox war chief Black Hawk, and "counted coup" (symbolically claiming victory over enemies) on a giant bulldozer. Bresette participated in a 28-day encampment in 1996 with the Bad River Ogitchida (Anishinaabe word for "Protectors of the People") on the Wisconsin Central Railroad tracks. They blocked a train carrying 64,000 gallons of sulfuric acid to be mixed with water and poured into the tunnels of a copper mine at White Pine, Michigan, only five miles from Lake Superior. Bresette had been a member of the Environmental Justice Committee of the Environmental Protection Agency (EPA) but had quit when the EPA approved this project without a hearing or clean-up plan. The EPA did do another review of the project, but while it was in process, the mining company withdrew its project. In 1998, thanks in part to efforts of Anishinaabe Niijii, the Wisconsin legislature passed a moratorium on mining in the state.

Convinced that the government should be obligated to protect common property at least to the same extent that it protects private property, Bresette authored the Common Property or Seventh Generation constitutional amendment, which would require lawmakers to think seven generations (or 150 years) ahead when considering laws that pertained to public lands, air, and water. During the summer of 1998, Bresette led a month-long walk from Red Cliff to Madison to promote the amendment.

Bresette received many awards for his activism, including the Eagle Feather Award from the Red Cliff Cultural Institute in 1985, the Wisconsin Labor Farm Party's Antinuclear Waste Organizing award in 1986, and the Progressive Social Change Award in 1987 from the Wisconsin Community Fund. *Walleye Warriors* received the Council for Wisconsin Writers book-length nonfiction award in 1993. Bresette worked on the All My Relations public radio project, which included conversations with people from Canada and the United States on the value of life and land, and appeared in the film "Wisconsin Powwow." He cofounded and served on the boards of directors of many organizations, including the Indigenous Environmental Network; the Midwest Treaty Network; Project Underground; the Anishinaabeg Millennium Project, which reclaims and redefines a vision for the future of the Anishinaabe nation; and many more.

Bresette died of a heart attack in Duluth, on February 21, 1999, while visiting friends. He left four children, Nicholas, Claudia, Katy, and Robin.

BIBLIOGRAPHY

Brain-Box Digital Archives, *Maawanji'iding: Ojibwe Histories and Narratives from Wisconsin* (multimedia CD), 1999; Bresette, Walt, "'We Are All Mohawks,'" *Green Letter*, 1990; Gedicks, Al, *The New Resource Wars, Native and Environmental Struggles against Multi-*

national Corporations, 1993; "Indigenous Environmental Network," http://www.alphacdc.com/ien/; "Midwest Treaty Network," http://www.alphacdc.com/treaty/; "Protect the Earth: Stories, Memories, and Condolences about Walt," http://www.protecttheearth.com/about-walt.html; Whaley, Rick, and Walt Bresette, *Walleye Warriors: The Chippewa Treaty Rights Story*, 1999 (available through ENFIELD@CONNRIVER.NET).

Brower, David

(July 1, 1912–)

Director of the Sierra Club; Founder of Friends of the Earth, League of Conservation Voters, and Earth Island Institute

David Brower, proclaimed "archdruid" of conservation by his biographer, John McPhee, is perhaps the most famous environmentalist of the twentieth-century United States. An intrepid mountaineer credited with over 70 first ascents, Brower has headed several of the country's most effective environmental groups. He directed the Sierra Club from 1952 to 1969, transforming it from a group of 2,000 mountaineers to a powerful political lobby representing 77,000 members, and then went on to found several more environmental groups, including the John Muir Institute, Friends of the Earth (FOE), the League of Conservation Voters, and Earth Island Institute. Brower can be credited for keeping Dinosaur National Monument and the Grand Canyon free of dams and for helping establish Kings Canyon, North Cascades, and Redwood National Parks, as well as the Point Reyes National Seashore. His credo is now inscribed on the National Aquarium in Washington, D.C.: "We do not inherit the Earth from our fathers, we are borrowing it from our children."

David Ross Brower was born in Berkeley, California, on July 1, 1912. His parents took their four children on frequent trips to Lake Tahoe and on hikes in Berkeley's hilly backdrop. As a child, Brower once broke open a chrysallis and discovered that the wings of the butterfly inside had not yet developed. Immediately he realized that his interference in its life cycle had doomed it, and this revelation transformed him into a conservationist. Brower began studying forestry and biology at the University of California at Berkeley in 1929 but dropped out during his sophomore year in 1931. In a 1994 interview, Brower claimed that his later success was due to this early escape from academia. "I hadn't been educated to know what you couldn't do," he told *Progressive* writer David Kupfer. After quitting school, Brower worked as a clerk at a candy company during the year and spent all his free time in Califonia's Sierra Nevada.

In 1933, Brower joined the Sierra Club and in 1935 published his first article in the Sierra Club *Bulletin*. Brower joined Sierra Club activists on a campaign to have Kings Canyon signed into the National Park System in 1940. He produced a publicity film about Kings Canyon and worked with the Sierra Club Press to

David Brower (courtesy of Anne Ronan Picture Library)

publish ANSEL ADAMS's book *The Sierra Nevada: The John Muir Trail.* With this publications experience, Brower took a job as an editor at the University of California Press. He met his wife, editor Anne Hus, at the Press and married her in 1943.

When World War II broke out, Brower enlisted in the 10th Mountain Division and taught U.S. troops mountaineering skills in the Italian Apennines. His commitment to conservation was reinforced after witnessing the war's environmental devastation of that scenic area.

Brower returned home after the war to become a vocal protagonist in several conservation struggles, including a drive against logging in Olympic National Park and a controversial plan to build roads through Kings Canyon. The Sierra Club was divided on the latter issue: some members felt that accessibility to the beauty of Kings Canyon should not be a privilege of only those fit enough to hike in while others believed that more roads through Kings Canyon would lead to overuse. Brower's side, against road construction, prevailed. He was chosen in 1952 to serve as executive director of the Sierra Club, because club members were realizing that political action would become increasingly necessary to save the natural wonders of the West.

Brower's first major political fight, in which he successfully convinced the federal government to scrap a dam project in Dinosaur National Monument in exchange for a promise not to oppose a dam at Glen Canyon, is one that has marked his work ever since. He favored the Glen Canyon dam simply because Dinosaur was so beautiful and he had never seen Glen Canyon. But later, once he visited Glen Canyon, he was convinced he had made a terrible mistake. Since then he has decried compromise, and in recent years Brower has joined a coalition lobbying for the destruction of Glen Canyon dam.

Working together with HOWARD ZAHNISER of the Wilderness Society, Brower's Sierra Club was instrumental in the passage of the 1964 Wilderness Act, which designated nine million acres of wild country in the United States as unexploitable wilderness. During the mid-1960s, Brower coordinated publicity campaigns to keep dams out of the Grand Canyon. His full-page newspaper ads, designed by JERRY MANDER's ad agency Freeman, Mander, and Gossage, were sufficiently political that the Sierra Club lost its tax-exempt status. This angered more conservative elements of the Sierra Club, who deposed Brower in 1969 after a bitter internal struggle over the siting of a California nuclear power plant. The PG&E power company intended to build its plant in the coastal Nipomo Dunes area. Since this was an ecologically rich site, the club's board of directors negotiated with PG&E for an alternative site. If PG&E would make Nipomo Dunes available for purchase as a state park, the Sierra Club would not protest the siting at the Diablo Canyon area. Brower reared at this compromise, similar to the compromise he considered the most serious mistake he had ever made. He published a "half-bulletin" about the controversy, including only his own opinion since the opposition did not submit its opinion in time for publication. A major conflict erupted, and Brower was ousted as director.

In 1968, Brower founded the John Muir Institute, which promoted environmental research and education, and in 1969 he founded the lobbying group Friends of the Earth along with its partner organization, the League of Conservation Voters. Brower served as president of FOE for 10 years, before founding Earth Island Institute in 1982, whose board of directors he continues to chair. Earth Island Institute provides institutional support and networking for some 30 projects that focus on environmental problems throughout the world. Certain projects, such as the Rainforest Action Network and Urban Habitat, have bloomed into independent organizations.

Brower has written a two-volume autobiography and has written or edited dozens of volumes on the nation's wild places. His most recent book, cowritten with Steve Chapple, *Let the Mountains Talk, Let the Rivers Run* (1995), calls for emergency CPR therapy for a stricken earth: Conservation, Preservation, and Restoration in massive doses. Among Brower's and Chapple's prescriptions: limit development and ring cities with green belts, link protected areas with wildlife corridors and expand them with buffer zones, and use alternative natural resources (for example, expand the production of electric cars, develop more ways to cook with solar energy, and use tree-free paper such as kenaf paper for publishing—the book is printed on this paper, as is *Earth Island Journal*).

Brower is still married to Anne Hus and with her has four children, Kenneth, Robert, Barbara, and John. He has received many conservation and publications awards during his life, including two nominations for the Nobel Peace Prize.

Bibliography

Brower, David, *For Earth's Sake, the Life and Times of David Brower*, 1990; Brower, David, *Visions of the Environmental Movement*, 1993; Brower, David, *Work in Progress*, 1991; McPhee, John, *Encounters with the Archdruid*, 1971; Strong, Douglas H., *Dreamers & Defenders: American Conservationists*, 1988.

Brown, Janet

(September 20, 1931–)
Director of Environmental Defense Fund, Political Scientist

Director of the Environmental Defense Fund (EDF) from 1979 to 1984, Janet Welsh Brown is an environmental policy advocate and expert on global environmental politics. She is the coauthor of two important books on global politics: *Bordering on Trouble (1986)*, with Andrew Maguire, and *Global Environmental Politics (1996)*, with Gareth Porter. She has served in a number of executive positions in policy and political organizations and continues to be an influential voice in Washington, D.C., environmental circles.

Janet Welsh Brown was born on September 20, 1931, in Albany, New York. She attended Smith College, earning a B.A. in government in 1953. She received an M.A. in 1955 from Yale University in Southeast Asian studies and a Ph.D. in international relations from American University in 1964. She taught at Sarah Lawrence College from 1956 to 1958, at Howard University from 1964 to 1968, and at the University of the District of Columbia from 1968 to 1973. In 1973 Brown became director of the Office of Opportunities in Science for the American Association for the Advancement of Science, where she developed a program to increase opportunities in the natural and social sciences for women, minorities, and the physically handicapped. She also served as president of the Federation of Organizations for Professional Women. From 1975 to 1976, she was president of the Scientific Manpower Commission. During this period Brown earned a reputation as an expert on science education. She served as consultant to the National Science Foundation, the Department of Health and Human Services, and other federal agencies.

In 1979, Brown was hired as executive director of the EDF (now known as Environmental Defense). She inherited the organization during a period of antienvironmental backlash. The energy crisis of the 1970s had led to high inflation and economic recession, and environmental regulation became a scapegoat for the economic crisis. In her first message as director in EDF's annual report, she emphasized the importance of the energy crunch: "All of us, in 1979, have felt the powerful pressures of inflation and the drive to develop new energy sources, however dangerous." Ronald Reagan's election in 1980 accelerated the fear of environmental rollbacks. EDF was preoccupied with protecting the Clean Air Act, the Marine Mammal Protection Act, the Clean Water Act, and other landmark legislation under threat of repeal by the Reagan administration. It was in this atmosphere that the "Group of Ten" environmental leaders first met at the Iron Grill in Washington, D.C., in January 1981, in an effort to combine forces to face attacks on established environmental policy. Brown was at that meeting, representing EDF. The organization developed into one of a new breed of professional, staff-based environmental agencies, distinct from and sometimes opposed to grassroots, activist-based groups. Brown's leadership took this professional, corporate mode. Brown celebrated the teams of scientists and lawyers who did much of EDF's work. These experts were consulted by elected officials, policy makers, and the media, and Brown nurtured the image of a knowledgeable, professional staff. During Brown's tenure as director, EDF focused broadly on energy, toxic chemicals, water resources, and wildlife issues, using both research and litigation to bring about policy changes. Brown also emphasized the importance of economics in formulating sound environmental policy, saying in her 1982 annual message to members, "EDF understands what the President doesn't—that good environmental practices are good business, and that there are at least as many jobs in caring for the environment as in destroying it." Brown resigned from EDF in 1984.

In 1985 Brown became a senior associate of the World Resources Institute, an organization devoted to research about global resources and environmental politics. As a research fellow Brown authored several studies of global environmental politics. *Bordering on Trouble: Resources and Politics in Latin America*, an anthology of essays by economic and political analysts coedited with Andrew Maguire, was published in the midst of the 1980s crises in Nicaragua and El Salvador. The authors of the essays attempt to shift debates about Latin America away from the old paradigm of "communist" versus the "free world" to an understanding of the way resource management and distribution shape political change. In their introduction to the collection, Brown and Maguire outline four major themes running through the essays: longstanding U.S. involvement in resource exploitation in Latin America, closeness of cultural ties between the United States and Latin America, damaging effects of U.S. support of repressive Latin American regimes, and the tendency of U.S. policy to focus on short-term economic and security interests rather than the long-term health of the economy and environment at home and abroad. *Global Environmental Politics* looks at the state of environmental politics across the world. Brown and Porter describe major trends in environmental policy, important actors in the arena (from governments and international state organizations to nongovernmental organizations and corporations), a broad array of case studies, and the politics of international development. The book serves as a primer to the current state of global environmental politics and is often used as such in college classrooms.

Brown has published a number of other works, including *In the U.S. Interest: Resources, Growth, and Security in the Developing World* (1990). She has served on a number of boards and committees, including the editorial board of *Society and Natural Resources*, the CARE Advisory Committee on Population and Development, and a term as president of the board of Friends of the Earth. She resides in Washington, D.C.

Bibliography

"Environmental Defense Fund," http://www.edf.org; Gottlieb, Robert, *Forcing the Spring*, 1993; "World Resources Institute," http://www.wri.org.

Brown, Lester

(March 28, 1934–)
President of Worldwatch Institute, Economist

Lester Brown is president and founder of the Worldwatch Institute, an interdisciplinary, global think tank that publishes the annual *State of the World* reports. He is an internationally influential agricultural economist and author, who sees food production and population growth as two of the most urgent environmental issues facing society today.

Lester Russell Brown, the oldest child of Calvin C. Brown and Delia Smith Brown, was born on March 28, 1934, in Bridgeton, New Jersey. He was raised on a farm in Bridgeton where his parents grew tomatoes. He was an active member of the local 4-H club and, later, the Future Farmers of America. When Brown was 14, he and his brother Carl bought a small plot of land and an old tractor and began farming tomatoes themselves. Their farming operation was producing 1.5 million pounds of tomatoes per year by the time Brown received his B.S. degree in agricultural science from Rutgers University in 1955, a yield that put the Browns in the most productive 2 percent of East Coast tomato farmers.

Upon graduation Brown had every intention of farming for the rest of his life. However, the year after he graduated, he was given the opportunity to live and work for six months in a small farming community in India, through the International Farm Youth Exchange program of the National 4-H Club Foundation. His experiences in India opened his eyes to the pervasive problem of world hunger, and suddenly "the idea of just growing tomatoes for the next forty years no longer seemed very challenging," he told Jim Patrico in *Top Producer* in 1987. In 1959, after completing an M.A. degree in agricultural economics at the University of Maryland, Brown became an agricultural analyst with the Foreign Agricultural Service, a branch of the U.S. Department of Agriculture (USDA). At the USDA, Brown produced a study that was the first to link the issue of food supply to rapid population growth. It received quite a bit of attention and was even featured on the cover of the January 6, 1964, issue of *US News and World Report.* This study led to Brown's first book, *Man Land and Food: Looking Ahead at the World's Food Needs*, published in 1963.

Lester Brown (courtesy of Anne Ronan Picture Library)

In 1963, Brown left the USDA and earned a master's degree in public administration from Harvard University. Orville L. Freeman, who was at that time secretary of agriculture, appointed Brown adviser on foreign agricultural policy in 1964. In this post, Brown was able to convince U.S. government officials to demand dramatic changes in India's agricultural policies in exchange for food shipments from the United States. These policy changes helped to avert large-scale famine. In 1966, Brown was named administrator of the USDA's International Agricultural Development Service, where he was responsible for overseeing projects being conducted in more than 40 countries. Brown stayed in this position for three years, until 1969, when he left government service altogether. At that time, along with James Grant, Brown formed the Overseas Development Council, a private, nonprofit organization whose purpose was to analyze economic and political issues affecting relations between the United States and developing countries.

In 1973, at a chance meeting with William Dietel, then executive vice president of the Rockefeller Brothers Fund, Brown learned that Dietel shared many of his concerns about the environment and was willing to support the formation of a research institute. In 1974, with a startup grant of $500,000 from the Rockefeller Brothers Fund, Brown founded the Worldwatch Institute. Since then, he has served as president and as one of the Institute's senior researchers. The purpose of Worldwatch is "to raise public awareness of environmental threats to the level where it will support an effective public policy response." The Institute conducts interdisciplinary, global research on everything from air pollution to biodiversity to "jobs in a sustainable society." It publishes two yearly reports, *Vital Signs: The Trends that Are Shaping Our Future* and *State of the World*, and a bimonthly magazine, *World Watch.*

The State of the World reports are broad in scope, and they differ from the reports of other think tanks in that they are sold (all of Worldwatch's publications are sold, rather than given away). Brown believes that the fact that the reports are purchased is proof of their usefulness and value; in addition, their sale helps pay the Institute's operating expenses. The reports are translated into many different languages and address agricultural and economic trends as well as population growth in developing countries. They provide up-to-date statistical information to the public, especially to policy makers.

Brown has written many books, including among others *World Without Borders* (1972), *In the Human Interest: A Strategy to Stabilize World Population* (1974), and *Building a Sustainable Society* (1981). In his books, Brown addresses population growth, soil erosion, overfishing, rain forest depletion, and the disappearance of oil reserves. He calls for a greater degree of international cooperation between scientists and governments, and he challenges people to change their consumptive habits and attitudes toward the environment.

In an interview with *Geographical Magazine* in 1995, Brown stated, "We live in a highly industrialized, urbanized society, and we take food supply for granted because it has been abundant for the past four decades. I think we have lost touch with nature and assumed that if food production has tripled in the last four decades, it will do so again in the next four." Brown believes not only that the types of changes that need to happen must occur at a local, individual level but that also "the whole system must change through political action." He believes that people need to start making choices between quality of life and quantity of life, that the earth simply cannot support six billion people all living a suburban U.S. lifestyle. The only viable solution he can foresee will occur when we reset "the environmental balance sheets" and act on the principle that if we cannot recreate what we destroy, then we should not destroy it.

Brown has received numerous service and literary awards over the course of his career, including the National Wildlife Federation Special Conservation Award in 1982, the Lorax Award, Global Tomorrow Coalition in 1985, and a Gold Medal from the Worldwide Fund for Nature, 1989. He was named one of "100 who made a difference" by *Earth Times* in 1995, and he is one of the Audubon Society's 1998 "100 Champions of Conservation." Brown lives in Washington, D.C. He and his ex-wife, Shirley Ann Woolington, have two grown children, Brian and Brenda Ann.

Bibliography

Lawson, Trevor, "The Voice of Reason," *The Geographical Magazine*, May 1995; "Lester Brown," *Current Biography Yearbook*, 1993; "Worldwatch Institute," http://www.worldwatch.org.

Brown, Michael

(March 5, 1952–)
Journalist

Journalist Michael Brown has contributed to the nation's environmental awareness through his investigative reporting on industrial pollution and toxic waste disposal. While working as a reporter at the *Niagara Gazette*, he tackled the issue of toxic contamination at Love Canal and helped bring it to national attention. He was nominated four times for a Pulitzer Prize for his coverage of the story, which he later chronicled in a book called *Laying Waste: The Poisoning of America by Toxic Chemicals* (1980). He also wrote *The Toxic Cloud: The Poisoning of America's Air* (1987), in which he examined the sources of pollutants in the air over the United States and the potentially disastrous health consequences of allowing industries to ignore their responsibility to those who live downwind.

Michael Harold Brown was born on March 5, 1952, in Niagara Falls, New York. His father, Harold Brown, was an accountant, and his mother Rose (Mento) Brown was a teacher. He attended Fordham University, where he graduated with a bachelor's degree in 1974. Having always wanted to be a writer, he began working three years later for the *Niagara Gazette* as a reporter. Soon after starting his job, he learned about the toxic waste leakage problems in the Love Canal neighborhood of Niagara Falls and began reporting on it for the *Gazette.* Residents of the neighborhood had first begun noticing something was wrong in 1976, when heavy rains were causing foul-smelling chemicals to seep into the local schoolyard and the basements of homes in the area. They discovered that the school had been built over an old, dry canal, which had been used by the Hooker Chemicals and Plastics Corporation to bury 20,000 tons of chemical wastes in the 1940s and 1950s.

Brown undertook an uncompromising investigation of the leak, which eventually led to more than 100 articles for the *Niagara Gazette.* The municipal government and Hooker Chemical attempted to downplay the issue, and at no time tried to warn the people who lived in the area about the waste dump or any health risks it could pose. Brown, dissatisfied with the way things were being handled, continued his investigations. Along with LOIS GIBBS, a resident who was quickly becoming an impassioned toxic wastes activist, Brown conducted informal health surveys of area residents that turned up abnormally high rates of cancer, miscarriages, birth defects, and nerve damage. Partly as a result of people like Brown and Gibbs urging officials to take responsibility—and Brown's refusal to let the issue die down in the press—governmental agencies began a detailed chemical analysis of the leak. They eventually isolated over 180 compounds, many of which were carcinogenic or capable of damaging the nervous system—this included dioxin, which is so toxic that its potency is measured in parts per trillion.

Besides exposing the problem through the newspaper, Brown took further action by continually pressuring Hooker Chemical and the federal Environmental Protection Agency to remedy the situation by cleaning up the leak and compensating the victims. Hooker Chemical refused to accept responsibility for what happened, claiming that it had followed standard procedures back in the 1940s and 1950s when it buried the waste. Finally, Pres. JIMMY CARTER declared Love Canal a national emergency and more than 900 families were evacuated from the area. State and federal authorities filed $760 million in damage suits against Hooker Chemical to recover costs and compensate displaced residents. Some of the lawsuits were settled when Hooker Chemical agreed to take some remedial steps and pay a portion of the cleanup costs. One result of the Love Canal incident is that the Environmental Protection Agency began to pay more attention to waste disposal sites. In 1978 the agency began surveying other hazardous waste sites; it found 32,000 that were deemed a risk to human health and the environment and at least 18,000 other sites that were unregulated.

For his reporting work on the Love Canal toxic waste leak, Brown earned four Pulitzer Prize nominations. In 1979 he received an award from the Environmental Protection Agency; in that same year he left the *Niagara Gazette* and began work on a book that would chronicle the entire Love Canal story. In *Laying Waste: The Poisoning of America by Toxic Chemicals*, published in 1980, Brown also looks beyond Love Canal to give an account of some of the other known toxic chemical dumps around the country. He discovered that only a small fraction of hazardous wastes are disposed of properly—even the Environmental Protection Agency estimated that as of 1979, less than 7 percent of the millions of tons of toxic wastes produced in the United States were legally disposed of. Brown concludes that the corporate irresponsibility behind this neglect leaves the rest of the population to pay the price and that federal, state, and local regulatory agencies have not provided adequate protection against the problem.

Continuing his mission to uncover the causes of environmental contamination, Brown wrote a book about airborne pollutants, *The Toxic Cloud: The Poisoning of America's Air* (1987). He found that pollution could travel great distances from its point of origin. In the 1970s, impurities were detected in an isolated lake ecosystem in Isle Royale in the northern part of Lake Superior, which was completely cut off from all other water sources except rain. Scientists found pesticide residues there that were traced back to cotton fields in the southern United States. Brown also describes the smothering petrochemical pollution along the Gulf coast and how the chances of contracting cancer there are one in three. In *The Toxic Cloud*, Brown reveals that no place in the country is free from unseen toxic particles blown by winds, causing potential health risks. He emphasizes that much of the emissions from industrial sources are unregulated, and many of the gases and chemicals that are released are untested—especially for long-term effects or for how they interact when combined. Once again Brown blames governmental agencies for policies that seem to favor industry over clean air and challenges them to take responsibility.

In addition to his books, Brown has written articles for various magazines such as *Atlantic Monthly*, *Science Digest*, *New York Times Magazine*, and *Reader's Digest*. More recently he wrote *The Search for Eve* (1990), a look at paleoanthropology and the evolutionary origin of humans. He lives in New York City.

Bibliography

Brown, Michael, *Laying Waste: The Poisoning of America by Toxic Chemicals*, 1980; Brown, Michael, *The Toxic Cloud: The Poisoning of America's Air*, 1987; Cohen, Richard, "Laying Waste: The Poisoning of America by Toxic Chemicals," *New Republic*, 1980; Kaiser, Charles, "Hell Holes," *Newsweek*, 1980.

Browner, Carol

(December 16, 1955–)
Administrator of the Environmental Protection Agency

Attorney and natural resources expert Carol Browner was nominated by Pres. Bill Clinton as administrator of the Environmental Protection Agency (EPA) in 1992 and has served longer in her position than any other EPA administrator. Browner has worked to reform a notoriously bureau-

Chinese Premier Zhu Rongji (left) and U.S. Vice President Al Gore (right) watch the signing of an environmental agreement by Xie Zhenhua, China's minister of state for environmental protection, and U.S. Environmental Protection Agency administrator Carol Browner, 9 April 1999 in Washington, D.C. (AFP/Corbis)

cratic and corrupt agency and has attempted to work out solutions to environmental problems through compromise among business, consumer, and environmental protection interests. At the same time that she has collected the largest fines from major polluters in the EPA's history, she has tried to reward compliant industries by streamlining the regulatory process in a way that has saved them time and money.

Carol Browner was born on December 16, 1955, in Miami, Florida, to Isabella Harty-Hughes and Michael Browner, the first of their three daughters. As a child, she would often bicycle into the wilderness of the Everglades very near her south Florida home, an activity that she now believes was a source of motivation for her career choice. Both of Browner's parents were university professors; her father taught English and her mother social sciences. Browner graduated from the University of Florida at Gainesville in 1977 with a B.A. in English, enrolled directly in the

University's law school, and earned her law degree in 1979.

Browner began her legal career in 1980, working for a year as general counsel for the Committee on Governmental Operation of the Florida House of Representatives. In 1983 she became the associate director of the Washington environmental lobbying group Citizen Action. Browner held that position until 1986, when Florida senator Lawton Chiles named her his senior legislative aide for environmental issues. During her tenure at that post, Browner performed her official duties with uncommon vigor. In 1987, while pregnant, she dived into Florida coastal waters to view the environmental effects of oil drilling firsthand. That experience later drove her to help negotiate a ban on drilling for oil and gas off the Florida Keys. Her nonconventional approach earned her notoriety and eventually a place as counsel for the U.S. Senate Committee on Energy and Natural Resources, a position that she accepted in 1989. That same year Browner left her charge with Chiles to join the staff of Tennessee senator AL GORE, JR. as a senior legislative aide. Browner worked with Senator Gore until 1991, at which time she left Washington at the request of then-governor Chiles to become secretary of Florida's Department of Environmental Regulation, the third-largest environmental agency in the country.

Exemplary of her desire to accommodate the interests of business while protecting the environment, one of Browner's first official acts as secretary was to help win passage of Florida's Clean Air Act, which streamlined the process by which businesses obtain permits to develop wetlands and expand manufacturing plants, thereby reducing the amount of money and time that the process was costing both business and the department. Browner's practical yet controversial approach earned her more praise than criticism, however, in a land-preservation compromise agreement with the Walt Disney Company in 1992. When Disney filed for the permits it needed to build on 400 acres of wetlands that it already owned, Browner brokered a deal that provided Walt Disney with the necessary means to proceed with its development plans in return for a $40 million commitment by the company to buy and restore 8,500 acres of endangered wetlands south of Orlando.

As secretary, Browner was also the primary negotiator in the settlement of a federal lawsuit brought against the state of Florida in 1988 for violating its own environmental laws. The suit alleged that the state allowed sugarcane growers and other farmers to siphon water from the Everglades for agricultural use, harming the ecosystem, and that it did not require them to filter the contaminated runoff water that flowed into the swamp. Browner took the side of the federal government and, in so doing, took on the region's wealthy sugarcane families. A settlement reached in 1991 resulted in the largest ecological restoration project ever attempted in the United States. It included the construction of filtering marshes to purify and restore the natural flow of water to the Everglades, the cost of which was shared by the sugarcane farmers and by state and federal governments.

In 1992, Vice President Al Gore asked Browner to join President Clinton's transitional team for environmental issues, and in January 1993, the president appointed Browner EPA administrator.

Browner's career at the EPA began precariously. She was thought of as inexperienced and unqualified for such an important position, and many worried that she would unfairly favor the White House because of her connection to Vice President Gore. Browner had to contend with the most antienvironmental Congress in recent history, the 104th Congress, which tried to roll back the major environmental gains of the last quarter century. The 104th Congress undercut the EPA's funding and its regulatory powers and denied the Clinton administration the votes necessary to raise the EPA to cabinet status.

Browner persevered through her first difficult years in the post and has focused on water quality, cleanup of toxic waste sites, children's health, and more efficient regulation of industry during her tenure. Browner lobbied for the Safe Drinking Water Act, which was signed by Clinton in 1996. Under her leadership, more Superfund sites have been cleaned up than had been since the Superfund program was established in 1980. Browner has prioritized the clean up of "brownfields" (contaminated lots in urban areas) as well, envisioning the conversion of these areas into revitalized places for new urban development. She announced new stringent regulations of future hazardous waste incinerators and increased community involvement in incinerator sitings and has issued new standards for air pollution. Browner has made children's health a focal point, introducing new policies that would always take children's health risks into consideration in the permit process. She introduced what she calls a Common Sense Initiative, by which regulations are designed for specific industries, and companies are asked to meet and even exceed environmental regulations in ways that are both creative and cost effective.

Browner's aims for the remainder of her tenure include improving water quality and food safety and support of the Clean Water, Safe Drinking-Water, and Resource Conservation and Recovery Acts, as well as adding 170 chemicals to the Toxic Release Inventory. Browner's primary objective as administrator is to renew people's confidence in the EPA, while seeking a balance between the interests of business and environmental protection.

Browner is married to Michael Podhorzer, who directs Citizen Action. They live in Maryland with their son, Zachary. She received the Mother's Day Committee's 1997 Mother of the Year Award and was recognized by *Working Mother* magazine in 1998 as one of the 25 most influential working mothers.

Bibliography

Breton, Mary Joy, *Women Pioneers for the Environment*, 1998; "Carol M. Browner," http://www.epa.gov/swerosps/bf/bf98/cf/biosnabs/brownerc.htm; Wilkinson, Francis, "The Sinkable Carol Browner," *Rolling Stone*, 1993.

Bullard, Robert

(December 21, 1946–)
Sociologist, Environmental Justice Movement Activist

Robert Bullard is the nation's foremost scholar specializing in issues of environmental justice. His research on patterns of environmental racism as well as the grassroots groups combating it has provided a solid academic foundation for the environmental justice movement. Through his technical assistance to grassroots environmental justice groups via the Environmental Justice Resource Center that he founded at Clark Atlanta University and the *People of Color Environmental Groups Directory* networking tool, of which he has edited three editions, Bullard has been as much an actor in the movement as a student of it.

Robert Doyle Bullard was born on December 21, 1946, in Elba, Alabama, to Nehemiah and Myrtle Bullard. As a youth in the segregated South, reminders of racism were everywhere, and he found inspiration in such African American leaders as W.E.B. Dubois, Sojourner Truth, and Malcom X. Bullard attended Alabama A & M University in Huntsville, Alabama, graduating with a B.S. in government in 1968. He spent two years with the United States Marine Corps as a communication specialist immediately following graduation, and after his honorable discharge in 1970, he resumed his studies, earning an M.A. in sociology in 1972 from Atlanta University (now Clark Atlanta University) and a Ph.D. in sociology in 1976 from Iowa State University. Bullard accepted a position as assistant professor at Texas Southern University in Houston in 1976.

Bullard's research into environmental racism began in Houston in 1979, when his wife, lawyer Linda McKeever Bullard, asked for his help in researching her case *Bean v. Southwestern Waste Management.* She was representing a group of African American homeowners of the Northwood Manor suburb who were suing the city of Houston, the state of Texas, and Browning-Ferris Industries for environmental discrimination in siting a municipal landfill in their neighborhood, whose population was 82 percent Black. This was the first environmentally oriented lawsuit filed under the 1964 Civil Rights Act. Bullard's research revealed that in Houston, all of the city-owned garbage dumps, six of the eight city-owned garbage incinerators, and three of the four privately owned municipal landfills were in Black neighborhoods. Despite the evidence that the city of Houston and private landfill companies were deliberately targeting Black neighborhoods, the homeowners lost their suit and the landfill was built. The publicity surrounding the case did have an effect, however. In 1980 the Houston city council passed a resolution that prohibited city trucks from dumping at the Northwood Manor landfill; the next year it passed an ordinance prohibiting landfill construction near schools and other public facilities. The state Department of Health rewrote its landfill permit applications to require detailed land use, economic, and sociodemographic data for its proposed sites. In the years since the Northwood Manor landfill was built, no new landfill in Houston has been sited in a Black neighborhood.

Robert Bullard (courtesy of Anne Ronan Picture Library)

Bullard's findings drove him to continue his research on the phenomenon of locally unwanted land uses (LULUs) in Black communities. Bullard found that Black communities in the South—regardless of their economic status—were much more likely than White communities to be targeted for LULUs. His case studies of Dallas and Houston, Texas; Emelle, Alabama; Institute, West Virginia; and Alsen, Louisiana, make up *Dumping in Dixie: Race, Class, and Environmental Quality* (1990), a primer on environmental racism. Bullard's later books, *Confronting Environmental Racism: Voices from the Grassroots* (1993) and *Unequal Protection: Environmental Justice and Communities of Color* (1994) examine environmental racism as it manifests itself in different areas of the country against people of color and how grassroots groups have organized to fight it. These two books have joined *Dumping in Dixie* as classics in the field of environmental justice.

Bullard has gone further than simply describing and decrying environmental racism; he has actively served as a networker between the environmental justice groups that he has come into contact with through his research. In 1990 he began to compile a list of grassroots environmental justice organizations run by people of color. In an attempt to help them network, Bullard and a planning committee consisting of Dana Alston, Patrick Bryant, Benjamin Chavis Jr., Donna Chavis, Charles Lee, and Richard Moore convoked the First National People of Color Environmental Summit held in Washington, D.C., October 24 to 27, 1991. More than 600 people from virtually every state in the United States as well as Canada, Central America, Puerto Rico, and the Marshall Islands gathered for the very first time to discuss environmental and social justice activism in communities of color. The four-day gathering resulted in the articulation of 17 principles of environmental justice (published in their entirety on the Clark Atlanta University Environmental Justice Resource Center web site http://www.ejrc.cau.edu/). These principles reaffirmed the strong connection that people of color have with the earth and set up guidelines for how environmental justice groups would collaborate both among themselves and with mainstream environmental organizations.

One result of the publicity surrounding the Summit was the decision of the U.S.

Environmental Protection Agency (EPA), under Bush appointee William Reilly, to create an Office of Environmental Equity. As part of a team of environmental justice experts on the Clinton-Gore transition team in 1992, Bullard cowrote a position paper that convinced Pres. Bill Clinton to expand the Office of Environmental Equity, appoint a National Environmental Justice Advisory Council (NEJAC) to advise the EPA on matters of environmental justice, and sign an executive order on environmental justice in 1994 that ordered every federal agency to identify and address "disproportionately high and adverse human health or environmental effects of its programs, policies, and activities on minority populations and low-income populations." Bullard served on NEJAC from 1994 to 1996, chairing its Health and Research Subcommittee.

During the 1980s and early 1990s, Bullard taught consecutively at Texas Southern University, Rice University, University of Tennessee, and University of California at Berkeley, Riverside, and Los Angeles, respectively. In 1994, he moved to Clark Atlanta University (CAU), where he holds an endowed chair as the Ware Professor of Sociology. In September 1994, he founded CAU's Environmental Justice Research Center (EJRC), a clearinghouse that generates scientific, technical, and legal research on environmental justice issues and provides this information to environmental justice activists.

With support from the EJRC, Bullard has prepared testimony, testified, and served as expert witness for dozens of cases having to do with LULUs and environmental racism. Recently, Bullard worked with the communities of Forest Grove and Center Springs, Louisiana, to fight the nation's first privately owned uranium enrichment plant. The plant was to be built between these two small, predominantly Black communities, less than a mile from either one. Residents would not benefit from the highly paid jobs, yet they were to bear increased traffic, noise, and threats to their health. They formed Citizens Against Nuclear Trash (CANT) and sued the company, Louisiana Energy Services. Bullard provided expert testimony for CANT, and a federal court ruled in its favor. The 1997 victory was the first major environmental justice lawsuit to be ruled upon favorably.

Bullard's most recent focus has been the intertwined issues of transportation and urban sprawl. With EJRC research associate GLENN S. JOHNSON, Bullard edited *Just Transportation: Dismantling Race and Class Barriers to Mobility* (1997), which describes how current transportation policy favors the wealthy and ignores the needs for public transit of poor people of color. Bullard's and Johnson's most recent collaboration, in which they are joined by EJRC geographical information system (GIS) specialist Angel O. Torres, is *Sprawl City: Race, Politics, and Planning in Atlanta* (2000), a study of the unrestrained spread outward of Atlanta, one of the nation's fastest-growing cities. Contributors to this book treat urban sprawl as an issue of environmental justice, in that it contributes to increasingly severe socioeconomic stratification.

Bullard continues to edit the *People of Color Environmental Groups Directory*, which was published by the Charles Stewart Mott Foundation first in 1992 and was updated in 1994 and 2000. Bullard serves on the board of directors for several scholarly and environmental organizations and has won numerous awards, including the Citizens Clearinghouse for Hazardous Waste's Environmental Justice Award in 1993, the National Wildlife Federation's 1990 Environmental Achievement Award, and several awards for his books. He directs the EJRC and is a widely sought speaker on issues of environmental justice. Bullard resides in Atlanta with his wife, Linda, and his children, Robert Jr. and Kai.

Bibliography

Bullard, Robert, "Overcoming Racism in Environmental Decisionmaking," *Environment*, 1994; "Clark Atlanta University Environmental Justice Resource Center," http://www.ejrc.cau.edu/; Johnson, Glenn S., "Robert D. Bullard: Godfather of Environmental Justice," *Environmental Activists*, John Mongillo and Bibi Booth, eds., 2000; Motavalli, Jim, "Dr. Robert Bullard," *E Magazine*, 1998.

Burroughs, John

(April 3, 1837–March 29, 1921)
Naturalist, Nature Writer

John Burroughs is a giant among natural historians, the creator of the literary genre of the nature essay as it exists in its modern form. He published many collections of nature essays, popularizing, in the process, natural history and a general interest in the natural world.

John Burroughs was born on April 3, 1837, in the Catskill Mountain town of Roxbury, New York. His parents, Chauncey and Amy Burroughs, were well off. They owned and worked a 300-acre farm, keeping a herd of dairy cows and growing such field crops as hay, oats, buckwheat, and potatoes. Burroughs would later describe his father as a "good farmer, a helpful neighbor, a devoted parent and husband." Of his mother, Burroughs wrote that from her he inherited "my temperament, my love of Nature, my brooding introspective habit of mind." Burroughs was their seventh child. He grew up helping out with chores around the farm and attending the local one-room schoolhouse. He spent Sundays wandering around the countryside, fishing, swimming, and exploring. As a teenager, he earned money for books by making maple sugar and selling cakes of it in town for as much as two cents apiece.

Burroughs left home in 1854, at the age of 17. For three years he held a series of teaching positions in small country schools, interspersing them with writing courses and attempts at writing newspaper articles for a living. After 1857, when he married Ursula North, the daughter of a farmer he had boarded with while teaching in Tongore, New York, he alternated between more temporary schoolteaching posts, half-hearted business ventures, and short stints on his family's farm.

Naturalists John Burroughs (left) and John Muir (right) (Bettmann/Corbis)

In 1860, Burroughs published an essay entitled "Expression" in the *Atlantic Monthly.* This piece was philosophical and theoretical in nature and borrowed heavily from the writings of Ralph Waldo Emerson. The publisher even held up its publication in order to compare it with Emerson's work to ensure that it had not been plagiarized. Shortly after its publication, Burroughs realized that such writing was not his strong point, and he began to focus on writing about the rural world he knew best. An occasional column of his called "From the Back Country" began appearing in the *New York Leader* later in 1860. It featured essays about such topics as making butter, maple sugar, and building stone walls.

In 1863, Burroughs and his wife moved to Washington, D.C. Burroughs found employment working as a clerk for the Treasury Department. This job actually inspired many of his nature essays because it was so bland. Later he would write, "During my long periods of leisure I took refuge in my pen. How my mind reacted from the iron wall in front of me, and sought solace in memories of the birds and summer fields and woods!" Essays that he wrote during this time include "With the Birds," "The Snow Walkers," and "In the Hemlocks." Also, during this time, Burroughs became friends with the famous poet Walt Whitman. They would remain lifelong friends, with Whitman having a significant influence on Burroughs's writing. In 1867, Burroughs published *Notes on Walt Whitman as Poet and Person*, a book for which Whitman himself wrote one chapter and which he helped edit.

In 1871, Burroughs's second book, *Wake-Robin*, was published. It was a collection of nature pieces, mostly about birds, some of which had appeared in slightly different forms in the *Atlantic Monthly*. *Wake-Robin* received positive reviews and sold well. It established Burroughs as a master of the nature essay, a form of essay based on field observations and marked by anecdotes and emotional interactions with the subject matter. At the end of December 1872, Burroughs resigned his post with the U.S. Treasury (he had, by that time, worked his way up to the position of special bank examiner) to take a position with a bank in Middletown, New York, just a few miles from his native Catskills.

Burroughs purchased property on the west bank of the Hudson River, near Poughkeepsie, where he designed and built a house that he dubbed "Riverby." Burroughs and his wife would live here for the rest of their lives. Burroughs worked part-time as a bank receiver. His idle weeks were spent writing essays and exploring the countryside. In 1875, his third book, *Winter Sunshine*, was published. It included such essays as "The Fox," "The Snow Walkers," and "The Apple." *Locusts and Wild Honey* (1879) and *Pepacton* (1881), more collections of his nature essays, appeared soon thereafter. Burroughs would publish, on average, one book every two years for the remainder of his life. In 1878, a son, Julian, arrived in the Burroughs household, an arrival not without some controversy, for Ursula was not the baby's mother. She was, however, convinced to adopt the child and the days at Riverby were brightened. "The baby," Burroughs wrote, "is a refuge."

Though very much a homebody, during his years at Riverby, Burroughs did travel extensively, visiting Canada, California, Hawaii, and Alaska. He made friends in some fairly high places during this time as well, Pres. THEODORE ROOSEVELT, Henry Ford (who in 1913 sent Burroughs a gift of a new Model T), and JOHN MUIR among them. Above all, he continued to write. He published a total of 54 books before his death in 1921. In the process he established and popularized the genre of nature essays and helped to foster a widespread interest in natural history in the American people. He believed that a nature essay should be based upon accurate information about the physical and biological environment. He also believed in the integration of humans into natural processes. He did not "observe" nature so much as he "participated" in it. And through his nature essays he helped his readers to do so as well.

Burroughs died on March 29, 1921, in a rail car on his way home from a trip to California.

BIBLIOGRAPHY

Black, Ralph W., *American Nature Writers*, 1996; Burroughs Kelley, Elizabeth, *John Burroughs: Naturalist, the Story of His Work and Family*, 1989; Kanze, Edward, *The World of John Burroughs*, 1999; Renehan, Edward, *John Burroughs: An American Naturalist*, 1998.

Butcher, Devereux

(September 24, 1906–May 22, 1991)
Executive Secretary of National Parks Association, Editor, Photographer, Writer

As executive secretary of the National Parks Association (NPA; now called the National Parks and Conservation Association [NPCA]) at a time in which U.S. national parks were threatened by budgetary problems and wartime needs for timber and mineral resources, Devereux Butcher worked to assure their continued preservation. He held firm throughout his life that national parks must also be defended from commercial development, and he worked to establish standards for different classes of protected land. In addition, Butcher is remembered for his photographs, his paintings, and his series of picture books on national protected areas.

Devereux Butcher was born in Devon, Pennsylvania, on September 24, 1906, to Henry Clay and Constance Devereux Butcher. His parents were wealthy and encouraged their son's interest in art. Butcher studied painting and photography at the Philadelphia Academy of Fine Arts, and upon graduation in 1928, he traveled to California to photograph its Spanish colonial missions. Butcher married Mary Taft in 1935, and the couple built their own stone cabin in the Delaware River valley. This experience deepened Butcher's interest in nature and conservation, and he began to submit articles, art, and photographs to *American Forests*, the publication of the American Forestry Association. He worked as an editorial assistant for *American Forests* from 1941 to 1942.

Butcher's work at *American Forests* caught the notice of Robert Sterling Yard, founder of the NPA. Founded in 1919, three years after the creation of the National Park Service (NPS), the NPA supported national parks by lobbying Congress for their protection, raising private funds that the financially strapped NPS could not provide on its own, and promoting parks and conservation to the American public. Butcher was recruited as executive secretary of the NPA in 1942, a time of crisis for the national parks. The wartime effort was making demands on the nation's natural resources, and some members of Congress were pressuring the Department of the Interior to allow the exploitation of timber and minerals within national parks. The NPA worked hard to oppose these forces and was supported by NPS director Newton Drury, who kept the NPA well-informed of challenges to NPS. Because NPS headquarters had been moved to Chicago and its budget reduced at the beginning of World War II, Drury depended upon the NPA, based in Washington, D.C., to maintain a close watch over Congress and lobby on its behalf. At the same time that Butcher was expected to defend national parks from violations of this type, he also led the NPA to demand a clear definition of standards for varying classes of protected land. New national parks, in Butcher's opinion, should be designated only in places of spectacular beauty and particular vulnerability to development. Other areas, in the opinion of Butcher and the NPA, should be designated for other types of protection.

In addition to his administrative duties, Butcher also edited NPA publications. Since the founding of NPA, Yard had served as the editor of *National Parks Bulletin*, a simple newsletter for members. But Yard, who previous to his work with NPA had edited highly popular picture books on the national parks, knew that the publication held great potential. Once Butcher took the reins of the publication, he quickly revamped it. It was renamed *National Parks Magazine*, and it began coming out as a regular quarterly. He recruited well-known contributors, including Olaus Murie and Sigurd Olson. Rather than addressing only NPA-related issues, the magazine became an attractive showcase for the national parks. Its increased printing budget allowed for better reproduction of beautiful photos—many by Butcher himself—of national park scenery. During the 1940s and 1950s, public and school libraries throughout the country subscribed to *National Parks Magazine*, giving many readers their first view of the most beautiful places in the nation. NPA membership under Butcher rose to its highest level in the organization's history.

In 1947, Butcher published *Exploring Our National Parks and Monuments*, a picture book that included reprints of features from the magazine and was republished eight times. The book fulfilled the vision of Yard, who early in his tenure with NPA had proposed such an album. When Butcher decided in 1950 that editing and directing NPA was too much for one person and that his publications work was more fulfilling to him than his administrative work, he requested that the board reassign him as field representative. This allowed him more time to travel the country with his wife and son, Russell, visiting national parks and preparing material for the magazine. Between 1950 and 1957, when he left NPA, Butcher continued to publish *National Parks Magazine* and edited two more books: *Exploring the National Parks of Canada* (1951) and *Seeing America's Wildlife in Our National Refuges* (1955).

During the 1950s, Butcher was a key player in the defense of Dinosaur National Monument against the dams proposed in Echo Park and Split Mountain. He took many of the photographs that revealed the area's beauty and swayed public opinion against the dams. Butcher also was recruited by Wilderness Society's Howard Zahniser to help draft and lobby for the Wilderness Bill. Yet as the decade wore on, Butcher became increasingly disillusioned by the growing commercialism in national parks. He was disgusted by the "tramways, chair lifts, swimming pools, golf courses and honky-tonks" infiltrating the parks, he wrote in a 1954 edition of *National Parks Magazine*, and he began to criticize the executive committee members of NPA for their refusal to fight this type of development. To censor what he wrote on the issue, the NPA executive committee appointed an editorial advisory board for the magazine. This was too much for Butcher, and he resigned from NPA in 1957.

Butcher and his wife published their own magazine, *National Wildlands News*, for three years, 1959 to 1962. In 1963, Butcher became director of Hawk Mountain Sanctuary Association, the organization that preservationist Rosalie Edge had established to administer her Hawk Mountain Reserve in Pennsylvania; he remained at that post until his retirement in 1980. Butcher wrote two more picture books, *Exploring Our National Wildlife Refuges* (1963) and *Our National Parks in Color* (1964), and cowrote the series *Knowing Your Trees*.

Butcher's son, Russell, followed in his father's footsteps, working as a conservation journalist during the 1960s and then becoming the southwestern regional representative for the NPCA in 1980. Devereux Butcher had initially sponsored his son in this position but withdrew his sponsorship shortly after, disagreeing with NPCA leadership about its stand on hunting in Alaska parks.

Devereux Butcher died on May 22, 1991, at his home in Gladwyn, Pennsylvania, of complications following a fall.

Bibliography

Butcher, Devereux, *Exploring Our National Parks and Monuments*, 8th ed., 1985; "Devereux Butcher, 84, A Park Preservationist," *New York Times*, 1991; Miles, John C., *Guardians of the Parks*, 1995.

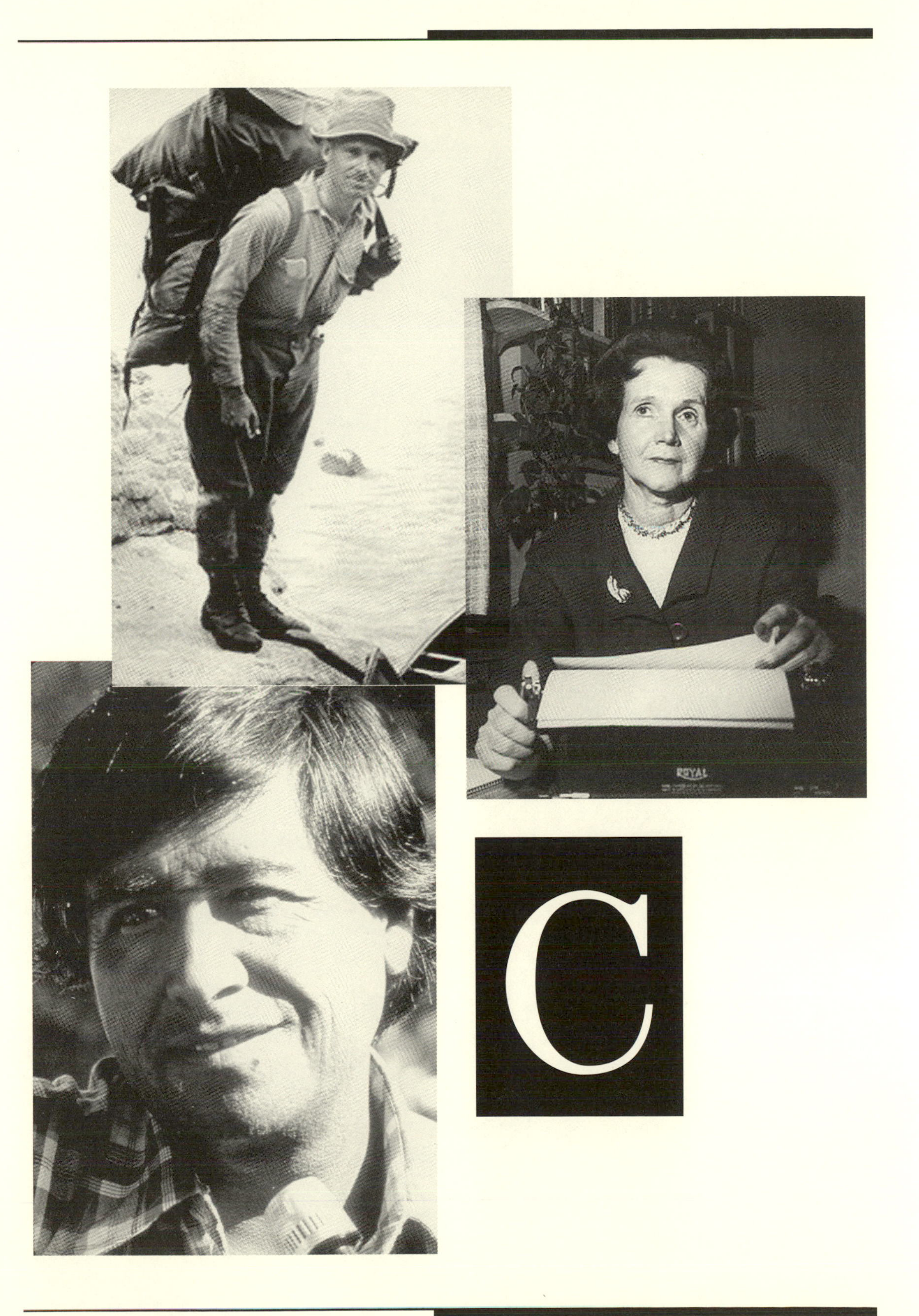
ROYAL
C

Cade, Thomas

(January 10, 1928–)
Ornithologist

Tom Cade has made a career of studying birds and how they are adapted to live within their particular environments; he is especially known as a champion of peregrine falcons. He has conducted field studies in Alaska, Africa, Arabia, Central America, Mauritius, and the southwestern United States and has published numerous articles and scientific papers. His efforts toward reestablishing peregrine falcons, which were nearly extinct by 1970, have been outstanding and have ranged from documenting their decline due to dichlordiphenyltrichlor (DDT) to establishing the Peregrine Fund, a conservation organization devoted to the preservation of falcons and other birds of prey. Cade's work on peregrine falcons has resulted in one of the first truly successful recovery efforts for an endangered species.

Thomas Joseph Cade was born January 10, 1928, in San Angelo, Texas. He spent some of his early years on a homestead in southern New Mexico and then moved with his father and mother to Glendale, California, in 1943. There he graduated from high school and spent two years at Occidental College. He studied biology at the University of Alaska, receiving his B.A. in 1951. He earned his master's degree in 1955 in zoology at the University of California, Los Angeles, and his Ph.D. in zoology from the same university in 1958. He taught at Syracuse University in the Department of Zoology, then at Cornell University, where he was professor in the Section of Ecology and Systematics and served as research director at the Laboratory of Ornithology. In 1988 he moved to Boise State University, Idaho, where he developed a graduate program in raptor biology.

In 1968 Cade and some colleagues published a paper in the *Condor* entitled "Peregrines and Pesticides in Alaska." At this time, peregrine falcons had begun to show reproductive failures resulting in catastrophic population declines over most of North America and Europe. Working in Alaska, where peregrines were still common and seemed to be unaffected by the decline, the researchers looked at the possibility that pesticides could be responsible for the dwindling numbers elsewhere. In the course of their study, they found residues of organochlorine pesticides such as DDT, dichlordiphenylethylene (DDE, a metabolite of DDT), dichlordiphenyldichlor (DDD), and dieldrin in the tissues of adult peregrines in concentrations 100 times greater than in the prey of these birds. While the Alaskan population appeared to be holding its own despite this contamination, Cade and his colleagues concluded their paper by warning that peregrines could be nearing a threshold level of pesticide poisoning that could lead to reproductive malfunction. Population declines in Alaskan peregrines subsequently occurred in the 1970s.

While researchers were beginning to uncover the link between the increased use of DDT that began just after World War II and the dramatic reductions in peregrine falcons and other species, the actual mechanism was not fully known. The emerging picture became much clearer in 1971 when Cade and others published a paper in *Science*, "DDE Residues and Eggshell Changes in Alaskan Falcons and Hawks." Cade and his coworkers found that the organochlorine pesticide residues in the Alaskan peregrines, which had previously shown little or no declines, were starting to have an effect. They found that eggshell thickness in these falcons was reduced after exposure to DDT, as reported earlier in Great Britain; in fact, there was a highly significant negative correlation between eggshell thickness and DDE content. This meant that a contaminated female would lay eggs that were so fragile that they would break during incubation, leaving the female incapable of reproducing.

Concerned over what he was finding and what it meant for the future of peregrines and other species, in 1970 Cade founded a conservation organization at Cornell University called the Peregrine Fund. This allowed for a focused effort on the recovery of the peregrine falcon, and Cade felt confident that his group could develop the techniques to do it. In the meantime, a new public awareness was emerging over DDT and its lethal effects on wildlife, and birds of prey such as the peregrine falcon figured prominently in this concern. Thanks in part to scientists like Cade who provided evidence on the threat of organochlorine pesticides, DDT was banned in 1972 by the Environmental Protection Agency.

At the time of the Peregrine Fund's inception, only a very small number of peregrines had ever been bred in captivity. Cade began a project to develop a captive breeding stock of peregrines to release to the wild. In 1973 the first successful hatch occurred, and the first release took place a year later—a breakthrough in endangered species research. Since then the Peregrine Fund has released over 4,000 captive-bred peregrines in 28 states, one of the first successful attempts at reestablishing an endangered species. The recovery of the peregrine has been so complete that in August 1999 the Department of the Interior announced its decision to remove the peregrine falcon from the Endangered Species List.

The organization that Cade founded in 1970 has greatly expanded and is now involved with restoration efforts in over 35 countries on five continents. The Peregrine Fund has moved its headquarters to the World Center for Birds of Prey in Boise, Idaho, where research and conservation projects continue. Cade is currently retired but remains involved with the organization as its founding chairman.

Bibliography

Cade, Tom J., *Ecology of Peregrine and Gyrfalcon Populations in Alaska*, 1960; Cade, Tom J., *The Falcons of the World*, 1982; Cade, Tom J., James H. Enderson, Carl G. Thelander, and Clayton M. White, eds., *Peregrine Falcon Populations, Their Management and Recovery*, 1988; Cade, Tom J., Jeffery L. Lincer, Clayton M. White, David G. Roseneau, and L. G. Swartz, "DDE Residues and Eggshell Changes in Alaskan Falcons and Hawks," *Science*, 1971; Cade, Tom J., Clayton M. White, and John R. Haugh, "Peregrines and Pesticides in Alaska," *The Condor*, 1968.

Caldwell, Lynton

(November 21, 1913–)
Political Scientist

Lynton Caldwell helped to author the National Environmental Policy Act (NEPA) of 1969, one of the best-known and most significant pieces of national environmental legislation. He is a political scientist and has written copiously on public administration and human-environment relations.

Lynton Keith Caldwell was born on November 21, 1913, in Montezuma, Iowa. He attended the University of Chicago, receiving a B.A. degree in English in 1935. He finished a master's degree in history and government three years later at Harvard, before returning to the University of Chicago, where he earned a Ph.D. in political science in 1943.

Caldwell began his career at Indiana University as an assistant professor of government and as director of the South Bend Center, where he worked from 1938 to 1944. In 1944, he took a position with the Council of State Governments, then in Chicago. He remained in this position until 1947, when he became professor of political science at the Maxwell Graduate School of Citizenship and Public Affairs at Syracuse University. In 1954 he joined the United Nations Technical Assistance Programme for one year as codirector of the Public Administration Institute for Turkey and the Middle East. In 1955 he was visiting professor of political science at the University of California at Berkeley. In 1956 he returned to Indiana University as professor of political science and coordinator of the university's training programs for public administration in Southeast Asia. Today he is the Arthur F. Bentley Professor of Political Science emeritus and Professor of Public and Environmental Affairs.

Caldwell began publishing in 1944 with a book entitled *The Administrative Theories of Hamilton and Jefferson.* His early writings did not focus on environmental topics. Over the course of the next 20 years he published books such as *The Government and Administration of New York* (1954) and *Improving the Public Service through Training* (1962), before shifting his focus to environmental issues. It was in 1962 that he had what he

Lynton Caldwell (courtesy of Anne Ronan Picture Library)

describes as an environmental "revelation." He was in Hong Kong for the Chinese New Year, and being that this is a family-oriented holiday, there was not much for him to do. He took a tram ride to the top of Victoria Peak, which overlooks the heavily populated island, and sat there in the early evening thinking about the future. Having traveled extensively in his role as an expert on public administration for the U.S. foreign aid program and the United Nations, Caldwell had become worried about the carelessness with which humans were treating the environment. He decided that he wanted to find a way to combine his expertise on management and public policy with a concern for the environment.

This decision led to a 1963 paper entitled "Environment: A New Focus for Public Policy?" for which he received the William E. Mosher Award from the American Society for Public Administration. This article has been credited for literally creating a new subfield of scholarly inquiry, one that suggests that governments should purposefully act to modify the interactions between humans and their environment. The article literally created the field of environmental management. Since 1964, Caldwell has written, edited, and contributed to many books that focus on environmental issues. Among those he has written are *Environment: A Challenge to Modern Society* (1970), *Man and His Environment: Policy and Administration* (1975), *International Environmental Policy Emergence and Dimensions* (1984), and *Between Two Worlds: Science, the Environmental Movement, and Policy Choice* (1992). Caldwell also contributed to such journals as *American Behavioral Scientist*, *Human Ecology*, *BioScience*, *Yale Review*, *Environmental Conservation*, *Public Administration Review*, *National Resources Journal*, *Unesco Courier*, *International Review of Administrative Science*, and numerous law reviews.

Caldwell is best known for the role he played in helping to bring about the NEPA, one of the United States' best-known and most significant pieces of national environmental legislation. Acting as a consultant to the Senate Committee on Interior and Insular Affairs in 1968, he composed *A Draft Resolution on a National Policy for the Environment*. The report created a statement of intent and purpose for Congress and examined the constitutional validity for a national environmental policy. Many of the concepts introduced by Caldwell in this draft were incorporated into the final bill, which was passed into law in 1969. NEPA created the Council on Environmental Quality (CEQ) and required environmental impact assessments to be carried out on "all major federal actions significantly affecting the quality of the human environment." These assessments were to be accompanied by environmental impact statements (EISs). NEPA placed responsibility on the federal government to act as an environmental trustee for future generations and to achieve a balance between population and resource use, allowing for a high standard of living. It has been argued, however, that because of the high-minded, sometimes vague wording of the act, it has been difficult to implement without active presidential initiative.

Caldwell has been an ardent supporter of NEPA since its enactment, despite its limited enforcement capabilities, pointing out that its passage marked the first time that any modern state had enacted a comprehensive commitment toward responsible custody of its environment. He sees the act as a yet to be fully utilized weapon for strong environmental protection, and he believes that under the provisions of NEPA, the executive branch of the U.S. government is in possession of great protective powers. In Caldwell's pamphlet, *Population and Environment: Inseparable Policy Issues* (1985), he writes that the NEPA could bring about a renewed "relationship of understanding and trust between the society-at-large and the government" when it comes to environmental issues. He envisions a situation where the president would work to identify crucial problems, while a network of locally based councils would work toward solving them.

Caldwell has also been active in international environmental issues. He acted as codirector of the Public Administration Institute for Turkey and the Middle East for the United Nations and on special assignments has provided technical assistance in Colombia, Pakistan, India, the Philippines, Thailand, and Indonesia. He is a strong believer in population control and sees environmental protection and population control as one issue rather than two separate ones. Caldwell has received many awards for his writings on public administration and the environment, including: National Science Foundation grants from 1965 to 1976, the Laverne Burchfield Award from the American Society for Public Administration in 1972, the Marshall E. Dimrock Award from the American Society for Public Administration in 1981, the John Gaus Award in 1996 from the American Political Science Association, and the Award of Achievement, Natural Resources Council of America, 1996. He is a United Nations Environment Programme Global 500 Laureate as of 1991. He lives in Bloomington, Indiana, with his wife, Helen Walcher, whom he married in 1940. They have two grown children, Edwin Lee and Elaine Lynette.

BIBLIOGRAPHY

Erdman, Karen, "Preventing Catastrophe," *Progressive*, 1985; Kareiva, Peter, "A Science with the Answers, but Too Little Influence?" *Ecology*, 1993; Metzger, Linda, ed., *Contemporary Authors*, 1984.

Callicott, J. Baird

(May 9, 1941–)
Environmental Philosopher

J. Baird Callicott is a founder and seminal thinker in the modern field of environmental philosophy. He is best known as one of the leading experts on the writings of ALDO LEOPOLD's land ethic and in interpreting and applying Leopold's land ethic to such modern resource issues as wilderness designation and biodiversity protection. As a trail-blazer in environmental philosophy,

Callicott has succeeded in communicating his ideas in a way that is both valuable to academics and accessible to the general public.

John Baird Callicott was born May 9, 1941, in Memphis, Tennessee, to Burton (an artist) and Evelyne Baird Callicott. Callicott completed his B.A. in philosophy with honors at Rhodes College in Memphis in 1963. He then went to Syracuse University for graduate studies in philosophy, completing his M.A. in 1966 and his Ph.D. in 1971.

Callicott's interest in the environment did not evolve from any intense childhood experience in or with nature. Rather, his environmental philosophies were born from his participation in the late 1950s and 1960s as a foot soldier in the movement led by Martin Luther King Jr. In 1966, after completing his class work at Syracuse, Callicott returned home to teach at Memphis State University (now the University of Memphis). He served as faculty sponsor/adviser to the Black Students' Association and participated in the civil rights actions in 1968, which brought Martin Luther King Jr. to Memphis for his last and fatal campaign. In the deep reflection brought about by King's assassination, Callicott began to consider that the environment was the ultimate object of oppression. Studying during the Age of Ecology of the 1960s, Callicott became aware that the existing environmental crisis challenged the most fundamental assumptions of Western thought. Whereas the philosophy animating civil rights was centuries old—though still recognizably unrealized in political practice—a philosophy animating a liberation of nature was hitherto nonexistent. The age-old problems of the essence of nature required reexplanation. As one of the founders of environmental philosophy, Callicott devoted his career to addressing these problems.

In 1971, Callicott accepted an appointment at the University of Wisconsin at Stevens Point. Until 1995, he taught in the Department of Philosophy where, in 1971, he designed and taught what is widely acknowledged as the nation's first college course in environmental ethics.

In 1979, Callicott contributed an article to the first issue of the *Journal of Environmental Ethics*. His essay, "Elements of an Environmental Ethic: Moral Considerability and the Biotic Community," elaborated on Aldo Leopold's argument for an environmental ethic. Two subsequent books on Leopold have made Callicott one of the nation's leading scholars of the author of *A Sand County Almanac*. His 1987 volume, *Companion to A Sand County Almanac*, which he edited and to which he contributed, is the first interpretive and critical discussion of Leopold's classic text. In 1991, Callicott coedited, with Susan Flader, *The River of the Mother of God and Other Essays by Aldo Leopold*, a collection of heretofore unpublished essays written by Aldo Leopold that chart his life from New Mexico Forest Service employee to author of one of the great conservation manuals of the twentieth century. This second book is especially valuable, as it illustrates the evolution of Leopold's ecological thinking, culminating with his advocation of a land ethic.

Callicott suggests that we live on the verge of a profound paradigmatic shift concerning human interactions, perceptions, and attitudes toward nature and the natural world. Much of his work has sought to address or discover an ecologically and philosophically valid concept of sustainability. Callicott's concept revolves around dynamic human/nonhuman mutualism; the suggested (or imaginary) dichotomy between culture and nature is only detrimental to the progress of ecosystem conservation and its restoration. In 1995, Callicott returned to the South to join the Department of Philosophy at the University of North Texas in Denton, where he currently resides.

Bibliography

Callicott, J. Baird, *Beyond the Land Ethic: More Essays in Environmental Philosophy*, 1999; Callicott, J. Baird, *In Defense of the Land Ethic: Essays in Environmental Philosophy*, 1989; Callicott, J. Baird, and Roger T. Ames, eds., *Asian Traditions of Thought: Essays in Environmental Philosophy*, 1989; Callicott, J. Baird, and Eric T. Freyfogle, eds., *For the Health of the Land: Previously Unpublished Essays and Other Writings by Aldo Leopold*, 1999; Callicott, J. Baird, and Michael P. Nelson, eds., *The Great New Wilderness Debate*, 1998.

Carhart, Arthur

(September 18, 1892–November 30, 1978)
Landscape Architect

As a young U.S. Forest Service (USFS) employee in 1919, Arthur Carhart was instrumental in convincing the USFS to preserve a portion of its land for public recreation. Until that time, the USFS had loyally followed the precept of its founder Gilbert Pinchot, who held that all public forest land should be available for managed use by the logging, mining, and grazing industries. Carhart went on to write over 5,000 articles and books, many of them promoting the preservation of wilderness for primitive recreational use.

Arthur Hawthorne Carhart was born in Mapleton, Iowa, on September 18, 1892. He gained an appreciation for tree-filled landscapes from his grandfather, who had acquired his Iowa farm under the Tree Claim Act, which gave land to settlers who planted trees on it. Carhart studied at Iowa State College, earning that institution's first degree in landscape design in 1916. Upon graduating, Carhart served as a nurseryman in the United States Army during World War I. After the war, in 1919, he was hired by the U.S. Forest Service for a recently developed recreational engineer position and was dispatched to Denver. His role was scorned by many of the forest rangers, who nicknamed him "Beauty Engineer" or "Beauty Doctor," but Carhart proceeded undaunted.

Carhart's first assignment was to spend the summer of 1919 at Trappers Lake in northwestern Colorado. The largest lake in Colorado, and "one of the three finest mountain lakes in the American West," he later wrote, Carhart was to survey the lake's shore for homesites and for a road that would encircle the lake. He bunked at Scott Teague's fishing camp and spent his evenings talking to guests. One guest, Paul J. Rainey, a well-known big-game hunter, conversed at length with Carhart

Arthur Carhart (courtesy of Anne Ronan Picture Library)

about the value of the plan to develop such a scenic public area. Eventually Carhart too doubted the ethics of the plan. Constructing private homes there would ruin the area's spectacular natural scenery. When Carhart submitted his report in the fall of 1919, which recommended that the area not be used for homesites, his supervisor, Carl Stahl, suggested he meet with another USFS employee, ALDO LEOPOLD, who was then supervisor of Carson National Forest in New Mexico. Leopold and Carhart no doubt were both encouraged to find that they shared the same hopes for USFS wilderness. (Two years later, in 1921, Leopold published his landmark article, "The Wilderness and Its Place in Forest Recreation Policy" in the *Journal of Forestry* and proposed that a 500,000-acre area of wilderness in Gila National Forest be set aside for recreational use only.) The memorandum that Carhart wrote about his meeting with Leopold outlined four types of USFS land that should be left undeveloped: the most beautiful places; mountain ridges and other areas that cannot support development; areas of use to groups, such as lakeshores, medicinal springs, and streams; and those so spectacular that God must have meant that they never be altered. After three years of consideration by the USFS, Carhart's proposal to scrap the road and shoreline homes at Trappers Lake was accepted. The land surrounding Trappers Lake was preserved in 1932 as the Flat Tops Primitive Area, to be used only for nonmotorized recreation.

During the interim, Carhart worked on two more projects. In 1920, he worked to preserve San Isabel National Forest in southern Colorado and develop the recreational possibilities there. Supported by an enthusiastic group of local public officials and business leaders, Carhart designed the first public campground in a national forest. The San Isabel Public Recreation Association raised funds to finance the improvements that the USFS would not pay for; several more campgrounds and recreational sites were built near the major towns bordering the national forest. Carhart surveyed the Quetico-Superior area of Superior National Forest in northern Minnesota in 1921 and recommended that it be left in as wild a state as possible, since the recreational potential there was immense. Now that area of lakes and streams is the preserved Boundary Waters Canoe Area.

Despite his success in promoting preservation at the USFS, Carhart left in 1922, disenchanted with its lack of greater support for recreation. He established his own landscape architecture firm, one of whose contracts was with the city of Denver. Carhart's plan was to design several large parks for Denver, each devoted to a different type of recreation. He worked especially on mountain parks in the foothills that border Denver to the west.

Carhart is also known for the 5,000 articles and books that he wrote after leaving the USFS. Some of these were popular western novels that he wrote either under his own name or his pen names, Hart Thorn and V. A. VanSickle. Others were guides to outdoor activities, such as *Hunting North American Deer* (1946) and *Fresh Water Fishing* (1949). Yet others were important contributions to the growing body of conservation scholarship. *Timber in your Life* (1955), for example, was a detailed description of the forestry industry. *National Forests* (1959) was a guide to the geological and anthropological history of land held by the USFS. *Planning for America's Wildlands* (1961), which came out during the long campaign to pass the Wilderness bill, consisted of a clear philosophical basis and concrete recommendations for the protection of wilderness. He delineated the types of protection that were necessary for different categories of wilderness and insisted that extractive industries could not be allowed in lands set aside for primitive (nonmotorized) recreation.

One of Carhart's most prized contributions to conservation was the Conservation Library Center at the Denver Public Library, which he conceived of and helped establish in 1961. This collection of important material relating to conservation includes his own papers, as well as those of Wilderness Society director Howard Zahniser, the Soil Conservation Service's Hugh Bennett, and population expert and writer William Vogt.

Carhart's wife, Vera VanSickle, died in 1966. They had no children. Carhart died on November 28, 1978.

Bibliography

Martin, E. J., "A Voice for the Wilderness: Arthur Carhart," *Landscape Architecture*, 1986; Nash, Roderick, "Arthur Carhart: Wildland Advocate," *Living Wilderness*, 1980; Stroud, Richard, *National Leaders of American Conservation*, 1985.

Carlton, Jasper

(March 16,1940–)
Director of Biodiversity Legal Foundation

Jasper Carlton directs the Biodiversity Legal Foundation (BLF), a science-based nonprofit organization that he and several colleagues founded in 1991 to protect imperiled ecosystems. The BLF differs from other conservation organizations in that it chooses ecosystems that others generally ignore and works on behalf of all of the species that inhabit them, from the bottom of the food chain on up—not just the so-called charismatic species on which mainstream conservation organizations often focus. Due to its commitment to impeccable science and the skill of its staff, the BLF's litigation to hold federal agencies responsible for upholding environmental legislation such as the Clean Water Act, the National Forest Management Act, the Migratory Birds Treaty Act, the Endangered Species Act, and others is successful about 90 percent of the time. The BLF has contributed to the listing of about one-third of the 1,200 species currently listed as endangered or threatened.

Donald Conrad "Jasper" Carlton was born in Picayune, Mississippi, on March 16, 1940. His father, an adventuresome horticulturist, had worked in Africa before Jasper was born and passed a strong conservation ethic on to his children. When he was four, Carlton moved with his family to the Amazon rain forest for two years, where his father was assigned by the U.S. government to extract rubber from the jungle for the war effort. Spending his formative years in a tropical rain forest awoke what has been a life-long interest in and love for nature. The family moved to the coast of Maine when Carlton was six years old, so that he and his sister could receive what their parents considered a proper education, and he spent the remainder of his youth there, frequently exploring Baxter State Park and the backwoods of Maine with naturalist friends.

Worried that his son was spending too much time in the woods, Carlton's father urged him to study business. Carlton did so at the University of Florida, working his way through college and graduating in 1962. He was hired by the international division of the Genesco Corporation, the world's leader in apparel, and spent the next eight years opening up new export markets all over the world. He had become one of the youngest international trade executives in the country by 1968.

But by 1970, he became dissatisfied with what he came to feel was the "futility and meaninglessness of spending one's life making a living." So he returned to college, studying botany, behavioral psychology, and law at the University of Tennessee and the National Testing Lab Institute for Applied Behavioral Science in Bethel, Maine. He worked for two years as assistant director of a school for autistic children in Nashville and in his free time became increasingly involved in Tennessee conservation movements. Carlton joined the Sierra Club and Friends of the Earth, holding Tennessee state chapter office positions for both organizations. He worked on campaigns in the Tennessee Valley to protect native fisheries, freshwater mollusks, and the valley's last free-flowing rivers.

Carlton moved to Oregon in 1972, where he led a legal fight to defeat an oil platform manufacturing plant at the mouth of Oregon's Columbia River. He then spent several years in Idaho, where he began working in a naturalist's capacity on mountain caribou and grizzly bears in the Selkirk ecosystem. These animals and others were negatively impacted by the Forest Service roads transecting the area, and Carlton eventually sued Interior secretary James Watt to obtain the emergency listing as an endangered species of the woodland caribou and the closure of the Forest Service roads. For these accomplishments, Carlton was awarded the Wildlife Society's Public Service award in 1982. Moving east to Montana, Carlton directed the Montana Woodland Caribou Ecology Project during the mid-1980s, during which time he helped force a court order to end sport hunting of grizzly bears in Montana's wilderness areas.

By the late 1980s, Carlton had become an expert in the integration of biology and law. He had developed his ecosystem approach to conservation, which combines both scientific and legal perspectives, and was well-known for his reliance on only the best science and his objectivity. In 1991, Carlton along with a number of attorneys and conservation biologists founded the Biodiversity Legal Foundation, based in Boulder, Colorado. The BLF is a sleek, efficient organization, with just five full-time employees and between six and a dozen volunteers. Its office space is donated, and its computers and other equipment are not new. Its annual budget is less than $100,000. From its very beginning the BLF chose not to solicit funds from major foundations, in order to protect its freedom to work on the vanguard and choose controversial cases that corporate-funded foundations might not like.

The BLF works in the following way. First, it chooses imperiled ecosystems that are ignored by other conservation organizations. Then the BLF engages expert scientists (who usually work pro bono) to share their knowledge about all of the ecosystem's component species, from the bottom of the food chain on up, how they interact, and what their individual habitat requirements are. The individual species that are most imperiled are identified, and reviews of their status are conducted. The federal agencies charged with protecting these species—the U.S. Fish and Wildlife Service, the U.S. Forest Service, the National Marine Fisheries Service, the Bureau of Land Management, the Bureau of Reclamation, and many more—are then approached and are forced, through administrative and sometimes legal actions as well, to protect the species in question. Because the science that BLF bases its actions on is irrefutable, and the BLF's pro bono attorneys are highly skilled at educating judges about the intricacies of ecosystems, BLF prevails in about 90 percent of its legal suits.

Carlton has traveled the country to teach environmental activists the steps necessary to successfully combine biology and law for conservation of species and their ecosystems. The BLF itself serves as a classroom for interns or staff who typically work for a few years to learn the process, then go on to begin their own similar organizations in other regions of the United States.

Because Carlton sees the limitations of environmental litigation as it is currently possible, he and his colleague REED NOSS, a conservation biologist, are in the process of drafting a Native Ecosystem Protection Act. Carlton estimates that some 9,000 species of U.S. plants and animals should probably be considered biologically threatened or endangered, yet as of 1999 only 1,200 plus U.S. species were listed, because the listing process is so slow and costly. An ecosystem protection act of this dimension, following the models set by the Endangered Species Act, the Clean Water Act, the National Forest Management Act, and other great conservation legislation, would preserve this country's most endangered ecosystems. A body of scientists such as the National Academy of Sciences would identify ecosystems that were endangered, and then those ecosystems would be evaluated to determine which sections of them should be completely closed to human activity, which could support passive, nonmotorized recreation such as hiking and cross country skiing, and which would tolerate more active or motorized recreation. Carlton believes that the ecosystem-based approach is the only way to preserve ecosystems that are still relatively healthy and that such an act would go further to stem the dangerous extinction trend currently in progress.

Although many of his colleagues at other conservation organizations doubt that any Congress would be courageous enough to pass such powerful legislation, Carlton responds that many environmental bills were discussed for decades before Congress was able to enact them. He points out that the establishment of Yellowstone National Park in 1872 was controversial, with many detractors, but that today no one would dare claim it was a mistake.

Carlton received the Deep Ecologist of the Year award by the Foundation for Deep Ecology in 1994. He currently lives in Louisville, Colorado.

Bibliography

Frankowski, Eric, "Defending Animals vs. Mankind Is His Battle," *Longmont Daily Times-Call*, 1999; Noss, Reed, *A Citizen's Guide to Ecosystem Management*, Biodiversity Legal Foundation Special Report, 1999; Worland, Gayle, "He Walks with the Animals," *Westword*, 1999.

Carr, Archie

(June 16, 1909–May 21, 1987)
Zoologist, Writer

Archie Carr, a prominent and dedicated pioneer of conservation biology, is credited for shedding light on the previously unknown life history of sea turtles and stimulating an international demand to protect them and their nesting grounds. Many of his books, including *The Windward Road* (1956) and *So Excellent a Fishe* (1967), have become natural history classics and have won great acclaim both for their high literary quality and for contributing to the awareness of the ecology of sea turtles and their tropical habitats. Carr went beyond the research of these animals and fought for their preservation at a time when turtle hunting was rampant. He convinced the government of Costa Rica to protect the sea turtle nesting site at Tortuguero beach as a national park, and thanks to his efforts, sea turtle populations have shown dramatic increases.

Archie Fairly Carr Jr. was born on June 16, 1909, in Mobile, Alabama, to Archibald Fairly and Louise (Deaderick) Carr. His father, a Presbyterian minister, moved the family away from Alabama, first to Fort Worth, Texas, and later to Savannah, Georgia. Archie Carr Sr. enjoyed duck hunting and had a collection of natural history books, which young Archie pored over. Early on, Archie asserted an interest in snakes and other reptiles, an interest that horrified his mother and especially his grandmother, who forbade anyone to talk about snakes after three o'clock in the afternoon for fear that she would dream about them. Both parents encouraged him to do a lot of reading as a child, and his love of words followed him for the rest of his life.

Carr began attending Davidson College in 1928, intending to study English. But his interest in the natural world won out, and he soon switched to the University of Florida in Gainesville to study biology, earning his B.S. in 1932. He continued studying there and in 1934 received his master's degree. His 1937 Ph.D. was the first granted in zoology by the University of Florida. On January 1, 1937, Carr married MARJORIE HARRIS CARR, who became a well-known conservationist herself, and they eventually raised five children together.

Carr started teaching at the University of Florida in 1938 but spent his summers on a fellowship at Harvard University's Museum of Comparative Zoology, honing his skills in taxonomy and evolutionary biology. Working with another scientist from Harvard, he coauthored a monograph on the turtles of Cuba, Haiti, and the Bahamas, *Antillean Terrapins.* At the time, there was substantial scientific knowledge on freshwater turtles, but information on sea turtles was almost nonexistent and seemed based mainly on rumor and folklore. Carr formulated a goal for himself: to become the world's expert on sea turtles. In order to study sea turtle populations from both the Atlantic and the Pacific, he arranged for a leave of absence from the University of Florida, and in 1945 he and his wife and two children moved to Honduras. He taught biology there for five years at the Escuela Agrícola Panamericana while also studying sea turtles. During those years he filled in some of the gaps of scientific knowledge about sea turtles and began creating an international reputation for himself.

Carr and his family returned to Florida in 1949, and he was made a full professor by the university. They bought a house, near Micanopy on an alligator pond, that became their permanent home. In 1952 Carr published a second monograph called *The Handbook of Turtles: The Turtles of The United States, Canada, and Baja California*, which identified 79 species and subspecies of sea turtles. In addition to this and other scientific papers reporting facts of the sea turtles' natural history, Carr published a book that would become a conservation classic and profoundly affect his career. *The Windward Road: Adventures of a Naturalist on Remote Caribbean Shores* (1956) told stories of exotic Caribbean wildlife and the interactions he had with the local people as he searched for sea turtle nesting areas. Although it reads almost like an adventure book, it also educates readers about the life history of sea turtles and the imminent dangers facing them and their habitat. The book created worldwide concern over the plight of sea turtles and helped to launch a campaign to protect their nesting areas. Part of the popularity and critical acclaim for the book came from Carr's famous and abundant sense of humor and from the high quality of his writing. One chapter of the book, titled "The Black Beach," was published separately in *Mademoiselle* and won an O. Henry Award as one of the best short stories of 1956. Such was Carr's growing fame that the University of

Florida relieved him of his teaching duties and made him a graduate research professor in 1959, which allowed him to concentrate on researching and writing books.

One especially avid fan of *The Windward Road*, Joshua Powers, formed an informal organization called the Brotherhood of the Green Turtles. An offshoot of this group was the Caribbean Conservation Corporation (CCC), which set about to preserve and protect sea turtles and their habitat in the tropics. Carr was the founding scientific director of CCC, a post he held until his death. Many of his research projects were funded by the CCC; he conducted most of them at Tortuguero, a beach on the coast of Costa Rica. His studies also led him on numerous expeditions to other sea turtle habitats, including Brazil, South Africa, West Africa, the Azores, every part of the Gulf of Mexico, Jamaica, Portugal, and Pacific Central America, to name but a few. But the study site at Tortuguero, the last nesting grounds of the green sea turtle in the Americas, was where Carr concentrated most of his efforts—resulting in one of the most intensive and longest-lasting studies of an animal population ever conducted. His studies, almost all of which have implications for conservation, elucidated aspects of sea turtles' migration, nesting behavior, nest physiology, nutrition, demography, and more. During the 1960s, when Tortuguero beach was plagued with turtle hunters, Carr began working on a project called Operation Green Turtle, to distribute green turtle eggs and hatchlings to historical nesting grounds around the Caribbean and the Gulf of Mexico, in an attempt to reestablish hatcheries. He eventually convinced the Costa Rican government to declare the beach at Tortuguero a national park in 1975 and to enforce laws against turtle hunting. Tortuguero National Park also became one of the first successful ecotourism destinations, as thousands of people traveled to witness the turtles that Carr had made famous in his books. Carr's conservation efforts there became evident during the 1980s with dramatic increases in the nesting green turtle population.

Carr's conservationism was not limited to sea turtles. In 1964 he published two books about Africa and its wildlife: *Ulendo: Travels of a Naturalist in and out of Africa* and *Land and Wildlife of Africa*. In both books he repeatedly expresses his concern for the future of these unique animals and ecosystems. By the end of his life, he had written ten books and more than 120 scientific papers and magazine articles on sea turtles and other natural history subjects. He was the recipient of many awards and honors for his conservation work, including the World Wildlife Fund Gold Medal (1973) and the National Audubon Society's first Hal Borland Award (1984) for contributing to the understanding and protection of nature. He continued his research and writing up until his death and also continued to serve as chairman of the Marine Turtle Specialist Group of the International Union for the Conservation of Nature, as he had done since the 1960s. Carr died of cancer on May 21, 1987, at his home on Wewa Pond near the town of Micanopy, Florida.

BIBLIOGRAPHY

Carr, Archie, *So Excellent a Fishe: A Natural History of Sea Turtles*, 1967; Graham, Frank, Jr., "What Matters Most: The Many Worlds of Archie and Marjorie Carr," *Audubon*, 1982; Maslow, Jonathan, *Footsteps in the Jungle: Adventures in the Scientific Exploration of the American Tropics*, 1996.

Carr, Marjorie Harris

(1915–October 10, 1997)
Biologist, Founder of Florida Defenders of the Environment

Marjorie Harris Carr, a biologist and conservationist, was a dominant figure in the protection of Florida's inland waters. She was the founder and longtime president of Florida Defenders of the Environment (FDE), a group she formed to help stop construction on the Cross-Florida Barge Canal, a shipping canal that was being excavated across the entire state to connect the Gulf of Mexico with the Atlantic Ocean. When she and her group were successful in halting the canal before it could be completed, she then worked to undo the environmental damage caused by the ill-advised project, which had included damming of one of Florida's most beautiful rivers, the Ocklawaha. The 110-mile Cross-Florida Greenway, a recreational trail system consisting of governmental lands formerly set aside for the canal, has been named after Carr, whose name will always be synonymous with conservation in Florida.

Marjorie Harris was born in 1915 in Boston, Massachusetts. In 1923, her father, a lawyer and schoolteacher, bought land in Florida as an escape from the bitter winters in New England. The family moved into a house near Bonita Springs, where her father planned to raise oranges in his retirement. Marjorie grew up with few neighbors but surrounded by a fascinating landscape of plants and animals, sparking a lifelong interest in natural history. She also enjoyed having parents who could answer her questions about the natural world and who encouraged her love of knowledge. By the age of nine she could identify more birds and flowers than most people know in a lifetime. Though the death of her father in 1931 and the subsequent years of the Depression had left her and her mother with little money, a bequest of $500 from a maiden aunt allowed Marjorie to begin attending the Florida State College for Women (now Florida State University) in 1932. She began a course of study that was highly unusual for a woman at that time—a major in zoology with a minor in bacteriology.

After graduating with a B.S. in zoology in 1936, she hoped to continue her studies at the University of North Carolina, but her fellowship fell through. So she took a job with the Resettlement Administration, a federal program established during the Depression, and became the state's first female wildlife technician at a fish hatchery in Welaka, Florida. Her supervisor, uncomfortable with a woman biologist, gave her a busywork assignment: to figure out what was wrong with some sick quail. She took the birds and sought advice at the laboratory at the University of Florida, where she met a doctoral student in zoology named Archie Carr. Three months later, in January 1937, they were married.

When her new husband was offered a summer fellowship at Harvard's Museum of Comparative Zoology, the Carrs began a routine of spending winters in Florida and summers in Massachusetts. Marjorie Carr worked at the Museum for several summers but was not assigned the kind of challenging research she wanted. When she applied to Cornell to do graduate work on her first love, ornithology, the director told her that there was no place for a woman in that field. She was finally able to earn her master's degree from the University of Florida in 1942, with her thesis, "The Breeding Habits, Embryology, and Larval Development of the Large-mouthed Bass in Florida." During the course of her graduate research she became the first scientist to discover cases of social parasitism (the type commonly associated with cowbirds, who leave their chicks in the nests of other bird species to raise) in freshwater fish.

During the 1940s and early 1950s, Carr moved to Honduras and assisted her husband with his studies on sea turtles, contributed a paper to the *Wilson Bulletin* on swifts, and had five children. In 1949 the Carrs returned to Florida and bought a house on Wewa Pond near Micanopy. Carr settled into raising a family and teaching biology at nearby Gainesville High School. She was also a charter member of the Alachua Conservation Society and served on the conservation committee of the Alachua Audubon Society. In the early 1960s Carr began investigating a proposal announced by the Army Corps of Engineers to resume construction on the Cross-Florida Barge Canal. The controversial project, intended to connect the Gulf of Mexico with the Atlantic Ocean by way of a 12-foot-deep, 150-foot-wide canal, was initiated in the 1930s but had stalled shortly thereafter, only to be revived in 1964. Carr opposed the shipping canal for many reasons, but her main cause for alarm was that the canal would necessitate damming the Ocklawaha River—flooding a 16-mile section of the free-flowing river, threatening wildlife, converting 9,000 acres of hardwood forest into a shallow weedy reservoir, and ultimately destroying a uniquely beautiful semitropical stream. Carr accumulated documents and scientific information on the river, and in 1965 she began an unwavering lobbying effort to save the river and stop the canal. She besieged members of Congress and state leaders with letters and recruited scientists, lawyers, and economists to help her cause. Nevertheless, the Corps of Engineers continued construction and erected the Rodman Dam on the Ocklawaha in 1968.

By the following year, the Corps of Engineers had completed about a third of the canal, and Carr rallied other conservationists and formed FDE, a nonprofit citizens' group. The organization concentrated on publicizing undisclosed facts about the canal project and educating policy makers about the issue—emphasizing the facts that the canal threatened to contaminate the freshwater aquifer and that many aquatic species were already disappearing as a result of the dam. At Carr's request, attorneys from the Environmental Defense Fund (EDF) joined in legal action to halt the canal, and in 1971 FDE and EDF won a federal court injunction and persuaded President Nixon to issue an executive order halting construction. Though construction on the Cross-Florida Barge Canal ended, it was still an authorized federal project, and Carr spent the next two decades fighting to get the project officially deauthorized. She set other goals as well—to restore the Ocklawaha to its natural meandering course by breaching the Rodman Dam and to have the federal land that was granted to the canal turned over to the state as a conservation area. By 1990 the canal was finally deauthorized with a bill signed by President Bush.

The fight to tear down the dam has been less successful, hampered by a handful of local constituents who claim that the reservoir is a prime bass fishing spot. But the rest of the old canal route was turned over to the state in 1992, and Carr and FDE came up with a management plan for the land as a conservation area. The state legislature endorsed the plan, and the 110-mile-long corridor, named the Marjorie Carr Cross-Florida Greenway, has become a model recreational trail system.

Carr influenced other conservation efforts in the state as well. She was instrumental in establishing the Payne's Prairie State Preserve south of Gainesville and helped with restoration in the Everglades and with protecting the endangered Florida panther. In 1997 she was inducted into the Florida Women's Hall of Fame. Carr died at her home in Micanopy on October 10, 1997, of emphysema, just months before state agencies began considering the permit application of the Florida Department of Environmental Protection to demolish the Rodman Dam.

BIBLIOGRAPHY

Barash, Leah, "People Who Made a Difference: Marjorie Carr/Jack Kaufman," *National Wildlife*, 1992; Breton, Mary Joy, *Women Pioneers for the Environment*, 1998; Graham, Frank, Jr., "What Matters Most: The Many Worlds of Archie and Marjorie Carr," *Audubon*, 1982; Pittman, Craig, "Digging Ourselves into a Hole," *St. Petersburg Times*, 1999.

Carson, Rachel

(May 7, 1907–April 14, 1964)
Biologist, Writer

Biologist and writer Rachel Carson is often credited as the founding mother of today's environmental movement. Her early books, *Under the Sea Wind* (1941), *The Sea around Us* (1951), and *The Edge of the Sea* (1958), opened a world that had previously been unknown to most landbound humans. They established Carson's reputation as an eloquent writer who could explain science both clearly and poetically. Her best-known book is *Silent Spring* (1962), an exposé that revealed the grave dangers posed by synthetic pesticides. These "elixirs of death," as she called them, were first used during World War II and after the war were introduced to civil society to combat insect and weed pests. A best seller, *Silent Spring* raised much public concern in the United States and resulted in the first federal and state laws that regulated pesticide use.

Born on May 7, 1907, in Springdale, Pennsylvania, to Robert Warden and Maria Frazier (McClean) Carson, Rachel Louise Carson was a child fascinated by nature. Her chief pastimes included writing and illustrating stories about the wildlife surrounding her western Pennsylvania home. By the age of 12 she had published several stories in a national children's magazine and had won three

Rachel Carson (Corbis/Bettmann)

prizes for her writing. Although her family was of modest means, and most women in her generation did not pursue a college education, Carson studied English and biology at Pennsylvania College for Women (now called Chatham College). After realizing that the natural world could provide great inspiration for her literary work, she opted to major in biology. Carson graduated with a B.A. in 1929. Her biology professor and mentor, Mary Skinker, encouraged Carson to pursue postgraduate studies in biology, and she did, earning an M.A. in marine zoology from Johns Hopkins University in 1932. She probably would have worked toward a doctorate in aquatic biology had the sudden deaths of her father and sister not left her the family's sole breadwinner. To support her mother and young nieces, Carson began to work for the U.S. Bureau of Fisheries, first editing transcripts for radio broadcasts and then, starting in 1936, as an aquatic biologist. Once that bureau merged with the U.S. Biological Survey to become the U.S. Department of Fish and Wildlife in 1940, she became the editor in chief of its publications. Carson published 12 pamphlets with a strong conservationist message, called the Conservation in Action series.

Carson always supplemented her work for the government with independent natural history writing, which was published in magazines and as books. She published an essay entitled "Undersea" in the *Atlantic Monthly* in 1937, which she expanded into her first book, *Under the Sea Wind* (1941). This work provided a scientifically astute and poetic portrait of the sea and its inhabitants, a world most people at that time knew little about. *Under the Sea Wind* was issued shortly before the bombing of Pearl Harbor, and the ensuing national crisis hindered its sales. However, it received favorable reviews and established Carson's literary reputation. It took her 11 more years to publish *The Sea around Us* (1951), a *New York Times* best seller and winner of the National Book Award. The success of this book led to the reissue of *Under the Sea Wind* in 1952. That same year, finally earning enough income solely from her writing, Carson left the Department of Fish and Wildlife to pursue writing, biology, and life with her adopted son, her orphaned grandnephew Roger. Carson next wrote *The Edge of the Sea*, a book about life on the ocean shore, which also met with high acclaim when it came out in 1955. Raising Roger, Carson became aware of the importance of helping children appreciate nature. She wrote a short article for *Women's Home Companion* in 1956, "Help Your Child to Wonder," which she hoped to expand into a full-length book.

In the years following World War II, Carson became increasingly concerned about a newly emerging environmental problem. Chemical companies were developing a host of new synthetic pesticides that on the recommendation of the U.S. Department of Agriculture were being used rashly throughout the United States, on farms, in cities and the suburbs, even in wildlife refuges. When a birdwatcher named Olga Owens Huckins wrote Carson that the government had sprayed dichlordiphenyltrichlor (DDT) in her private wildlife refuge, resulting in the deaths of most of the refuge's bird life, Carson was inspired to begin a new writing project, which was to become her most famous and influential. *Silent Spring* (1962) was the product of painstaking, voluminous research into the problem. There was scientific evidence that such common pesticides as DDT, chlordane, dieldrin, aldrin, and others were harmful to many more organisms than the pests they were designed to kill, but the studies were scattered and published mostly in specialized academic journals. Carson culled the evidence and consolidated it into an eloquent and concise yet horrifying wake-up call to a society becoming increasingly dependent upon these toxic substances. Carson indicted chemical companies and the government for exaggerating the threat of insect pests and withholding information on the dangers posed by overuse of pesticides.

The book had an enormous public impact. Within days of its publication, a U.S. Senate hearing on the dangers of pesticide was convoked, and Pres. John F. Kennedy was compelled to respond to questions of pesticide abuse at press conferences. The chemical industry lashed out against Carson, spending a quarter of a million dollars to defame her. The industry withdrew its sponsorship of a television talk show she appeared on and pulled its advertisements from the *New Yorker* magazine, when it serialized excerpts of *Silent Spring*. Nevertheless, Carson's clarion held the public's attention, and in the years following the book's publication, 42 bills in state legislatures around the country were introduced to curb widespread use of insecticides, many of them becoming law. Carson herself campaigned for many of the bills, and she was awarded the Conservationist of the Year award from the National Wildlife Federation in 1963.

Ironically, during the writing of *Silent Spring*, Carson was struck by breast cancer (which her research revealed to be linked to pesticide exposure). Carson endured frequent trips to the hospital for radiation treatment during her four years' work on the book. Carson made herculean efforts to continue publicizing her cause during her last years and finally succumbed to the cancer on April 14, 1964. Her article "Help Your Child to Wonder" was reprinted in book form as *A Sense of Wonder* in 1965. In that year, the Rachel Carson Council, Inc., was founded by Carson's friends and colleagues to further Carson's research on pesticides. The Coastal Maine Wildlife Refuge was renamed for her in 1969, and she was posthumously awarded the Presidential Medal of Freedom in 1980.

Bibliography

Brooks, Paul, *The House of Life: Rachel Carson at Work*, 1972; Gartner, Carol B., *Rachel Carson*,

1983; Lear, Linda, *Lost Woods: The Discovered Writing of Rachel Carson*, 1998; Lear, Linda, *Rachel Carson: Witness for Nature*, 1997; McCay, Mary A., *Rachel Carson*, Twayne's United States Authors Series, edited by Frank Day, 1993; "Rachel Carson Council, Inc.," http://members.aol.com/rccouncil/ourpage/rcc_page.html; *Rachel Carson's Silent Spring*, PBS Video, 1992; Strong, Douglas H., *Dreamers & Defenders: American Conservationists*, 1988.

Carter, Jimmy

(October 1, 1924–)
Governor of Georgia, President of the United States

As governor of Georgia and president of the United States, Jimmy Carter initiated significant environmental protection. While governor, he created Georgia's Department of Natural Resources and enacted tough regulations on polluters. As president, he enacted the Superfund toxic waste cleanup legislation in the form of the Comprehensive Environmental Response Compensation and Liability Act and signed the Alaska Lands Act, the largest designation of wilderness in the history of the United States.

Born in Plains, Georgia, on October 1, 1924, James Earl Carter was the first child of James Earl and Lillian (Bessie) Carter. His father ran a farm supply store and an office that purchased peanuts from local farmers. He also, eventually, obtained 4,000 acres of farmland in Archery, Georgia, just outside of Plains, where he added the bagging and selling of fertilizer to his other lines of business. Carter followed in his father's footsteps and developed a capacity for hard work and entrepreneurship. As an adolescent, he would go in to town on Saturdays to sell peanuts, ice cream, and hot dogs with a cousin. Carter attended high school in Plains, where he is remembered as a bookworm and an excellent student. He played basketball and participated in debate.

Carter graduated from high school in 1941, at the age of 16. He studied at Georgia Southwestern College in Americus and the Georgia Institute of Technology, before entering the United States Naval Academy in 1943. At the Naval Academy, he continued to earn high grades and graduated in the top 10 percent of his class, in 1946. He spent his next two years working on battleships in the Navy. In 1948, he transferred to submarine service, applying in 1952 for admission to the nuclear submarine program. Carter was assigned to duty in Schenectady, New York, where he studied nuclear physics and engineering at Union College. However, when his father died in 1953, he abandoned his naval aspirations. Carter returned to Plains and took over the floundering family businesses, rebuilding and expanding them.

Carter's political career began in 1962, with a bid for the Georgia state senate. He was narrowly defeated in the primary but was selected as the Democratic candidate after it was discovered that his op-

Jimmy Carter (Charles Rafshoon/Archive Photos)

ponent had stuffed the ballot boxes. Carter won this election and was reelected two years later, serving in the Georgia senate from 1963 to 1966. He sought the Democratic nomination in the Georgia gubernatorial race of 1963 but did not receive it. It was at this point in Carter's life that he was "born again" as a Christian. This did not occur because of any one specific experience but rather emerged as a result of a cumulative period of spiritual growth for Carter. By nearly all accounts, being "born again" had little external effect on Carter's life. He did not go through any dramatic transformation as a politician, and he retained his confident, comfortable style of interacting with his constituents.

In 1970, he again sought the Democratic gubernatorial nomination. This time he was successful. He was elected the 76th governor of Georgia and served from 1971 to 1975. As governor, Carter involved women and minorities in Georgia's government, introduced a merit system for cabinet and judicial appointments, and improved prison rehabilitation programs. One of his most significant accomplishments was structural in nature. He consolidated the 300 existing state agencies into a total of 22 agencies. He also instituted a system known as zero-base budgeting, which required each department to justify its annual budget from scratch.

Carter's period as governor was marked by significant environmental protection as well. He wrote in his book *Why Not the Best?*, published in 1975 as he was campaigning for the 1976 presidential election, that as governor he spent more time on preserving natural resources than on any other issue. Out of all of the newly reorganized state agencies, the Department of Natural Resources was perhaps the most successful. It gave environmentalists considerably more political clout, as well as a larger share of the annual state budget. In addition, Carter proposed several measures for the prevention of erosion and sedimentation and for the regulation of flood hazard areas. He also signed legislation that strengthened environmental enforcement procedures and provided civil penalties for violations of water, air, and surface mining protection laws. By the end of his term, nearly all industrial air and water polluters in Georgia had met federal and state pollution control regulations.

In December 1974, just before his first and only term as governor was about to end, Carter announced his candidacy for the Democratic presidential nomination. He campaigned across the country for the next two years and won the nomination at the Democratic National Convention in July 1976. He won the presidency with his open, populist style of campaigning, defeating incumbent Gerald Ford by a narrow margin. His first act as president was to sign an executive order granting full and unconditional pardons to Vietnam War draft resisters. His term of presidency was characterized by considerable achievements in foreign policy. Carter established diplomatic relations with China and negotiated a successful agreement between the leaders of Egypt and Israel, virtually ending the strife between these two countries. He signed an agreement to relinquish control of the Panama Canal to Panama in 2000, and arranged for an agreement with the USSR, enacting the Strategic Arms Limitation Talks II arms reduction treaty. Carter had a more difficult time at home, however, suffering poor relations with Congress and being unable to stimulate a U.S. economy that was undergoing a period of both inflation and unemployment.

As in his tenure as governor, though, Carter's presidency had a strong environmental focus. One of his first acts as president elect was to nominate Idaho governor and conservationist, Cecil Andrus, for the position of secretary of the interior. Carter was in staunch opposition to the pork-barrel federal dam projects that were then very popular with Congress. And he attempted, with a moderate degree of success, to bring an end to the dam-building frenzy in the western United States. He issued executive orders protecting wetlands, flood plains, and desert ecosystems. He created a Department of Energy. He was a believer in alternative energy sources, and entering office when he did, in the midst of an energy crisis, he advocated the use of natural gas and solar power and coordinated, in his first year in office, a comprehensive long-range energy policy. In response to the Love Canal contamination crisis, Carter initiated Superfund legislation that mandated collection on chemical manufacturers' insurance policies in order to clean up dumping grounds for toxic-waste. He also sent to Congress one of the most sweeping pieces of environmental legislation in U.S. history: the Alaska Lands Act, which in one bill doubled the size of the national parks in the United States and nearly tripled the amount of U.S. land designated as wilderness. The legislation was, according to author Douglas Brinkley, "hailed as a miracle by environmentalists."

Carter was not reelected to office in the 1980 election. He lost in a landslide to Ronald Reagan. Since leaving office, Carter has remained active in international and humanitarian affairs through the Carter Center of Emory University in Atlanta, Georgia, which he founded in 1982. The Center serves as a forum for the discussion of such issues as human rights and democracy. Carter has also been heavily involved since the mid-1980s with Habitat for Humanity, an organization that provides housing for low-income, needy families in the United States and internationally. Carter found himself again in the international spotlight in 1994 when he helped North Korea to negotiate a dispute over the production of nuclear weapons; he also journeyed to Bosnia in 1994, where he aided in bringing about a four-month cease-fire in the conflict between Serbia and Kosovo. Carter has been married to Rosalynn (Smith) Carter since 1946. They have four grown children, Jack, Chip, Jeff, and Amy Lyn.

Bibliography

Brinkley, Douglas, *The Unfinished Presidency*, 1998; Glad, Betty, *Jimmy Carter, in Search of the Great White House*, 1980; Moritz, Charles, ed. *Current Biography Yearbook*, 1977.

Carver, George Washington

(1860–1864 (?)–January 5, 1943)
Botanist, Agricultural Scientist, Inventor

George Washington Carver developed and promoted conservationist agricultural practices during the 46 years that he worked at the Tuskegee Institute in Alabama. His work focused on the most practical aspects of agricultural science, and he directed his efforts toward the people he most wanted to help, poor rural African American farmers of the Deep South.

George Washington Carver was born in Diamond Grove, Missouri. His exact birthdate is unknown, but his biographers estimate that he was born sometime between 1860 and 1864. His mother, Mary, was a slave owned by Moses and Susan Carver, and his father was most likely a slave too, killed before or shortly after his birth. When George Carver was just a year old, a gang of slave raiders kidnapped him and his mother. Mary was never found, but the Carvers tracked down George and raised him and his older brother, Jim, on their farm. Although he showed a precocious interest in learning and knew so much about plants that even as a child local farmers sought him out to diagnose and cure ailing plants, he was not allowed to attend the local elementary school because it accepted only White children. He left the Carver family when he was 14 years old to attend school in Neosho, Missouri. Once he realized that he knew more than the teacher, he moved with a local family to Fort Scott, Kansas. He fled Fort Scott horrified after witnessing the lynching of a Black man and spent the next decade wandering the Midwest, supporting himself by doing odd jobs and even spending a short period of time homesteading on 160 acres in arid western Kansas.

Carver graduated from high school in Minneapolis, Kansas, in the early 1880s and sought admission to Highland College in northeastern Kansas in 1885. He was prohibited from attending there because of his race but was accepted in 1890 at Simpson College in Indianola, Iowa. He studied art and supported himself by taking in laundry but was dissuaded from completing his degree by a professor who believed that an African American man—even one as artistically gifted as Carver—could not make a living as an artist. Carver moved to Ames, Iowa, to study agriculture at Iowa State College, where he rose to take charge of the school's greenhouse, assist botany professors in their research, and teach freshman botany courses. The school's first African American graduate, he earned a B.S. in 1894 and an M.S. in 1896 from Iowa State.

In 1896, Carver was offered a position as head of the new agricultural experimental station at Booker T. Washington's Tuskegee Normal and Industrial Institute in Alabama, an educational and research institution run by and for African Americans. Carver accepted the offer, replying to Washington that "it has always been the one ideal of my life to be of the greatest good to the greatest number of my people." His duties included running an experimental station, sharing his findings with local farmers, and teaching a variety of disciplines, including botany, chemistry, agricultural science, and mycology.

Upon his arrival in 1896, Carver dove immediately into the serious agricultural problems of the Deep South, which included soil depletion and pest infestation. Farmland had been used for intensive cultivation of cotton and tobacco for many generations, and it was severely eroded. Carver developed and promoted composting techniques, and he discovered that the soil could be nourished by rotating such alternative crops as peanuts, sweet potatoes, and black-eyed peas. Alternating these crops with cotton and tobacco helped discourage plant disease; Carver also experimented with hybridization to increase plant resistance to common pests. Through these techniques, he was able to produce impressive yields without the use of commercial fertilizer.

To disseminate his findings and advice, Carver wrote a series of 44 instructional bulletins on different topics over the years. They were notable for their readability and were printed in large quantities to be distributed to small farmers. Some focused on agricultural techniques; No. 6, for example, was entitled *How to Build Up Worn Out Soils* and described how he built up an acre of depleted farmland that initially produced $2.40 per acre, to the point that it yielded a net profit of $94.65 per acre. Others served to convince farmers of the usefulness of the new crops that Carver recommended. No. 5, *Cow Peas*, heralded this legume as a "nutritious and palatable food for man and beast" and provided 25 recipes that featured cow peas; No. 17, *Possibilities of the Sweet Potato in Macon County, Alabama*, covered every aspect of cultivation and use of this tuber and was such a popular edition that it was reprinted several times.

Despite the demand for Carver's bulletins, they did not entirely fulfill Tuskegee's mission to improve the lives of the southern farmer. Because many local farmers were illiterate, Carver traveled through the countryside to give demonstrations and lectures at church after Sunday services or in town squares. Educating farmers in this direct way was formalized in 1906, with the "movable school," a wagon that Tuskegee students equipped with demonstration materials and exhibits. This school on wheels promoted Carver's methods of scientific agriculture to some 2,000 farmers per month during its first summer and served as a model for the U.S. Department of Agriculture extension program.

Carver is best remembered for his promotion of the peanut. Carver had discovered that peanuts effectively added nitrogen to the soil and protected it from erosion, but he understood that farmers would not cultivate them unless there was a market for them. Bulletin No. 31, *How to Grow the Peanut and 105 Ways of Preparing It for Human Consumption*, published in 1916, suggested a tremendous variety of uses for peanuts and their by-products, including peanut butter, peanut oil, peanut flour, shaving cream, face cream, ink, cardboard, dyes, and shoe polish. His later work with the peanut resulted in almost 200 more uses for it. Once a strong peanut industry developed, it adopted Carver as its spokesman and asked him to address Congress in 1921 during its campaign to raise import duties on peanuts raised in the Orient. Carver impressed the congressmen with all of the samples of peanut products that he had brought, and they subsequently voted an import tax of four cents per bushel to save the southern peanut industry.

After this presentation Carver became an instant legend. Articles in national publications extolled his miraculous rise from slavery, his deep religious faith, his humble lifestyle and genuine unpretentiousness. Henry Ford and Thomas Edison both tried to convince Carver to work for them at much higher salaries than Tuskegee could ever offer him, but Carver insisted that he preferred to remain at Tuskegee and serve his own people.

Carver received many awards during his lifetime, including his induction into Great Britain's Royal Society of the Arts in 1916 and the Spingarn Medal from the National Association for the Advancement of Colored People in 1923. There have been many more after his death. Carver died of heart failure at Tuskegee on January 5, 1943.

Bibliography

Adair, Gene, *George Washington Carver*, 1989; Elliott, Lawrence, *George Washington Carver: The Man Who Overcame*, 1966; Holt, Rackham, *George Washington Carver*, 1943; Kremer, Gary, *George Washington Carver in His Own Words*, 1987; McMurry, Linda, *George Washington Carver, Scientist and Symbol*, 1981.

Castillo, Aurora

(1914–April 30, 1998)
Cofounder and Director of Mothers of East Los Angeles

Aurora Castillo was a community activist who helped found and direct Mothers of East Los Angeles (MELA), a grassroots organization that works to improve the environments of poor, largely Chicano neighborhoods in southern California. Known as "La Doña" of East Los Angeles, Castillo received a 1995 Goldman Environmental Prize for her work.

Born in 1914, along with her twin sister, Bertha, Aurora Castillo was the daughter of Frances and Joaquin Pedro Castillo, both of whom were laundry workers. Aurora Castillo was the great-great-granddaughter of Augustine Pedro Olvera, for whom Los Angeles's famous Olvera Street was named. Castillo credited her father for her fighting spirit, saying she always carried with her his advice: "Put your shoulders back, hold your head high, be proud of your heritage and don't let them buffalo you." When she was in high school, Castillo wanted to study accounting, though she was discouraged by the anti-Latino prejudice of her teachers. Castillo did go to business school, as well as study drama and voice at Los Angeles City College. In 1940 she spent three months as a translator for the movie *Across the Wide Missouri*, starring Clark Gable and the Mexican actress María Elena Márquez. She eventually landed a job as secretary at Douglas Aircraft, after scoring high on a business law test; she worked for Douglas Aircraft for most of her career. Castillo never married or had children of her own, and she remained close to her family, including her many nieces and nephews, throughout her life.

Castillo's career as an activist began in 1984, when John Moretta, the priest of the local East Los Angeles Church of the Resurrection, asked women parishioners to protest the construction of a state prison in the neighborhood. The prison would have been the eighth in the area, and the women united to form MELA. Moretta later said he was surprised at how effective the group was, because its members had no training or experience in political organizing. The women were determined to protect their children, fearing possible escapes from the facility, and were no longer willing to see their neighborhood used as a dumping ground for state problems. Castillo said she decided to "fight like a lioness for the children of East Los Angeles." The group informed the community about the threat, using church crowds to spread the word. For two years, the mothers held protest marches every Monday at noon. The marches grew in size, the largest involving over 3,500 participants. MELA united with other groups in the Coalition Against the Prison in East Los Angeles, and the prison was finally relocated in 1992.

As MELA grew in size and experience, it began to work on a broad range of environmental issues. In 1987 MELA began a successful fight against the Lancer Project, a municipal waste incinerator that was to have been built in East Los Angeles. In 1988 the community organized against another toxic incinerator, this time to be built in Vernon, a small industrial community in the area. MELA developed a phone list of over 400 women who could be mobilized very quickly, and Castillo guarded this list fiercely, refusing to sell or give the names to any other group. In 1989 MELA united with Huntington High School students to stop a chemical treatment plant. The Chem-Clear Plant Project, which was to process cyanide and other hazardous chemicals, would have been located across the street from the high school, the largest in the district. The group also successfully rerouted an oil pipeline, which would have gone directly beneath a middle school. One notable defeat was the failure to stop malathion spraying for the Mediterranean fruit fly, commonly referred to as the Medfly. The group currently participates in the Water Conservation Program, which provides low flush toilets to area residents; leads a Lead Poison Awareness Program, which employs high school students to go door to door to educate the community about the dangers of lead poisoning; sponsors higher education scholarships; and directs a Graffiti Abatement Program.

In 1995 Castillo was awarded a Goldman Environmental Prize, which carried a no-strings-attached award of $75,000, the largest of any environmental award. Castillo was the first Latina and the oldest person ever to win the award, at 81. In an interview with the *Los Angeles Times* on receiving the award, Castillo said, "We may not have a Ph.D. after our names, but we have common sense and logic and we are not a dumping ground. We are not the sleeping giant people think we are. We're wide awake and no way will anything be put over on us." Castillo died of leukemia in Los Angeles on April 30, 1998.

Bibliography

Berger, Rose Marie, "Women Heroes of Environmental Activism," *Sojourner's Magazine*, 1997; "Madres del Este de Los Angeles Santa Isabel," http://clnet.ucr.edu/community/intercambios/melasi/; Quintanilla, Michael, "The Earth Mother," *Los Angeles Times*, 1995; Schwab, Jim, *Deeper Shades of Green*, 1994.

Catlin, George

(July 26, 1796–December 23, 1872)
Painter

George Catlin is best known for his large collection of paintings of Native Americans, pictured at work in their villages or in traditional ceremonies. He is known as a conservationist for his proposal, in 1841, of a "nation's Park" that would preserve wild America—"man and beast, in all the wild and freshness of their nature's beauty!"—as he had witnessed it during his painting expeditions. This call was sounded 30 years before the first national park, Yellowstone, was established by Congress. Although some now complain of the racist tinge to Catlin's call for parks as a place where Indians would be on exhibit for visitors, environmental historians cite Catlin as the first American to move from lamenting the destruction of nature during westward expansion to a plan for preserving wild places.

George Catlin was born on July 26, 1796, in Wilkes Barre, Pennsylvania, the fifth of 14 children. His mother, who had been captured by the Iroquois when she was eight years old, told her children stories about Indian ways of life, and Catlin heard more from the many white visitors who stayed at their home on their way to or from Indian lands. Catlin studied and practiced law professionally and painted as a hobby until 1823, when he decided to paint full time. He painted portraits of the wealthy until about 1830, when he discovered his calling. On a visit to Philadelphia he saw a delegation of a dozen Indians who seemed to him awesome and dignified in their traditional dress, and he was struck by the thought that when their lands were conquered by Whites, their culture and traditions would disappear. Catlin later wrote that seeing the delegation inspired his decision "to use my art . . . in rescuing from oblivion the looks and customs of the vanishing races of native man in America."

Catlin took off west in approximately 1831 with a fur-trapping expedition to paint Plains Indians. He painted or drew whenever he could, spending sometimes

George Catlin (courtesy of Anne Ronan Picture Library)

just a few minutes doing a rapid pencil sketch that he later filled in or, if he had more time, painting with oils on canvas. His first expedition was three months long, and for the next eight years, Catlin would travel extensively with explorers and frequently with William Clark, the U.S. Superintendent of Indian Affairs. He, his wife, Clara Gregory, and their son would use Saint Louis as their base, and he did the finishing work on his paintings there.

During his time with the Indians, Catlin developed a familiarity and respect for them. He became convinced that the Plains Indians were superior beings who were able to manage their land well and live comfortably without depleting their main source of meat, the buffalo. Catlin was worried, however, as were other thinkers of his time, most notably author James Fenimore Cooper, that Indian culture would be destroyed by the droves of Whites to come. Instead of merely bemoaning what he feared was inevitable, however, Catlin proposed the creation of "a nation's Park," which would preserve an area of the Great Plains as it was, complete with buffalo and Indians, and keep out White settlers. Catlin wrote his proposal in "Letter—No. 1," a chapter in his 1841 book *Letters and Notes on the Manners, Customs, and Conditions of the North American Indians:*

> And what a splendid contemplation too, when one (who has traveled these realms, and can duly appreciate them) imagines them as they might in future be seen, (by some great protecting policy of government) preserved in their pristine beauty and wildness, in a *magnificent park*, where the world could see for ages to come, the native Indian in his classic attire, galloping his wild horse, with sinewy bow, and shield and lance, amid the fleeting herds of elks and buffaloes. What a beautiful and thrilling specimen for America to preserve and hold up to the view of her refined citizens and the world, in future ages! A *nation's Park*, containing man and beast, in all the wild and freshness of their nature's beauty!
>
> I would ask no other monument to my memory, nor any other enrollment of my name amongst the famous dead, than the reputation of having been the founder of such an institution.

Catlin was not to be the founder of national parks; he spent the rest of his life painting and exhibiting his work and trying unsuccessfully to make a living from his art. He suffered financial ruin and traded his first 600 paintings in 1852 to Joseph Harrison for the cash he needed to repay his debts. He set out once again for the western United States and to South America as well and by 1870 had completed 600 more paintings. Catlin had made a friend at the Smithsonian Institution and exhibited his work there in 1872, in the hopes of convincing the U.S. Congress to purchase his "Indian Gallery," as he called his collection. While he was in Washington, Congress voted to establish Yellowstone National Park, the first national park in the United States.

Catlin died on December 23, 1872, in Jersey City, New Jersey, after a sudden decline in his health. The heirs of Joseph Harrison donated Catlin's work to the Smithsonian Institution after Harrison's death, and Catlin's later collection of paintings was purchased by the American Museum of Natural History and the New York Historical Society.

Bibliography

Dippie, Brian W., *Catlin and His Contemporaries: The Politics of Patronage,* 1990; Mitchell, Lee Clark, *Witness to a Vanishing America,* 1981; Weber, Ronald, "I Would Ask No Other Monument to My Memory: George Catlin and a Nation's Park," *Journal of the West,* 1998.

Caudill, Harry

(May 3, 1922–November 29, 1990)
Writer, Attorney

Harry Caudill, whose roots trace back to the earliest white settlement of southeast Kentucky's Appalachia region, was an eloquent spokesman for his homeland and its people in their struggle against the exploitations of the coal-mining industry. For years he practiced law in mountain courthouses, and he served three terms in the Kentucky state legislature—striving to loosen the grip of the out-of-state coal companies that have wreaked such destruction in the Appalachian region in their pursuit of coal and huge profits. Eventually Caudill began writing books in an attempt to make the abuses of the industry a national issue and to sound the warning that in disregarding coal mining's cost in human and environmental misery, a precedent is set that the whole country must contend with. His books, *Night Comes to the Cumberlands: A Biography of a Depressed Area* (1963), *My Land Is Dying* (1971), *Watches of the Night* (1976) and others, many of which draw an unyielding portrait of a land rich in resources yet impoverished by neglect and misuse, were extremely influential in bringing attention to the problems in Appalachia.

Harry Monroe Caudill was born in Whitesburg, Kentucky, on May 3, 1922, to Cro Carr and Martha (Blair) Caudill. Cro Carr worked in the coal mines until he lost an arm in an accident in 1917. Caudill and his siblings were raised in Whitesburg on the Cumberland Plateau in southeastern Kentucky, a mountainous region of ridges and hollows, which contains dense forests and rich veins of coal. In 1943 during World War II the U.S. Army sent Caudill to fight in Italy; he eventually sustained a severe leg injury and was posted home. By 1948 he had earned a bachelor of law degree from the University of Kentucky and was admitted to the bar of the State of Kentucky. He began practicing law in a private practice in mountain courthouses in the Appalachian region, which he would continue to do for 28 years. In 1949 he married Anne Frye, with whom he would raise a family of two sons and a daughter. Three times he represented Letcher County in the Kentucky legislature, starting in 1954.

In the spring of 1960, Caudill accepted an invitation to give the commencement speech at an eighth-grade graduation in a coal camp school. Only seven students were graduating from the ramshackle two-room schoolhouse, and only one of the students' parents had steady work. From the window Caudill could see a hill of mining slag and was pained to hear the students singing "America the Beautiful." The irony of that setting inspired him to write his first book, *Night Comes to the Cumberlands: A Biography of a Depressed Area*, which was published in 1963. This book, which would later come to influence the formation of public policy, rendered a harsh indictment of the coal industry's abuses. Caudill articulates the irony he perceived in that two-room coal camp schoolhouse: the Appalachian region, endowed with unmatched deposits of coal, petroleum, and minerals, holds several of the poorest counties in the United States. Coal barons managed to extract vast fortunes out of the coal veins in the hills and left nothing behind but political corruption and a wasted landscape.

In *Night Comes to the Cumberlands* Caudill traces the history of mining in his homeland, highlighting some of the worst atrocities, such as the "broad-form deed," which the earliest coal speculators devised to chisel the coal out from under the landowners without paying a fair price for it. Coal company agents would approach the people living in the hillsides and offer a sum of money in exchange for the right to mine there someday, using whatever methods were "convenient or necessary" to extract the coal. It wasn't until years later that these deeds, many of which had been signed by illiterate farmers, would come back to haunt their descendants. Strip mining had become the mode of choice among coal operators, being the fastest and cheapest way, yet it left the surface land and even the landowners' houses destroyed and left the landowners with no legal recourse or remedy. Those who worked in the mines suffered as well, from rock-bottom wages and negligent safety standards. The lax governmental regulations allowing coal companies to operate in destructive and oppressive ways set a pattern for later industrial ventures everywhere else in the country. *Night Comes to the Cumberlands* eventually stirred things up in the rest of the nation: the general public began to take an aversion to strip mining, and the plight of Appalachian people began to gain publicity. Even President Kennedy took notice and began making an effort to alleviate some of the evils of poverty that plagued the region.

In *My Land Is Dying*, published in 1971 complete with photo illustrations, Caudill focuses on surface mining and the trauma it causes the environment. He writes with pride about the Appalachian woodlands where he grew up and the complex diversity of tree species there—a greater variety of trees than anywhere else in the Northern Hemisphere. Yet by the beginning of 1970, over a million acres in the region had been strip-mined, with operations accelerating every year. In the process, strips of mountainside were peeled away, denuding the area of all vegetation, and the coal was mined out of the cuts. The resultant spoil was dumped down the side of the mountain, filling the hollow below with silt and allowing the creek to turn to acid mud. Anyone who raised concerns for the wrecked land was just plowing straight into heavy seas, writes Caudill, since no politician from a coal state wanted to challenge the power of the coal cabal.

In 1976, *Watches of the Night* was published, perhaps Caudill's angriest book of all. In frustration he writes that in the 13 years since *Night Comes to the Cumberlands* there had been no significant changes in the federal effort to regulate coal strip mining or to finance reclamation efforts. He continues the warnings of earlier books and describes the results of a coal boom in the early 1970s as the "murder of a land."

Caudill's work earned him many rewards, including a Kentucky Statesman award (1968), a Tom Wallace Forestry award (1976), and honorary degrees from Tusculum College (1966), Berea College (1971), and the University of Kentucky (1971). In addition to writing books and attending conferences on public affairs around the country, Caudill taught Appalachian studies at the University of Kentucky from 1977 to 1990.

Suffering from Parkinson's disease, Caudill died of a self-inflicted gunshot wound to the head on November 29, 1990, in Whitesburg, Kentucky.

BIBLIOGRAPHY

Caudill, Harry M., *My Land is Dying*, 1971; Caudill, Harry M., *Night Comes to the Cumberlands: A Biography of a Depressed Area*, 1963; Caudill, Harry M., *The Watches of the Night*, 1976; Mitchell, John G., "The Mountains, the Miners, and Mister Caudill," *Audubon*, 1988.

Chafee, John

(October 22, 1922–October 24, 1999)
U.S. Senator from Rhode Island

John Chafee was a moderate Republican who served as governor of Rhode Island for 6 years and as a senator for Rhode Island for 22 years. He was a successful statesman, largely due to his consensus-building capabilities. During his tenure in the U.S. Senate, which lasted from 1977 to 1999, he was significantly involved in the passage of nearly every piece of environmental legislation.

John Hubbard Chafee was born in Providence, Rhode Island, on October 22, 1922, to John and Janet (Hunter) Chafee. His father was a tool manufacturer and was descended from a family that had been in Rhode Island since the seventeenth century. On his mother's side he had two relatives who had served as governors of Rhode Island: his great-grandfather and one of his great-uncles. However (as he is quoted in a 1994 *Washington Monthly* article), "in those days you were governor of Rhode Island for a year, so everybody and his brother had served as governor. I never met a politician before I went to college." Chafee attended the Providence public elementary schools, then the Providence Country Day School, and finally Deerfield Academy in Deerfield, Massachusetts, from which he graduated in 1940.

Chafee attended Yale University, where he captained the undefeated freshman wrestling team. During Chafee's sophomore year at Yale, the United States entered World War II. Chafee left school and enlisted in the U.S. Marine

Corps as a private. He served from 1942 to 1945, first as a soldier, landing with the first troops at Guadalcanal. He then served in Australia before being ordered back to the United States in 1943 to attend Officers Candidate School. He was commissioned a second lieutenant, U.S. Marine Corps Reserve, in June 1944. At the beginning of 1945, he was sent to Guam, and he served in the battle of Okinawa with the 6th Marine Division. He left active duty in December 1945.

Returning to Yale at the end of the war, Chafee resumed his studies. At Yale, he rubbed elbows with prominent politicians and future statesmen, including George Bush. Like Bush, Chafee was "tapped" into the distinguished Skull and Bones secret society in his junior year. He received his B.A degree in 1947. Chafee then attended Harvard Law School, graduating with an LL.B. in 1950. He was admitted to the Rhode Island bar later that year, and he opened a practice in Providence, only to have his career interrupted by the Korean War. He served as commander with a rifle company in Korea, and then with the Marine Corps legal office at Pearl Harbor for the years 1951 to 1953. A longtime aide of Chaffee's would later observe that he thought the military, with its diverse array of people from different backgrounds gathered for a common purpose, had a profound effect on Chafee's attitude toward people. It made him curious about the lives of others.

When Chafee returned from Korea, he resumed his law practice and became involved in local politics. In 1952, he served as an aid in the unsuccessful Providence mayoral campaign of Christopher Del Sesto. In 1956, Chafee sought and was elected to a seat in the Rhode Island House of Representatives from the third district, Warwick. He was a member of the house from 1957 to 1962 and was minority leader from 1959 on. In 1962, Chafee ran for governor. Being a Republican in a heavily Democratic state, Chafee was not favored to win. However, thanks to his keen political sense and his promises to impose no new taxes and to bring more jobs and higher wages to Rhode Islanders, he defeated the incumbent by the narrow margin of 398 votes. Serving a total of three terms as governor, from 1963 to 1969, Chafee worked successfully with a Democratic legislature, signing a comprehensive medical aid program for the aged and authorizing the acquisition of large tracts of land for seven woodland and waterfront state parks. In 1969, Chafee was defeated in his bid for a fourth term as governor, probably due to his frank advocacy of a new, unattractive tax on Rhode Islanders (which his opponent instituted despite his campaign promises to the contrary). By this time, Chafee had risen to prominence as a leader in the national Republican Party. After he failed to be reelected as governor, Chafee was appointed to the post of secretary of the navy by Richard Nixon in 1969, a position he held until 1972.

During the years 1973 through 1976 Chafee practiced law in Providence, an occupation he found to be not nearly as exciting as public service. In 1976, he was elected to the U.S. Senate. He served in the Senate from 1977 until his death in 1999. Chafee was a moderate Republican who was well respected for his consensus-building skills and for his willingness to compromise. From his post on the Environment and Public Works Committee (which he occupied from his arrival in the Senate; he was its chairman from 1995 to 1999) he was involved in nearly every piece of important environmental legislation that went through the Senate during his 22-year tenure. In 1980, he authored the Superfund program that was created under the Comprehensive Environmental Response Compensation and Liability Act (CERCLA), which funded and directed the cleanup of hazardous-waste dump sites across the country. In 1982, he helped to create the National Estuary Program and the Coastal Barrier Resources Act, a piece of legislation that cataloged shoreline areas to be protected from development. Chafee was also involved in the passage of the Clean Water Act of 1986 and in the Oil Pollution Act of 1990, which demanded that polluters pay for oil cleanup and compensate victims. He also played a central role in 1990 in amending the Clean Air Act, which had first been passed in 1967.

Chafee was a defender of the Environmental Protection Agency, protecting it from the cuts and budget riders with which it was assailed by the conservative Republican Congress throughout the mid-1990s. He carried a lifetime 70 percent rating from the League of Conservation Voters and enjoyed the support of such organizations as Defenders of Wildlife. Chafee received many awards for his environmental efforts. He received a National Environmental Quality award from the Natural Resources Council of America in 1995, and in 1999 the League of Conservation Voters presented him with a Lifetime Achievement Award for his "successful leadership in strengthening the Clean Air and Safe Drinking Water acts and his tireless efforts to preserve open space and conserve America's natural resources."

Chafee died October 24, 1999, from heart failure at Bethesda's National Naval Medical Center at the age of 77. He is survived by his wife of 49 years, Virginia, and by five of their grown children (a sixth died in 1968).

Bibliography

Benenson, Bob, "Environment and Public Works," *Congressional Quarterly Weekly Report*, 1994; Pope, Charles, "In the Lull after Chafee's Death, A Sigh of Uncertainty," *Congressional Quarterly Weekly Report*, 1999; Shenk, Joshua Wolf, "An Endangered Species," *Washington Monthly*, 1995.

Chapman, Frank

(June 12, 1864–November 15, 1945)
Ornithologist, Editor

Frank Chapman was a self-educated ornithologist who rose to curate the Department of Birds at the American Museum of Natural History. He founded and edited the ornithological publication *Bird Lore*, later to become the National Audubon Society's magazine *Audubon*. Chapman's work encouraged both amateur bird-watchers and professional ornithologists. From his

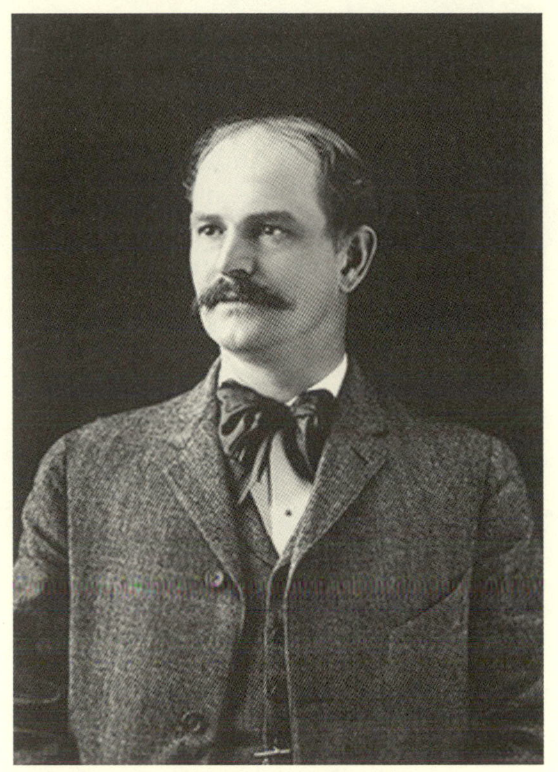

Frank Chapman (Library of Congress)

position of prestige and influence, he called for strict federal protection of birds and their habitat.

Frank Michler Chapman was born on June 12, 1864, in Englewood, New Jersey. When he graduated from high school in 1880, he immediately began working for the American Exchange National Bank, with which his late father had been associated. Chapman worked there for six years, spending all of his free time bird-watching. He befriended local birders, experimented with taxidermy, and worked with the U.S. Biological Survey on bird counts. By 1886, he reportedly could no longer stand to spend so much time away from birds, and he quit the bank. An inheritance from his father funded a year of independent study in Florida.

When Chapman returned from Florida with a large collection of bird skins, he visited the American Museum of Natural History in New York City and received permission to compare his specimens with those in the museum's laboratories. Within the year his talent and interest became apparent to the staff, and he was offered a position as assistant to the head of the Department of Mammals and Birds, Dr. J. A. Allen. Chapman soon became assistant curator of that department, and when it was divided in 1920, he was named curator of the Department of Birds. He retained that position until his retirement in 1942.

Chapman made several innovations to the way birds had traditionally been exhibited: rows of stuffed specimens on perches. He divided the museum's bird collection by region and designed exhibits that incorporated the birds' natural habitats. His trademark designs included a painted background and a foreground with natural elements such as vegetation and rocks. Because these were far more expensive to mount, Chapman raised the money for them himself. Chapman also encouraged local bird-watchers by providing a special exhibit of birds local to New York City and a rotating New York City "Birds of the Month" exhibit.

A board member of the newly fledged Audubon Society, Chapman founded and edited its official publication *Bird Lore*, again financing it himself. In addition to its studies written for the experienced ornithologist, *Bird Lore* included notes from Audubon Society meetings and editorials calling for more stringent government protection of birds. After the passage of the first law to protect birds, the Lacey Act of 1900 that prohibited interstate commerce of birds killed in violation of state laws, Chapman accompanied the first National Audubon Society president, William Dutcher, on surveys of New York clothing and millinery shops in search of illegal plumes. This vigilance was crucial to the success of the Lacey Act. Chapman reportedly persuaded Pres. Theodore Roosevelt to proclaim Florida's Pelican Island the country's first federal bird reserve in 1904. He also established Audubon's annual Christmas bird count, in which bird-watchers spend a 24-hour period during the Christmas season recording species and numbers of all the birds they see within an area 15 miles in diameter.

Chapman was an important behind-the-scenes actor in the controversies that rocked the Audubon Society during its first few decades. When Society president Gilbert Pearson was challenged by several influential members at annual meetings from 1929 to 1933 for what they felt was inappropriate collaboration with the hunting industry, Chapman excluded any mention of it in his meeting notes for *Bird-Lore.* He likewise published nothing about the Emergency Conservation Committee, formed by Audubon members Rosalie Edge, Willard Van Name, and Irving Brant to reform the Audubon Society and oust Pearson. But when Pearson's decision to lease trapping rights to Audubon's Louisiana Paul Rainey Wildlife Refuge and his exorbitant annual salary became public knowledge, Chapman played a critical role in finally deposing Pearson.

Chapman handed the reins of *Bird-Lore* to the National Audubon Society in 1934. The name of the publication was changed to *Audubon* in 1941; it continues to be an influential conservation magazine to this day. Chapman authored 17 books, including bird guides and accounts of his expeditions. For his scientific work he was awarded the Brewster Medal, the John Burroughs Medal, the Roosevelt Medal, and the very first Elliot and Linnaean Society Medals.

Upon his retirement from the Museum in 1942, Chapman moved to Florida where he could study birds year-round. His wife, Fannie Bates, whom he had married in 1898 and of whom he once wrote that "she made it the chief object of her life to advance the aims of mine," died in 1944. Chapman, survived by his son Frank Chapman Jr., died in New York City on November 15, 1945.

Bibliography

Griscom, Ludlow, "Frank Michler Chapman, 1864–1945," *Audubon*, 1946; Murphy, Robert Cushman, "Frank Michler Chapman, 1864–1945," *The Auk*, 1950; "National Audubon Society Christmas Bird Count," http://www.audubon.org/bird/cbc/; Zimmer, John T., "Frank Michler Chapman,"*American Naturalist*, 1946.

Chappell, Kate, and Tom Chappell

(October 17, 1945– ; February 17, 1943–)
Entrepreneurs

In 1970, Kate and Tom Chappell founded Tom's of Maine in Kennebunk, Maine, for the purpose of selling environmentally safe household products. The company has become one of the most successful and influential players in the "greening" of the American marketplace. Tom's of Maine currently grosses over $20 million a year, mainly from the sale of toothpaste and other personal hygiene products. The Chappells are credited with helping to create a market for ecologically sound products. The success of Tom's of Maine demonstrated that environmentally and socially responsible business practices are compatible with profit.

Kate Cheney Chappell was born on October 17, 1945. She attended Chatham College, Sarah Lawrence, and the Sorbonne and graduated from the University of Southern Maine with an A.B. in communications in 1983. Tom Chappell was born on February 17, 1943, and attributes his commitment to the environment to the time he spent during his boyhood in the countryside near Pittsfield, Massachusetts, and on the coastal islands of Maine. His father ran an unsuccessful textile company. When the business failed, the family lost its home, an experience that shaped Tom's business acumen and understanding of the need to attend to the bottom line. He graduated from Trinity College in 1966, with a B.A. in English. The Chappells married in 1966.

In 1968 Kate and Tom Chappell moved from Philadelphia, where Tom was working in the insurance industry, to Kennebunk, Maine. They were ready to leave the pressures of corporate and urban life for a quieter, more rural world. In 1970 they founded Kennebunk Chemical Center, with the aim of combining a business with their ecological concerns. They wanted to develop and market products that would be safe for the consumer and the environment, using natural, nonpolluting ingredients. The first two products were a cleaner for dairy equipment and a phosphate-free detergent, ClearLake. The company really began to grow when the Chappells met PAUL HAWKEN, who owned a chain of natural food stores called Erewhon. Hawken needed natural soap to sell to his customers, so the Chappells developed one and called it "Tom's." By 1981, the company, which had been renamed Tom's of Maine, dominated the personal-care sections of health and natural food stores nationwide and boasted annual revenues of over $1.5 million. In 1983 the Chappells decided to expand the business. They raised new capital, hired experienced business advisers, and began to sell their products in larger, more mainstream drug and grocery stores.

As the business grew, the Chappells maintained a controlling interest in it but began to give themselves the freedom to explore other pursuits. In addition to completing her bachelor's degree, Kate devoted time to her painting and became a successful professional artist. In 1987 Tom enrolled in the Harvard Divinity School and earned a master's degree in theology in 1991. His work at the Divinity School brought new focus to Tom's of Maine. This new sense of mission is outlined in his two books, *The Soul of a Business: Managing for Profit and the Common Good* (1993) and *Managing Upside Down: Seven Intentions for Values-Centered Leadership* (1999). The books offer a vision of a company unified by values, in its relationships to customers, employees, shareholders, and community. The company's mission statement, written collaboratively by the employees and board of directors, offers eleven goals, including: "to respect, value, and serve not only our customers, but also our coworkers, owners, agents, suppliers, and our community"; "to be distinctive in products and policies which honor and sustain our natural world"; and "to be a profitable and successful company while acting in a socially and environmentally responsible manner." In practice, the company fulfills its mission in a variety of ways, including the kinds of ingredients it uses, using recycled packaging materials, refusing to test on animals, and respecting its own workers and those who supply the raw materials. The company tries to use natural ingredients, produced in sustainable ways, and faithfully donates 10 percent of its pretax profits to environmental, human need, arts, and education organizations. Some nonprofit organizations supported by the company include the Rainforest Alliance and the Jane Goodall Institute. Tom's of Maine also frees its employees to devote 5 percent of their on-the-job hours to local volunteer projects.

Tom's of Maine is recognized as a leader in the world of "green" commerce and is often compared to Ben and Jerry's, the Body Shop, and Stonyfield Farm, which have succeeded as socially responsible enterprises. These companies have helped create consumers attentive to the environmental consequences of their purchases, by marketing their own commitment to social welfare. Though these companies are not all industry giants, they have influenced larger companies to include the environment as an important factor in decision making. Tom's of Maine has created a niche for itself and altered the surrounding commercial climate at the same time. The Chappells have often been honored for their business practices, winning recognition from *Working Mother* and *Child* magazines. In 1992 Tom's of Maine received the Corporate Conscience Award for Charitable Contributions from the Council on Economic Priorities. In 1993, the Chappells were presented with the New England Environmental Leadership Award.

The Chappells still control the company, Tom as president and Kate as vice president. Tom functions as the company's chief executive officer, while Kate is more involved with product creation and research. The Chappells live in Kennebunk, Maine, and have five children.

BIBLIOGRAPHY

Barasch, Douglas, "God and Toothpaste," *New York Times*, 1996; Hart, Stuart, "Beyond Greening: Strategies for a Sustainable World," *Harvard Business Review*, 1997; Kontzer, Tony, "The Greening of American Marketing," *E Business Magazine*, 1998; "Tom's of Maine," http://www.tomsofmaine.com.

Chávez, César

(March 31, 1927–April 23, 1993)
Cofounder of United Farm Workers

As a labor leader whose constituency was endangered by exposure to carcinogenic chemical pesticides and fertilizers, César Chávez and his organizing team at United Farm Workers (UFW) fought California's powerful agribusiness interests for healthier working conditions and better salaries. The table grape and wine boycott that the UFW maintained from 1965 to 1970 achieved health and safety regulations that had previously seemed impossible for agricultural workers. Chávez continued to serve as a visible, influential advocate for farm laborers throughout the 1970s, 1980s, and early 1990s, always calling for more humane and healthier working and living conditions.

César Estrada Chávez was born in Yuma, Arizona, on March 31, 1927. After his father was swindled out of his small business and the government seized his grandfather's farm, Chávez's family moved to California and joined the migrant farm labor force. Chávez attended 37 schools before graduating from the eighth grade, at which time his father suffered an accident that left him incapacitated. To support his family, Chávez began to work full time as a farm laborer until he enlisted in the Navy at the age of 17.

First subjected to racial and linguistic prejudice in the English-only, skin-color-segregated California schools, Chávez grew even more aware of racism in the United States grew during his two-year stint in the Navy. While on leave, Chávez went to a movie theater in Delano, California, and was arrested for refusing to respect its segregationist seating policy. Chávez married Helen Fabela in 1948 and settled with their growing family—they eventually had eight children—in Delano, California.

Chávez was introduced to community organizing in 1952 in the Sal Si Puedes neighborhood of San Jose, California, where he was living temporarily as a migrant farm worker. Fred Ross and Father Donald McDonnell of the Community Service Organization (CSO) quickly identified Chávez as a charismatic leader and recruited him to work for the

César Chávez (Archive Photos)

CSO on a voter registration drive. McDonnell served as a mentor to Chávez, introducing him to the nonviolent organizing techniques of Mahatma Gandhi and the respect for all life forms of St. Francis of Assisi. In 1958 Chávez became the national director of the CSO. His major victory while at CSO was forcing the government to investigate the corrupt hiring practices of the government's Farm Placement Service, which favored low-paid Mexican *braceros* over migrant farm workers who were U.S. citizens.

By 1961, Chávez wanted to focus exclusively on problems faced by farm workers, so he left the CSO and in 1962 founded the National Farm Workers Association (NFWA). Chávez was soon joined by former CSO organizer Dolores Huerta, and together they chose grape pickers as their first group of workers to organize. The first major action of the NWFA was in 1965, when it joined the Agricultural Workers Organizing Committee (AWOC) in a strike. Inspired by the AWOC success, the NWFA initiated its first strike in September 1965 after the grape pickers at Lucas and Sons' farm walked off the job. When the strike was declared illegal, the NFWA resorted to another tactic, a five-year nationally coordinated boycott of table grapes and wine.

Chávez and Huerta knew that concern for farm worker welfare would not be enough to convince American consumers to boycott grapes and wine. Farm laborers, among the lowest paid workers in the United States, were a population mostly ignored by mainstream Americans. Pesticides, however, did concern the American public, especially in the years following Rachel Carson's *Silent Spring*. By focusing on pesticides, the NWFA was able to appeal to consumer self-interest and at the same time push growers on an important safety issue for workers. The NWFA ran farm worker clinics throughout California that were tracking pesticide poisonings among farm workers. Two nurses, Peggy McGivern and Marion Moses, collected data, worked with medical experts to devise a proposal for a better pesticide policy, and toured the country to educate boycott organizers on the dangers of pesticides in the food stream. The NFWA's legal team, working with California Rural Legal Assistance, began to sue individual growers for exposing workers to dangerous levels of pesticides and withholding information about which pesticides were applied. Even as late as 1969, one California judge declared that he saw no good reason to release documents that described the types of pesticides used and the dates and locations where they were applied, nor did he understand why it would be beneficial to require growers to post notices in fields where pesticides had recently been applied.

With a boycott in full swing and an ever-growing union membership (at its peak about 50,000 members), Chávez decided to undertake a water-only, 25-day fast in 1968. A powerful symbol, the fast allowed Chávez a poignant way to declare his solidarity with the suffering of farm workers and to declare a stance of noncooperation with supermarkets, which continued to stock boycotted grapes.

By 1970—when 75 percent of grape pickers were NWFA members, such major political figures as Robert F. Kennedy and Walter Mondale had sided with the NWFA, and the boycott was cutting deeply into grape and wine sales—

the majority of growers finally signed a contract that protected agricultural workers in ways they had never enjoyed before. Among the health and safety gains: certain pesticides, including dichlordiphenyltrichlor (DDT) and parathion, would no longer be used; the union could have free access to records of pesticide application; any application of organophosphates had to be approved by a union committee; and whenever the highly toxic organophosphates were used, workers were given cholinesterase tests to measure their organophosphate blood levels. Thanks in part to the impetus of the NWFA, DDT was banned by the Environmental Protection Agency (EPA) in 1972.

Grape workers organized and protected, the NWFA moved on to organize a lettuce boycott. When courts ordered that the lettuce boycott be lifted, the NWFA refused, and Chávez was jailed for 14 days. NWFA supporters maintained a 24-hour vigil outside the jail for the duration of his incarceration.

The NWFA was chartered by the American Federation of Labor–Congress of Industrial Organizations (AFL-CIO) and became the United Farm Workers in 1972. The UFW by no means won all of its battles. When the 1970 contracts with the UFW expired in 1972, many growers signed with the Teamsters Union, which did not include a health and safety clause in its contracts, and threatened those growers who maintained contracts with the UFW. The UFW declared a new boycott against producers who signed with the Teamsters.

Throughout the 1980s, Chávez continued to fight against pesticide abuse in agroindustry. In 1987, he and RALPH NADER jointly called for a boycott of all grapes sprayed with pesticides that the EPA had declared hazardous. Chávez fasted for 36 days in 1988 to bring attention to the children of farm workers dying of cancers most likely caused by pesticide exposure.

Chávez led the UFW until his death, April 23, 1993. He died in San Luis, Arizona, during his sleep. Before his death, in 1990, Chávez had been awarded Mexico's *Águila Azteca*, that country's highest civilian award. Posthumously, he was given the Presidential Medal of Freedom in 1994, and also in 1994, his birthday, March 31, was designated a California state holiday.

BIBLIOGRAPHY

Conord, Bruce, *César Chávez*, 1992; Faistein, Mark, *César Chávez*, 1994; Levy, Jacques E., *César Chávez: An Autobiography of La Causa*, 1975; Mattheissen, Peter, *Sal Si Puedes: César Chávez and the New American Revolution*, 1969; Pulido, Laura, *Latino Environmental Struggles in the Southwest*, 1991; Ross, Fred, *Conquering Goliath: César Chávez at the Beginning*, 1989; United Farm Workers, http://www.ufw.org/.

Chavis, Benjamin

(January 22, 1948–)
Civil Rights and Environmental Justice Activist, Executive Director of the National Association for the Advancement of Colored People

Former executive director of the National Association for the Advancement of Colored People (NAACP) Benjamin Chavis coined the term *environmental racism* in 1982, when, while working for the United Church of Christ's Racial Justice Commission, he decried the frequent practice of siting toxic dumps and industries in areas settled predominantly by racial minorities. After he cosigned a letter to the ten largest national environmental organizations in 1990 asking why there were no people of color on their boards of directors and why they were ignoring the problem of environmental racism, most of the "Group of 10" began to work more actively with minority populations and the environmental problems they face.

Benjamin Franklin Chavis Jr. was born on January 22, 1948, in Oxford, North Carolina. He was one of four children of Benjamin F. Chavis Sr., a bricklayer and orphanage administrator, and Elizabeth Chavis, a schoolteacher. His parents told him early in his life of his great-great-grandfather, John Chavis, a freed slave who was the first Black graduate of Princeton University and worked for White slaveholders in North Carolina as a private teacher for their children. When he was discovered teaching slave children to read, a violation of a state law, he was beaten to death by Whites. Chavis grew up with an awareness of racial injustice and led a reform movement at the age of 13 when he entered Oxford's Whites-only library and attempted to borrow a book. The librarians told him to leave, Chavis asked why, the librarians called his parents, other people got involved, and soon the library was open to all townspeople, regardless of race. As a teenager, Chavis joined the NAACP and the Southern Christian Leadership Conference (SCLC). He went to Washington, D.C., with the local NAACP youth council for the March on Washington in 1963 and heard Martin Luther King Jr. give his renowned "I Have a Dream" speech.

Chavis entered the all-Black St. Augustine's College in Raleigh, then transferred to the University of North Carolina in Charlotte, where he was one of 14 Black students in the university, and the only one majoring in sciences. He organized his fellow Black students and campaigned for a Black student union and a Black studies department. Chavis worked as the western North Carolina coordinator for the SCLC, and during the summers he worked for the Raleigh office of the United Church of Christ's Commission for Racial Justice (CRJ). He graduated with his B.A. in chemistry in 1969 and taught high school chemistry for a year in Oxford, before moving on to full-time civil rights work at the CRJ.

In 1971 Chavis was asked by the CRJ to go to work in the newly desegregated schools in Wilmington, North Carolina, to work with Black students who were being discriminated against by White students and school administrators. The tension in Wilmington quickly accelerated, and a White-owned grocery store was burned. Chavis and nine others were indicted with arson and conspiracy charges. The jury, composed of ten White people, some of them known members of the Ku Klux Klan, and two Black people, found Chavis guilty and he was sentenced to 34 years in prison. Chavis served four years, beginning in 1976. During his time in prison, Chavis studied at Duke University and received a degree in divinity, taught a seminar in Black church studies at Duke, worked in leadership roles for the National Alliance Against Racism and Political Repression and the Southern Organizing Committee for Economic and Social Justice, and wrote a book of prayers, *Psalms from Prison.* Amnesty International and the United Church of Christ, in the meantime, were working on his behalf, since the trial had been flawed and he was considered a political prisoner. In 1980 he was freed after three important witnesses recanted their testimony, claiming that they had been pressured into testifying against the "Wilmington Ten" by local police.

Upon his release, Chavis was ordained as a minister in the United Church of Christ, and resumed working for the CRJ. In 1982, he became involved in the struggle against the siting of a toxic polychlorinated biphenyl (PCB) landfill in Warren County, North Carolina, just east of his native Oxford. Warren County was one of the poorest counties in the state, with a population that was 64 percent Black. The county was a poor choice for the landfill because the water table was high and most residents used wells for their drinking water. Local citizens organized themselves and were supported by several nationally known civil rights and labor leaders, including Chavis, yet the state began hauling the PCB-contaminated soil to the dump site. In September 1982, Chavis was arrested along with more than 400 other protesters for blocking the dump trucks' access to the site. Environmental justice scholar ROBERT BULLARD says that this protest "marked the first time anyone in the United States had been jailed trying to halt a toxic waste landfill."

Chavis commissioned CRJ director of research Charles Lee to perform a national study, "Toxic Wastes and Race in the United States," which showed that polluting industry and waste processing sites were predominantly sited in areas inhabited by racial minorities. Chavis is credited with coining the term *environmental racism* to describe the problem. He was such an effective spokesman for the movement for environmental justice—which works for an end to environmental racism—that the Bill Clinton–AL GORE team appointed him to their transition team.

In 1990, Chavis, along with several other representatives of the growing environmental justice movement, cosigned a letter addressed to the ten largest mainstream environmental organizations in the United States: the Sierra Club, the National Wildlife Federation, the Audubon Society, the Natural Resource Defense Council, the National Parks and Conservation Association, the Wilderness Society, the Izaak Walton League, the Environmental Defense Fund, the Environmental Policy Center, and Friends of the Earth. The letter chastised the "Group of 10" for their overwhelming whiteness and for their lack of action on issues of environmental racism. More money and effort were being spent on struggles in other developing countries than on combating environmental threats to poor racial minorities in this country, the signers complained, and they demanded action. Most of the "Group of 10" acknowledged that their boards and staffs were too White, and they responded with recruitment drives for minority employees and board members. During the 1990s there was a notable increase in attention from many of these groups to problems of environmental racism.

Chavis left the CRJ in 1993 when he was named executive director of the NAACP. He held that post for 16 months, reaching out to young Black people and to other Black organizations, including the Nation of Islam, in an attempt to increase membership. That he did; 160,000 new members signed up during his first year. But Chavis was fired in August 1994 after the board discovered that he had made unauthorized use of NAACP funds to settle with an employee who had filed a sexual discrimination suit.

Chavis joined the Nation of Islam shortly after leaving the NAACP, changed his name to Benjamin Mohammed, and serves as a top adviser to Louis Farrakhan. He has six children, four from his first marriage and two from his second, to Martha Chavis.

Bibliography

Anaya, Toney, and Benjamin Chavis Jr., "Race and the Environment: Protecting the Have-Nots," *Atlanta Constitution*, 1991; "Benjamin F. Chavis," *Current Biography Yearbook*, 1994; Bullard, Robert, *Dumping in Dixie*, 1990; Kotlowitz, "A Bridge Too Far? Benjamin Chavis," *New York Times*, 1994; Lewis, Neil A., "Seasoned by Civil Rights Struggle," *New York Times*, 1993.

Clark, Anita

(April 11, 1946–)
Education Materials Recycler

Child development expert Anita Clark directs Creative Exchange, a nonprofit material resource exchange in Denver, Colorado, that recycles business and industry discards by selling them at a low price to art teachers and organizers of community art projects. The project promotes children's and teacher's creativity, saves schools lots of money on art supplies, and diverts hundreds of thousands of cubic feet of discards annually from landfills.

Anita Mae Clark was born Great Falls, Montana, on April 11, 1946. Growing up in a family of ranchers, she was aware early on of the need to respect the land and use everything available efficiently, in as many ways possible, before discarding it. She recalls her disbelief when a sixth-grade teacher told her that the West was blessed with unlimited natural resources. "It never made sense to me, and from that point on I was very aware and environmentally focussed," she said. Clark's involvement in the Girl Scouts reinforced environmental stewardship, and later, as a mother, she saved everything that seemed even remotely useful for her children's art projects.

Clark studied nursing at the University of Arizona, the University of Southern California, and California State University and early childhood education at Metro State College in Denver, Colorado. She earned her B.A. in human resources and organizational management at Colorado Christian University in Lakewood, Colorado, in 1988 and followed that with three years of graduate work at the University of Denver in applied communications. In 1996, Clark obtained an M.Ed. in curriculum and instruction with a specialization in creative arts and learning at Lesley College in Cambridge, Massachusetts, in a program based on the theory of multiple intelligences conceived by Howard Gardner. Gardner's theory acknowledges the many facets and functions of the brain and advocates providing children with a variety of outlets for creative expression.

Clark always combined her studies with work. She operated a home day care and ran an early learning center from 1976 to 1984; managed home health services for two medical centers from 1985 to 1989; managed Medicare services for an insurance company from 1990 to 1994; worked as a national consultant for a large corporate child-care company from 1994 to 1997; and during all of this, consulted and spoke to national professional groups about early childhood.

Clark chose to focus her career on creativity and arts education after completing her M.Ed. degree. With help from her daughter Rebecca, also an arts educator and graduate of Lesley College, Clark brainstormed the idea for a materials exchange facility that would bridge five areas: arts, education, the environment, nonprofits, and for-profit industry. The Clarks, along with Mary Simon, founder of the thriving Bay-Area Resource Area for Teachers materials exchange facility, researched 200 materials exchange facilities throughout North America, culling ideas from the best models.

Creative Exchange, founded in 1998, collects reusable by-products of business and industry—cardboard tubes, fabric strips, circuit boards, injection molded plastic, wooden dowels, to name just a few—and sells them inexpensively to teachers who either pay an annual membership fee or buy a one-day pass for their visit to the Creative Exchange warehouse/distribution center. Creative Exchange offers workshops for teachers on developing curricula for use with Creative Exchange's materials, and classes can also make field trips to Creative Exchange. Craft ideas include rainsticks, pinhole cameras, solar ovens, poetry books, costumes, and infinitely more.

Creative Exchange's board of directors includes childhood education experts, arts administrators, environmental administrators, representatives of various cultural organizations, and business leaders. Clark's entire family works for Creative Exchange: her husband, Rick, works as warehouse transportation manager; daughters Julie and Kari are materials and business office coordinator, respectively; and eldest daughter Rebecca is associate director.

Clark's Creative Exchange, at this writing still in its nascent phase, is expanding rapidly to serve educators and children in all corners of the state of Colorado and the Rocky Mountain region. Clark resides in Littleton, Colorado.

Bibliography

Davila, Erika, "200 Children Take Recycling for a Spin," *Denver Post*, 1998; Ellingboe, Sonya,

"Creative Exchange: Arts, Education, Environment," *Littleton Independent*, 1998; Go, Kristin, "Creative Exchange Makes Trash Treasure," *Denver Post*, 1999; Human, Katie, "Denver Firm Recycles Trash for Art," *Boulder Daily Camera*, 1999.

Clawson, Marion

(August 10, 1905–April 12, 1998)
Agricultural Economist, Public Lands Policy Analyst, Director of Resources for the Future

Agricultural economist Marion Clawson had a significant impact on public lands policy over the course of his 70-year career. He is well known for his insightful critical analyses of agriculture and recreation on federal lands. He served as director of the Bureau of Land Management from 1948 to 1953, directed and held various offices at the nonprofit organization Resources for the Future (RFF), and wrote many books on the topic of federal land use policy. He was one of the first analysts to apply the principles of social science to forestry, outdoor recreation, and agriculture and was a pioneer of the "multiple use" concept used in public lands management.

Son of William Ennes and Agnes Thompson Clawson, Marion Clawson was born on August 10, 1905, in Elko, Nevada, where he grew up. Clawson graduated from the University of Nevada in 1926 with a B.S. in agriculture. Three years later, in 1929, he earned an M.S. in agriculture and economics, also from the University of Nevada. During the time he was working on his master's degree he was employed as an agricultural economist at the University of Nevada's experimental agriculture station. Upon receiving his master's degree, Clawson moved to Washington, D.C., to become an agricultural economist at the Bureau of Agricultural Economics. He remained in Washington until 1938, when he relocated to Berkeley, California, and continued to work for the bureau as an agricultural economist. He also became involved in land use on the Columbia Basin and in California's Central Valley during this time, acting as head of research and planning for both of these areas from 1940 to 1942 and 1942 to 1945, respectively. Clawson earned a Ph.D. in economics from Harvard University in 1943.

In 1947, Clawson joined the U.S. Department of the Interior. He worked as regional administrator in San Francisco, California, for a year before moving back to Washington, D.C., in 1948. Clawson served as the second director of the Bureau of Land Management (BLM) from 1948 to 1953. When he arrived at the BLM, the agency was primarily concerned with the management of ranch land and with extractive industries. "The bulk of our land was grazing land and our administration of that land was of great importance to the ranchers who used it," he stated in a 1989 interview with the *Journal of Forestry*. The BLM was also heavily involved with oil and gas developers and, in the West, the timber industry. There was literally no official recognition of the value of recreation and conservation on the federally owned lands that the BLM was responsible for managing. The BLM had been attempting to establish itself as an important natural resource "production" agency, along much the same lines as the U.S. Forest Service during this same time period, and the policies created to manage the federal lands were, according to Clawson, "totally unrelated to reality and much criticized by the field people." He changed this by introducing the concept of economic policy analysis to the BLM. This was the first instance in which a federal land management bureau utilized this tool, which measures the effectiveness of a given policy based on economic factors. At the time, Clawson did not think of it as policy analysis, so much as he considered it to be a matter of practical efficiency. Through this efficiency, he was able to "find" up to 25 percent more funding for his district managers.

Clawson was discharged as BLM administrator in 1953 when the newly elected president, Dwight Eisenhower, appointed Douglas McKay to the post of secretary of the interior. Clawson spent the next two years in Israel as a member of the BLM economic advisory staff. He returned to the United States in 1955, accepting a position with a nonprofit research and educational organization, Resources for the Future, where he stayed until his death in 1998. Clawson acted as director of the land use and management program for 18 years and then served as the organization's acting president, vice president, and consultant, consecutively. He became a senior fellow emeritus in 1979. Clawson believed that RFF's greatest influence was exerted through the graduate students who were exposed to the institution's work and who then brought that experience to government and private businesses. And while the specific policy results brought about by the activities of RFF are difficult to pin down, they manifest, Clawson believed, in the actions of policy makers influenced and inspired by the organization's work.

Clawson wrote extensively on the subject of public lands policy, particularly focusing on national forests, national parks, and recreation. He published more than 30 books on these issues, including *Uncle Sam's Acres* (1951) and *Federal Lands: Their Use and Management* (1957), which are considered classic primers on public lands history and administration. He also wrote such important works as *Economics of Outdoor Recreation* (1966), *Forests for Whom and for What?* (1975), *Federal Lands Revisited* (1983), and in 1987, his autobiography, *From Sagebrush to Sage: The Making of a Natural Resource Economist.* He also contributed to many professional journals, *Journal of Forestry*, *Journal of Business Administration*, and *Journal of Forest History* included.

Clawson will be remembered for his contributions to the fields of agricultural economics and public lands administration. He was one of the first analysts to apply the principles of social science to forestry, outdoor recreation, and agriculture, analyzing natural resource policy with a consideration for ecological, economical, and sociocultural factors.

Clawson was a member of the American Agricultural Economics Association, the American Society for Range Management, and the American Academy of Arts and Sciences. He married three times, first to Clara Partridge in 1931, second to Mary Montgomery in 1947, and third to

Nora Roots in 1972. He had four children, two each from his first two marriages. Clawson died on April 12, 1998, in Washington, D.C., during surgery for a hernia.

Bibliography

Clawson, Marion, *From Sagebrush to Sage: The Making of a Natural Resource Economist,* 1987; Dowie, Mark, *Losing Ground: American Environmentalism at the Close of the Twentieth Century,* 1995; Healy, Robert G., and William E. Shands, "A Conversation with Marion Clawson: How Times (and Foresters) Have Changed," *Journal Of Forestry,* 1989; Norgaard, Richard B., "From Sagebrush to Sage: The Making of a Natural Resource Economist," *Quarterly Review of Biology,* 1990.

Cobb, John B.

(February 9, 1925–)
Theologian

A theologian who has worked primarily in academic contexts, John Cobb Jr. has, since the early 1970s, called for an integration of Christian and environmental thought. Cobb is a "process" theologian, who emphasizes the importance of interdisciplinary and interfaith work. He has articulated a revision of Christian theology that responds to the pressing social issues of our time and reimagines humankind's relation to the natural order. In particular, Cobb has stressed the importance of shifting Christian thought away from human dominion over the earth toward an ethic of stewardship.

John B. Cobb Jr. was born in Japan on February 9, 1925. The youngest of three children, he lived in Japan with his parents, who were Methodist missionaries, until 1940, when Americans were urged to leave the country because of the outbreak of World War II. Cobb moved to Georgia, the home state of both his parents, where he lived with his grandmother and went to Emory Junior College at Oxford, Georgia. In 1944, before he finished his studies, Cobb joined the United States Army, where he was assigned to the Japanese language program. His army experiences had a profound shaping effect on Cobb's thought because he was exposed for the first time to non-Protestant intellectuals, primarily Jews and Irish Catholics. He came to understand the limits of his Protestant beliefs and sought a wider understanding of twentieth-century ideas. Upon leaving the service, Cobb enrolled at the University of Chicago, eventually settling in the Divinity School.

While there, Cobb's thought was shaped primarily by two people: Richard McKeon and Charles Hartshorne. McKeon led Cobb to a stance of relativism, that is, to the belief that all systems of philosophical thought are capable of arriving at the truth. Hartshorne introduced Cobb to the ideas of Alfred North Whitehead. Whitehead reversed the Platonic celebration of the ideal, by emphasizing the importance of process. In Whitehead's thought, actuality is determined by process, and anything outside process is perceived as an abstraction and therefore outside of what is real. Whitehead's thought thus values the life of the world over the realm of ideal forms. Cobb's career has been in large part an effort to adapt Whitehead's thought for Christian theology. Cobb graduated from the University of Chicago with a Ph.D. in 1952 and took a position at Emory University's School of Theology. In 1955, Cobb was invited to take a position in California's Claremont University School of Theology, where he has taught ever since.

Cobb's engagement with environmental issues began in 1969, when he read PAUL EHRLICH's *The Population Bomb*. In the introduction to *Sustainability: Economics, Ecology & Justice (1992)*, Cobb says that Ehrlich opened his eyes to current and impending ecological disasters and made him reorder his priorities accordingly. During this time Cobb also read LYNN WHITE JR.'s essay "The Historical Roots of the Environmental Crisis," which taught him that Christian theology fostered attitudes that supported exploitation of the natural world. Cobb's ecological awakening led him to organize in 1970 a conference on "theology of survival," to establish a chapter of Zero Population Growth, and to serve as chair of the ecojustice task force of the Southern California Ecumenical Conference. Cobb also began studying the ideas of architect PAOLO SOLERI and economist HERMAN DALY. In 1972, he organized a conference, "Alternatives to Catastrophe," at which both Soleri and Daly spoke, and published his own book, *Is It Too Late? A Theology of Ecology*. In this book Cobb returns to the tradition of process philosophy and finds strands of thought useful for "ecological thinking." In particular, Cobb discovers within process philosophy ideas to counter anthropocentric and dualistic thinking, both of which lead to ecologically destructive relationships between humankind and the natural world.

In 1973, Cobb, together with David Griffin, founded the Center for Process Studies at Claremont University. Cobb became convinced of the value of process theology for solving social and political crises and founded the center to encourage increased interdisciplinary work. Cobb sees the separation between disciplines as contributing to destructive social and environmental practices and has worked to build dialogues between disparate fields. His own work has exemplified this principle. In 1982, he collaborated with biologist Charles Birch to write *The Liberation of Life: From the Cell to the Community*, which challenged the traditional separation of science and theology. Cobb also returned to his early involvement with Japanese ideas in his 1982 *Beyond Dialogue: Toward a Mutual Transformation of Buddhism and Christianity*, again challenging a separation between Eastern and Western thought. *For the Common Good: Redirecting the Economy toward Community, the Environment, and a Sustainable Future*, cowritten in 1989 with Herman Daly and Clifford Cobb, offered a profound critique of traditional economics and proposed a new, more humanistic and environmentally sustainable approach.

Cobb's career has been devoted to creating a Christian theology responsive to present-day political concerns. In recent years he has joined efforts to address Christianity's antigay and misogynist strands. He has helped lead the Mobilization for the Human Family, a group of liberal theologians organized to counter the power of the Christian right. He retired from Claremont's School of Theology in 1990. He continues to direct the Center for Process Studies and lives in Los Angeles.

Bibliography

"Center for Process Studies," www.ctr4process.org; Griffin, David Ray, and Joseph C. Hough, *Theology and the University: Essays in Honor of John B. Cobb, Jr.*, 1991; Kim, JeeHo, "John B. Cobb, Jr.," http://www.bu.edu/wwildman.

Colborn, Theo

(March 28, 1927–)
Zoologist

Our Stolen Future, written in 1996 by Theo Colborn with coauthors John Peterson Myers and Dianne Dumanoski, alerted the world to a serious yet largely unrecognized chemical contamination problem. Present on earth since their introduction by industry as early as the 1920s, many chlorine- and petroleum-derived chemicals, chemical byproducts, and chemical breakdown products proliferating the food web are toxic in the most minute quantities. They have already affected the sexual development and the endocrine and immune systems of many animal species at the top of aquatic food chains and have caused neurodevelopmental and neuromuscular problems in a sizable segment of human children. Colborn, a senior scientist at World Wildlife Fund (WWF), coordinates research efforts by scientists all over the world and works with environmental groups, government agencies, and the chemical industry toward the phaseout of these so-called endocrine-disrupting chemicals.

Theodora Decker Colborn was born on March 28, 1927, in Plainfield, New Jersey. From a very early age, she was attracted to water, and she spent as much time as she could playing in a creek that flowed past the farm where she lived. Her mother loved gardening, flowers, and birds and instilled a love for all of these in her daughter. Colborn attended Rutgers University, earning a degree in pharmacy in 1947. She and her husband, Harry Colborn, whom she met at Rutgers, took over the Colborn family pharmacy in New Jersey and soon opened two more. They had four children before deciding in 1962 to escape the hassle and fast pace of their life in New Jersey and move to Colorado. After a brief stint in Boulder, where Colborn did a semester of graduate work in pharmacy, the family settled in the valley of the North Fork of the Gunnison River in western Colorado, and began to raise sheep.

The pastoral beauty of the valley was disturbed in the early 1970s, when large-scale coal mining compounded the environmental damage already being caused by the region's molybdenum and uranium mining activity. Colborn and two friends founded the Western Slope Energy Research Center (WSERC) to promote sustainable mining. Once recognized by the

Theo Colborn (courtesy of Anne Ronan Picture Library)

Sierra Club as one of the most efficient grassroots organizations in the country, WSERC nonetheless lost many of its battles. It was through this experience that Colborn learned the importance of impeccable—not anecdotal—scientific information. In 1978 she began a master's degree program in freshwater ecology at Western State College of Colorado in Gunnison. She spent summers at the Rocky Mountain Biological Laboratory, studying aquatic insects in the area's streams to find out if they were accurate indicators of ecosystem health and meeting world-renowned scientists who would later provide crucial support for her work. While at Western State she organized a series of conferences on water issues, the first national and international meetings of their type at the college, which attracted scientists, policy makers, environmental activists, and officials from government agencies.

In addition to the heightened effectiveness that association with an academic institution granted her, Colborn found herself fascinated by ecological science. Free at this point in her life of the constraints of motherhood, since her children were all grown, she entered the University of Wisconsin at Madison's Ph.D. program in zoology in 1982. Her professor there was Stanley Dodson, whom she had studied under at the Rocky Mountain Biology Laboratory; he helped her design a distributed program of study focusing on water quality. After she graduated in 1985, Colborn was awarded a fellowship with the congressional Office of Technology Assessment (OTA) in Washington, D.C.

After two years at the OTA, Colborn was invited to join a team at the Conservation Foundation (later merged with the World Wildlife Fund), also in Washington, D.C., to provide the science for a book on the state of the Great Lakes ecosystem, which is contaminated by massive quantities of industrial pollutants. Colborn found that animals in the lakes were disappearing but not because they had cancer. Instead, animals that ate fish were having difficulty reproducing, and those that were able to reproduce were bearing offspring that suffered a suite of health effects that suggested that their endocrine, immune, and reproductive systems were being undermined before they were born, preventing them from surviving to adulthood. As she searched for the chemical contaminants responsible for this problem, Colborn discovered that many of the possible culprits shared a damning characteristic: they somehow jammed animals' endocrine systems or mimicked some of the hormones, which interfered with their ability to transmit intricate hormonal messages that control how an individual develops. Other scientists who had become aware of this problem ahead of Colborn, including psychologists Sandra and Joseph Jacobson of Detroit, had already started to look at the effects on baby rats who had been fed Great Lakes fish and human babies whose mothers ate Great Lakes fish.

Her work quite promising yet still far from conclusive, Colborn received support from the W. Alton Jones Foundation in 1990, which allowed her to focus entirely on it. The W. Alton Jones Foundation's director, zoologist John Peterson Myers, was intrigued by Colborn's work and became a close collaborator. Together, Colborn and Myers convoked the first Wingspread Work Session in July 1991, which brought together 21 scientists to share findings and contemplate the implications of this research. The scientists collectively came up with the term *endocrine disruption* at this session, and at its conclusion signed a consensus statement recommending the phaseout of endocrine disruptors. They also contributed peer-reviewed papers to a technical volume entitled *Chemically Induced Alteration in Sexual and Functional Development: The Wildlife/Human Connection*, which was edited by Colborn and her assistant at World Wildlife Fund, Coralie Clement, and published in 1992. This book caught the attention of industry and the public health community but was too technical to be accessible to most lay readers. Colborn organized five more Wingspread Work Sessions between 1993 and 1996, each of which resulted in better-coordinated research and further consensus statements.

Knowing that the threats of endocrine disruptors would be ignored unless more people learned about them, Colborn collaborated with Myers and Dianne Dumanoski on *Our Stolen Future*, published in 1996. This book traces Colborn's steps as she pieced together the endocrine disruptor thesis, explaining the complexities of it for lay readers. The book describes the effect of endocrine-disrupting chemicals on wildlife and explores what humans suffer as well. There is already evidence that sperm count is declining, there is a higher rate of abnormalities in the development of sexual organs, and testicular cancer is increasing, to name a few of the problems that studies on the human population have revealed. As dangerous as the organochlorines that bioaccumulate (or become more concentrated each step up the food chain), so too are other types of chemicals that are found in food or products that people encounter every day at home, work, school, and in their cars. Bisphenol A, for example, is an estrogen-mimicking plastic monomer that was first used in the 1920s and occurs in many man-made products. Although estrogen disruptors can penetrate the wombs of female lab animals, little is known about what they do in humans. What may be most troubling to readers of *Our Stolen Future* and policy makers is that there is no safe dose for some of these chemicals, and there is no safe haven from them. Unfortunately, governments have not yet supported research to design screens and assays to test chemicals for endocrine-disrupting effects.

As those familiar with the history of Rachel Carson's 1962 *Silent Spring* might have predicted, *Our Stolen Future* was met with a virulent counterattack by the chlorine industry. According to *Sierra* writer David Helvarg, the multimillion-dollar smear campaign targeting the book was coordinated and funded by the Chlorine Chemistry Council (CCC), an industry group representing this country's major chlorine manufacturers. The CCC still publishes a weekly newsletter with its spin on every aspect of the endocrine disruptor issue. Despite this discrediting attempt, the book has been translated into sixteen languages and has been through many printings. The endocrine disruptor theory has claimed many prestigious and influential adherents. The Environmental Protection Agency, the National Academy of Sciences, and the President's National Science and Technology Council have all prioritized the issue.

Colborn continues to direct WWF's Wildlife and Contaminants Program, which gathers and synthesizes research on endocrine disruptors, facilitates contact between scientists working on the issue, and monitors endocrine disruptor–related work of federal agencies. Colborn's husband, Harry, died in 1983; she resides in Washington, D.C., and owns a home in Paonia, Colorado, which she visits as often as possible.

Bibliography

Burger, Alyssa, "Sex Offenders," *E Magazine*, 1996; Helvarg, David, "Poison Pens," *Sierra*, 1997; Lerner, Michael, "Crossed Signals," *Whole Earth Review*, 1997; Wapner, Kenneth, "Theo Colborn Studies Waterways and Wildlife," *Amicus Journal*, 1995. "World Wildlife Fund: Global Campaign to Reduce the Use of Toxic Chemicals," http://www.worldwildlife.org/toxics/.

Colby, William

(May 28, 1875–November 9, 1964)
Secretary and Director of the Sierra Club

Few Sierra Club members have been more enthusiastic or more boundless in their energy when it came to conservation than William Colby. A key member of the Sierra Club after John Muir's death in 1914, Colby served as a director for 49 years. During his lifetime, he contributed substantially to the saving of redwoods, the enlargement of Sequoia National Park, and the establishment of Kings Canyon and Olympic National Parks. However, Colby is probably best remembered for his participation in the battle for the Hetch Hetchy Valley, where his ability as a determined and eloquent protector of nature came to the fore.

William E. Colby, born May 28, 1875, in Benecia, California, was one of five children of Gilbert and Caroline (Smith) Colby. Their untimely deaths left Colby an orphan at age six, and he was brought up by a legal-minded aunt. Influenced by his aunt, Colby, who had once expressed interest in becoming a naturalist like his hero and future comrade-in-arms, John Muir, took an early interest in the law.

Colby started at the University of California but was forced to drop out owing to financial problems. He began teaching in an Oakland preparatory school, but the ambitious Colby would not be denied his goal. He made two round-trips daily across the bay, one to attend an early morning class at Hastings Law School and another to attend more classes in the afternoon after his teaching duties were done for the day. He graduated in 1898 from Hastings Law School. Tired from the rigor of work and study, Colby gladly accepted a post with the Sierra Club as its representative in the Yosemite Valley. Colby's enthusiasm for the mountains and the many Sierra Club friends he had made on his first trip into the Sierra a few years earlier won him his appointment, and in 1900, he became the club's recording secretary. In 1901, Colby initiated the annual high trips that began the club's popular outings program; he also led these trips until 1929. As Colby gained prominence within the Sierra Club, he started his career as an attorney with a Bay area law firm. During the course of that career, he achieved considerable respect as an attorney who specialized in mining and water law, which served him well in his conservation work.

By 1905, Colby's principal activity in the Sierra Club had become the protection of the Hetch Hetchy Valley. Plans to dam the Tuolumne River at the bottom of the valley to provide water for the growing city of San Francisco were made more urgent after the city's devastating earthquake in April 1906. The plan was met by stern conservationist opposition; the Hetch Hetchy Valley lay within the confines of Yosemite National Park. Damming protected land would set a dangerous precedent.

During the struggle to preserve Hetch Hetchy, Colby worked in close relation with John Muir, the leading figure and spokesperson in the efforts to prevent the dam from being built. Together, they wrote countless letters and pamphlets advocating the protection of Hetch Hetchy. Colby's careful technical arguments complemented Muir's stirring rhetoric to create emotional but rational pleas in favor of the valley. Often Muir and his arguments were dismissed as misanthropic and impractical. Colby, with his experience as a mining lawyer, worked to control Muir's zeal and to deal with the technical problems a dam would create in Yosemite National Park. Interestingly, the law firm where Colby worked was one of the leading proponents of the proposed dam, so Colby often had to pull strings from a distance so as not to jeopardize his job.

Division within the Sierra Club regarding the Hetch Hetchy Valley led Colby to establish a separate organization called the Society for the Preservation of National Parks in 1907 in order to ease the growing rift within the outdoors club. With Muir as president, this organization was composed of a network of Sierra Club council members from all over the country, making the Hetch Hetchy campaign national and relieving much of the pressure from the San Francisco members of the Sierra Club.

Ultimately, Hetch Hetchy, which became a national *cause célèbre*, was approved for damming in 1913 after almost a decade of struggle. While this case did set the dangerous precedent of violating a national park, Colby was at least able to appreciate one positive outcome of the experience: the conscience of the nation had been awakened. As Colby wrote 30 years later in the first *Sierra Club Handbook*, "While this particular battle was lost . . . it has deterred others from attempting similar inroads."

Nevertheless, the loss of the Hetch Hetchy was enough to kill John Muir, who died the following year. After his death, Colby's prominence in the Sierra Club grew. He served as a director for the next 49 years. During this time, he was active in leading directives to protect forests and establish national parks. In 1927, he became the first chairman of the California State Park Commission. Colby also suggested the creation of the John Muir Trail in honor of his friend and the Sierra Club's inspirational first president.

After a lifetime of service to the Sierra Club and the cause of conservation, Colby became the first recipient of the John Muir Award, the Sierra Club's highest recognition for achievement in conservation. Colby died November 9, 1964, at his home in Big Sur, California.

Bibliography

Cohen, Michael P. *The History of the Sierra Club, 1892–1970*, 1988; Jones, Holway R., *John Muir and the Sierra Club: The Battle for Yosemite*, 1965; Wolar, Glynn Gary, "The Conceptualization and Development of Pedestrian Recreational Wilderness Trails in the American West, 1890–1940: A Landscape History," Ph.D. Dissertation, University of Idaho, 1998.

Cole, Thomas

(February 1, 1801–February 11, 1848)
Landscape Painter, Founder of the Hudson River School of Painting

Thomas Cole, the founder of the famous Hudson River School, the first American painting movement, portrayed the grandeur of wilderness in the United States through his landscape paintings. In doing so, he began an artistic tradition that helped to define the ways that Americans interact with and appreciate nature.

Thomas Cole was born on February 1, 1801, in Bolton-le-Moor, Lancashire, England, to James and Mary Cole. His father was an unsuccessful muslin manufacturer, who was unable to keep up with the trend toward mechanization that was occurring in that region, the center of England's textile manufacturing trade and the very cradle of the industrial revolution. Cole's parents encouraged his artistic interests but could not pay for an artistic education. He did attend a boarding school for a short time in Chester and at the age of 14 was apprenticed to a designer of calico prints at a calico printworks in Chorley. Later, he apprenticed with a wood engraver in Liverpool. In 1818, after James Cole's business failed in the depression following the Napoleonic Wars, the Cole family emigrated to the United States, arriving in Philadelphia in 1819. Here, Cole worked as a wood engraver. After a year in Philadelphia, Cole and his family moved to Steubenville, Ohio, where Cole worked as a woodblock carver in his father's wallpaper factory for two years.

It was in Ohio that Cole had something of an awakening after making the acquaintance of a traveling portraitist by the name of Stein. Cole was inspired, and although he had had no formal instruction in painting, he set out in February 1822 to become an artist himself. He traveled around the countryside painting portraits to survive. He had always had a great appreciation for nature, taking long hikes as a child and later exploring the Ohio River. The turning point in his artistic career occurred in May 1823, when he decided to take his notebook into the woods and work from nature. He drew a picture of a gnarled tree trunk, his first nonhuman subject. Once winter fell, Cole returned to Philadelphia to live in a bare, unheated room. He used the tablecloth his mother had given him when he left home as a blanket; it had been the only item she had had that could provide some warmth. Cole haunted the Pennsylvania Academy of Fine Arts with the single-minded purpose of learning how to translate his passion for the natural world into paintings. Cole studied the paintings on view, including landscapes painted by Thomas Birch and Thomas Doughty, two prominent landscape artists of the time.

In 1825, Cole moved to New York and began to exhibit his work. He displayed three of his Hudson River Valley landscapes in a store window. They were well reviewed, and more importantly for Cole, were noticed and purchased by three influential painters, John Trumbull, Asher Durand, and William Cullen Bryant. The prominent Trumbull, who was at that time president of the American Academy of the Arts, is said to have praised Cole for accomplishing in his landscapes what he himself had been trying in vain to do for years. Trumbull arranged for the exhibition of Cole's work in the American Academy and introduced him to a network of wealthy patrons and collectors.

Cole's landscape paintings were revolutionary. Until this point, nature had been primarily depicted as a backdrop for historical paintings featuring famous people acting out well-known situations. Cole changed all of this by making nature itself the powerful, dynamic subject of his paintings. According to author James Thomas Flexner, "Cole was developing a landscape style that did not so much break with the vernacular tradition as transform it." He was painting with an exuberance and sense of importance that had never before been brought to the American landscape. Cole believed in the primacy of the subject. To him, techniques such as use of color and shading were to remain subservient to the subject and were never to be ends in and of themselves. In his portrayals of vast landscapes he deliberately created visual and symbolic oppositions—storm and sunshine, wilderness and human society—to produce a type of exhilaration similar to that experienced when looking at actual landscape panoramas. In the process, he literally helped to invent the way that Americans view wilderness.

The three paintings that he sold in 1825, including the well-known *Lake with Dead Trees (Catskill)*, with their moodiness and power, established the style that would become Cole's signature. Their sale marks the birth of the Hudson River School of Painting and Cole's initiation as its founder. The Hudson River School was the first American movement in painting. It would eventually include such painters as Frederic Church, Asher B. Durand, and Albert Bierstadt. These artists dominated painting in the United States from the 1850s to the 1870s, creating large-scale works that captured the beauty and splendor of the wilderness.

In 1829, on a grant from a Baltimore art collector, Cole traveled to Europe. He visited his native England and Italy. While in Italy he had the idea for a series of five paintings that would track the course of a civilization from its beginning to its end. After he returned to the United States in 1832, a New York art collector named Luman Reed commissioned Cole to paint the series, which he called *The Course of Empire*. These paintings, completed in 1836, were not as well received publicly as his pure landscapes had been, owing to their moral undertones.

Cole settled in Catskill, New York, in 1833, where he boarded with the family of John Alexander Thompson at Cedar Grove, a house surrounded by 88 acres of woods on three sides. Cole married Maria Bartow, one of Thompson's nieces, in 1836. They continued living at Cedar Grove and had four children together, Theodore, Mary, Emily, and Thomas. Cole continued to paint during this time, producing several hundred paintings while at Cedar Grove. When the Catskill-Canajoharie Railroad cut through the woods north toward Albany, Cole became one of the first conservationists in the United States before the word *conservation* even existed with its present meaning. In an article for *National Parks* magazine, Lynne Bertrand quotes him as having said, "Beauty should be of some value among us that where it is not necessary to destroy a tree or a grove, the hand of the woodsman should be checked."

Cole lived at Cedar Grove for 12 years. He died, apparently of pleurisy, February 11, 1848, at the age of 47. Some of his most significant works include: *Lake With Dead Trees (Catskill)*, 1825; *The Oxbow*, 1836; and *The Voyage of Life*, 1839–1840.

Bibliography

Baigell, Matthew, *Thomas Cole*, 1981; Bertrand, Lynne, "The American Canvas," *National Parks*, 1989; Flexner, James Thomas, *That Wilder Image*, 1962.

Collom, Jack

(November 8, 1931–)
Poet, Teacher

Jack Collom is a poet from Colorado who teaches poetry and ecology up and down the Rocky Mountains, from Taos, New Mexico, to Salmon, Idaho. He has written tirelessly about his environs, both outer and inner, since his first book of verse, *Blue Heron & IBC*, was published in 1972. At Naropa University, where he has taught in the Jack Kerouac School of Disembodied Poetics since 1986, Collom has developed eco-lit curricula as well as pedagogies for elementary and secondary schools to activate poetry and ecology simultaneously in the classroom.

Jack Collom was born John Aldridge Collom on November 8, 1931, in Chicago, Illinois. His parents were secondary school teachers who encouraged young Jack's interest in nature. A bird-watcher from the age of 11, Jack Collom has always been a nonviolent outdoorsman, preferring birding and trekking over hunting and trapping. In 1947, when he was 16 years old, his family moved to the small cold town of Fraser, Colorado. He graduated from high school in a class of four, going on to forestry school at what is now Colorado State University at Fort Collins.

After a stint in the Air Force in Germany and North Africa, Collom spent the late 1950s and early 1960s learning his chops at verse writing in and around New York City and other places. He began working in factories in Connecticut to support his poetry habit and his family. He edited the literary magazine *The* from 1966 to 1977.

By 1974 he was working in "Poetry in the Schools" programs in Colorado, Wyoming, and Nebraska, and in 1980 he became a full-time poet in the public schools of New York City. The early teaching approaches that he developed are presented in his book, *Moving Windows* (1985). A variety of writing experiments that are especially appropriate in conjunction with ecological study are included in *Poetry Everywhere* (1994). In that book, and as a first assignment for any group, Collom asks students to write a list of "Things to Save," including both personal possessions and the wonders of our planet earth. Other forms Collom is wont to assign include "Compost-based Poems," "Place Poems," "Recipes" (how to make a horse, a planet, for example), and "Talking to Animals."

Collom always has researched the animals treated in his poems, and he has a penchant for small vertebrates. Animals of particular interest to Collom are mice, fox, passenger pigeons, the blue heron, as well as migratory birds.

In addition to his 16 books of poetry, Collom has written essays on eco-poetics in which he explains his unique and creative approach to impending ecological collapse. He wrote in his 1994 essay, "On, at, around Active Eco-Lit," "Perhaps some poetry, in its eco-response, can reach forward in time, and humanize or enchant scientific prophecy." Like many poets, he is annoyed by pure science's tendency to quantify all things. In the same essay, he advised, "'Out there' is the sense of an endless supply, & endlessness equals unpredictability. Once things can be counted, they're almost gone."

Collom resides in Boulder, Colorado, with his wife, Jennifer Heath. He has four grown children, Nat, Chris, Franz, and Sierra.

Bibliography

Collom, Jack, *Arguing with Something Plato Said*, 1990; Collom, Jack, *Blue Heron & IBC*, 1972; Jack Collom, "On, at, around Active Eco-Lit," *Annals of the Jack Kerouac School of Disembodied Poetics*, Anne Waldman and Andrew Schelling, eds. 1994.

Commoner, Barry

(May 28, 1917–)
Biologist, Activist, Founder of the Center for the Biology of Natural Systems

One of the founding fathers of the modern U.S. environmental movement, Barry Commoner has since the 1950s advocated the correction of environmental problems at their source. Commoner's Center for the Biology of Natural Systems at Queens College, New York, has researched and proposed solutions to a variety of environmental threats, including air pollution, nuclear power, soil and water contamination, and waste disposal. Commoner's focus on problems that affect the public, together with his desire to publicize his common-sense solutions, has motivated him to write five books and even to run for president on the Citizen's Party ticket in 1980.

Barry Commoner was born on May 28, 1917, in New York City. An uncle gave him a microscope as a child, and Commoner spent his free time gathering and examining specimens from Brooklyn's Prospect Park. Commoner studied zoology at Columbia College, graduating in 1937 with honors. He then moved on to Harvard University, where he earned an M.A. in 1938 and a Ph.D. in 1941, both in biology. After a year of teaching at Queens College, Commoner was drafted by the Navy Reserve. An early assignment was to design a method to spray Pacific island beaches with dichlordiphenyltrichlor (DDT) before Navy troops invaded them, and when he tested the procedure on the New Jersey shore, a huge fish kill resulted. This was Commoner's first personal experience with a phenomenon he has seen repeated many times: that society learns about the pitfalls of a product or technology only after it has adopted it and become dependent upon it.

Following his discharge from the Navy and a two-year stint as associate editor of *Science Illustrated* magazine, Commoner accepted a position in the botany department at Washington University in Saint Louis, Missouri. By 1953, Commoner had risen to full professor of botany and became the department's chair in 1965. His research at that point in his career was on the tobacco mosaic virus. Commoner felt that some of the discoveries he was making about the virus might be applied to viruses occur-

Barry Commoner (courtesy of Anne Ronan Picture Library)

ring in humans, and his work was funded in part by a pharmaceutical company and the American Cancer Society. Commoner was responsible for some of the earliest findings about the association between free radicals and cancer.

At the same time that his academic career was blossoming, Commoner's conscience drew him toward research that would address some of the environmental problems that had emerged since the end of World War II. Nuclear fallout was an urgent concern, since the United States and the Soviet Union in the early 1950s were making frequent atmospheric tests of nuclear bombs. Commoner, together with Linus Pauling and Margaret Mead, convinced their scientific colleagues to form study groups and lobby the government about this and other problems. Commoner helped form and for 17 years served on the board of directors of the Scientists' Institute for Public Information, which made scientific information available to the public.

By 1966, Commoner had decided to wrap up his studies of tobacco viruses and devote all of his energies to solving environmental problems. With a $4.25 million grant from the U.S. Public Health Service, Commoner established the Center for the Biology of Natural Systems (CBNC) at Washington University, which would coordinate a multidisciplinary group of researchers, including economists, biophysicists, sanitary engineers, anthropologists, and the like, to study the "real-world problems" resulting from society's dependence on electricity, automobiles, detergents, insecticides, plastics, and so on.

The lessons that Commoner has drawn from CBNC's studies and history are that environmental problems are best solved at their source. For example, in a 1990 interview in *Mother Earth News*, Commoner decries the regulatory approach used by most legislation. The Clean Air Act of 1970, for example, required a 90 percent reduction of carbon monoxide, hydrocarbons, and ozone and set a seven-year deadline for that goal. It was not achieved by 1977, nor by 1982, nor even by 1987; in fact levels started to rise after 1982. The only dramatic improvement in air quality was an 86 percent drop in lead content, a result of the banning of leaded gasoline. Commoner recommends the same tactic with pesticide contamination of soil and water: a ban on pesticides. When CBNC studied the productivity of organic farms in the midwestern United States, it found that they were 8.5 percent less productive but that farmers made the same amount of money per acre because they did not have to purchase expensive pesticides. To halt air pollution caused by fossil fuel–fired electric plants, Commoner says that at the very least, natural gas should be burned to generate electricity, and it would be even better to use methane gas produced by manure or coastal algae, or solar energy gathered by photovoltaic cells.

Commoner's ideas, by his own admission, would be difficult to implement in an economic system such as ours that encourages short-term gain even at the expense of the environment. He is a proponent of more government control over certain industries. In his view, for example, the government should mandate what types of fuels can be used in electric plants or automobiles. Because his ideas could be implemented more effectively if the government assumed more control over environmental issues, Commoner agreed to run for president on the

Citizens Party ticket in 1980. The Citizens Party advocated public control of the energy industry and a sharp reduction in defense spending. Ronald Reagan won that election, but Commoner did garner 250,000 votes.

In 1981, CBNC moved to Queens College at the City University of New York. Commoner stepped down from teaching at the age of 70 in 1987 but continues to direct CBNS. Commoner has been awarded numerous honorary doctorates; he serves on the Vietnam Veterans of America Foundation Council on Dioxin and the Committee for Responsible Genetics. He has two children from his 1946 marriage to Gloria Gordon and has been married to Lisa Feiner since 1980.

Bibliography

"Center for the Biology of Natural Systems," http://www.qc.edu/CBNS/index.html; Commoner, Barry, "Environmental Democracy Is the Planet's Best Hope, *Utne Reader*, 1990; Commoner, Barry, *Making Peace with the Planet*, 1990; Nixon, Will, "Barry Commoner: Earth's Advocate," *In These Times*, 1991; "The Ploughboy Interview," *Mother Earth News*, 1990; Strong, Douglas, *Dreamers and Defenders, American Conservationists*, 1988.

Connett, Ellen, and Paul Connett

(June 21, 1943– ; October 20, 1940–)
Editor of *Waste Not;* Professor of Chemistry

Since 1985, Paul and Ellen Connett have provided solid scientific information and networking support to communities throughout the world that are fighting municipal solid waste incinerators. Paul Connett, a chemistry professor, has made more than 1,600 presentations to activist groups and in public hearings about the hazards of incineration, including the health risks of dioxin and other toxic emissions. He has produced dozens of videotapes on various aspects of waste management that are distributed worldwide. Since 1988, the Connetts have published *Waste Not*, a newsletter for citizen activists that summarizes important waste issues not covered by other media.

Ellen Langle was born on June 21, 1943, in Brooklyn, New York, and was raised in Dumont, New Jersey. She moved to New York City during the 1960s, worked for an advertising agency, and in her free time became involved in the American Committee to Keep Biafra Alive. Through this work she met Paul Connett.

Paul Connett was born on October 20, 1940, in Sussex, England. He graduated with honors from Cambridge University, earning a B.A. in natural sciences in 1962. After teaching high school for four years, he moved to the United States and commenced a doctoral degree program in biochemistry. Quickly Connett was swept up by the anti–Vietnam War movement, and he left the program. In New York City, at a meeting about the situation in Biafra, Connett met Ellen Langle. They began working together on political issues and married in 1970.

The couple moved to England in 1970. They worked for one year opposing the war in East Pakistan, now Bangladesh. Ellen Connett and Roger Moody, an editor at *Peace News* in London, were cofounders of Operation Omega, a group that defied the blockade by delivering much-needed supplies to the besieged country. She and Operation Omega member Gordon Slaven were captured by the West Pakistan army, charged with illegal entry, and sentenced to two years in prison. They served two months and were released in December 1971 when the Indian army liberated Jessore, East Pakistan, where the prison was located. After the war the Connetts lived in London, where Paul Connett taught general studies at a technical college and Ellen Connett raised their three sons. They were particularly concerned about nuclear power plants and worked against the proposed expansion of the Windscale nuclear plant (now known as Stellafield). The Connetts relocated to the United States in 1979, when Paul Connett entered a Ph.D. program in chemistry at Dartmouth College. He graduated in 1983 and that same year accepted the position he still holds today, as professor of chemistry at Saint Lawrence University in Canton, New York.

The Connetts believed that their days of activism had ended and that they could expect the quiet life of an academic family. Paul Connett was looking forward to devoting himself to teaching, and he began hosting a weekly program at the local public radio station.

In 1985, however, they learned of a plan to site a municipal waste incinerator in their rural county. Ellen Connett, who was then working at St. Lawrence University as a periodicals clerk, searched the library's journal archives, and the couple discovered that the incinerator would present a serious health risk. In 1977 it was discovered that dioxin, a highly toxic substance, for which there is no safe dose, is created as a by-product when chlorinated substances are burned. Chlorine is prevalent in municipal waste, occurring in low levels in any bleached cotton or paper product and at significantly higher levels in polyvinyl chloride (PVC) plastic. A health risk assessment was performed for the proposed incinerator in 1985, which assessed only inhalation as the route of exposure to dioxin. Paul Connett believed that ingestion might present a greater risk. He and Tom Webster, who worked with BARRY COMMONER at the Center for the Biology of Natural Systems, published what became the first study on the uptake of dioxin in cows' milk from the dioxin emitted by municipal solid waste incinerators. That study showed that dioxin emitted by trash incinerator smokestacks near grazing cows presented a far greater route of exposure than inhalation, owing to the direct deposition of dioxin onto the pasture and its uptake by the grazing cows. Thus, anyone consuming beef or milk from the area (the county was then the second-largest milk producer in the state) was potentially exposed to very high levels of dioxin. Connett and Webster went on to publish six more scientific papers on dioxins, all of which were published in *Chemosphere* and presented at the International Symposia on Dioxin and Related Compounds. It is now accepted that the major route of exposure to dioxin is through the ingestion of milk, meat, and fish. Of concern is that the fetus and breast-fed baby are exposed to the highest levels of dioxin in the human species.

During the mid-1980s hundreds of incinerators were proposed throughout the country. The Connetts began to share information with other communities through their newsletter *Waste Not*, which by 1988 came out weekly and was sent to hundreds of citizen activists. Each two-page issue of *Waste Not* documented incinerator proposals countrywide, tracked industry proponents as they traveled from community to community, described their arguments, offered scientifically sound responses to these arguments, hailed victories against incinerators, and mourned defeats.

Paul Connett gained a reputation for his articulate, scientifically sound testimony, and he began receiving invitations from grassroots activists to attend public hearings on incinerators all over the country. Since 1985 he has made more than 1,600 presentations to a wide variety of audiences, usually pro bono, in 49 states, 5 Canadian provinces, and 43 foreign countries. He has devoted three sabbaticals and most of his vacation time to being on call for activists in need. Paul and Ellen Connett, together with Tom Webster and Billie Elmore, launched the Citizens Conferences on Dioxin. The first two conferences were held in 1991 and 1994. In 1996, they worked with several other groups, including the Citizens Clearinghouse on Hazardous Waste, to coorganize the Third Citizens Conference on Dioxin.

The St. Lawrence County municipal solid waste incinerator debate lasted five and a half years; the proposal was finally defeated by a narrow margin of county legislators in July 1990. During the period between 1985 and 1993, incinerator proposals were defeated in more than 300 communities across the nation. The last municipal solid waste incinerator was built in the United States in 1995. The incinerator industry has moved abroad, however, and incinerators are being proposed in Asia, Latin America, and some European countries. Most of Paul Connett's current presentations are done abroad at the request of local grassroots activists.

Although the national fight against municipal incinerators has wound down, the Connetts are still concerned about the solid waste problem. The U.S. Environmental Protection Agency and the New York Department of Health collaborated on two studies (1997 and 2000) that estimate that two to eight barrels of burning garbage in back yards release as much dioxin as one 200-ton-per-day trash incinerator. Ellen Connett has become involved with this major rural issue. The Connetts believe that long-term alternatives to chlorine must be pursued, including a phaseout of the use of chlorinated chemicals in manufacturing and more emphasis on the "3 Rs": reducing consumption, reusing, and recycling. They believe that if society can solve its waste problem, it will be much closer to solving other global environmental issues.

The Connetts still publish *Waste Not* but have changed the format to one 20-page issue per month. Their new field of interest is the toxicity of fluoride. Paul Connett currently works with a large Internet-based coalition of scientists, dentists, and doctors called the Fluoride Action Network. A 1997 *Waste Not* article by Joel Griffiths and Chris Bryson, "The Atomic Bomb, Teeth and Fluoride," won recognition from Project Censored, which each year calls attention to the top

25 stories that the mass media failed to cover.

Paul Connett, together with Robert Baily, a professor of fine arts at St. Lawrence University, produced 40 videos that documented the dangers of incineration and safer alternatives for Video-Active Productions between 1985 and 1994. Of special note is a ten-part series on dioxin. Paul Connett now produces videos for Grass Roots and Global Video. The videos' topics range from the poisoning of a South African community from the operations of a medical waste incinerator and a hazardous waste landfill to video interviews with fluoride researchers. His son, Michael Connett, is coproducer of the fluoride series.

The Connetts have received many awards for their work. They shared the Conservationist of the Year award in 1990 from the Environmental Planning Lobby in Albany, New York. Paul Connett received a Leadership Award from the National Campaign Against the Misuse of Pesticides in 2000. Ellen Connett was appointed in 2000 to the board of the Citizens Environmental Coalition, based in Albany, New York.

The Connetts reside in Canton, New York.

Bibliography

Connett, Paul, "The Disposable Society," *Ecology, Economics, and Ethics: The Broken Circle*, F. H. Borman and S. R. Kellert, eds., 1991; Connett, Paul, and Ellen Connett, "Municipal Waste Incineration: Wrong Answer to the Wrong Question," *The Ecologist*, 1994; Schwab, Jim, *Deeper Shades of Green*, 1994; "Fluoride Action Network," http://www.FluorideAlert.org.

Conte, Lisa

(March 24, 1959–)
Entrepreneur

Lisa Conte is founder, president, and corporate executive officer of Shaman Pharmaceuticals, Inc., which in 1999 became ShamanBotanicals.com. The company has set a new standard for the pharmaceutical and dietary supplement industries with its environmentally and culturally astute practices. Shaman's researchers work with indigenous healers in rain forests around the world to identify key compounds from tropical plants with a history of medicinal use. In return for this head start on developing new products, Shaman contributes funds, goods, and services for community development projects in the healer's village. To provide long-term support for this type of project, Conte also set up a nonprofit conservation organization, the Healing Forest Conservancy, to promote the survival of the biological diversity of tropical forests and the cultural diversity of tropical forest peoples. After the U.S. Food and Drug Administration (FDA) delayed Shaman Pharmaceuticals' plans to market its first drug, Provir, in 1999, an action that threatened to bankrupt the company,

Conte successfully transitioned Shaman into the dietary supplement business and an Internet/cable content company.

Lisa Ann Conte was born on March 24, 1959, in Manhasset, New York. She grew up in Great Neck, New York, and as a teenager clerked at a local retail pharmacy. During summer vacations, Conte worked as a horse wrangler in Alaska for her adopted godfather, who was a master guide in Alaska. Conte attended Dartmouth College, receiving her A.B. in biochemistry in 1981, and followed that with an M.S. in physiology/pharmacology at the University of California at San Diego in 1983 and an M.B.A. from the Amos Tuck School of Business Administration at Dartmouth in 1985.

Following her graduation from business school, Conte worked for Strategic Decisions Group in Menlo Park, California, conducting risk and strategy audits for technology companies funded by venture capital. In 1987 she moved to Technology Funding, a venture capital fund, and was responsible for recommending investments in health care companies. In the waiting room of one biotechnology firm, Conte read an article in *Smithsonian* magazine about rain forest destruction and ethnobotanists' fears of losing what could potentially be thousands of medicinal plants. Conte decided that she wanted to be part of the solution, and in 1989 she founded a new type of pharmaceutical company.

Conte provided seed capital for Shaman Pharmaceuticals, Inc., by maxing out her credit cards, and she then raised $200 million from private, public, and corporate investors. The company incorporated the best of ethnobotanical and conservation techniques. In contrast to the usual pharmaceutical testing practice, in which artificially generated molecules are randomly screened for biological activity, Shaman's physicians and ethnobotanists, working in 30 countries in Asia, Africa, and South America, gathered specimens of plants that local healers used, notated exactly what they were used for, and sent them back to the Shaman labs for testing. In exchange for receiving the traditional knowledge of the healers, which in many cases had been passed through uncountable generations of healers, Shaman provided up to $8,000, or its equivalent in goods and services, to the village of each healer who helped its research teams. And it promised the communities profit-sharing agreements if medicines from any of the plants were marketed. In addition, in 1990 Conte founded the Healing Forest Conservancy, a nonprofit organization associated with Shaman that would channel Shaman's profits to all communities and cultures that contributed to the development of products. The Healing Forest Conservancy also works with other organizations that promote sustainable development and environmentally friendly employment opportunities in the countries where Shaman's products originate. These exemplary practices garnered Shaman recognition by the executive office of the Convention on Biological Diversity.

Over its first ten years, Shaman gained the attention of industry observers and environmentalists for its innovative approach. Its findings seemed promising as well. Shaman isolated some 30 compounds that seemed useful for diabetes because they lowered blood sugar levels and began testing Provir, an antidiarrhea drug that promised to be useful for people living with acquired immunodeficiency syndrome (AIDS). Shaman had high hopes for Provir, derived from the sap of the *Croton lecheri* tree that grows in rain forests in South America and has been used traditionally by indigenous people there as a remedy for diarrhea and cholera. When initial tests indicated its high level of effectiveness, Shaman invested in agroforestry projects that would assure a sustainable harvest from the *Croton lecheri* tree from which it was derived.

Provir underwent two years of clinical testing and proved to be safe and effective enough to garner fast-track status from the FDA, requiring only a single final phase-III trial instead of the usual two. But in early 1999, the FDA requested a second clinical test that could delay the drug an additional two years and cost the company tens of millions of dollars. Stockholders felt that this was too risky, and in December 1998, Conte responded by canceling Shaman's drug research and announcing that the company would focus instead on the dietary supplement market. Conte downsized Shaman and created a wholly owned subsidiary, ShamanBotanicals.com.

The first product of ShamanBotanicals.com to be marketed is similar to Provir in origin and use. Normal Stool Formula (NSF) is being marketed to those who suffer from diarrhea as a result of traveling, irritable bowel syndrome, or AIDS or people who experience diarrhea as a side effect of medication or chemotherapy.

Conte has also led Shaman into the field of cable and Internet content production, to generate more income for the company.

Conte has served on many boards of directors, including that of the AIDS Treatment Data Network, the Biotechnology Industry Organization, the Bay Area Bioscience Center, the Amazon Conservation Team, the California Healthcare Institute, the University of California at San Francisco Foundation, Dartmouth College President's Leadership Council, and California governor Pete Wilson's Council on Biotechnology. She received Ernst & Young's prestigious 1994 Entrepreneur of the Year award. She resides in San Francisco and spends her free time on high-adventure travels and playing piano. She is married and the mother of three children.

Bibliography

Abramovitch, Ingrid, "Mother Nature Inc.," *Success*, 1993; "Can a West African Plant Slow the Scourge of Diabetes?" *Life*, 1998; "Healing Forest Conservancy," http://www.shaman.com/Healing_Forest.html; "Shaman Botanicals," http://www.shamanbotanicals.com/.

Cooper, James Fenimore

(September 15, 1789–September 14, 1851)
Novelist

James Fenimore Cooper earned his place as one of the founders of American fiction when he wrote the Leatherstocking Tales—stories about Natty Bumppo, a rugged frontiersman who lived in the late eighteenth and early nineteenth centuries. Generations of readers have since enjoyed these tales of adventure set in the backwoods and forests of the early United States. Cooper wrote many other novels as well and became known for his detailed descriptions of a beautiful natural world and for his resounding criticism of wasteful and destructive pioneering practices in the United States. His novels helped to define the American identity of the times and to document the excesses of civilization on the new frontier.

James Fenimore Cooper was born in Burlington, New Jersey, on September 15, 1789. His father, Judge William Cooper, was a wealthy landowner who founded Cooperstown in central New York State, where the family moved when James was one year old. At the age of 14, Cooper entered Yale, but he was thrown out for misconduct after a year. Following that, he became a sailor for a year in preparation for entering the Navy, which he did in 1808. After three and a half years of service, the last year of which was spent on furlough, he resigned from the Navy and began a career as a gentleman farmer. He married Susan Augusta DeLancey in 1811 and eventually settled at Angevine Farm near Scarsdale, New York, and had five daughters and two sons. Cooper's ambitions as a gentleman farmer were proving difficult, because although his father left an immense estate when he died, he also left sizable debts, and Cooper and his family had to struggle with financial crisis.

It was only after these various trials and travails that Cooper began his literary career. He was 31 years old when he wrote his first novel, *Precaution* (1820), and it was a flop. The next book he wrote was a patriotic tale of the American Revolution called *The Spy* (1821). It appealed to the American sense of adventure and national loyalty and met immediate success. Cooper's career took off. With his next book, *The Pioneers* (1823), which takes place in 1793–1794 near Lake Otsego, New York, he introduced to the world the character of Natty Bumppo, also known as Leatherstocking. By the end of his career, Cooper had written four more tales about Natty Bumppo, including *The Last of the Mohicans* (1826), *The Prairie* (1827), *The Pathfinder* (1840), and *The Deerslayer* (1841), known collectively as the Leatherstocking Tales.

Leatherstocking claims a unique place in literary history as the first American fictional hero. The United States in the late eighteenth and early nineteenth centuries was stepping toward autonomy and working hard to become established as a nation of power. This involved a national mentality of conquering and civilizing the wildness of the physical environment and the native people who lived in it. Settlements were expanding farther and farther into the frontier, uprooting everything in their path. The encroachment of civilization cramped Leather-

James Fenimore Cooper (Archive Photos)

stocking, who lived in a rough hut by Lake Otsego, close to the town of Templeton (modeled after Cooperstown). Leatherstocking's way of life was to live off the land, never taking more than he needed to sustain himself. Like the Native Americans he interacted with, he was extremely adept at reading the woodlands for subtle cues. He exemplified a way of life that was in tune with his natural surroundings, in sharp contrast to what he called the "wasty ways" of the settlers living nearby. There are endless descriptions in this book and other Leatherstocking Tales of the senseless destruction that was common on the frontier: frenzied killing of passenger pigeons, heedless cutting of trees, netting fish in such huge quantities that most of them went to waste. Most of the settlers at this time fell into the all-too-common American habit of mistaking abundance for inexhaustibility, and Leatherstocking made a lonely stand for a more disciplined approach. In the end, though, unable to bear the sound of axes felling trees and unable to trust that the people around him would learn to be less selfish, he moved halfway across the continent to escape. Cooper clearly wished that more Americans would follow Leatherstocking's moral path, and his stories have left us with questions just as pertinent today: can a society bent on progress survive its wasteful and destructive ways?

In the midst of writing these and other novels, Cooper and his family moved to Europe, where they lived from 1826 until 1833, mainly in Paris and Italy. His novels gave Europeans a portrait of American life, and during his time there he was a literary spokesperson for the United States and defended it in his writing. However, on his return to the United States he became disgusted with the abuses of democracy and what he considered to be the tyranny of the majority. He began writing satire and nonfiction, much of which contained candid political criticism—something that made him very unpopular and earned him many enemies. This phase of his career brought on increasing controversy and struggles with a libelous press, though he did manage to resolve much of it and return to writing fiction. By the end of his life he had written over 50 books.

James Fenimore Cooper died on September 14, 1851, in Cooperstown, New York.

Bibliography

Ringe, Donald A., *James Fenimore Cooper*, 1962; Schwarz, Frederic D., "Cooper's Coup," *American Heritage*, 1998; Spiller, Robert E., *Fenimore Cooper, Critic of His Times*, 1931; Taylor, Alan, "Fenimore Cooper's America," *History Today*, 1996.

Costle, Douglas

(July 27, 1939–)
Administrator of the Environmental Protection Agency

Douglas Michael Costle is a lawyer, civil servant, and activist. His environmental career has spanned nearly 40 years and includes such accomplishments as helping to create the Environmental Protection Agency (EPA) in 1970 and serving as its head from 1977 to 1981.

Douglas Michael Costle was born on July 27, 1939, in Long Beach, California.

Soon thereafter, his parents, Michael and Shirley Joan, moved the family to Seattle, Washington, where Costle spent his adolescent years, attending Jane Addams Junior High School and Lincoln High School. After graduating from high school, Costle entered Harvard University, where he earned a B.A. degree in 1961. Three years later, in 1964, he received a J.D. degree from the University of Chicago.

Costle began his career as a trial attorney with the civil rights division of the U.S. Department of Justice in Washington, D.C. After a year in this capacity, he transferred to the Department of Commerce, where he remained until 1967, serving as an attorney for the Economic Development Administration. Costle entered into private law practice in 1967 as an associate with Kelso, Cotton, Seligman and Ray, a San Francisco law firm. In 1968, he became a senior associate with another San Francisco law firm, Marshall, Kaplan, Gans and Kahn.

Costle returned to Washington in 1969, accepting a position as senior staff associate for environment and natural resources of the President's Advisory Council on Executive Organization. It was in this position that Costle contributed significantly to the creation of the Environmental Protection Agency, an independent agency of the U.S. government responsible for the protection and maintenance of the environment for future generations. The EPA effectively replaced the former Environmental Health Service, focusing on the control of air and water pollution through the integration of research, monitoring, and enforcement. Costle lobbied to be appointed as assistant administrator of the new agency but was not awarded the position by President Nixon. Costle maintained an association with the EPA for another year, however, acting as a consultant and also serving on the President's Council on Environmental Quality.

In 1971, Costle was a fellow at the Smithsonian's Woodrow Wilson International Center for Scholars. In 1972, he became deputy commissioner for the Department of Environmental Protection in Connecticut. He was appointed commissioner in 1973 by Connecticut governor Thomas Meskill. As commissioner, Costle gained a reputation as an effective and fair administrator. He instituted a penalty program, referred to as the "Connecticut Plan," that required polluters not only to equip their plants with the proper means to prevent pollution, but also to pay fines equal to what they would have paid had they complied with environmental regulations in the first place. Both environmental and industrial interests praised the plan as fair. Costle resigned from the Connecticut Department of Environmental Protection in 1975 and accepted a position as assistant director of the Congressional Budget Office in charge of natural resources and commerce. He kept this position for a year and a half before being appointed by Pres. Jimmy Carter to the post of administrator of the EPA. Costle acted as EPA administrator from 1977 to 1981.

Costle's relatively late appointment as head of the EPA in the Carter administration was questioned by some who saw him as something of a "compromise candidate." A top EPA official wondered, in a 1977 *Newsweek* article, whether or not Costle had "the horsepower and the guts" to stand up to industry and enforce the country's growing number of environmental regulations effectively. However, under Costle's direction, the EPA grew to hire 600 new employees, and it increased its budget from $2.7 billion to $5.5 billion. Costle emphasized the use of scientific data in creating and enforcing environmental policy, leading the agency in a cautious, deliberate manner. He made a controversial decision to delay the imposition of new automobile emission standards and allowed work to continue on a nuclear power plant project in Seabrook, New Hampshire. He also negotiated what he referred to as "the largest environmental agreement in the history of the steel industry" with United States Steel, which reduced air and water pollution from several large Pittsburgh steel facilities. While with the EPA, Costle endorsed the "bubble concept," which treats polluting plants as if they are under a single bubble that covers the entire facility and allows pollution reductions to be made in the manner most cost effective for the polluter. Costle declared that the EPA was a public health agency and was instrumental in the adoption of the famous Superfund toxic waste cleanup legislation created by the Comprehensive Environmental Response Compensation and Liability Act (CERCLA).

Since leaving the EPA in 1981, Costle has remained active in environmental affairs. He cofounded and acted as executive committee chair for the Environmental Certification Corporation and was chair of the United States/People's Republic of China Environmental Protection Protocol. He served as dean of the University of Vermont Law School from 1987 to 1991. In 1993, he joined the Institute for Sustainable Communities (ISC) as chair of the board of directors. ISC is a nonprofit organization that provides technical training and financial support to communities in an effort to foster environmental protection and economic and social well being. It was founded by former Vermont governor Madeleine M. Kunin in 1991, and it focuses on encouraging civic participation in strong democratic institutions that encompass a diverse array of interests in decision-making processes. Costle believes strongly that economic development and environmental protection are not mutually exclusive concepts but in fact actually reinforce each other. The ISC works on both fronts through its four foci: community development and environmental action; democracy building; environmental training; and environmental and civic education. Costle served on the board of this organization until his retirement in 1999. Today he is chair emeritus of the ISC and serves on the board of directors of the National Audubon Society.

Costle married Elizabeth Holmes Rowe in 1965; they have two grown children, Douglas Michael and Caroline Elizabeth.

Bibliography

"Institute for Sustainable Communities," http://www.iscvt.org/ischome3.htm; Langway, Lynn, "The EPA's New Man," *Newsweek*, 1977; Moritz, Charles, ed., *Current Biography Yearbook*, 1980.

Cowles, Henry

(February 27, 1869–September 12, 1939)
Botanist, Ecologist

In 1899 Henry Cowles published "The Ecological Relations of the Vegetation of the Sand Dunes of Lake Michigan," a work now regarded as one of the founding moments of modern ecology. Cowles's work in this and in later publications helped transform botany from a system of classification and catalog to the scientific study of plants in their environment. Cowles studied ecological succession and pioneered the view that ecology is dynamic, shaped by relationships among species, individuals, and their environment.

Born in Kensington, Connecticut, on February 27, 1869, Henry Chandler Cowles attended local public schools and graduated from New Britain High School. He went to Oberlin College in Ohio and graduated in 1893. In 1894 he began teaching natural science at Gates College in Nebraska but left after one year to study geology at the University of Chicago. There he worked with Thomas Chamberlin and Rollin Salisbury, both of whom practiced a dynamic approach to geology. Cowles eventually switched his field to botany, attracted by John Merle Coulter, who had recently been recruited by the university to start a department of botany.

Coulter introduced Cowles to the work of European botanist Eugenius Warming, who was evolving an ecological approach to the study of plants. Cowles combined his work in botany with his background in geology and began his doctoral research on the succession of plants on the Indiana Dunes at the southern end of Lake Michigan. The dunes provided an ideal site for Cowles's study because their advance caused the rapid succession of plant species. Cowles came to see succession not simply as the product of competition between species but as a more complex process involving the struggle of individuals with their environment. Cowles's work challenged the view that succession was always progressive—the result of strong species squeezing out the weak—and argued that sometimes succession had destructive, negative results. Cowles finished the dissertation, "An Ecological Study of the Sand Dune Flora of Northern Indiana," in 1898 and received his Ph.D. in botany from the University of Chicago. In 1899 he published his results in the *Botanical Gazette*, under the title "The Ecological Relations of the Vegetation of the Sand Dunes of Lake Michigan."

This work initiated a school of ecology known as "dynamics," that is, focused on describing and understanding processes of change in ecosystems. Cowles argued that dynamics were a key component of arriving at accurate botanical classification. His ideas were part of a growing scientific interest in dynamics, derived in part from Darwin. A central idea of the dynamic school is the "climax" community, a stage of equilibrium reached in a particular ecosystem in which plant communities are stable and self-sustaining. Cowles argued for a subtle, complex view of the climax stage. His work showed that what looks self-sustaining can in fact prove self-destructive if, for example, one species crowds out neighboring species necessary to the health of the whole ecosystem.

Cowles followed his first studies on the Indiana Dunes with work on the surrounding ecosystem, publishing his results in 1901 in a paper entitled, "The Physiographic Ecology of Chicago and Vicinity." This paper is given credit as the first formal study to systematically employ the concepts of succession and climax. He also made comparative studies of dunes on Cape Cod. In 1911, Cowles became full professor at the University of Chicago and in 1925 was appointed chair of the botany department. He held the position until his retirement. As a teacher, Cowles was remembered for his use of photography and other methods that took students out of the classroom and into the field. His photographs—slides and prints—were recently rediscovered and are on display at the University of Chicago. He is credited with generating enthusiasm for the emerging field of ecology and created a large and loyal following of students who went on to careers in a variety of scientific fields.

Cowles worked with a number of influential scientists, including Sir Arthur George Tansley, who first coined the term and concept of the ecosystem, and ecologist Frederick Clements. His work on the dunes of Lake Michigan also attracted the attention of landscape designer Jens Jensen. In 1914, Cowles, along with several of his students, founded the Ecological Society of America. In 1926 he was appointed editor of the *Botanical Gazette*. Cowles used his influence to help establish the Illinois state parks system and the forest preserves of Cook County. He was a charter member of the Illinois State Academy of Science. He served as president of the Chicago Academy of Science, the Association of American Geographers, the Botanical Society of America, and the phytogeography and ecology section of the International Botanical Congress.

Cowles retired from the University of Chicago in 1934. He died at his home in Chicago on September 12, 1939, after a long illness.

Bibliography

Cooper, William, "Henry Chandler Cowles," *Ecology*, 1935; Golley, Frank, *A History of the Ecosystem Concept in Ecology*, 1993; Yoe, Mary Ruth, "The Once & Future Scenes," *University of Chicago Magazine*, 1997.

Cox, Paul

(October 10, 1953–)
Ethnobotanist

Ethnobotanist Paul Cox shared a Goldman Environmental Prize in 1997 with Fuiono Senio, high chief of the village of Falealupo, Western Samoa, for their cooperative role in saving a 30,000-acre rain forest outside Falealupo from imminent destruction. The Falealupo reserve is a model of sustainable use: villagers harvest plants for medicinal or traditional uses, an arboreal

Paul Cox (Courtesy of Paul Cox)

walkway used by ecotourists and researchers provides the village with an extra source of income, and the forest continues to flourish. Cox performs ethnobotanical research throughout the world and currently directs the National Tropical Botanical Garden in Hawai'i.

Paul Alan Cox was born in Salt Lake City, Utah, on October 10, 1953, to a family long devoted to conservation. His great grandfather helped establish Arbor Day in Utah; his grandfather founded wildfowl preserves and fish hatcheries in the West and played a lead role in establishing the elk preserve in Jackson Hole, Wyoming. Cox's mother, a fisheries biologist, and his father, a conservation officer, both worked for the U.S. Fish and Wildlife Service; and his father also worked for the National Park Service and in the state park system of Utah as superintendent. Cox spent much of his childhood outdoors, accompanying his parents on their duties, and they both encouraged their son's early interest in plants. When he became fascinated with a rare plant, *Darlingtonia californica*, at the age of ten, for instance, they drove him to the California coast near Crescent City, where one of two natural populations grow.

When Cox was 19 years old, he was called to serve as a missionary by the Church of Jesus Christ of Latter-day Saints. He began his two years of service in the small, remote village of Safune, on the island of Savaii, Western Samoa. The village's high chief, Aumalosi, taught Cox the chiefly version of Samoan, a dialect reserved for chiefs and almost never taught to *palagi*, or foreigners. Cox still does not understand why he was chosen for such instruction, but he proved a facile learner and now speaks the dialect well. This fluency has been of great use in his ethnobotanical studies and conservationist efforts in Samoa, since both entail complex interactions with Samoan people.

Cox returned to Utah in 1975 and resumed studies at Brigham Young University (BYU), graduating with a B.S. in botany in 1976. He then earned two master's degrees in 1978, one from the University of Wales in ecology and the other in biology from Harvard University. Harvard awarded him a Ph.D. in 1981. At Harvard, Cox was deeply influenced by Prof. Richard Evans Schultes, who upon learning that he was fluent in Samoan, encouraged him to begin ethnobotanical studies there. After a two-year stint at Miller Institute for Basic Research in Science at the University of California, Berkeley, from 1981 to 1983, Cox was offered a professorship in the botany department at BYU.

Cox was devastated by the death of his mother, in 1984, of breast cancer. In his grief, he vowed that he would dedicate his life to searching for a cure to that disease. His training led him to believe that of the many medicinal plants growing in tropical rain forests, at least one might hold a cure for cancer. Two months after his mother's death, Cox learned that he had won a National Science Foundation Presidential Young Investigator Award to fund any research project of his choosing. So, in 1984, Cox was able to return to the island of Savaii in Western Samoa to begin what would become a long and fruitful career studying the medicinal plants of its lush rain forests. This time, he was accompanied by his wife, Barbara, and their four children. Living in a *fale* (an open-air Samoan-style home) on the beach in the village of Falealupo, the family readily integrated into the village's social life. Cox studied with local healers, and he gathered specimens of medicinal plants that he felt might offer cures for cancer to be screened by the National Cancer Institute (NCI) in Washington, D.C. During his stay in Falealupo that year, Cox made two discoveries that would fuel his later conservationist efforts: that the Western Samoan flying fox, a large flying fox bat of the rain forest, was facing extinction and that the village of Falealupo was being forced by the Western Samoa government to build a new school for its children, which was such an expensive endeavor that the village had to sell logging rights to its rain forest to pay for it.

Cox used funds from his National Science Foundation award to fund a study of the Western Samoan flying fox. He returned in 1988 to help his research team set up their study in Falealupo, and while he was there, he learned that the company that had been granted the logging rights to the forest in return for a down payment on the village school had just begun cutting trees. Cox and the researchers visited the logging site and were horrified. Cox writes in his 1997 memoir *Nafanua* that watching the rain forest fall reminded him painfully of watching his mother succumb to cancer. Just as he would have done anything for her comfort and survival, he felt compelled to save the Falealupo rain forest. A few days later he made a daring proposal to the village Council of Chiefs: that he himself would pay for the village school and pay off the logging company's down payment on the school, so that the forest could remain standing. After several days of debating the offer—controversial because Samoans have many times in their history been taken advantage of by power-hungry *palagi*—High Chief Fuiono Senio convinced the Council to accept it. Fuiono then ran three miles to the logging site, machete in hand, and forced the loggers to leave.

Cox returned to the United States and raised money to buy back the lease from the logging company and to pay for the new school. He and Barbara had decided to sell their own home to raise the necessary funds, but that became unnecessary when other donors contributed to the fund. Soon Cox was able to return and sign the official contract in which the villagers promised to protect the rain forest for 50 years, extracting forest products in a sustainable manner only for medicines and other traditional uses. They also vowed to protect the endangered Western Samoan flying fox. He later returned to help construct an aerial walkway and observation platform 100 feet above ground. Designed for researchers and tourists who pay entrance fees, this provides additional income to the village, where the average per capita yearly income is under $100. For his leadership in saving the village's rain forest, the Council of Chiefs bestowed a unique honor on Cox, the title of Nafanua, or Goddess of War, which he still holds today and which entails a lifelong responsibility for the well-being of the village. Cox and Fuiono Senio were jointly awarded a 1997 Goldman Environmental Prize for saving the Falealupo rain forest.

In addition to this achievement, Cox also lobbied successfully for an act setting up a national park in American Samoa, which protected habitat of the flying fox. It was signed into law in 1988. Although he had been unable to persuade the U.S. Fish and Wildlife Service that the flying fox merited listing as an endangered or threatened species, Cox and several members of his research team attended the 1989 conference of the Convention on International Trade in Endangered Species (CITES), whose delegates voted to prohibit commercial traffic of the Western Samoan flying fox and six other species of Pacific Island flying foxes. One of the plants from the Falealupo forest that Cox sent to NCI yielded prostratin, a new anti–acquired immunodeficiency syndrome (AIDS) drug, and in accordance with his contract with NCI, the Samoan people will share half of NCI's royalty income. Cox has signed a similar agreement with Nu Skin International, which has a line of botanical products that Cox helped formulate. Nu Skin funded the aerial walkway in Falealupo, a water catchment scheme, and other projects in Samoa and other islands. These projects of Nu Skin's "Force for Good" campaign are funded via contributions to Seacology, a nonprofit organization that seeks to preserve the ecosystems and cultures of islands throughout the world and on whose scientific advisory board Cox sits.

Cox continues to visit Samoa at least once a year for his on-going ethnobotanical research. He serves as the King Carl XVI Gustaf Professor of Environmental Science at the Swedish Biodiversity Centre, where he works closely with the king and studies medicinal plants used by the Lappish reindeer herders above the Arctic Circle. Cox directs the congressionally chartered National Tropical Botanical Garden in Hawai'i and was recently named distinguished professor by BYU-Hawai'i.

Bibliography

Balick, Michael, and Paul Cox, *Plants, People, and Culture: The Science of Ethnobotany*, 1996; Cox, Paul, *Nafanua: Saving the Samoan Rain Forest*, 1997; Hallowell, Christopher, "Rainforest Pharmacist," *Audubon*, 1999; "National Tropical Botanical Garden," http://www.ntbg.org/; "The Seacology Foundation," http://www.seacology.org; Willis, Monica Michael, "The Plant Detective," *Country Living*, 1998.

Craighead, Frank, and John Craighead

(August 14, 1916–)
Wildlife Biologists

Wildlife biologists Frank and John Craighead became famous for their research on the grizzly bears of Yellowstone National Park and their commitment to sharing their research with the general public via articles for *National Geographic* magazine and several educational and documentary films. Through their research, they pioneered the use of radio tracking collars and satellite telemetry in the 1960s and early 1970s. In addition to their work with wildlife, they were particularly concerned about preserving the pristine rivers of the Northern Rockies and provided the impetus and much of the wording for the 1968 Wild and Scenic Rivers Act.

Frank Cooper Jr. and John Johnson Craighead, identical twins, were born on August 14, 1916, in Washington, D.C. Their father, Frank Craighead Sr., was chief of the Bureau of Forest Entomology in the U.S. Department of Agriculture in Washington, D.C. Upon his retirement to Florida, Frank Sr. authored books on the orchids, air plants, and trees of southern Florida and published many papers about the region's changing ecology. As teenagers, Frank and John became interested in falconry, and at the age of 21, they cowrote "Adventures with Birds of Prey" for *National Geographic*, the first of their 20 articles for that magazine. Their first book, *Hawks in the Hand*, was published in 1939. The Craigheads attended Pennsylvania State University, each receiving a B.A. in science in 1939. They each earned M.S. degrees in ecology and wildlife management in 1940 from the University of Michigan. In 1940, with a National Geographic Society grant, they traveled to India, and as guests of Prince Dharmakumarsinhji, they trained and flew falcons, coursed cheetah at antelope, and participated in the pomp and ceremony of the royal family. They served as lieutenants in the Naval Reserve during World War II, organizing the Navy's Land Survival Training Program and writing the manual *How to Survive on Land and Sea*. This garnered them a special citation in 1946 from the secretary of the navy. The brothers returned to the University of Michigan after the war to earn their Ph.D.s in vertebrate ecology in 1950. Their doctoral research, done with fellowships from the Wildlife Management Institute, compared the annual predation of complete raptor communities in Michigan and Wyoming. It resulted in the 1956 publication of *Hawks, Owls and Wildlife*, which set a new standard for the scientific study of raptors and their role in ecosystems.

John Craighead worked for the U.S. Fish and Wildlife Service as a wildlife biologist, led the Montana Cooperative Wildlife Research Unit, and was professor of zoology and forestry at the University of Montana from 1952 to 1977, earning the Outstanding Educator of America award in 1973. During this same period, Frank Craighead managed the Desert Game Range for bighorn sheep in Nevada, worked as a wildlife biologist for the U.S. Forest Service, founded the Outdoor Recreation Institute in 1955, and took a post as senior research associate and adjunct professor at the State University of New York (SUNY) at Albany. The brothers carried out research together, individually, and with other biologists in a variety of areas, including water quality in western rivers, aquatic insects, air quality in Yellowstone, and such wildlife species as elk, Canada geese, the golden eagle and other raptors, mountain lions, magpies, and pheasants. Together the Craigheads wrote the Peterson field guide *Rocky Mountain Wildflowers* in 1964.

The Craigheads were invited in 1959 to study grizzly bears in Yellowstone National Park, home to the largest remnant population of that species. They spent 12 years in this endeavor, pioneering the use of radio tracking (and later satellite tracking) to study the physiology and ecology of wild grizzlies. Their study compiled the most extensive population data on the grizzlies, which led to a much more complete understanding of the grizzlies' interaction with their ecosystem. When Yellowstone officials announced plans to rapidly close down the park's refuse dumps, the Craigheads warned that this would put the grizzlies in danger. They had discovered that the bears did not distinguish between natural food sources and nonnatural ones and they correctly predicted that the sudden disappearance of a major food source resulted in bears' becoming nuisances in nearby areas inhabited by humans. The Park Service not only did not heed the Craigheads' warning, forced the Craigheads to leave their research base in the park in 1971, and the study of the Yellowstone grizzly came under strict government control.

The Craigheads' 12 years of research yielded important biological data and evidence of misguided management that led to the 1975 listing of the grizzly as a threatened species. During the 1980s, their controversy with the National Park Service began to heal, and they were able to collaborate with Park Service biologists and other scientific groups trying to facilitate the recovery of the species. Their work with the grizzlies resulted in dozens of technical papers, several articles for *National Geographic* and other popular magazines, and the National Geographic Society documentary film *Grizzly* (1967), which featured the brothers. Frank Craighead's *Track of the Grizzly* (1979) was published by the Sierra Club. John Craighead wrote two technical books. *A Definitive System for Analysis of Grizzly Bear Habitat and Other Wilderness Resources* (1983) explains how Landsat multispectral imagery using on-the-ground botanical data and computer technology can be used to map and evaluate the vegetation complexes and landform characteristics of large expanses of remote wilderness. His second book, *The Grizzly Bears of Yellowstone: Their Ecology in the Yellowstone Ecosystem, 1959–1992*, was coauthored with Jay Sumner and John Mitchell in 1994. Both books won The Wildlife Society's publication of the year award. In his acceptance speech for the award to the latter book, John Craighead summed up how grizzly bear protection fits into wilderness conservation as a whole: "To preserve the grizzly bear in its natural state, we must keep, intact, the entire spectrum of biodiversity present within its public-land habitat in the Northern Rockies. Concomitantly by preserving the grizzly we automatically preserve the great biodiversity that is its environment."

When the Craigheads first traveled west early in their careers, they were impressed by the region's wild rivers. They boated many of them and advocated classifying them according to their natural assets and recreational potential and preserving the most pristine in their wild state. They published a classification system, listing specific data necessary for classification purposes. They drafted legislation that eventually became the Wild and Scenic Rivers Act. With a strong lobbying effort from Stewart Brandborg's Wilderness Society, the bill was signed into law in 1968. This dedication was recognized by the National Geographic Society, which featured the Craigheads in its 1970 documentary film *Wild River.*

Once the Craigheads retired from their respective academic positions, they each devoted themselves to their own nonprofit wildlife research institutes.

Frank Craighead's Outdoor Recreation Institute changed its name in 1980 to the Craighead Environmental Research Institute (CERI). It is currently codirected by his sons Frank Lance Craighead and Charles Craighead and his daughter-in-law, April Hudoff Craighead. Frank is an assistant professor of biology at Montana State University, Charles is a wildlife biologist who authors books and produces conservation documentaries, and April is a biologist. The mission of CERI is "to increase humankind's understanding, appreciation, and protection of our natural environment; particularly wildlife populations and wild landscapes," and it works in four program areas: nature reserve design that accommodates movement through corridors; conservation biology and genetics; Geographic Information Systems for conservation; and conservation education. In 1994, Frank Craighead published *For Everything There Is a Season: The Sequence of Natural Events in the Grand Teton–Yellowstone Areas,* which describes the seasonal changes for plants and ecosystems that he observed during his 50 years in the field.

Upon John Craighead's retirement in 1977, he founded and directed the Wildlife-Wildlands Institute in Missoula, Montana, which plans, coordinates, and conducts research projects on wildlife and wildlands. Through the institute, John Craighead and his colleagues developed a system for botanically describing and mapping wilderness ecosystems and conducted a variety of research programs. The institute was renamed the Craighead Wildlife-Wildlands Institute, Inc., in 1987. John Craighead served as its president from 1987 to 1990 and currently serves as the chairman of its board of directors, responsible for setting the institute's goals and direction, developing policy, and soliciting financial support. His sons, John Willis and Derek, are director designate and president, respectively.

The Craigheads have been honored for their contributions to conservation through innovation and leadership in scientific inquiry, public education, and promotion of science-based environmental policy. They have shared many accolades for their work, including the 1970 Distinguished Alumnus Award and the 1973 Alumni Fellow Award from Pennsylvania State University; the National Geographic Society's John Oliver La Gorce gold medal in 1979; and the 1988 National Geographic Centennial Award, which recognized the "fifteen individuals who symbolize the best in their fields." John Craighead received the Aldo Leopold Award from The Wildlife Society in 1998.

Frank Craighead resides in Moose, Wyoming; John Craighead resides in Missoula, Montana.

Bibliography

"Craighead Environmental Research Institute," http://www.grizzlybear.org/; Devlin, Sherry, "John and Frank Craighead," *Missoulian's 100 Montanans, Our Pick of the Most Influential Figures of the 20th Century*, 2000; Puckett, Karl, "John and Frank Craighead," *Great Falls Tribune 100 Montanans of the 20th Century*," 1999; Stroud, Richard, *National Leaders of American Conservation*, 1986; Weaver, John L., "John and Frank Craighead," *Wildlife Society Bulletin*, 1996.

Cronon, William

(September 11, 1954–)
Environmental Historian

One of the leading American environmental historians of his generation, William Cronon has written extensively on a variety of topics ranging from indigenous land use in New England to problems with the modern environmental movement. As Frederick Jackson Turner Professor of History, Geography, and Environmental Studies at the University of Wisconsin at Madison, one of the elite universities for environmental studies, Cronon's academic work in environmental history demands the respect of his peers and influences both social and natural scientists whose focus is environmental studies.

William John Cronon was born September 11, 1954, in New Haven, Connecticut, to E. David, a historian, and Mary Jean Hotmar Cronon, a nurse. He grew up in Madison, Wisconsin, and studied at the University of Wisconsin at Madison, where he majored in English and history. He graduated with honors in 1976 and won a Rhodes Scholarship to study at Oxford University for two years, between 1976 and 1978. Returning to the United States, he completed his M.A. and M.Phil. at Yale University in 1979 and 1980 respectively. Between 1976 and 1982, Cronon was a Danforth Fellow. Cronon then went back to Oxford University, where he earned his D.Phil. in 1981. Back in the United States, Cronon taught western American and urban history at Yale University and earned his Ph.D. in 1990. He taught at Yale for over a decade before becoming the Frederick Jackson Turner Professor of History, Geography, and Environmental Studies at the University of Wisconsin at Madison in 1992, a position he still holds.

Cronon has written extensively. His first book, *Changes in the Land: Indians, Colonists, and the Ecology of New England* (1983), was a study of how the New England landscape changed with the arrival of Europeans to the region and their colonization. In 1984, the work was awarded the Francis Parkman Prize of the Society of American Historians. In 1991, Cronon's *Nature's Metropolis: Chicago and the Great West*, was published. A study of Chicago's relationship with the natural-resource-rich American West, the book won several prizes, including the *Chicago Tribune*'s Heartland Prize for the best literary work of nonfiction published during the previous year (1992), the esteemed Bancroft Prize for the best work in American history (1992), the George Perkins Marsh Prize from the American Society for Environmental History (1993), and the Charles A. Weyerhauser Award from the Forest History Society (1993). *Nature's Metropolis* was also one of three nominees for the Pulitzer Prize in history in 1992.

His 1995 essay, "The Trouble with Wilderness or Getting Back to the Wrong Nature," sparked considerable controversy among academics and environmentalists alike. Cronon argued that wilderness was nothing but a cultural construction and examined the implications of different cultural ideas of nature for modern environmental problems. Poet and naturalist Gary Snyder, among others, took issue with Cronon's contention that nature and wilderness were cultural constructions. Cronon is currently at work on several projects, including a local history of Portage, Wisconsin, which will explore ways of integrating environmental and social historical methods with nontraditional narrative literary forms. He is also completing a book about writing nonfiction and an anthology of first-person accounts describing life on the land in different parts of the United States.

Along with numerous teaching prizes, Cronon has also served as president of the American Society for Environmental History and continues to serve as general editor of the Weyerhauser Environmental Books Series for the University of Washington. During the spring semester of 1994, Cronon organized and chaired a faculty research seminar on the theme of "Reinventing Nature" at the University of California's Humanities Research Institute in Irvine, California. In January 1996, he became director of the Honors Program for the College of Letters and Science at the University of Wisconsin at Madison, and he has most recently become the founding faculty director of the new Chadbourne Residential College at the University of Wisconsin at Madison.

Avocationally, Cronon enjoys hiking, backpacking, swimming, and cross-country skiing. He is also active in the Wilderness Society. Cronon continues to live and teach in Madison, Wisconsin.

Bibliography

Cronon, William, "A Place for Stories: Nature, History, and Narrative," *The Journal of American History*, 1992; Cronon, William, "The Uses of Environmental History," *Environmental History Review*, 1993; Cronon, William, ed., *Uncommon Ground: Rethinking the Human Place in Nature*, 1995.

D

Daly, Herman

(July 21, 1938–)
Economist

Economist Herman Daly has challenged many assumptions of traditional economics regarding the idea that supply is infinite and that growth can continue indefinitely. He is author or coauthor of several influential books, including *Steady State Economics* (1977), *For the Common Good: Redirecting the Economy towards Community, the Environment, and a Sustainable Future* (1989), and *An Introduction to Ecological Economics* (1997). He believes that economic restrictions should be imposed that would limit resource consumption, discourage waste, and ultimately lead to economically and ecologically sustainable societies.

Herman E. Daly was born on July 21, 1938, in Houston, Texas, where, as a child, he was exposed to the rich cultural diversity of that city. He developed a special interest in Latin American culture and decided he wanted to help poorer, developing countries achieve a standard of living equal to that of the United States. He felt that the field of economics would be the ideal vehicle for him to help to bring about this change. He attended Rice University, receiving a B.A. degree in 1960, and continued his education at Vanderbilt University, where he earned a Ph.D. in 1967. Meanwhile he had accepted a position as assistant professor of economics at Louisiana State University in Baton Rouge. Daly taught at Louisiana State University, first as an assistant professor and later as full professor, from 1964 until 1988.

In 1967, Daly traveled to the northeastern Brazilian state of Ceará as a Fulbright scholar and Ford Foundation Visiting Scholar to teach economics at the University of Fortaleza. It was while in Brazil that Daly experienced an economic and ecological awakening. His eyes were opened to the real and dramatic danger posed by population growth in Brazil, and he came to the realization that, just as surely as there is a limited number of bodies the earth can support, there must be a similar limit to the amount of sustainable commodities available as well. He realized, in short, that there must be a limit to economic growth. This was a revolutionary realization, coming as it did from a person who up until this point in his life had been a "traditional economist," which is to say, an economist who believes that the solution to every problem is growth. In a 1980 interview with *The Mother Earth News*, Daly stated that "most economists do agree on the basic theory that the economy is a machine that continually needs to be fueled and made bigger." Daly's idea that economic growth cannot be sustained indefinitely is in direct opposition to the traditional economic perspective.

Daly returned to the United States in 1968, and in 1973 he edited his first book, a collection of essays entitled *Toward a Steady State Economy*. This book attempted to expose the fundamental fallacy inherent in traditional, as Daly put it in his 1980 interview, "growthmania" economics. Daly expanded further on this theme in 1977 with the publication of

Steady State Economics. Daly's version of economics takes into account such factors as sunshine, population growth, and the first and second laws of thermodynamics (energy cannot be created or destroyed, and everything in the universe trends toward disorder). Daly uses these limiting factors to suggest practical policy changes that would create an economy based on sustainability rather than growth. To achieve sustainability, he believes, we should establish economic and social limits that reflect the natural world's limits. For example, we should recognize solar power as the ultimate source of all energy, calculate what it costs to produce a given unit of usable solar energy, and use that figure to determine the cost of energy derived from depletable sources such as coal and oil. He finds it completely illogical that the current economy makes solar power uneconomical in comparison with the rapidly extracted "sunshine of Paleozoic summers" (fossil fuels), when in reality, solar power is the most economically and environmentally sound power source we have at our disposal.

Daly left Louisiana State University in 1988 for a position with the World Bank as senior economist. Given his visionary and heretical views on economics, many thought that Daly would not last long at the World Bank. However, he stayed for six years, until 1994. While at the World Bank, Daly introduced the idea that natural resources such as forests, soils, clean water, and clean air should be considered "natural capital" and should not be "spent" by developing countries in order to make a quick profit. He criticized the Gross National Product as an inaccurate indication of a country's prosperity, introducing the concept of the Index of Sustainable Economic Welfare as a more accurate indication of a country's economic well being.

In a speech he made just before leaving the World Bank in 1994, Daly prescribed a series of remedies for what he saw as the 50-year-old institution's "middle-aged infirmities." He recommended that labor and income be taxed less, because they ought to be encouraged, but that the consumption of energy and the depletion of natural resources be taxed more, in order to discourage them. We should, in other words, make it less expensive to earn and more expensive to waste. Another of his recommendations was that we should move away from the ideology of international free trade and, instead, begin to think in terms of national production for internal markets. Daly also suggested that natural capital be treated as monetary capital, that it be invested in and increased.

Since 1994, Daly has served as a senior research scholar at the University of Maryland's School of Public Affairs. He is the author of several important books in addition to *Steady State Economics.* In 1980 *Economics, Ecology and Ethics* came out, and in 1989 he published *For the Common Good: Redirecting the Economy towards Community, the Environment and a Sustainable Future* (coauthored with JOHN B. COBB JR. and Clifford W. Cobb). This book, directed toward a general audience, offers a profound critique of traditional economics and advocates a more environmentally sustainable and humanistic approach. In 1997, he wrote *An Introduction to Ecological Economics,* which was coauthored with several others. This book describes the emergence of the field of ecological economics as a "transdisciplinary" science whose practitioners dedicate themselves to understanding the interrelationships between natural and human systems in an attempt to encourage sustainable human activities. Daly thinks of economics as a branch of biology, since both fields focus on the exchange of resources between organisms. Ecology and economics, he explains, are both derived from the same Greek root word of *oikos*, which means "household." Where ecology is the study of the "household," economics is its management.

Daly is married and has two children.

Bibliography

Logan, William Bryant, "What Is Prosperity?" *Whole Earth Review*, 1995; Meadows, Donella, "Daly Medicine: The End of the World Bank as We Know It?" *Amicus Journal*, 1994; Stone, Pat, "Herman E. Daly Steady State Economics," *Mother Earth News*, 1980.

Darling, Jay Norwood "Ding"

(October 21, 1876–February 12, 1962)
Cartoonist, Cofounder of the National Wildlife Federation

Pulitzer Prize–winning cartoonist Jay Norwood "Ding" Darling is known in conservation history for his efforts to restore wildfowl habitat during the 1930s and as one of the founders of what is now the National Wildlife Federation.

Jay Norwood Darling was born on October 21, 1876, in Horwood, Michigan. The son of a minister who moved his family to various towns in Michigan, Indiana, and Iowa, Darling grew up exploring the countryside and hunting ducks in lush midwestern wetlands. He was seduced by the power of cartoons as an eight-year-old, when his father received a postcard with a friend's hand-penned cartoon lampooning the ministry. Darling watched his father laugh harder than he had ever seen him laugh before. The boy copied the cartoon and filled many notebooks with cartoon exercises from a correspondence course. As a student at Beloit College in Beloit, Wisconsin, Darling was suspended for a year after the school yearbook published his cartoon of the faculty as young ballerinas lined up at the barre. Darling did graduate with a B.A. from Beloit in 1900; later the school also bestowed on him an honorary Doctorate of Letters.

After graduating from Beloit, Darling immediately went to work for the *Sioux City Journal* as a reporter, photographer, and cartoonist. Quickly he distinguished himself as a talented cartoonist. He moved to the *Des Moines Register and Leader* in 1906, introducing himself with a cartoon about that city's serious air pollution from the soft coal burned in furnaces: a monk (*moine* in French means "monk") smoking a "soft coal" pipe and blowing big black smoke rings. Apart from a short episode in New York City with the *Globe* syndicate, Darling remained with the *Register and Leader* (which later became the *Register)* for the rest of his cartooning career, selling car-

Jay Norwood "Ding" Darling (Bettmann/Corbis)

toons to syndicates and appearing, at his peak, in 300 newspapers. Darling's pen could be acid, his opinions strong. His ability to distill national problems into simple attention-grabbing images made him a force that all major politicians had to recognize and reckon with. Though his subjects were mostly political, Darling frequently delved into the environmental issues of the day: erosion, industrial contamination of waterways, and greedy hunters and anglers. A survey of newspaper editors named him the country's best cartoonist in 1934, and he won two Pulitzer prizes, in 1923 and 1942.

Darling, a Republican, was critical of Pres. FRANKLIN D. ROOSEVELT's New Deal social policies as well as his lack of response to the crisis that threatened Darling's beloved ducks. The dustbowl conditions of the early 1930s had dried up the wetlands that waterfowl used as breeding and feeding grounds, and their numbers diminished drastically. Although Darling supported outspoken prowildlife advocates such as WILLIAM TEMPLE HORNADAY and ROSALIE EDGE, who called for shortening the hunting season and cutting bag limits, he restrained himself from blatant criticism, because responsibility for this problem fell to Roosevelt's secretary of agriculture, Henry Wallace, a fellow Republican and a close friend. In 1933, Wallace invited Darling to serve on a special Presidential Commission on Wildlife Restoration, charged with devising a strategy for saving the ducks. The committee, which included wildlife management expert ALDO LEOPOLD, issued recommendations to budget $17 million to purchase several million acres of submarginal farmland for conversion into wetland waterfowl refuges. A few months later, in the spring of 1934, Darling was invited to become chief of the U.S. Biological Survey and try to implement this recommendation. Despite the huge pay cut entailed (Darling was then earning $100,000 for his cartoons, and the Biological Survey chief earned $8,000), Darling accepted the challenge.

Conservation historians still speculate that Roosevelt offered Darling this post to quell Republican conservationist critics. While that might have been the case, during his 20 months in the post Darling was allotted $14.5 million to establish 19 major wildfowl refuges and 13 secondary ones, on a total of 840,000 acres of reclaimed farmland. Roosevelt, who at the same time was trying unsuccessfully to wrestle Congress for money to rebuild inner-city tenements and fortify New Deal welfare programs, was envious of

Darling's congressional support. But for Darling's ambitious save-the-duck plan to succeed, this was not enough. He raised more money through his one-dollar duck stamps that hunters had to buy and affix to their hunting licenses. The gun and ammunition industries offered more support, with a pledge of 10 percent of their annual sales to raise a ten-million-dollar endowment fund for wildlife protection. Darling applauded, but Roosevelt declined the offer because it was conditional on canceling the 10 percent federal excise tax on all guns and ammunitions sold in the United States. Disappointed, Darling resigned and returned to cartooning in the fall of 1935.

Darling remained active in wildlife conservation. He helped form the American Wildlife Institute (AWI) as a lobby for all industries that benefited from hunting, including the gun and ammunition industries, automobile manufacturers, and oil and railway companies. Chaired by the top executives of these companies, the AWI favored restoration of wildlife habitat and game management and gave research money to 10 land-grant colleges to promote scholarship in these fields. The AWI, with support from Roosevelt's staff and endorsements from 36,000 wildlife groups, organized the largest-ever wildlife conference in February 1936 in Washington, D.C. The conference resulted in the founding of the General Wild Life Fund (GWLF), whose new membership immediately elected Darling as president. He led the GWLF as it successfully pushed Congress to pass the 1937 Pittman-Robertson bill, essentially the same plan as the one the guns and ammunition industries had proposed two years earlier. GWLF changed its name to the National Wildlife Federation (NWF) in 1938. Disenchanted with the NWF because it acted too often as a lobbying arm for the gun industry, Darling moved on to become more active in the National Audubon Society. During the remainder of his life, Darling focused on two conservation projects in particular: the preservation of the Lewis and Clark Trail and the protection of Captiva and Sanibel Islands off the Gulf Coast of Florida, where he spent his winters with his wife, Genevieve Pendleton, and their two children. A wildlife refuge on Sanibel Island was named for him after his death.

Darling continued his cartooning until illness debilitated him in 1949. He died in Des Moines on February 12, 1962, after a series of strokes.

BIBLIOGRAPHY

Fox, Stephen, *The American Conservation Movement: John Muir and His Legacy*, 1981; Lendt, David, *The Life of Jay Norwood Darling*, 1989; Mahoney, Tom, "How to Be a Cartoonist," in John E. Drewry, ed., *More Post Biographies*, 1947.

Dawson, Richard

(August 2, 1935–)
Educator

Richard Dawson is considered a founding father of environmental education, especially in Kansas City, where he has taught biology and environmental science since 1958. Recipient of numerous awards for excellent teaching, Dawson is an innovator. He set up outdoor ecological laboratories for high school students in the 1960s, designed high school curricula in the areas of world futuristics and bioethics, and is known for a unique approach to science teaching whereby he asks his students to hone their observational skills by writing poetry.

Richard Glen Dawson was born on August 2, 1935, in Columbia, Missouri. An artist uncle gave him a book of John James Audubon's bird paintings while Dawson was in grade school, which helped inspire his lifelong fascination with nature. As a high school student he led field trips for the Burroughs Nature Club, Kansas City's Audubon Society, and developed a nature center at the local Camp Lake of the Woods, where he fell in love with teaching children about "critters."

Dawson attended Carleton College in Northfield, Minnesota, which is endowed with a 700-acre arboretum. He transformed the footpaths through the "Arb" into self-guiding nature trails, with numbered stakes and a monthly descriptive booklet to point out seasonal plants and other wildlife, and he painted interpretive habitat displays that were erected along the trails. Dawson met his wife, Ellie, also a nature educator, when she came to a meeting of the Carleton Natural History Club. Dawson, who was president of the club, invited her to accompany him on setting up the next month's guide. They married in 1959 and later had two daughters, Andrea and Carolyn.

Following his graduation from Carleton with a B.A. in biology (1957), Dawson earned an M.S. in biology from the University of Michigan (1958) and returned to Kansas City to teach at Shawnee Mission North High School. By the mid-1960s, Dawson was already distinguishing himself as a tireless and inventive biology teacher. He established an Environmental Science Camp at a 300-acre outdoor recreation site in Kansas City's Swope Park in 1964, the first of its kind in the region, and directed summer programs there for 18 years until the tax money that funded it dried up. Dawson's unique summer camp programs combined regular camp activities such as horseback riding, archery, cookout/campouts with ecology activities, with staff-led habitat studies and campers' individual research projects culminating in presentations for parents. In 1965, Dawson led an effort to convert a former restaurant into the Lakeside Nature Center, Kansas City's first public environmental education center. During the school year, Dawson held class on the ecologically valuable vacant lands near Shawnee Mission North and invited students on camping trips in Swope Park on weekends. Dawson proposed a formal outdoor laboratory composed of 24 acres of forest, fields, pond, and stream adjacent to Shawnee Mission South High School in 1968. When the school board agreed,

Dawson transferred to Shawnee Mission South and set up the Shawnee Mission Environmental Science Laboratory.

In addition to his work setting up areas for outdoor environmental education, Dawson is recognized for the classroom curricula he has developed. In his "Science and Survival" course, students study contemporary science and environmental issues, form opinions about pertinent public policy, and then write letters to elected officials and newspapers. The "World Futuristics" course that Dawson designed in 1974 was the first of its kind in the United States. Asking students to imagine the type of world they wanted to bequeath to their children, Dawson encourages long-range planning skills and works with issues such as resource availability, pollution, and population dynamics, using computer-generated forecasting graphs, simulation games, trend extension, cross-impact matrices, and more techniques. In 1992, Dawson was invited by the Woodrow Wilson Fellowship Foundation to work with a team at Princeton University to design curricula for incorporating bioethics issues into biology courses to teach critical thinking and decision making in areas such as genetics, reproduction, and environmental problems.

Dawson's innovations and teaching methods have earned him numerous professional awards, ranging from Outstanding Biology Teacher Award of the Midwest region from the National Association of Biology Teachers in 1968, to Kansas Master Teacher Award from Emporia State University in 1986, to a spot in the Mid-America Education Hall of Fame at Kansas City Kansas Community College (1998).

Nominators for the Hall of Fame recognition describe another of Dawson's approaches to teaching. He asks his students to write poetry. "The young people are encouraged to examine leaves, insects, everything amazing in the world of nature, and then to express the emotions evoked," writes literature teacher Rowena Unger Turk on Dawson's nomination form. Dawson publishes booklets of student poetry, and he shares his own poetry with his students as well. From his booklet *Buffalo Chip:*

Black Oak
Summer's warm rushing winds
Twist and crack the brittle trunk
Until the oak's crown crashes.
Vulnerable now, the fractured bole
Softens to the fungus touch
For woodpecker carpentry.
Pewee, Nuthatch, Chickadee
Inheriting these holes
Protect unbroken trees.

Bibliography

Dawson, Richard, *Buffalo Chip and Other Biological Droppings*, 1993; Dawson, Richard, "A Natural History Camp for Pre-Teens," *The American Biology Teacher*, 1967; Hoskins, Alan, "Environmental Pioneer Dawson Selected to Education Hall of Fame," *Kansas City Kansan*, 1998.

DeBonis, Jeff

(March 18, 1951–)
Forester, Founder of Association of Forest Service Employees for Environmental Ethics and Public Employees for Environmental Ethics

As a U.S. Forest Service employee who spoke out in 1989 against the ecologically destructive forest management practices of the agency, Jeff DeBonis was a pioneer whistle-blower. Since then, he has organized two nonprofit organizations, the Association of Forest Service Employees for Environmental Ethics (AFSEE) and Public Employees for Environmental Ethics (PEER), both of which organize and support environmental activism and whistle-blowing among public employees.

Jeffery Nicholas DeBonis was born on March 18, 1951. He grew up in Bedford, Massachusetts, where the Shawsheen River ran behind his house. He spent a significant amount of time on and along the river as a child. When his family arrived in Bedford, the river was healthy and full of native fish such as eastern pickerel, shiners, sunfish, and brook trout. DeBonis remembers that pollution had killed the river by the time he was in 11th grade.

DeBonis attended Colorado State University in Fort Collins, Colorado, graduating in 1974 with a B.S. degree in forestry. He spent his next three years in El Salvador with the U.S. Peace Corps, where he worked with El Salvadoran government agencies, nongovernmental organizations (NGOs), and private citizens on reforestation and soil conservation projects. He lived in remote villages, instructing small groups of farmers on how to utilize appropriate soil conservation and forest management techniques on their land. He also helped to develop a large-scale experimental reforestation project in conjunction with the United Nations and NGOs. The purpose of this project was to reintroduce native hardwood species into degraded, clear-cut forest lands. His experiences with the massive clear-cuts of El Salvador made a lasting impression on him.

After returning to the United States, DeBonis began his career with the U.S. Forest Service in 1978. He would remain with the forest service until 1991. His career as a forester has been well documented in Todd Wilkinson's *Science under Siege*, in a chapter revealingly entitled "Confessions of a Timber Beast." Wilkinson writes, "During a career that spanned thirteen years, DeBonis had a role in delivering billions of board feet of lumber into the laps of local timber mills in the Pacific Northwest. . . . His decisions brought down tree trunks a millennium old, killed fish, displaced grizzly bears, and spotted owls, carved up mountainsides into rectangles of visual blight and cost taxpayers at least tens of millions of dollars in losses through publicly subsidized road construction." During his time with the U.S. Forest Service, DeBonis was a timber-sale planner. It was his job to decide how big the clearcuts would be and where the trees would fall.

DeBonis, however, will not be remembered for his role in harvesting national forests in the United States. He will be remembered for speaking out against the ecologically unsound practices of the U.S. Forest Service and for attempting to

Jeff DeBonis (Dan Lamont/Corbis)

reform this bureaucracy from within. According to DeBonis, the beginnings of his apprehension about the U.S. Forest Service management policies date back to 1978 when he was a forester trainee on the Kootenai National Forest out of Troy, Montana. The clear-cutting of forests, the resulting erosion, and the decimation of trout fisheries reminded him of El Salvador. He was also concerned with the Forest Service management imperative to produce timber outputs, creating a system that rewarded those who cut down more trees.

The turning point in DeBonis's career came in 1989. After attending a seminar in Eugene, Oregon, on ancient forests, he wrote a two-page memorandum and distributed it throughout the Forest Service using the agency's computerized communications system, its equivalent of e-mail. In the memo, DeBonis attacked the Forests Service's position as "an advocate of the timber industry's agenda," calling for greater attention to conservation and stewardship. He also stated that the Forest Service should look to the "conservation community" for assistance "in developing a strategy which will contribute to an ecologically sustainable lifestyle in the 21st century."

This memo caused quite a stir throughout the agency and in the timber industry. DeBonis had broken an unspoken law of the Forest Service in speaking out against its policies. He was not alone, however, in feeling that the Forest Service should reevaluate its practices. In the months following the distribution of his memo, DeBonis was contacted by hundreds of fellow Forest Service employees who wanted to speak out but were afraid. The Association of Forest Service Employees for Environmental Ethics grew out of his organizational efforts during this time. DeBonis had become a self-described "whistle-blower," intent on exposing (in his own words) "the Forest Service's willful violation of the spirit and intent of environmental laws that resulted in extensive over-cutting of national forests." He organized his fellow employees around these issues, encouraging diverse opinions and freedom of speech in an agency not known to foster such behavior.

DeBonis acted as executive director of AFSEE (which has dropped the *A* and is now simply known as FSEE) until 1991, when he left the Forest Service to concentrate on environmental advocacy through nonprofit organizations. In 1992, he left AFSEE and founded another nonprofit group, Public Employees for Environmental Ethics. PEER is an advocacy-oriented organization that works to protect the environment through organizing and supporting public employees. PEER has supported dozens of environmental whistle-blowers throughout the United States and has helped to organize hundreds of public employees as environmental activists in federal and state enforcement and land management agencies. Under the direction of DeBonis, PEER also worked on many environmental issues, challenging enforcement of national and state environmental protection laws, as well as combating overgrazing and overcutting on federal lands, threats to endangered species, toxic pollution, and corporate timber theft. PEER has also sued and won numerous court decisions on behalf of employee activists and the environment, conducted advocacy surveys of entire agencies, and disclosed environmental damage and illegal activities in numerous federal and state agencies. PEER is unique among environmental organizations in that it provides inside information, often from anonymous sources, the sources themselves being the ones required by inappropriate agency policies to cause environmental degradation. PEER has provided a forum where these policies can be brought to light by those most familiar with their ill effects and has given a voice and an opportunity to act to countless environmentally concerned public employees.

DeBonis resigned his position as executive director of PEER in 1997. Since then he has spent his time renovating a house in Hood River, Oregon, with his wife, Susan, and acting as an organizational consultant to nonprofit organizations in the environmental community. DeBonis has received many awards for his advocacy and environmental accomplishments. Included among these are the National Conservation Achievement Award from the National Wildlife Federation, 1996; the Environmental Leadership Award from the California League of Conservation Voters, 1995; the Alliance for the Wild Rockies Conservation Award, 1989; and the Giraffe Award for Environmental Whistleblowing from the Giraffe Foundation, 1989.

Bibliography

Ervin, Keith, "Voice of Doubt Sickened by Timber Practices Condoned by the Government, Forester Spoke Out," *Seattle Times*, 1990; McLean, Herbert, "A Very Hot Potato," *American Forests*, 1990; Nilsen, Richard, "Reforming the Forest Service from Within," *Whole Earth Review*, 1989; Wilkinson, Todd, *Science under Siege*, 1998.

Desser, Chris

(September 24, 1954–)
Attorney, Cofounder of Muir Investment Trust, Project Director for Migratory Species Project

Chris Desser has been active in the environmental movement for her entire professional career. She cofounded the Muir Investment Trust and has practiced land use and environmental law, acted as project director for the

Migratory Species Project since 1993, and served on numerous boards of directors and advisory boards for foundations and nonprofit organizations with environmental foci. As executive director of Earth Day 1990, she redefined grassroots environmental organizational strategies and opened the eyes of environmentalists to the possibilities afforded by effective use of the media. As coordinator of the Funders Working Group on Technology, she is deeply engaged in the health, safety, environmental, and policy issues implicated in biotechnology.

Christina Louise Desser was born on September 24, 1954, in Los Angeles, California. She was the first of two children born to Alan and Shirley Desser. She has a younger brother, Jim. As a child and young adult her experiences in the natural world shaped her perspective and commitment to environmental activism. She skied, rock climbed, and backpacked extensively in the Sierra Nevada and in the Rocky Mountains, swam in mountain lakes, and body surfed in the Pacific Ocean. Desser traveled frequently during her formative years, with her family and on her own. She still does. She has had some penetrating and heartbreaking experiences in her travels. During the 1960s in coastal Mexico, she saw the remains of slaughtered sea turtles floating in the Sea of Cortez. In Yalta, in the 1970s, she took a midnight swim in the Black Sea; when she returned to the same spot the next day, she found an oil sheen polluting the water she had been swimming in.

Desser was in high school during the first Earth Day celebration in 1970. This event underscored for her the threats facing the environment and demonstrated to her the importance of the political process to environmental change. In 1972, motivated to help the effort to end the Vietnam War, Desser joined George McGovern's presidential campaign. She was the youngest member of the national staff. In this campaign she began to develop her political organizing and media skills. She remained politically active throughout her time at the University of California at Berkeley, where she studied philosophy and rhetoric. In 1978 she took two years off from school to work at *Rolling Stone* magazine and to serve as assistant to the director of ACTION, a government agency, now defunct, that oversaw the Peace Corps and Volunteers in Service to America (VISTA). She graduated from Berkeley in 1978 with honors. Desser studied law at the McGeorge School of Law at the University of the Pacific. During this time she interned at the Environmental Defense Fund and wrote a manual about California's pesticide regulations. She received her J.D. degree in 1983 and went to work with Manatt, Phelps, Rothenberg and Tunney in Los Angeles, practicing land use and environmental law until 1987. She then served as deputy city attorney for land use in San Francisco until 1989. According to Desser, practicing environmental law deepened her understanding of how the political process can be brought to bear on environmental issues.

In 1989, Desser left law to organize Earth Day 1990. Earth Day afforded her an opportunity to use the power of the media and political-campaign-style organizing to mobilize a large environmental constituency. This approach, novel at the time, was dramatically successful. Earth Day 1990 culminated in the largest grassroots demonstration in history and confirmed the potential of political organization to those in the environmental movement, which, owing to Earth Day 1990's example, has since become more savvy about the power and uses of media and organizing by constituency.

In 1991, Desser cofounded the Muir Investment Trust, the first socially responsible and environmentally sound municipal bond mutual fund. She remained with the Muir Investment Trust until 1992. Since 1993, she has served as project director of the Migratory Species Project in San Francisco. This project focuses on promoting the understanding of interconnection and interdependence by linking communities along important migratory routes. Desser became the coordinator of the Funders Working Group on Biotechnology in early 1999. This group concerns itself with the myriad environmental and health issues raised by biotechnology, as well as its threat to democracy and dissent. Also in 1999, California governor Gray Davis appointed Desser to the California Coastal Commission, a twelve-member commission responsible for managing the conservation of California's coastal resources through a comprehensive planning and regulatory program. In her capacity as a coastal commissioner, Desser is committed to limiting coastal development and preserving the coastal and marine environment.

Desser has said that she experiences violence to the environment in an almost physical way, a way that literally requires action of her. When she walks through a clear-cut forest or stands before an oil-fouled ocean, it pains her so deeply, and with such awareness, that she cannot not do something. She lives in San Francisco, California, with her husband, Kirk Marckwald, whom she married in 1986. She has written numerous articles. She served as a delegate to the United Nations PrepCom for the United Nations Conference on Environment and Development (UNCED) in 1991 and continues her decade-long membership with the International Forum on Globalization.

Bibliography

"California Coastal Commission, Commissioner Biographies," http://ceres.ca.gov/coastalcomm/bios.html; Desser, Christine, *California's New Pesticide Regulations and You. A Guide to What They Say. What Your Rights Are. What You Can Do*, 1982; Desser, Christine, "Making the Connections," *Inquiring Mind*, 1996; Desser, Christine, "Transgenics: Unnatural Selection or Bad Choice," *Wild Duck Review*, 1999; "Migratory Species Project," http://www.well.com/user/suscon/esalen/participants/Desser/statement.html; "Sustainability Consciousness, Chris Desser," http://www.well.com/user/suscon/esalen/participants/Desser/.

Devall, Bill

(December 2, 1938–)
Deep Ecologist, Activist

Bill Devall has dedicated his life and work to the protection of the environment, both actively and intellectually. Since his move to California in the late 1960s, Devall has been a regular participant in countless nonviolent demonstrations to preserve redwood forests, establish wilderness areas, and protect coastlines. Intellectually, Devall is one of the leading proponents in the United States of deep ecology, a biocentric movement that challenges practical and philosophical contemporary anthropocentric conceptualizations of resource management.

William Bert Devall was born December 2, 1938, in Kansas City, Kansas, to William and Marie (Culp) Devall at the tail end of the Great Depression. His father worked in the steel industry, and his mother, the first person in her family to receive a college education, was a schoolteacher who became a housewife. Devall was raised in the suburbs of Kansas City and was educated primarily during the quiet decade of the 1950s. Growing up outside of such cities as Kansas City and Denver and never spending much time away from the manicured suburbs, Devall's early impressions of nature bordered on the romantic and the mystical. His parents taught him to admire nature but from a distance. Their first family vacation, in 1947, was an auto trip through the Colorado Rockies. They looked at the scenery from the windows of the car but did not hike into the mountains or sleep outside under the stars.

Upon completion of his B.A. in sociology from the University of Kansas in 1960, Devall earned an M.A. in sociology at the University of Hawaii in 1962. He wrote his doctoral dissertation—on the Sierra Club—at the University of Oregon, where he received his Ph.D. in 1970.

Devall's first involvement in the conservation movement came when he went to Humboldt County, California, in 1968 to teach at Humboldt State University. There, he was inspired by DAVID BROWER, who was then executive director of the Sierra Club. He was also inspired by local Sierra Club activists who faced great adversity as they fought for the establishment of Redwood National Park. Devall's first real activism came two years later, in 1970, when he helped to organize local participation for the first Earth Day. From that point on, Devall was committed to conservation activism and has worked on almost every campaign for the protection of California wilderness, including its coastal land and forests. His activism has always been based on principles of nonviolent, direct action.

While developing his own thoughts on social science, philosophy, and ecology, Devall was heavily influenced by the writings of Paul Shepard, ALDO LEOPOLD, and JOHN MUIR. With his discovery of Norwegian philosopher Arne Naess's short, but highly influential, paper on a new theory called deep ecology, "The Shallow and the Deep, Long-Range Ecology Movement" (published in *Inquiry* in 1973), Devall found direction and a philosophical framework for his own ideas and thoughts. Along with GEORGE SESSIONS, with whom he has worked closely, Devall became an advocate for the deep, long-range ecology movement in the United States.

During the 1980s, Devall was active with the Earth First! movement through all of its various metamorphoses. He served as president of the Earth First! Foundation (later renamed the Fund for Wild Nature) from 1984 to 1989. As an academic, however, he was never really accepted by the self-proclaimed "rednecks for wilderness," the urban anarchists, or the feminists who influenced the movement during the latter half of the decade. A close friend and associate of DAVE FOREMAN, Devall resigned from the

Earth First! Foundation in 1990 at the same time that Foreman left the movement. Around the same time, he and Foreman were invited by Doug Tompkins, an environmentalist and outdoors clothing manufacturer, to discuss the need for a foundation to promote the deep, long-range ecology movement. Tompkins then established the Foundation for Deep Ecology, and Devall's association with the foundation led to his work on many projects sponsored by the foundation, including the 1993 publication of *Clearcut: The Tragedy of Industrial Forestry.* Devall is presently working as webmaster for the deep ecology web site, www.deep-ecology.net.

Devall retired from the Department of Sociology at Humboldt State University in 1995 but has remained active on several projects. His current work includes a review of the deep, long-range ecology movement, from 1960 to 2000, which will be published in a forthcoming issue of *Ethics and the Environment.* Devall is also working on a manuscript on forest fire, particularly how a 138,000-acre forest fire that burned much of the Trinity Alps Wilderness Area in northern California during the summer and fall of 1999 helped change perspectives and deepen the local community's understanding of ecology.

Devall dwells in Trinidad, California.

BIBLIOGRAPHY

Devall, Bill, *Bioregion on the Edge*, forthcoming; Devall, Bill, *Living Richly in an Age of Limits*, 1992; Devall, Bill, *Simple in Means, Rich in Ends*, 1988; Devall, Bill, and George Sessions, eds., *Deep Ecology*, 1985.

Devoto, Bernard

(January 11, 1897–November 13, 1955)
Writer, Historian

Bernard Devoto was a Pulitzer Prize–winning author, historian, and social critic. From his position as editor of "The Easy Chair" in *Harper's Magazine* from 1935 to 1955, he contributed greatly to the early conservation movement, fueling public sentiment that would eventually result in the preservation of Dinosaur National Monument and in such legislation as the 1964 Wilderness Act. He was named an honorary lifetime member of the Sierra Club.

Bernard Augustine Devoto was born January 11, 1897, in Ogden, Utah, to Florian and Rhoda Dye Devoto. His family was a microcosmic representation of the small community in which it lived. His mother was Mormon, his father Catholic, and Devoto, as it would later turn out, a serious practitioner of neither religion. Devoto's education began at a convent school and continued in public schools. He graduated from Ogden High School in 1914. Growing up, he experienced all of the freedoms that go along with a frontier childhood. He spent his time hiking, climbing, and camping in the wilderness of northern Utah.

Bernard Devoto (Bettmann/Corbis)

Devoto attended the University of Utah for one year (1914–1915), where he helped to organize a chapter of the intercollegiate Socialist Society, quickly disbanded by the university administration. This event, along with the dismissal of four unorthodox faculty members, disgusted Devoto, and he left for Harvard in 1915. His focus at this time was primarily on philosophy, though he studied writing and continued to act as an intellectual revolutionary. Upon the declaration of war, Devoto enlisted and was commissioned a lieutenant of infantry in 1918. He was not sent overseas. Instead, he spent the next two years, until the armistice, at Camp Perry in Ohio teaching marksmanship. In 1920, he returned to Harvard and received his degree "as of the class of 1918."

After graduating, Devoto returned to Utah. In 1921, he took a job as a teacher at the North Junior High School in Ogden. He was definitely not what people had come to expect from a junior high school teacher in Ogden, Utah. His sharp tongue and his willingness to use profanity fascinated his students but shocked and offended their parents and his fellow teachers.

He accepted a position teaching English at Northwestern University in 1922. He taught at Northwestern for five years, gaining a reputation as one of the best teachers on campus and attracting large numbers of students to his classes, with his sarcastic showmanship and tough but fair grading practices. While there, he met and married Helen Avis MacAir, with whom he would have two sons, Gordon King and Mark Bernard. Devoto refused to study for a Ph.D., but his success with the students and his growing prominence as a writer brought him an assistant professorship in 1927. His first book, *The Crooked Mile*, had been published in 1924 and his second, *The Chariot of Fire*, in 1926. During these years, he also wrote many articles for such magazines as *Harper's* and *The Saturday Evening Post.*

Finding it difficult to continue balancing his teaching with his writing, Devoto resigned his professorship in 1927 and moved with his family to Cambridge, Massachusetts, planning to do nothing more than write. He alternated between whipping up potboilers under the pseudonym John August to pay the bills and producing serious works of social history and biography such as *Mark Twain's America*, 1932, and *We Accept with Pleasure*, 1934. He was an instructor and lecturer at Harvard from 1929 to 1936. And, in 1935, he became the editor of the "Easy Chair" of *Harper's*, a position he held for 20 years.

It was in this capacity that Devoto had his greatest impact as a conservationist. He used this platform to express his views on public lands and the federal responsibilities for their management. He compiled these views into a series of conservation essays and published them in 1952 in his four-page column, "The Easy Chair." In 1953, Devoto wrote four "Easy Chair" columns in relation to public-lands issues, going so far in one of them as to suggest that Congress close the national parks until it saw fit to allocate enough money to allow them to run properly. He lit many fires with his articles in *Harper's*, fires that would help lead to the passage of the Wilderness Act in 1964 and to other major conservation legislation during the Kennedy and Johnson administrations.

Devoto also contributed to the movement that would eventually erupt into the struggle to save Dinosaur National Monument in 1956. The essays and articles he produced in the name of conservation were works that he was proud of, because they validated his position as a professional journalist and as a controversialist. They also gave his reforming impulses, which can be traced all the way back to his early college days, a cause and a purpose. Wallace Stegner writes, "His conservation writings record a continuous controversy unmarred by any scramble for personal advantage or any impulse towards self justification, a controversy in every way dignified by concern for public good and for the future of the West."

Devoto did not live to see many of the successes of the conservation movement he helped to fuel. He died of a heart attack on November 13, 1955.

Bibliography

Bowen, Catherine Drinker, Edith R. Mirrielees, Arthur M. Schlesinger, and Wallace Stegner, *Four Portraits of One Subject: Bernard Devoto*, 1963; Burrows, Russell, *Bernard Devoto*, Western Writers Series, 1997; Stegner, Wallace, *The Uneasy Chair*, 1973.

Dilg, Will

(1867–March 8, 1927)
Founder of Izaak Walton League of America

Publicist and advertising man Will Dilg led 53 fishermen and hunters to found the Izaak Walton League of America (IWLA) in 1922. Although his time at the helm of the new organization was short, Dilg's charismatic leadership gave the IWLA the push it needed to become the first nationwide conservation organization and the largest one of its time.

Little is known of the early life of Williamson H. Dilg, who was born during the fall of 1867 in Milwaukee, Wisconsin. He was an advertising man and a publicist by profession, a fisherman by vocation. He contributed articles to hunting and fishing publications. According to a story cited by writer Stephen Fox, Dilg was moved to dedicate his life to preserving opportunities for young boys to enjoy outdoor experiences after his own son died. Dilg and 53 other sportsmen founded the Izaak Walton League of America with the goal of saving "outdoor America for future generations," according to the IWLA's official history. The problem they were most concerned about was the degradation of the country's best fishing streams due to contamination by industry and raw sewage and to sedimentation caused by soil erosion. They named their organization after the seventeenth-century British fisherman who wrote the conservationist classic *The Compleat Angler.* The IWLA was modeled on fraternal orders such as the Kiwanis Club. Within a few years of its founding it boasted 100,000 members, most of whom lived in the Midwest.

In August 1922, Dilg founded IWLA's publication, the monthly *Outdoor America* (renamed *Izaak Walton League Monthly* during its first year). Dilg wanted *Outdoor America* to appeal to a large, mainstream audience and was able to woo many famous American writers to contribute free articles. Dilg's editorials decried industrial civilization and yearned for the calm of nature. *Outdoor America* became the conservation movement's largest publication and served as a powerful recruitment tool.

In 1923, Dilg learned of a private developer's plan to drain the river bottoms of the Upper Mississippi basin. This was Chicagoan Dilg's favorite place to fish—he spent two months every year fishing there. The river bottoms were also the best spawning grounds for black bass, Dilg's favorite game fish. Dilg mounted an energetic and successful assault on this plan. He moved to Washington, D.C., with an army of IWLA staff members and set up an office in a suite at the New Willard Hotel. His goal was to have a 300-mile stretch of the Upper Mississippi declared a federal wildlife refuge. Dilg and his staff wrote the Upper Mississippi River Wild Life and Fish Refuge bill, recruited two fellow sportsmen legislators to introduce it, and lobbied hard to assure its passage. With help from secretary of commerce and veteran angler Herbert Hoover, Dilg convinced Pres. Calvin Coolidge to sign the bill. No expense had been spared on this project; the conservation movement had never before seen a conservationist effort of this magnitude.

The IWLA annual meetings were huge occasions in which Dilg and other members of the IWLA's brotherhood expounded on the glory of their struggle. Conservationists from the Audubon Society, the Boone and Crockett Club, and other organizations who attended the IWLA meetings were astounded. Their members were primarily of the privileged elite classes of the eastern seaboard, staid and dignified. But at the IWLA, the crowd was uninhibited. Dilg was not a member of the privileged elite; he had made his money in the brash business of advertising. At the annual meetings, compared by some observers to revivalist meetings, religious metaphors flowed freely, reports Stephen Fox: members compared Dilg to St. Paul, to Stephen rousing a crusade, to David fighting Goliath. Dilg, for his part, frequently referred to "the God of Nature" as his savior and the object of his passion.

But within a few years, Dilg's weaknesses became apparent to other leaders of the IWLA. Despite the strength of the organization and the wealth and generosity of a number of the board members,

Dilg's activities hemorrhaged the organization's treasury. Dilg had outspent the organization's budget by 100 percent. His lobbying efforts for the Upper Mississippi protection bill had cost more than any similar conservationist effort before it. *Outdoor America* ran a deficit too. To replenish its accounts, *Outdoor America* began to accept large advertisements from the gun and ammunition industry, which Dilg strongly opposed. But he was not in a position to impose his commitment to independence. Further weakening him was throat cancer, which had been diagnosed in 1924.

Other leaders of the IWLA decided that Dilg had to be deposed for the organization to continue. Dilg was forced out of *Outdoor America* in 1925, and at the annual meeting of April 1926 he was removed from his post as president, by a vote of over two-thirds of the members. He tried to fight but at this point was too weak. Dilg dedicated the remainder of his life in Washington, D.C., to efforts to convince the president to establish a Conservation Department in his cabinet.

The IWLA has continued its advocacy for clean water and healthy forests and was an important actor in much of the landmark conservation legislation of the twentieth century.

Dilg died of throat cancer on March 8, 1927, in Washington, D.C.

BIBLIOGRAPHY

Fox, Stephen, *The American Conservation Movement: John Muir and His Legacy*, 1981; "Izaac Walton League of America," http://www.iwla.org/; Stroud, Richard, ed., *National Leaders of American Conservation*, 1985.

Dingell, John, Jr.

(July 8, 1926–)
U.S. Representative from Michigan

John Dingell Jr. has been a member of the U.S. House of Representatives since 1955. Over the past 45 years he has been involved with the creation of many important pieces of environmental legislation, including the National Environmental Policy Act of 1970, the Marine Mammal Protection Act of 1972, the Endangered Species Act of 1973, and the Clean Air Act of 1990.

John David Dingell Jr. was born on July 8, 1926, in Colorado Springs, Colorado, to John David Dingell and Grace Bigler Dingell. The family moved to Detroit, Michigan, in the late 1920s. Dingell's father, a New Deal Democrat and an advocate of a national health insurance program, was elected to the U.S. House of Representatives in 1932. Dingell attended private elementary and secondary schools in Washington, D.C., and served as a page in the House of Representatives from 1938 to 1943.

From 1944 to 1946, Dingell served as an infantryman in the United States Army. He was discharged with the rank of 2nd lieutenant following the completion of World War II. Later that year, he entered Georgetown University. He graduated in 1949 with a B.S. degree in chemistry, and in 1952 he earned a law degree from Georgetown. In 1952, Dingell accepted a position as research assistant to U.S. circuit judge Theodore Levine. Dingell became assistant prosecuting attorney in Wayne County, Michigan, which includes Detroit and its suburbs, a position he occupied from 1954 to 1955. In September of 1955, Dingell's father died in the middle of his term in the House of Representatives. Dingell ran in a special election to fill the vacated position and was elected, at the age of 29, to succeed his father as representative of the Sixteenth Congressional District in Detroit.

Dingell began his career as representative by supporting strong civil rights legislation: introducing a bill in 1956 that led to the creation of the Civil Rights Division in the Justice Department and proposing legislation that prohibited segregated hospitals from receiving federal aid. He was especially active in environmental issues as well. In 1963, Dingell organized a group of House representatives in criticizing the United States Public Health Service's apparent unconcern with the ability of streams and lakes to support wildlife. Two years later, in 1965, Dingell helped to create a bill that required rural water and sewage treatment plants that were built with federal funds to comply with all federal water pollution standards.

In 1966, Dingell was appointed to the position of chairman of the Fisheries and Wildlife Conservation subcommittee of the Energy and Commerce committee of the House of Representatives. From this position, Dingell was able to secure important pieces of environmental legislation. He introduced the House version of what was to become the National Environmental Policy Act of 1970. He was floor manager of the Marine Mammal Protection Act of 1972, and he helped to draft the Endangered Species Act of 1973, a bill that requires the U.S. Fish and Wildlife Service to list threatened or endangered species and work toward the recovery of viable populations and also prohibits any federal projects from destroying the habitat of an endangered species. Dingell became chairman of the Commerce Committee's energy and power subcommittee in 1975, a move that marked his increasing prominence in the House of Representatives. Energy was an important national issue in the mid-1970s, owing to the Arab oil embargo of 1973 and 1974, and Dingell was in a key position to help formulate energy policy. He backed government regulations and price controls to conserve U.S. energy resources, and he was an important participant in creating the Energy Policy and Conservation Act, a bill that provided the president the authority to set domestic oil prices, ration gasoline, and establish energy conservation plans. Dingell also supported Pres. JIMMY CARTER's efforts to reduce energy consumption through a combination of taxes and other federal regulatory actions.

Dingell assumed the chair of the Energy and Commerce Committee (now called the Commerce Committee) in 1980, also acting as chair of its Oversights and Investigations subcommittee. From these positions Dingell began to wield significant power in the House of Representatives. The Commerce Committee was one of the House's most powerful, because of its far-reaching jurisdiction. It was concerned with everything from health and the environment to communications, consumer protection, and trade. The Oversights and Investigations subcommittee is important for its role as a "whistle-blower" on all manner of issues. Investigations conducted by Dingell and his staff revealed cases of wrongdoing, including spurious "administrative cost" billings at Stanford University, falsified research data published in a paper by a Nobel Prize–winning biologist, and the mishandling of blood supplies by the American Red Cross. One important investigation conducted by this committee under Dingell's direction was into the criminal misconduct of Environmental Protection Agency (EPA) personnel in managing the toxic cleanup money made available in the "Superfund" clause of the Comprehensive Environmental Response Compensation and Liability Act (CERCLA). This 1983 investigation led to the resignation of Pres. Ronald Reagan's EPA administrator, Anne Gorsuch, and to a perjury conviction for one of her deputies, Rita Lavelle.

Dingell's district is one of the most heavily industrialized in the country. It includes a large number of autoworkers and union members and is home to the main Ford automobile plant. So, while Dingell has been an important player in creating environmental legislation, he has also been an ardent supporter of the auto industry. He long opposed stricter air quality standards, understanding that such legislation would have a negative impact on his constituents. It was not until 1990 that shifting congressional interests and power led him to support a clean air bill for the sake of his own political survival. Dingell did help to author, and was a supporter of, the Clean Air Act of 1990. Dingell served as chair of the Energy and Commerce Committee until 1995. In the 107th Congress (1997–1999), Dingell was the ranking Democrat on the Commerce Committee. He has been elected to more consecutive terms than any other member of Congress in the House of Representatives. He lives with his second wife, Debbie Dingell, in Virginia with two of his four children from his previous marriage.

Bibliography

Cifelli, Anna, "Capitol Hill's One-Man Gauntlet," *Fortune*, 1985; Hook, Janet, "By Shifting Tactics on Clean Air, Dingell Guarded His Power," *Congressional Quarterly Weekly Report*, 1990; Noah, Timothy, "Corporate Watchdog." *Newsweek*, 1987; Raine, Harrison, and Gary Cohen, "Congress's Most Feared Democrat," *U.S. News and World Report*, 1991.

Dittmar, Hank

(January 12, 1956–)
Sustainable Transportation Advocate, Executive Director of Surface Transportation Policy Project, Chief Executive Officer of Great American Station Foundation

Hank Dittmar is president and chief executive officer of the Great American Station Foundation, which promotes community economic development and encourages sustainable transportation by transforming historic railroad stations into improved centers of transportation and commerce. In his previous position as executive director of the Surface Transportation Policy Project (STPP), he managed a successful campaign for the passage of the Transportation Equity Act of the 21st Century (TEA-21), the landmark 1998 federal transportation bill that emphasizes economically efficient and environmentally sustainable transportation policy. The bill dramatically increased funding for public transit, bicycle and pedestrian facilities, and train stations and mitigated problems caused by highways, including runoff, wildlife mortality, and the loss of scenic vistas and open space to sprawl.

Henry Eric ("Hank") Dittmar was born on January 12, 1956, in Oklahoma City, Oklahoma, and was inspired to become active in environmental issues on the first Earth Day, in 1970. He attended Northwestern University, receiving a B.S. in communication studies in 1976, and went on to earn a master's degree in community and regional planning from the University of Texas at Austin in 1980, where he contributed to the development of Texas's Coastal Zone Management legislation.

Dittmar was responsible for coordination of public transit systems in the nine-county San Francisco Bay area from 1980 to 1983 and worked as a transit planner for the municipal bus service of Santa Monica, California, from 1983 to 1984. He directed the Santa Monica airport from 1984 to 1989, where he developed the nation's most advanced airport noise ordinance. While managing the departments of legislation and finance for the metropolitan Transportation Commission in Oakland, from 1989 to 1993, he helped San Francisco replace the earthquake-damaged Embarcadero Freeway with a surface boulevard and a streetcar line.

In 1993, Dittmar became executive director of the Surface Transportation Policy Project, a national, nonprofit coalition of more than 200 organizations dedicated to ensuring that environmental policy and government investments help to support a strong economy, promote energy conservation, protect environmental quality, and build more equitable and livable communities. Specifically, STPP focuses on documenting and publicizing the tremendous quality of life impacts of auto dependency and personal and societal solutions to the problem. STPP publishes *Transfer*, a biweekly e-mail newsletter, and *Progress*, a bimonthly forum for transportation and community issues. It also publishes numerous reports in tune with its mission statement that "we emphasize the needs of people, rather than vehicles, in assuring access to jobs, services, and recreational opportunities." Among its reports (available for purchase on its web site) are "Road Work Ahead: Is Construction Worth the Wait?" (which reveals that road construction delays are not usually offset by the later gains in travel time), and "High Mileage Moms" (which describes the heavy driving routine of the typical homemaker). The coalition's policy publications have been widely credited with providing Congress with the intellectual underpinnings for both the 1991 Intermodal Surface Transportation Efficiency Act (ISTEA) and the 1998 TEA-21 legislation.

Dittmar was appointed by President Bill Clinton to the White House Advisory Committee on Transportation and Greenhouse Gas Emissions in 1994 and to the

President's Council on Sustainable Development's Metropolitan Working Group in 1996, for which he was chair. His major emphasis during his five years at STPP was to manage its campaign to pass TEA-21. President George Bush had signed ISTEA in 1991, which acknowledged aspects of transportation policy that had not been included in previous legislation. ISTEA represented "a shift from a highway building era to an era of managing our transportation system in a way that balances mobility and accessibility concerns with environmental priorities," according to a 1997 STPP report. Despite its high-reaching goals, however, a study by the General Accounting Office found that the Department of Transportation's research agenda did not meet ISTEA goals, especially in the areas of sustainable development and intermodalism (the use of more than one mode of transportation during a trip, driving an automobile to a train station, then taking a train to complete the trip, for example).

Under Dittmar's leadership, STPP responded to this weakness by organizing a workshop for experts to develop a proposed research agenda that could be incorporated into the ISTEA reauthorization bill, TEA-21, and by working to establish a national network of regional activists who sought to force state departments of transportation to implement the progressive aspects of the new legislation. The workshop that STPP organized in November 1996 specifically recommended that TEA-21 further develop Intelligent Transportation Systems (ITS) that would maximize the number of cars that could use freeways by improving transit routing and dispatching, providing reliable, up-to-the-minute information on road conditions and transit service, and clearing traffic pile-ups promptly. ITS could be used as an alternative to highway widening and building new roads. The workshop also recommended discontinuing the funding of research on the Automated Highway System, which would have removed urban freeway lanes from ordinary use and restricted them to use by specially equipped cars that would have been electronically controlled by a central computer to operate in high-speed platoons as close as 18 inches apart. Recommendations were also made to research the relationship between travel behavior, land use, and transportation and to study how the boom in information technology has affected transportation and the sustainability of communities. The TEA-21 legislation responded to these recommendations by discontinuing the Automated Highway System and by calling for the creation of a National Transportation and the Environment Cooperative Research Program, now being designed by the National Research Council.

After the passage of TEA-21 in June, 1998, Dittmar left his position as executive director of STPP and moved to New Mexico to work on transportation issues at the grassroots level. He remains a member of the STPP board of directors and is president and chief executive officer of the Great American Station Foundation, which promotes the revitalization of historic train stations (and in turn the community centers surrounding them) by offering funding for their renovation. The Great American Station Foundation facilitated a grant for the renovation of the historic station in Las Vegas, New Mexico. That station will soon serve as the terminal for Greyhound and Amtrak and as the seat of the Las Vegas chamber of commerce.

Dittmar has served on numerous boards of directors of transit and planning-oriented groups, including the Institute for Location Efficiency, the Congress for the New Urbanism, the Center for Neighborhood Technology, the Environmental Leadership Project, the Biodiversity Project, and the League of Conservation Voters. Dittmar lives with his wife, Kelle, and their twin children, Cole and Clara, in a traditional village outside of Las Vegas. In addition to his work on transit issues, he is a published poet and short story writer.

Bibliography

Dittmar, Hank, "Road to Nowhere: The Automobile, Sprawl, and the Illusory Suburban Dream," http://www.newdream.org/transport/nowhere.html; "The Great American Station Foundation," http://www.stationfoundation.org/; Horan, Thomas A., Hank Dittmar, and Daniel R. Jordan, "ISTEA and the New Era in Transportation Policy: Sustainable Communities from a Federal Initiative," *Toward Sustainable Communities: Transition and Transformations in Environmental Policy*, 1999; "The Transportation Action Network," http://www.transact.org/; "Transportation Equity Act for the 21st Century," http://www.fhwa.dot.gov/tea21/index.htm.

Dombeck, Michael

(September 21, 1948–)
Chief of U.S. Forest Service, Fisheries Biologist

Michael Dombeck, named fourteenth chief of the U.S. Forest Service (USFS) by Pres. Bill Clinton in 1997, steered the Forest Service on a new course: away from its former promotion of timber extraction above all other uses and toward the broader goal of promoting the long-term ecological health of the land it manages.

Michael P. Dombeck was born on September 21, 1948, in Stevens Point, Wisconsin, and grew up in the Chequamegon National Forest in northern Wisconsin where his parents ran a general store. Dombeck's first summer job was as a logger, cutting pulpwood, but after two summers he turned to guiding fishermen, which he continued for 11 years until 1977. He earned his B.S. degree in biology and natural science from the University of Wisconsin at Stevens Point in 1971 and an M.S.T. teaching degree in biology education there in 1974. While at Stevens Point, Dombeck read Aldo Leopold's *A Sand County Almanac*, which still governs his thinking on natural resource management. Dombeck taught high school science classes; he continued his studies at the University of Minnesota, obtaining an M.S. in zoology in 1977, and at Iowa State University, where he was awarded a Ph.D. in fisheries biology in 1984.

During his time at Iowa State, Dombeck became a fisheries biologist for the USFS. He worked in the USFS's fisheries program as its Pacific Southwest Region manager from 1985 to 1987 and then as its national manager in Washington, D.C., until 1989, when he joined the Depart-

Michael Dombeck discusses the Forest Service's new road proposal during a news conference, 2 March 2000. (AP Photo/Douglas C. Pizac)

ment of the Interior's Bureau of Land Management (BLM). Dombeck rose through the ranks of the BLM, working first as science advisor and special assistant to the director, then in 1993 as acting assistant secretary in land and minerals management, and in late 1993 to 1994 as chief of staff to the assistant secretary of land and minerals management. President Clinton nominated Dombeck for director of the BLM in early 1994, but western Republican congresspeople, who generally favor extraction interests and the rights of ranchers to graze their livestock on government land for minimal leasing fees, successfully opposed the nomination because of Dombeck's commitment to conservation. President Clinton maintained Dombeck as acting director of the BLM, a position not requiring congressional confirmation, for three years until the resignation of thirteenth USFS director Jack Ward Thomas in 1996. Since congressional approval is not required for USFS chiefs, Clinton was able to appoint Dombeck to that position in December 1996.

Dombeck inherited an agency that some had nicknamed the "U.S. Timber Service" because of its blatant alignment with timber industry. From the beginnings of the USFS under founding director GIFFORD PINCHOT, its goal has been to facilitate extraction of timber from the national forests in the United States. Pinchot's vision was to manage extraction so that it would allow a sustainable yield over the long term. But over the decades the goal of sustainability was cast aside. The USFS evaluations of its employees were based on timber output in the regions they supervised, and USFS programs and salaries were funded by fees timber companies paid to log national forests. By the 1980s, the most recent extraction peak, timber companies were taking 11 billion board feet annually from national forests; clear-cuts were scarring even the steepest mountain forests, and the resulting erosion was contaminating streams and rivers and causing landslides. USFS managers were caught between lawsuits from environmentalists about the agency's violation of environmental protection laws, continuous pressure from industries to allow more extraction, and the agency's own muddled bureaucracy. President Clinton's naming of Dombeck as USFS chief in 1996 was a response to public protest about the crisis in national forests.

Dombeck, still influenced by the land ethic as articulated by Aldo Leopold, immediately called for a shift in USFS priorities. He declared that the USFS would no longer prioritize timber extraction over all other activity in national forests and that the long-term health of the land should be the basis on which decisions are made about what is allowed in national forests. He called for a new approach of "collaborative stewardship," which would seek public participation in an effort to balance the varied activities in national forests, including timber and mineral extraction, grazing, hunting and fishing, and other forms of recreation. Furthermore, USFS employees would no longer be evaluated on how many board-feet were taken from their regions. Instead their performance would be evaluated according to the health of the forests they were in charge of: conditions and health of stream banks and watersheds, water quality, stability of soil, management of noxious weeds, landscape management, and protection and habitat management of endangered species.

Some western Republicans, such as Rep. Helen Chenoweth and Sen. Larry Craig, both of Idaho, were not happy with Dombeck's work at the USFS and called him to testify before congressional committees almost monthly after he became USFS chief. Despite their attempts to stop him, Dombeck has been able to repair some key structural flaws at the USFS. He replaced all six deputy chiefs and replaced or transferred seven of the USFS's nine regional heads, moving in officials who agreed with his collaborative stewardship approach. He authorized and published a report that showed how the USFS actually lost money on its timber sales, owing to low fees charged the timber industry and huge overhead costs of building timber access roads. In 1998 Dombeck declared an 18-month moratorium on building new roads in the most remote areas of national forests until the USFS could determine a long-term road policy that would allow for the maintenance of existing, heavily used roads, the construction of those roads deemed necessary, and the designation of areas that would be kept roadless in perpetuity.

Although the *New York Times Magazine* called him "the most aggressive conservationist to head the Forest Service in at least half a century," Dombeck's taming of an agency notorious for its inefficiency is, as of early 2000, far from accomplished. However, Dombeck is known for his patience and diplomacy in dealing with lawmakers who oppose his reforms and with the residents of towns neighboring national forests who are despondent at the decline of the logging industry. Dombeck urges long-term thinking at the policymaking level and for towns whose economy has been dependent upon unsustainable resource extraction. To the general public, Dombeck advises a return to his intellectual mentor Aldo Leopold's ecological ethic. Specifically, Dombeck urges Americans to make conscientious decisions about resource consumption so that more stringent USFS policies in this country do not lead other countries with more lenient environmental protection policies to overharvest their natural resources to meet our demand.

Dombeck currently lives in northern Virginia with his wife and daughter.

BIBLIOGRAPHY

Anderson, H. Michael, "Reshaping National Forest Policy," *Issues in Science and Technology*, 1999; Annin, Peter, "Saving the Tall Timber: The U.S. Forest Service Turns a Bit More Green," *Newsweek*, 1999; Kriz, Margaret, "Fighting over Forests," *National Journal*, 1998; Lewis, Daniel, "The Trailblazer," *New York Times Magazine*, 1999; "United States Forest Service," http://www.fs.fed.us/.

Donovan, Richard

(August 6, 1952–)
Natural Resources Manager, Director of SmartWood

Richard Donovan directs the Rainforest Alliance's SmartWood program, which certifies wood and wood products grown and harvested in an environmentally responsible manner throughout the world. A founding member of the Forest Stewardship Council (FSC), the international body that accredits wood certification programs such as SmartWood, Donovan helped write the "Principles and Criteria for Natural Forest Management," the document upon which the FSC bases its decisions to accredit certification programs.

Richard Zell Donovan was born on August 6, 1952, in Englewood, New Jersey. His parents were both from northern Minnesota, where three generations of his mother's family ran a sawmill from the late 1800s until 1954. As a young man, Donovan learned how to work as a woodcutter and logger in the north woods, skills and practical experience that have served him well during his career. Donovan attended Wabash College in Crawfordsville, Indiana, from 1970 to 1972, before transferring to the University of South Florida and graduating with a double major in Latin American history and romance languages in 1974.

Upon graduation, Donovan moved to Maine and worked as an activist on forest issues in the region and with the Alaska Coalition on the Alaska Lands bill. Donovan enlisted in the Peace Corps in 1975 and was sent to Paraguay, where he spent three years working with Paraguay's National Environmental Sanitation Service. Upon his return to the United States, Donovan worked in northern Minnesota as a woodcutter and in 1979 entered the Antioch New England Graduate School to study natural resources management and administration. His major research project there was to assess the quality of forest management in Vermont's municipal forests and analyze their history; it culminated in recommendations to the public and the private sector about more effective ways that municipal forests could be managed for multiple purposes, including watershed protection, recreation, education, and timber and nontimber forest products extraction.

Donovan graduated from Antioch with an M.S. in 1981 and throughout most of the rest of the 1980s worked as a consultant on natural resource management projects in developing countries and North America. He spent three years in Costa Rica from 1987 to 1990 as a World Wildlife Fund (WWF) senior fellow, during which time he designed and implemented an integrated conservation/community development project in a recently settled buffer zone of Costa Rica's spectacular Corcovado National Park. The BOSCOSA Forest Conservation and Management project offered environmentally sustainable alternatives to *campesinos*, who generally practiced traditional, environmentally destructive slash-and-burn agricultural techniques. Through BOSCOSA, *campesinos* established and managed tree plantations and managed natural forest for timber and nontimber products, watershed protection, and ecotourism.

Donovan returned to Vermont in 1990, where he continued his work on conservation and development as a senior fellow with WWF and also taught in the University of Vermont's Environmental Program and School of Natural Resources. In mid-1992, he became director of the Rainforest Alliance's SmartWood Program. SmartWood, established in 1989, was the first timber certification program in the world and was the first of the Rainforest Alliance's ecological certification programs (the Rainforest Alliance also certifies tropical agricultural crops such as bananas, coffee, cocoa, oranges, and more through its ECO-O.K. program, cofounded by CHRIS WILLE).

SmartWood works for forest conservation by certifying wood and wood products that are grown and harvested responsibly from natural forests and tree plantations in tropical, temperate, and boreal forest regions. Its network of regional nonprofit certifying organizations assesses companies and awards a seal of approval to those that comply with SmartWood's stringent environmental, economic, and social criteria. Once a company's forest products have been certified by SmartWood, they can be marketed to customers who are often willing to pay more for them, and this provides an economic incentive that SmartWood hopes will motivate more forest managers to manage their forests in a more sustainable manner and earn certification. The SmartWood seal has become a mark of excellence in forestry, to the point where many public and private forestry operations seek to become certified for public recognition and peer credibility. Since 1992, the program has certified almost 100 different forest management operations on more than three million acres in Europe, throughout the Americas, and Southeast Asia, and SmartWood has also certified another 200 forest products companies to turn certified wood into certified products ranging from plywood to flooring to furniture to musical instruments.

SmartWood is proud of its many success stories on both the environmental and the socioeconomic fronts. Among them is one on which it collaborates with its Brazilian partner Imaflora, which established a guitar-making school in the Amazon outpost of Manaus, where poor children can go after their regular classes to learn to make guitars and other stringed instruments with nonendangered wood species of the Amazon rain forest. Not only has this helped protect the endangered woods with which guitars are traditionally made, it has taught at-risk children a marketable skill and has engendered in them an appreciation for conservation. Another success story is taking place in Bolivia's "Red Zone," where *campesinos* have cultivated the coca bush and sold its leaves to cocaine processors as a lucrative cash crop. The Bolivian military has tried to eradicate this practice, and the peasants have suffered severe economic setbacks as a result. SmartWood and its Bolivian partner Centro de Investigación y Manejo de Recursos (Resource Investigation and Management Center or CIMAR) are offering agrarian unions training in sustainable forest management, so that SmartWood–certified wood could become a new, stabilizing product.

With the ecocertification boom that began after the renewed public interest in the environment after Earth Day 1990, Donovan and others serious about certification realized that their efforts would be for naught if other certifying entities began offering less stringent certifications. During 1991 and 1992, Donovan worked with a number of international and regional organizations in Europe, the Americas, Melanesia, and Asia to develop the concept of an international body for reviewing and approving forest management certifiers. This idea led to the formation in 1993 of the Forest Stewardship Council, an internationally recognized organization based in Oaxaca, Mexico, that evaluates, monitors, and accredits certification bodies like SmartWood. Donovan cochaired the committee that wrote FSC's "Principles and Criteria for Forest Management," the seminal document upon which all FSC-approved certifications are based. Since its formation, the accredited certification organizations in the FSC system have certified approximately 1,000 forest management and forest products operations covering more than 40 million acres in some 25 countries.

Donovan directs SmartWood from its headquarters in Richmond, Vermont; he resides in the nearby town of Jericho with his wife, Karen, a special education teacher, and their two children, Andrew and Emily. The Donovan family continues their ancestral tradition of heating their home with firewood harvested from local forests.

BIBLIOGRAPHY

Ervin, J., R. Donovan, et al., eds., *Forest Products Certification: Opportunities and Constraints*, 1996; "Forest Stewardship Council," http://www.fscoax.org/; "Rainforest Alliance," http://rainforest-alliance.org/; "Smart Wood," http://www.smartwood.org/.

Douglas, Marjory Stoneman

(April 7, 1890–May 14, 1998)
Writer, Founder of Friends of the Everglades

Variously referred to as the Everglades' "patron saint," "empress," "champion," and "one of the true guiding lights," writer Marjory Stoneman Douglas started the movement to save the Florida Everglades from development with her best-selling *The Everglades: River of Grass*, published in 1947. Through this book and other writings and her fierce grassroots activism, Douglas fought for the preservation of this unique ecosystem and was one of the few environmentalists who lived long enough to enjoy the fruits of her labor. The year before she died, Everglades National Park was expanded significantly, and the federal government announced rehabilitation plans to undo the decades of damage that development and agriculture had inflicted upon it.

Marjory Stoneman Douglas was born in Minneapolis on April 7, 1890, to Lillian Trefethen Stoneman and Frank Bryant Stoneman. While she was still very young, her mother, a concert violinist, left Marjory's father and took her to her grandparents' home in Taunton, Massachusetts. Douglas was raised on stories told by her French grandmother and was encouraged by high school teachers to write. She attended Wellesley College, graduating with a B.A. in English composition in 1912. At Wellesley, Douglas was elected class orator and published her work in the college's literary magazine. In 1914, she married Kenneth Douglas, who was 30 years older than she and turned out to be an alcoholic check forger. She divorced him three years later and never remarried. In 1915, after her mother had died, Douglas moved to Miami, Florida, to be closer to her father, who had founded the *Miami Herald*. She worked as a reporter for the *Herald* until the United States entered World War I. She joined the Red Cross and worked in its publicity department, traveling throughout Europe to write articles about child refugee relief. When she returned to Miami in 1920, she worked as an assistant editor at the *Herald* for three more years before leaving to write short stories. Her stories, many of which were set in a region unfamiliar to most readers—southern Florida—met rapid success. Over a period of 15 years, 40 of her stories were published in the *Saturday Evening Post*, and other magazines such as *Collier's*, *Woman's Home Companion*, and *Reader's Digest* published many more. She won second place in the 1928 O. Henry Memorial Prize, and in 1937 her "Story of a Homely Woman" was included in an anthology of the best stories published in the *Post*.

An avid student of Florida's geography and history, Douglas was especially interested in the area west of Miami that most Floridians considered a useless, pest-infested swamp. The Everglades is a huge marshy area that originally extended south from Lake Okeechobee to the Gulf of Mexico, covering approximately one-third of the Florida peninsula. A shallow, slightly inclined pan of water covered with saw grass, it serves as a giant water purifier. Water originally drained out of Lake Okeechobee slowly through the Everglades, the entire trip to the ocean lasting a full year. The Ever-

Marjory Stoneman Douglas (Kevin Fleming/Corbis)

glades' ability to soak up excess rainfall gave it an important role in flood control, and it stored water that evaporated into north-blowing clouds that hydrate central and northern Florida. Farmers, developers, and industrialists did not recognize the ecological importance of the Everglades and as early as the mid-1800s sought to "reclaim" the land for other uses. The 1850 federal Swamp and Overflowed Lands Act turned over the Everglades to the state of Florida on the condition that it drain the area. More than 400 miles of drainage canals were built by the 1930s to divert water from the region; in later years the U.S. Army Corps of Engineers increased the canal network to 1,400 miles and built levees that locked up water in reservoirs for irrigation. Seven massive pumping systems were set up to protect crops on the reclaimed land from inundation. Douglas's father denounced this work in a *Herald* editorial in 1905, and his daughter also believed that the area deserved more respect for its natural features and its ecological significance. In 1927 she joined a citizen's committee that lobbied for the Everglades' protection as a national park.

In the 1930s, Douglas successfully pitched a nonfiction piece about the threats of wildlife poaching in the Everglades to the editor of the *Saturday Evening Post*, and by 1941 she had started work on her famous *The Everglades: River of Grass*, published first in 1947 and still in print. Douglas spent five years exploring the Everglades by canoe, row boat, and swamp buggy and on foot to research the book and in the process realized that her love for the area would convert her into a lifelong activist to protect it. The book reveals to lay readers not only the complexity of the water cycle and wetland ecology, but also the history of human interaction with the area—how indigenous people learned to live off the Everglades and the clumsy attempts of white colonists to convert them into the dry land that they knew how to use. This human history was the first documentation of its kind about the Everglades.

The book was an immediate success, with the first printing of 7,500 copies selling out in just a month. In that same year, 1947, Everglades National Park was established, protecting over a million areas of the Everglades. But the area remained under siege, because it was surrounded by agriculture and contaminated by pesticide- and fertilizer-laden runoff. Developers ate further into the Everglades to build housing and industry, and the east-west Everglades Highway impeded water flow.

The Everglades remained in print throughout the 1950s and 1960s, earning the Everglades a widening group of admirers. Local conservation groups continued to push for more protection, but it was not until 1969, when Douglas was 79 years old, that she became a full-time activist for the Everglades. In that year, the twin threats of an oil refinery on Biscayne Bay south of Miami and a jetport in the middle of the Everglades emerged. Douglas founded Friends of the Everglades, a local, grassroots group that anyone could join for a dollar, and became its first president. The group defeated the refinery and the jetport and lobbied to improve the north-south flow of water. In a 1999 article for *International Wildlife*, Douglas recalled that the proponent of the refinery really did the environmental movement a great deed: "His idea was so ridiculous and it stimulated such widespread opposition that many people who'd otherwise been sitting back were enlisted in the environmental movement right then."

Douglas devoted the last 30 years of her long life to the continuing struggle to protect the Everglades. She continued to address any group that invited her, despite her age and worsening blindness. Friends of the Everglades grew to 5,000 members and in concert with major national conservation organizations successfully lobbied for a major restoration project for the Everglades. Florida voters voted in 1996 for a constitutional amendment to clean up the Everglades but against a penny-a-pound tax on sugar that would have paid for it. The Clinton administration stepped in and agreed to buy 50,000 acres of sugarcane fields surrounding Everglades National Park. The U.S. Army Corps of Engineers and the South Florida Water Management District were assigned the task of restoring water flow in this area with $1.5 billion from the federal government. After Douglas's death two years later, Vice President AL GORE said that "Marjory was one of the true guiding lights. I am thankful she lived long enough to see the fruits of her good efforts." Douglas was honored for her work with a Presidential Medal of Freedom in 1993. A nature center in Key Biscayne, the Department of Environmental Protection building in Tallahassee, and several schools and parks in south Florida have been named after her.

Douglas lived alone for more than 70 years in a thatched cottage that she designed herself in the Coconut Grove neighborhood of Miami. She died on May 14, 1998, at home.

Bibliography

Breton, Mary Joy, *Women Pioneers for the Environment*, 1998; Douglas, Marjory Stoneman, "A Dollar for the Everglades," *International Wildlife*, 1999; Douglas, Marjory Stoneman with John Rothchild, *Marjory Stoneman Douglas: Voice of the River*, 1987; "Friends of the Everglades, founded by Marjory Stoneman Douglas," http://www.everglades.org/; Lim, Grace, and Patrick Rogers, "Lady Everglades," *People Weekly*, 1998; Severo, Richard, "Marjory Douglas, Champion of Everglades, Dies at 108," *New York Times*, 1998.

Douglas, William Orville

(October 16, 1898–January 19, 1980)
Supreme Court Justice

In 1939, at the age of 40, William Orville Douglas became the second youngest appointee to the U.S. Supreme Court. While his years on the Supreme Court were hardly free from controversy (twice his foes tried to impeach him), many observers now regard Douglas as one of the finest justices in the Court's history. Douglas was an ardent defender of the nation's wild areas when environmental cases were heard by the Supreme Court, writing eloquent minority opinions. He is also remembered for leading two major grassroots preservation battles in the 1950s and for his books on nature, which were widely read and highly acclaimed.

William Orville Douglas was born October 16, 1898, in Maine, Minnesota, to Julia Fisk Bickford Douglas and William Douglas, a Presbyterian minister. Douglas's family did not stay long in Minnesota, as the elder Douglas transferred to a new pastorate in Estrella, California, followed shortly by another move to Cleveland, Washington. In Washington, Orville, as he was known in his youth, first came across the awesome natural features of the Pacific Northwest that would influence much of his later career. The Cascade Mountains were 40 miles to the west of Cleveland, and the Columbia River was 30 miles to the south. Spending formative years in such spectacular country endowed him with a lifelong emotional and spiritual tie to that landscape.

In 1904, not long after the family's arrival in Washington, the elder Douglas died of complications following surgery for stomach ulcers. Julia Douglas moved her family of three young children to Yakima, Washington, where they bought a house with part of the life insurance money her husband had left behind. She invested the rest in a highly speculative irrigation project that soon failed. Housed but penniless, Orville Douglas took odd jobs washing store windows, sweeping floors, and picking fruit in the fertile Yakima Valley. This early experience instilled in him a deep sympathy for the poor, which later influenced a significant number of his decisions on the U.S. Supreme Court.

William O. Douglas (Archive Photos)

Although poverty undoubtedly influenced Douglas's views, another factor even more profoundly shaped the course of his life. While still in Minnesota, Douglas had contracted a minor case of polio. As a young boy, his legs remained weak, and he was bullied at school. Determined to rise above his peers, Douglas quickly climbed to the top of his class. He also strengthened his legs by regular walking in the foothills north of Yakima, which tightened his bond to the magnificent northwestern landscape.

Finishing school, Douglas earned a scholarship to Whitman College. To supplement the scholarship, Douglas worked summers as a farmhand, meeting radical members of the Wobblies, the Industrial Workers of the World, who recruited members from among farm laborers. His success at Whitman led to a short stint teaching in Yakima and then to Columbia Law School in 1922. He was admitted to the New York bar in 1926 but continued teaching at Columbia until 1928, when he accepted an offer at Yale University Law School. In 1932, he was named Sterling Professor of Law, a title he held until being appointed to the Supreme Court. In the midst of the Depression, in 1934, Douglas was asked to direct a protective committee study by the U.S. Securities and Exchange Commission (SEC). The study lasted two years, at the end of which Douglas was named commissioner of the SEC, and chairman in 1937. In 1939, he was appointed to the Supreme Court.

Pres. FRANKLIN D. ROOSEVELT wanted to appoint a westerner to the nation's highest tribunal to balance regional representation on the Court. Douglas's 1939 appointment was mildly controversial; despite growing up in Washington, Douglas had spent his entire adult career as a member of the Eastern intelligentsia and boasted a meteoric career. Nevertheless, Douglas quickly showed himself to be a genuine westerner, cultivating an image as a rugged individualist, expert outdoorsman, and determined conservationist.

The majority of Douglas's successful conservation battles were personal, however, not issues he ruled upon as Supreme Court Justice. His first publicized conservation battle involved the Chesapeake & Ohio Canal near Washington, D.C., which some had hoped to turn into a scenic highway. Douglas frequently hiked along the historic canal and enjoyed both its wildlife and its solitude. In 1954, he challenged the editors of the *Washington Post*, who had endorsed the idea of a parkway, to hike the 189-mile canal from Washington to Cumberland, Maryland. Douglas's weighty influence as a Supreme Court justice attracted the attention of the Wilderness Society and other environmentally concerned individuals. The editors of the *Post* accepted Douglas's challenge and ultimately changed their opinion. The well-publicized hike generated all kinds of interest and annual reunion hikes to maintain the issue's prominence. In 1961, Pres. Dwight D. Eisenhower declared the C & O Canal a national monument, and in 1971 it was declared a national historic park.

Douglas's next fight was in 1958, to prevent the building of a road along an ocean beach in Olympic National Park in Washington State. Local business interests were lobbying for a new road that would wind along the beach to make the area more accessible to automobiles. Having hiked the beach before, Douglas felt it possessed a special quality worthy of protection. Along the lines of his successful C & O Canal protest, Douglas and local activist POLLY DYER organized a three-day, 22-mile hike along the primitive beach to gain publicity for the beach's protection. This tactic succeeded again, and the road was never built.

Douglas's interests in preserving hiking or outdoors areas were not simply for aesthetic values or interests. In his first book of nature writing, *My Wilderness: The Pacific West*, published in 1960, Douglas explained the deeper issues involved in the 1958 fight. Providing both a natural and cultural history of the region, Douglas offered a number of ecological lessons relevant to the proposed road and its impact on both terrestrial and marine wildlife.

For most of Douglas's tenure on the Supreme Court, few environmentally charged cases came to the tribunal. When they did, Douglas likely dissented from the majority opinion in favor of environmental protection but was largely unable to change existing environmental law. Nevertheless, his activity in the public forum and his eloquence on environmental matters remain a significant aspect of his lasting environmental legacy.

Douglas died January 19, 1980, at Walter Reed Army Hospital in Washington, D.C., with his children and wife at his side.

BIBLIOGRAPHY

Douglas, William O., *My Wilderness: East to Katahdin*, 1961; Douglas, William O., *My Wilderness: The Pacific West*, 1960; Douglas, William O., *A Wilderness Bill of Rights*, 1965; Sowards, Adam M., "William O. Douglas: The Environmental Justice and the American West," in Regan Lutz and Benson Tong, eds., *The Human Tradition in the American West*, forthcoming .

Dowie, Mark

(May 20, 1939–)
Investigative Reporter, Editor, Publisher

Mark Dowie's journalism career, which includes working as publisher and editor of *Mother Jones* magazine, has earned him 17 major journalism awards for his coverage of important social and environmental issues. While working at *Mother Jones*, Dowie broke the stories on the Dalkon Shield and the Ford Pinto and helped launch provocative investigations of unethical corporate practices. Through the course of his investigative reporting, Dowie became critical of the current mainstream environmental movement, believing that it should abandon diplomatic tactics and attempts at "win-win" capitulation with corporate polluters. In his book *Losing Ground: American Environmentalism at the Close of the Twentieth Century* (1995), he holds out hope for a new form of environmentalism, one that is diverse in class, race, and gender and strongly linked to grassroots human rights movements.

Mark Dowie was born in Toronto, Ontario, on May 20, 1939, the son of Ian and Shan (Campbell) Dowie. When he was eight years old his family moved to Cleveland, Ohio. He attended Denison University, receiving his B.A. in English in 1962. From the time he graduated from college to the time he went into journalism, Dowie pursued several careers, many of which prepared him well for investigative reporting. He worked in metal prospecting and investment banking and then helped operate a cattle ranch in Wyoming in the early 1960s. He then moved to San Francisco, where he went to graduate school in economics at the University of California; he later coordinated long-range economic planning at Industrial Indemnity from 1966 to 1969. He eventually became involved in prisoners' rights and acted as executive director of Transitions, Incorporated, an employment project for ex-prisoners, from 1969 to 1974. In 1974, he published *Transitions to Freedom: Comparative Community Response to the Return of Imprisoned Convicts*, the first step in his career in journalism.

In 1976, Dowie began working for *Mother Jones*, a newly launched progressive magazine offering investigative articles and promoting activism and social change. By getting involved at such an early stage at the magazine, he was able to influence its scope and help the magazine break new ground. He wrote a landmark investigative article for *Mother Jones* in 1977 about Ford Motor Company's safety problems. Titled "Pinto Madness," the article stated publicly for the first time that Ford Motor Company had made unethical decisions regarding the production of the Ford Pinto by opting for maximum profit instead of safety. Dowie made the claim that in order to make the Pinto lightweight and inexpensive, certain safety measures were neglected, leaving the fuel tank unnecessarily vulnerable to rupture in the event of rear-end collisions. Furthermore, Ford engineers were aware of this defect through preproduction crash tests, but they proceeded with the manufacturing process anyway, since the assembly-line machinery was already set up for production. Dowie went on to expose the fact that Ford Motor Company spent eight years fighting against and effectively delaying the implementation of a governmental safety standard that would have forced the company to redesign the Pinto's unsafe gas tank. The article won a National Magazine Award from the Columbia University School of Journalism in 1978.

Dowie had also written an earlier article for the magazine revealing the corporate history of the Dalkon Shield, a poorly designed intrauterine device that killed at least 17 women and injured hundreds more before a federal investigation forced its manufacturers to take it off the market. In researching corporate documents on the case, Dowie had noticed that the total number of the devices far exceeded the sum of those sold and recalled in the United States. He discovered that one million of the dangerous Dalkon Shields had been exported to 42 countries, many of them poor, developing countries. Over the next three years, Dowie gathered information on many similar cases involving unsafe products banned in the United States, such as toxic pesticides, tainted foods, known carcinogens, and defective medical devices that were being "dumped" on other countries. *Mother Jones* assembled an investigative team to write up a special report on U.S. policies that allowed corporations to sell shiploads of these products, mostly to Third World nations. In an attempt to pressure the Carter administration and Congress into banning the practice, *Mother Jones* staff delivered copies of the issue with the investigative report on dumping to every embassy in Washington, D.C., to all delegations at the U.N., and to major newspapers throughout the world.

Dowie was publisher of *Mother Jones* from 1977 to 1980 and then editor from 1980 to 1985, and during his years at the magazine he was responsible for many revealing and environmentally relevant investigative reports. He left the magazine in the mid-1980s, but continued writing. In 1995, Dowie published *Losing Ground: American Environmentalism at the Close of the Twentieth Century,* a critique of the mainstream environmental movement. In the book, he makes clear his frustration that the movement has not lived up to its potential and is in fact "courting irrelevance." He follows the history of environmentalism, starting with conservation notions at the turn of the century, up to the debut of the idea of the "environment" as including both the human and the natural habitat, which occurred with the publication of RACHEL CARSON's *Silent Spring* (1962). But increasing motivation over the next two decades to preserve wilderness and enact environmental legislation was roadblocked during the 1980s by the Reagan administration. And currently, Dowie writes, environmentalism is being shackled by a bureaucratic attempt to find a happy medium between politicians and corporate polluters. The mainstream environmental groups are becoming too close to and too much like the industries they claim to fight, and phrases such as "nonadversarial dialogue" have become their buzzwords. The hope Dowie holds for the future of environmentalism lies in true grassroots movements that are linked to civil rights, involve men and women of all cultures and races, and include urban as well as rural issues. The diversity and activism of these groups will make them able to more powerfully assert society's rights to a satisfactory environment, Dowie contends.

In addition to his research and writing, Dowie serves as a national advisory board member to the Center for Investigative Reporting; Sustainable Urban Neighborhoods of Louisville, Kentucky; and Global Resource Action Center for the Environment (GRACE). After years of investigative reporting, Dowie has become more and more interested in history and now describes himself as an investigative historian. He currently has a book in peer review on the history of foundations in the United States and is also writing screenplays. He lives in Inverness, California.

Bibliography

Dowie, Mark, "A Case of Corporate Malpractice: The Corporate History of the Dalkon Shield Intrauterine Device," *Mother Jones*, 1976; Dowie, Mark, *Losing Ground: American Environmentalism at the Close of the Twentieth Century*, 1995; Dowie, Mark, "Pinto Madness," *Mother Jones*, 1977; Hochschild, Adam, "Dumping Our Mistakes on the World," *Mother Jones*, 1979.

Drayton, William

(June 15, 1943–)
Social Entrepreneur, Founder and President of Ashoka

William Drayton founded the international fellowship Ashoka in 1980 to enable "social entrepreneurs" (highly ethical, motivated people with revolutionary ideas for improving society, who possess the problem-solving ability to implement them) to devote themselves full time to putting into practice their vision of change. Ashoka's fellowships—which take the form of a living expense stipend for three years—generally come at the crucial point in which its fellows are just launching their projects. Ashoka support allows them to develop and disseminate their ideas and to build stable institutions through which they can continue to work after the financial support ceases. More than 1,000 Ashoka fellows work in 34 countries worldwide, on projects concerning the environment, alleviation of poverty, women's issues, and problems for disabled people.

Born on June 15, 1943, in New York City, William Drayton was a natural entrepreneur from an early age. One of the first entrepreneurial experiences he remembers was selling small items to his parents' dinner guests; another was publishing an elementary school newspaper with 60 pages of advertising. At Phillips Academy, Drayton founded the Asia Society, which became one of the school's most popular student organizations. While studying at Harvard, Drayton created the Ashoka Table, dinners at which prominent government, union, and religious leaders explained, off the record, the inner workings of their organizations. Drayton founded Yale Legislative Services at Yale Law School, which at its peak attracted one-third of the law school's student body. Drayton graduated with an A.B. from Harvard College in 1965; earned his M.A. from Balliol College, Oxford University, in 1967; and received his J.D. from Yale Law School in 1970.

Drayton began working for New York–based McKinsey & Company in 1970, helping a wide range of clients (government, corporations, foundations) solve management and policy problems. He taught at Stanford Law School from 1975 to 1976 and taught regulatory and management reform at Harvard's John F. Kennedy School of Government in 1977. He worked with President Jimmy Carter's transition planning team, designing regulatory and management reform programs for such areas as airline and trucking deregulation and civil service reform. During Carter's term in office, Drayton served as assistant administrator of the Environmental Protection Agency (EPA). Among his chief accomplishments was the design and implementation of a flexible, market-like approach to regulation that encouraged industry to meet environmental goals in efficient, yet enforceable ways. He proposed tradeable pollution rights, similar to what the presidential administration of Bill Clinton implemented 20 years later in its attempt to decrease emissions of ozone-depleting gasses. When President Reagan's administration took over in 1981 and tried to weaken the EPA, Drayton became president of Save EPA, an association of prominent environmental managers that worked to limit the damage. Save EPA was renamed Environmental Safety in 1983. It serves as a monitor of the EPA and offers pro bono counsel and analysis to senior executive branch and legislative leaders. Drayton still serves as its chair.

During the late 1970s Drayton and a group of friends began laying the foundation for what officially in 1980 became Ashoka. The organization was named for an Indian leader of the third century B.C. who, after successfully uniting India by force, was stricken by great remorse. Ashoka renounced violence and dedicated the rest of his life to promoting social welfare, economic justice, and religious tolerance. Drayton and his covisionaries spent their vacations during the late 1970s looking for people who embodied Ashoka's approach. They traveled in India, Indonesia, and Venezuela, countries selected for their variation in size and culture, seeking effective innovators who dedicated themselves to the public good. They amassed a database that included hundreds of "social entrepreneurs."

By 1981 Ashoka had raised enough funds to begin its second phase: offering fellowships to selected social entrepreneurs whose work in the fields of environment, poverty alleviation, women's issues, and disability was very promising and who were at the particularly vulnerable early stages of their innovative projects. With a goal similar to that of venture capital investors, Ashoka sought fellows whose innovations for the public good would yield great social "profit" if moderate funding came at the right time.

Ashoka fellows are selected through a rigorous process consisting of several screenings, reference checks, and interviews with in-country and out-of-country Ashoka representatives. The fellowships take the form of a stipend for living expenses for three years and are tailored to the fellow's needs, ranging from $2,500 to $20,000 per year. Ashoka does not fund projects; it expects its fellows to mobilize financing from local sources or to make the project self-financing.

In 1984, Drayton was awarded a $200,000 MacArthur Foundation grant. This enabled him to quit his job at McKinsey & Company and dedicate himself full-time to Ashoka. He has raised millions of dollars, mostly from U.S. and international foundations.

One of Ashoka's early fellows was Gloria De Souza, a teacher from Bombay, India, who had been experimenting with a teaching approach that emphasized inquiry and problem-solving to stimulate independent, creative thinking and environmental awareness. This was in direct opposition to the rote style of teaching that dominated India's public schools. It was criticized by officials wary of such experimentation, but it resulted in much higher reading and math test scores for the children. Ashoka's four-year fellowship for De Souza, amounting to a total of $10,000, allowed her to disseminate her approach throughout India. In 1985, the Bombay school system asked De Souza to introduce her methods through a pilot program, and three years later her environmental studies program was incorporated into the national educational curriculum. Her case illustrates the approach of Ashoka: contributing to substantial social change that affects a large number of people, by investing in a single individual with the outstanding qualities of a social entrepreneur.

Another of the more than 1,000 fellows working in 34 countries is Albina Ruíz Ríos, of Lima, Peru, who developed a successful program to pick up garbage in poor neighborhoods where the government had discontinued its garbage collection service. She set up several microbusinesses that hired employees from the neighborhoods as garbage collectors. Residents are encouraged to pay the monthly fee for the collection through incentives. Prompt payers in a neighborhood built on a barren hillside get a tree planted in their front yard and a sticker for their window that says "This house is clean." The household with the most prompt payment record gets gifts such as food baskets. These poor neighborhoods have better payment records than Lima's wealthier neighborhoods with government garbage collection. Currently, more than three million low-income Limeños are clients of Ruíz Ríos's companies.

One of Ashoka's criteria is that the work of fellows be replicable, so that it can be implemented elsewhere. Ashoka looks for people that Drayton says will leave their "scratch on history," people who will set or change national or regional patterns. After the initial fellowship ceases, support continues in the form of a lifelong association with Ashoka and its worldwide network of social entrepreneurs.

In addition to serving as President of Ashoka, Drayton works with several other organizations. He is chair of two groups. One is Youth Venture, which helps community groups provide youth with the support they need to undertake their own projects, ranging from producing their own radio programs to setting up a teen counseling phone service. The other is Get America Working, which works to create job opportunities for this country's unemployed. Drayton has received many awards for his work: The Yale School of Management gave him the 1987 Award for Entrepreneurial Excellence; he received the 1995 National Public Service Award from the National Academy of Public Administration and the American Society for Public Administration; and in 1996 he was elected Member of the American Academy of Arts and Sciences.

BIBLIOGRAPHY

"Ashoka," http://www.ashoka.org; Bornstein, David, "Changing the World on a Shoestring," *Atlantic Monthly*, 1999; Holmstrom, David, "Change Happens, One Entrepreneur at a Time," *Christian Science Monitor*, 1999; Parrish, Michael, "Hands-on Giving, Part II: Ashoka, Venturing Abroad," *Environmental News Service*, http://www.enn.com/enn-features-archive/1999/10/101999/capital2_6421.asp.

Drury, Newton

(April 9, 1889–December 14, 1978)
National Park Service Director

Generally castigated for allowing Bureau of Reclamation surveyors to study dam sites in Dinosaur National Monument in 1941 and for his involvement (or lack thereof) in the Echo Park controversy of 1949–1956, Newton Drury's positive work as director of the National Park Service (NPS) from 1940 to 1951 has gone largely unnoticed. The conventional criticism of Drury's tenure tends to ignore his first nine years as director, when he successfully arrested proposals to exploit park resources for the World War II effort and fought against the Bridge Canyon and Glacier View Dams. In an age of high development, Drury believed that the federal government had a duty to protect places where the aesthetic should have primacy over the economic. This commitment, which bucked the tradition of the period's often blind pursuit of economic interests, ensured that national parks in the United States would remain intact for future generations.

Newton Bishop Drury, the son of Wells, a newspaper editor in Virginia City, Nevada, and San Francisco, California, and Ella Bishop Drury, was born April 9, 1889, in San Francisco. Drury attended the University of California, Berkeley, and graduated in 1912. He worked at the university for several years following his graduation, teaching English and forensics and assisting the university's president, then served in World War I with the Army Balloon Corps. After the armistice, he returned to San Francisco and together with his brother Aubrey, who eventually wrote a highly acclaimed guidebook to California, ran an advertising and public relations agency. In 1919, Drury became the first executive secretary (the functional equivalent of executive director today) of the Save-the-Redwoods League. This organization, which was founded with the help of National Park Service director STEPHEN MATHER, managed during Drury's 21-year tenure to directly purchase some 50,000 acres of redwood groves and convince the state of California to buy nearly half a million acres more for state parks. Some historians consider this Drury's greatest contribution to conservation.

Drury declined appointment as director of the National Park Service when Horace Albright retired in 1933, because he opposed Albright's expansion into areas that Drury felt were more the concern of state and local governments. Like many preservationists, Drury had grown alarmed by the direction the NPS took during the New Deal. In an effort to create jobs for organizations such as the Civilian Conservation Corps (CCC), the size and nature of the agency's jurisdiction expanded to include not only new national parkland (some of which many preservationists saw as less than worthy of the name), but also a host of monuments, historic sites, and buildings. Drury, who was a purist, believed that the NPS should revert to its original mission of managing and preserving natural areas of truly spectacular natural beauty.

By 1940, when NPS director Arno Cammerer suffered a heart attack and resigned, the Park Service's reputation as a preservation agency had suffered greatly. The National Parks Association, a nonprofit organization that monitored the NPS and defended national parks from private incursions and commercial interests, had opposed some park proposals on the grounds that the land was not suitable for parks. Groups such as the Wilderness Society and the Sierra Club objected to CCC developments within the parks. In order to avoid further controversy, Secretary of the Interior Harold Ickes decided to look outside the Park Service for a new director whose preservationist credentials were impeccable. Drury's work with Save-the-Redwoods had gained him a national reputation, and his character was exactly what Ickes was looking for. Even though he had declined the position seven years earlier, Drury accepted this offer and on August 20, 1940, became the National Park Service's fourth director, the first without any prior Park Service responsibilities.

Less eager than his predecessors to expand the Park System, Drury opposed NPS involvement in areas he judged not to meet national park standards. Fewer than 20 new areas were added during Drury's 11 years as director, and most were small. The largest—Coulee Dam National Recreation Area (99,000 acres) and Theodore Roosevelt National Memorial Park (70,000 acres)—were both areas that Stephen Mather, the first director of the National Park Service, had opposed. That Drury's tenure coincided with World War II might also have contributed to Drury's less-than-eager approach to enlarge the National Park System.

Drury might have had a greater effect on the Park Service had his directorship not coincided with World War II. When the war began, the Park Service and other "nonessential" agencies were moved to Chicago. This move resulted in the agency's lack of contact with Congress and Congress's subsequent 50 percent reduction in park budgets and staffing to help fund the war effort. Nevertheless, Drury fought hard to prevent the exploitation of the natural resources in the national parks to aid the war effort. The problems only intensified after the war, as the agency did not return to Washington, D.C., until 1947. While its budget increased somewhat during the following two years, the Korean War in 1949 threatened further cuts. Albright, who continued to be an influential adviser to NPS directors, and conservation groups such as the Sierra Club and the Wilderness Society lobbied hard to prevent the threatened cuts. Weakened by his distance from Congress, Drury witnessed such groups becoming prominent in battling his causes for him. As such, Drury's work as director is often considered weak.

Nevertheless, Drury was successful in the fight against the Glacier View Dam, which the Army Corps of Engineers sought to build in Glacier National Park in the late 1940s. He also led the struggle against the construction of Bridge Canyon Dam, which would have affected Grand Canyon National Park and Monument. Such activity suggests that Drury was not the timid appeaser many historians have described.

Drury was also active, though arguably equally ineffective, in the controversy that led to his resignation. The Bureau of Reclamation wanted to build dams in Dinosaur National Monument, and in 1944 Drury approved a Park Service study of Dinosaur that suggested that dams could be constructed within the monument. To protect the purity of the national monument concept, the area would be downgraded to a national recreation area. The report was given to the Bureau of Reclamation, which began planning for the dams. In 1947, Drury visited Dinosaur National Monument for the first time. Impressed by the scenery, Drury immediately regretted having originally supported the dam proposal. In 1948, the Park Service submitted a report to the Bureau of Reclamation stating that a dam would be incompatible with Dinosaur National Monument. Feeling betrayed, the Bureau of Reclamation leaked the report to dam supporters, who demanded that Congress approve the dam. Under siege, Drury agreed to a compromise that would allow dams in the monument but at different, less damaging sites than had originally been proposed. But the situation was already too polarized, with other park officials opposed to any dams. Accused of hypocrisy by the Bureau of Reclamation, Drury was forced to resign by Interior Secretary Oscar L. Chapman in 1951, leaving other dam opponents, such as the Sierra Club, to win glory in preventing the dams after years of debate.

In 1951, Drury became director of the California Division of Beaches and Parks. During his tenure there, much of the State Park System's share of shore oil royalties, which had been suspended in 1947, began to flow once again. The extra income allowed Drury to expand the system so that, by 1959, when he retired at the age of 70, the California State Park System was composed of 150 beaches, parks, and historic monuments, which covered 615,000 acres.

Drury died in Berkeley, California, December 14, 1978. He was 89 years old.

BIBLIOGRAPHY

Harvey, Mark, *A Symbol of Wilderness: Echo Park and the American Conservation Movement*, 1994; Ise, John, *Our National Park Policy: A Critical History*, 1961; Neel, Susan Rhoades, "Irreconcilable Differences: Reclamation, Preservation and the Origins of the Echo Park Controversy," Ph.D. dissertation, UCLA, 1990; Pearson, Byron, "Newton Drury of the National Park Service: A Reappraisal," in *Pacific Historical Review*, August 1999; Shankland, Robert, *Steve Mather of the National Parks*, 1970.

Dubos, René

(February 20, 1901–February 20, 1982)
Microbiologist

René Dubos was a soil microbiologist who discovered the enzymes in soil microbes that were used as the world's first antibiotic drugs. In his work, he learned that the environmental conditions to which soil microbes are subjected will affect their many characteristics and capabilities. Transferring this observation to the larger world, Dubos discovered that all organisms are affected in this way by their environment: their environment helps determine which of their varied capabilities emerge. Dubos was disturbed by wanton environmental destruction and in the early 1960s became an outspoken advocate for the nascent environmental movement of that time.

René Jules Dubos was born in Saint-Brice, France, a small village outside of Paris, on February 20, 1901. A severe case of rheumatic fever at the age of eight robbed him of his aspirations to race bicycles or play professional tennis. His convalescence included long walks in the countryside, and Dubos developed an appreciation during this period of that area's beautiful pastoral landscape. As a teenager he read an essay by philosopher Hippolyte Taine about the effect that that particular landscape had on French fable-writer La Fontaine, whose stories most French children grow up hearing. The idea that an environment had a great effect upon its inhabitants was one that Dubos would return to throughout his professional life.

Dubos and his parents moved to Paris and opened a butcher shop just before World War I broke out. His father was soon called to military service and died of a war injury when Dubos was 18 years old. Dubos finished secondary school on scholarship and was set to attend a college that specialized in physics and chemistry, when his rheumatic fever recurred. Once he recovered, there was only one college in Paris that was still accepting new students so late into the school year, the National Institute of Agronomy. He enrolled there and graduated with a B.S. degree in 1921, but so disliked his courses that he determined to never again enter a laboratory. Dubos moved in 1922 to Rome, where he worked as associate editor for an agricultural journal. While in Rome, he read a paper by Russian soil bacteriologist Sergei Winogradsky that immediately made sense to him and inspired him to return to microbiology. The paper postulated that soil microbes should not be isolated and studied in laboratories but instead must be studied in their own environment so that their interactions with other soil bacteria can be observed.

Dubos traveled to the United States in hopes that he could study there. On board the ship, Dubos met Selman Waksman, a soil scientist from Rutgers University whom he had previously led on a tour of Rome. Waksman and another colleague whom Dubos had met in Rome, Jacob Lipman, set up Dubos at Rutgers, and within three years he had earned his Ph.D. in soil microbiology. Dubos's thesis was about how environmental conditions determine the ability of soil bacteria to digest cellulose. In 1927, after receiving his Ph.D., Dubos met Oswald Avery at the Rockefeller Institute for Medical Re-

René Dubos (courtesy of Anne Ronan Picture Library)

search. Avery was a physician who was researching pneumococcus, the bacteria that causes lobar pneumonia. Avery believed that the cure for pneumonia lay in discovering how to dissolve the polysaccaride capsule encasing pneumococcus. Dubos offered to test soil bacteria for such an agent—cellulose is a polysaccharide commonly found in soil, so he felt it was likely that another soil microbe might digest the pneumococcus polysaccharide. Avery helped Dubos obtain a fellowship to the Rockefeller Institute, and Dubos set out into the bogs of New Jersey in search of the right bacterium. Within three years, Dubos had discovered a bacterium that could eat the polysaccharide capsule and isolated the particular enzyme that digested it. Work on the enzyme as a cure for pneumonia would have continued had it not been for the emergence of sulfa drugs as an effective cure for the disease.

What was most interesting to Dubos about this project was that the bacterium in question produced the enzyme it needed to digest the polysaccharide only when there was no other food source available. This was an observation that Dubos realized had many implications: organisms have diverse capabilities, and many of these capabilities will emerge only in response to certain environmental conditions.

Dubos continued his work to develop new antibacterial substances during the 1930s. In 1939 he discovered tyrothricin, which contained the antibacterials gramicidin and tyrocidine. He insisted that the drugs he discovered be referred to as antibacterial or antimicrobial as opposed to "antibiotic" (or "antilife"), because he felt these were medicines that restored life.

In 1942, Dubos's wife Marie Louise Bonnet died of tuberculosis. His grief at her loss led him to look for a cure for the disease and also to question why the disease, which she had had as a child, reemerged at that time. Dubos had moved to Harvard in 1942 but returned to the Rockefeller Institute two years later to open a department dedicated to tuberculosis research. By 1947 he had discovered a way to breed large numbers of the tuberculosis bacilli without their mutating to become vastly different from the form that causes tuberculosis in humans. This enabled other researchers to develop the Bacille Capmette-guérin (BCG) vaccine, still used today. His research on the causes of tuberculosis also led him to believe that people's early environment and nutrition have a long-term effect on their health. He believed that a healthy child would have a much better chance for a long, healthy life and that social reforms that would improve environment and nutrition for all children were of moral necessity. Dubos's second wife, Jean Porter, whom he married in 1946, also contracted tuberculosis but managed to recover. Together, they wrote the 1952 exploration of the disease, *The White Plague: Tuberculosis, Man, and Society.*

The realization that illness is caused by more than just microbes, together with his deep humanism, led him to write prolifically for a general audience about medicine, human adaptability, and the environment. His three books *Man Adapting, Man, Medicine and Environment,* and *So Human an Animal* celebrate the ability of all organisms to adapt to varied environments but also reveal humanity's particularly awesome adaptation skills as a double-edged sword. Humans are so adaptable that they could very easily adapt to the worst manifestations of modern life: pollution, a landscape devoid of beauty, a joyless life. Despite this worry, however, Dubos parted from the emerging environmental movement of the 1970s with his insistence that people were not enemies of nature but rather had the potential to restore health to their environment. Dubos believed that people need to understand the value of a healthy natural environment and that they must protect it in order to live to their full potential. He believed that a healthy environment was a natural right. He was inspired by a grassroots movement in Queens, New York, to clean up Jamaica Bay and resist the expansion of Kennedy Airport onto a landfill in the bay. A peninsula into the bay is now named Dubos Point Wetland Park.

Dubos received many awards and honorary degrees for his work in health and the environment. After his retirement from Rockefeller University in 1971, he became professor of environmental studies at the State University of New York, College at Purchase, and was an adviser at Richmond College on Staten Island. Dubos and his wife divided their time between their apartment in Manhattan and their estate in Garrison, New York, where they planted many trees and gardens. René Dubos died from heart failure in New York City on February 20, 1982, his 81st birthday.

Bibliography

Moberg, Carol L., and Zanvil A. Cohn, "René Jules Dubos," *Scientific American,* 1991; Moberg, Carol L., and Zanvil A. Cohn, eds., *Launching the Antibiotic Era: Personal Accounts of the Discovery and Use of the First Antibiotics,* 1990; Piel, Gerard, and Osborn Segerberg Jr., eds., *The World of René Dubos: A Collection from his Writings,* 1990.

Durning, Alan

(November 7, 1964–)
Founder and Director of Northwest Environment Watch, Writer

Alan Durning is founder and director of Northwest Environment Watch, a private, nonprofit research organization that seeks to foster a sustainable economy and way of life in the Pacific Northwest. Durning has authored and coauthored numerous books on a wide range of socioenvironmental topics.

Alan Thein Durning, the son of Marvin and Jean Durning, was born November 7, 1964, in Seattle, Washington. Durning grew up in Seattle but came from a family that had always roamed widely. Consequently, he aspired to the life of a world traveler, so much so that after his name in his grandmother's guest book, Durning, at the age of 11, wrote the title "world traveler."

Durning attended Oberlin College in Oberlin, Ohio, where he graduated with high honors in 1986, studying environmental policy but majoring in philosophy. While there, he also earned a Bachelor of Music degree from Oberlin Conservatory in trombone performance (he no longer plays). Shortly after college, Durning moved to Washington, D.C., where he joined the staff of the Worldwatch Institute, a research center that monitors the world's social and ecological health. After a few years of 70-hour weeks, Durning was promoted and began traveling the world, studying everything from poverty to atmospheric chemistry. "It was urgent stuff," he wrote in his award-winning book, *This Place on Earth*, "documenting injustice, testifying before Congress, jet-setting on behalf of future generations."

During this "jet-setting" period, Durning experienced something of an epiphany in the Philippines, while interviewing members of remote hill tribes about their land and livelihood. He met with an old, traditional priestess who asked him about his homeland. Durning did not know what to say. Home—when he was there—was Washington, D.C., a city he did not hold dear to his heart; crime and poverty were rampant. Durning abashedly replied that in the United States, people had careers, not places. Durning recalls that the old woman

Alan Durning (courtesy of Anne Ronan Picture Library)

looked at him with pity. It was this encounter that persuaded Durning to focus on local issues rather than global ones.

Looking for a place to call home, Durning left Washington, D.C., and returned to Seattle. There, in 1993, he founded Northwest Environment Watch, a wholly independent research center designed to foster an environmentally sound economy and way of life in the Pacific Northwest—the biological region stretching from southeast Alaska to northern California and from the Pacific to the crest of the Rocky Mountains. Northwest Environmental Watch is predicated on the belief that if an environmentally sustainable economy cannot be created in the greenest part of history's richest civilization, it probably cannot be done. If it can be done, the Pacific Northwest will set an example for the world.

Northwest Environmental Watch serves as a monitor of the region's environmental conditions and a pathfinder for routes toward a lasting economy. Through action-oriented interdisciplinary research, Northwest Environmental Watch provides northwesterners with reliable information about what sustainable development is and how to achieve it.

According to its mission statement, Northwest Environmental Watch creates tools for reconciling people and place, economy and ecology. One of its major tools is its series of publications, for which Durning is a prolific writer. The books in the series inform both generalists and experts of cutting-edge findings on a wide range of topics, such as the current health of ecosystems, the relationship between cars and cities, and the creation of green jobs. Durning's own publications include *This Place on Earth*, published in 1996, which was the winner of the 1997 Governor's Writers Award. *This Place on Earth* is part autobiographical as it recounts Durning's return to Seattle with his wife and children, the evolution of his own bioregional thinking, and the development of Northwest Environment Watch. Durning's other recent publications include *Green-Collar Jobs* (1999), *Tax Shift* (1998), *Misplaced Blame* (1997), *The Car and the City* (1996), and *How Much Is Enough?* (1992), which has been translated into seven languages. As a general theme, Durning's work focuses on socially and ecologically sustainable living.

Durning has also coauthored seven of the Worldwatch Institute's State of the World reports and two of its Vital Signs reports. His articles have been published in *World Watch*, *Washington Post*, *Los Angeles Times*, *New York Times*, *Christian Science Monitor*, *International Herald Tribune*, *Foreign Policy*, *Sierra*, *Utne Reader*, *Technology Review*, and more than 100 other periodicals. A sought-after keynote speaker, he has lectured at the White House, major universities, and numerous conferences. In 1996, Durning was awarded a Building Economic Alternatives award by Co-op America for his work.

Durning continues to live and work in Seattle with his wife, Amy, and their three children, Gary, Kathryn, and Peter.

Bibliography

Durning, Alan, *Green-Collar Jobs: Working in the New Northwest*, 1999; Durning, Alan, and Yoram Bauman, *Tax Shift: How to Help the Economy, Improve the Environment, and Get the Tax Man off Our Backs*, 1998; Durning, Alan, and Christopher Crowther, *Misplaced Blame: The Real Roots of Population Growth*, 1997; Durning, Alan, and John C. Ryan, *Stuff: The Secret Lives of Everyday Things*, 1997; Ryan, John C., *State of the Northwest*, 1994.

Dutcher, William

(January 20, 1846–July 1, 1920)
Amateur Ornithologist, First President of National Audubon Society

William Dutcher, a successful insurance agent for Prudential Life and expert amateur ornithologist, was a leader for the bird-protection legislation movement of the late nineteenth century and early twentieth century and a pioneer of the National Audubon Society's sanctuary program.

William Dutcher was born on January 20, 1846, in Shelton, New Jersey, and went to work at age 13. With little formal education, Dutcher became a very successful businessman and expert birder. His acquaintances spoke of "extraordinary personal qualities" that offset his lack of formal education. His interest in birds was showcased through his interest in hunting birds and his impressive collection of bird specimens from the Long Island and New Jersey shores. He wrote numerous respected scientific papers about birdlife in the region.

The American Ornithologists Union (AOU), at the American Museum of Natural History in New York, was founded in 1883 for "the protection and study of birds." That same year, Dutcher was elected as an associate member. He served as its treasurer for 14 years and rapidly became one of the most active members of its Committee on Bird Protection. He was especially interested in legislation that would protect North American birds, and with the Committee, he drafted model state legislation for nongame bird protection in 1886. During the late 1800s, birds were in high demand for their plumage, which was used to decorate women's hats and considered high fashion at the time.

In 1898, Senator George F. Hoar of Massachusetts introduced a bill in Congress that would have outlawed the importation, sale, or shipment of millinery plumes in the United States. The bill did not gain the support of the AOU, and many claimed that this was a primary reason for its failure. Congressman John F. Lacey of Iowa introduced a similar bill, called the Lacey Act, that prohibited intrastate shipment of birds and other animals if it violated state law, using the commerce clause from the U.S. Constitution that allows the federal government to regulate state trade. The AOU joined the effort to lobby for the Lacey bill, and it was passed in 1900. However, only five states had bird-protection legislation at the time.

The bird-protection movement by 1900 was fragmented. There were Audubon bird-watching and protection groups in many states, and some were passing bird-protection legislation, but Audubon groups were not unified in a singular effort. Dutcher suggested forming a union of state Audubon societies to create an impressive national front. In 1901, at a meeting of state Audubon groups, members decided that a national committee should be formed, and Dutcher was elected chairman. By 1903, there were 37 state societies, most of which were part of the national committee.

With the passage of the Lacey Act as well as the passage of bird-protection legislation by a growing number of states, Dutcher recognized the need for enforcement of the law on the ground in the individual states. In 1900, he initiated

William Dutcher (Library of Congress)

an enforcement system that evolved into the successful Audubon sanctuary and warden system. With the fund-raising help of New Hampshire AOU member Abbott H. Thayer, the program began with the hiring of wardens to protect seabird colonies. The first wardens recruited were lighthouse keepers who were hired to post No Trespassing signs and watch the coast for bird poachers. By 1901, 27 wardens were working for the program. Dutcher organized questionnaires for the wardens to complete that provided a wealth of information about bird habitat, mating, and hatching. For example, the questionnaires included information about where the largest breeding colonies of each species were along the eastern seaboard.

By January 1903, the national committee of Audubon societies, under Dutcher's leadership, had fallen $700 in debt. An anonymous benefactor promised a legacy of $100,000 to be given to the Audubon Society if it incorporated immediately and promised to broaden its focus to include wild animals as well as birds. In 1905, the National Association of Audubon Societies for the Protection of Wild Birds and Animals was officially incorporated.

To help run the organization, Dutcher hired T. Gilbert Pearson as the first salaried, full-time Audubon executive. Together they supervised the warden system, lobbied for more bird-protection legislation, and initiated a bird education program for children. While Dutcher grew more fanatical in his belief for bird protection toward the end of his career (he eventually came to disagree with the taking of scientific specimens), Pearson served as a middle-of-the-road figure. Pearson was also effective as a fundraiser and public speaker for the association.

Dutcher's budgeting woes continued until 1907, when the association accumulated $8,000 worth of debt. Dutcher, responsible for the mismanaging, was reprimanded by the board of directors, and his control of the budget was limited. He resigned as president but was reinstated shortly after.

Dutcher remained dedicated to his work on bird-protection issues despite his managerial problems. Through education and legislation, Dutcher made significant progress in the work toward bird protection. As early as 1902, Dutcher insisted that educational programs would be essential to the success of the Audubon movement. He wrote educational leaflets for the National Committee of Audubon Societies until 1910. Each leaflet, focusing on one bird, consisted of four pages of text that described the bird's habitat and actions, accompanied by illustrations. At the end of the leaflet were tips for teachers and students on studying the bird. The education program did not do as well as Dutcher had hoped, largely for lack of funds, until 1906. At that time Margaret Olivia Slocum Sage, widow of Russell Sage, established the Russell Sage Foundation, which gave money for the education program. By 1915, there were 152,164 children enrolled in 7,728 Junior Audubon classes.

Toward the end of his career, in 1910, Dutcher was instrumental in the passage of the Audubon Plumage Bill in New York by Gov. Charles Evans Hughes. The bill prohibited the sale or possession of plumage of birds in the same family of any species protected in New York, including herons, egrets, gulls, terns, and songbirds. It gave milliners a year to use the prohibited feathers they had in stock. Although the law pertained only to New York, that state was the main center of the millinery activity. While Pearson lobbied for the bill, Dutcher was the mastermind behind it.

On October 19, 1910, Dutcher suffered a paralyzing stroke that prohibited him from speaking or writing again, although he lived for ten more years. Dutcher retained the title of president while Pearson succeeded him as executive officer of the Audubon Association. Dutcher died July 1, 1920.

BIBLIOGRAPHY

Fox, Stephen, *The American Conservation Movement, John Muir and His Legacy,* 1981; Graham, Frank, Jr., *The Audubon Ark,* 1990; Stroud, Richard, ed., *National Leaders of American Conservation,* 1985.

Dyer, Polly

(February 4, 1920–)
Wilderness Preservation Activist

Polly Dyer has been a major force behind wilderness protection efforts in the Pacific Northwest. She was one of the few women involved in the conservation debates of the 1950s and 1960s, and once her talent for organizing environmental activists was revealed, she quickly rose to coordinate protection campaigns for Olympic, Mt. Rainier, and North Cascades National Parks. She has cofounded four regional conservation organizations and two state chapters for national groups, and since 1964 she has organized Northwest Wilderness Conferences, initially for the Federation of Western Outdoor Clubs, and since 1984 under the auspices of the Northwest Wilderness and Park Conference.

Polly Dyer was born Pauline Tomkiel on February 4, 1920, in Honolulu, Hawaii. Her father was a career officer in the United States Coast Guard, and the family moved frequently. It was not until they moved to Ketchikan, Alaska, in 1940 that Dyer discovered wilderness and an appreciation and love for the natural world. She explored the natural areas near

Ketchikan, and one day climbed to the top of Deer Mountain, where she was awed by the beautiful, wild expanse in view. During another hike up Deer Mountain, she met John Dyer, a young chemical engineer and world-class rock climber who had moved to Ketchikan from the San Francisco Bay area in 1943. The two married in 1945. John Dyer had been an officer for the Bay Area chapter of the Sierra Club, and Polly Dyer quickly became interested in the club's conservation work. Together they were active in a local Ketchikan conservation group that John had helped found.

The Dyers remained involved in conservation causes when they moved to Berkeley, California, in 1947 and then to Auburn, Washington, in 1950. In Washington, they and another couple cofounded the Pacific Northwest chapter of the Sierra Club, the club's first definitive step toward becoming a national organization. They also joined the Mountaineers, a Washington-based hiking and conservation organization. The couple backpacked in the Cascade and Olympic mountain ranges, and Polly Dyer led a Girl Scout troop, emphasizing hiking over arts and crafts.

Dyer became secretary for the Mountaineers' Conservation Committee in 1952, and in 1953 she represented the Mountaineers on the Washington governor's Olympic National Park Review Committee, to study and make recommendations for the possible removal of the west-side forests from the park. Two other conservationist women served on the 14-member committee, which was dominated by men (and one other woman) sympathetic to the timber industry. The committee's majority preferred reducing the size of the park, but its report recommended instead that a study be made by a "higher authority." The conservationist forces, joined by two other members representing trade unions, wrote a minority report to oppose reduction. It was known that the governor agreed with the majority, but Dyer and the other conservationists had mobilized their organizations to send so many letters to the committee that he subsequently responded that there was no need for further study.

This was the first of what would be Dyer's many major organizing efforts on behalf of wilderness preservation. She continued to defend Olympic National Park from constant attempts during the 1950s by the timber industry to log the parks, even testifying before the U.S. Senate in 1955. Rayonier, a timber company owning substantial forest land on the Olympic Peninsula, tried to ridicule Dyer and other women preservationists in a full-page national magazine advertisement that showed women scolding loggers. The advertisement explained that the women proposed wasting potential timber because they believed that trees should "reach maturity, die, topple over, and rot."

In 1956, a scenic highway was proposed for the new coastal strip of Olympic National Park, the longest roadless portion of coastline in the continental United States. Brainstorming with colleagues from the Wilderness Society (TWS) about how to defeat this proposal, Dyer responded enthusiastically to TWS executive secretary Howard Zahniser's suggestion to invite Supreme Court justice William O. Douglas to lead a hike along the stretch. This hike would be similar to the well-publicized three-day hike he had led along the Chesapeake and Ohio Canal in

1954. Participants on that trek had been well-known conservationists and writers and editors from the *Washington Post.* The hike had convinced them to oppose plans to build an expressway there and instead to support Douglas's proposal to establish a national historic park. Douglas agreed to the Olympic Coast hike, and Dyer coordinated the 22-mile trip in August 1958. Seventy-two people participated, including the nation's best-known conservationists and several journalists. A 1964 reunion hike along the southern part of the park's coastal area drew more than 150 participants and once again drew public attention to the pristine, roadless beauty of the area.

In 1957, Dyer cofounded the North Cascades Conservation Council, the idea originating at a meeting in her living room. The council fought for national park designation for an area of mountainous national forest in northwest Washington that was slated for logging. Finally in 1968 the North Cascades National Park Complex, which includes Ross Lake and Lake Chelan National Recreation Areas, was designated. Dyer opposed the construction of the North Cascades Scenic Highway during the 1960s for its intrusion on unspoiled wilderness and was one of two women who accompanied Gov. Dan Evans on a 20-person horseback survey of the route.

Although Governor Evans supported the road project and presided at the opening of the road in 1972, he is a committed environmentalist, and he and Dyer remained allies. They had first worked together in the early 1960s, when he was a state legislator, on a bill that regulated billboards. Dyer participated in the lobbying effort for the Wilderness Act of 1964 and contributed some of its language. As governor, Evans appointed Dyer to the Forest Practices Board in 1974. She convinced him to add seven more miles of roadless coastline to Olympic National Park, including Shi Shi Beach and Point of the Arches. Once Evans was elected U.S. senator, Dyer worked with him to have areas within Olympic, Mount Rainier, and North Cascades National Parks designated as wilderness under the Wilderness Act, protecting them from logging, road building, new mining claims, dams, and off-the-road vehicles.

In addition to these major preservation accomplishments, Dyer has cofounded several state preservation groups. She cofounded Mount Rainier National Park Associates in 1958. She and three others from the governor's 1950s Olympic National Park Review Committee were invited to join the board of trustees of Olympic Park Associates. Both organizations act as watchdogs over their respective national parks. She has served on the boards of both of these groups and is currently president of Olympic Park Associates. In 1984 she cofounded and for two years served as president of the Puget Soundkeeper Alliance, which works with automotive shops, shipyards, and marinas along the sound to reduce water pollution. Dyer organized biennial wilderness conferences for the Federation of Western Outdoor Clubs from 1964 to 1974 and then in 1984 to celebrate the 20th anniversary of the Wilderness Act, under the auspices of the Northwest Wilderness and Park Conference. These conferences enthuse and rally activists, help them network more effectively, and bring in new recruits to learn more about wilderness and how they can contribute to the effort.

Although all of the work mentioned thus far has been done as a volunteer, Dyer worked from 1974 to 1994 as continuing environmental education director at the University of Washington's Institute for Environmental Studies. Dyer earned a B.A. in geography in 1970 from the University of Washington.

Dyer continues to live in northeast Seattle with her husband, John Dyer.

Bibliography

Breton, Mary Joy, *Women Pioneers for the Environment*, 1998; Kaufman, Polly Welts, *National Parks and the Woman's Voice*, 1996; Pryne, Eric, "A Fighter by Nature, Longtime Conservation Leader in the State Doesn't Plan to Slow Down in Retirement," *Seattle Times*, 1994.

E

Earle, Sylvia

(August 30, 1935–)
Marine Botanist, Oceanographer

Sylvia Earle has spent almost 6,000 hours underwater exploring and studying the life of the oceans, earning the nickname "Her Royal Deepness." Her explorations have given her a vast appreciation of the ocean as a living system, one upon which the health of the planet is utterly dependent. As the most famous woman marine scientist of our time, she is highly visible, and she has used this attention to speak out against the harmful effects of pollution, development, and overfishing on the earth's seas. She is leading a project called Sustainable Seas Expeditions, in which her goal is to dive in submersibles in all 12 of the U.S. marine sanctuaries to document the wildlife there and to try to understand some of the long-term effects of environmental abuse.

Born on August 30, 1935, in Gibbstown, New Jersey, Sylvia Alice Earle was raised with her two brothers on a small farm near Camden. Her interest in the oceans started early, on vacations to the New Jersey shore. Curious, she read all she could find about oceans and began recording her observations in a notebook. Although neither of her parents had been to college, they valued education and encouraged her love of exploring. When she was 12 years old her family moved to Clearwater, Florida, and suddenly the Gulf of Mexico was at her back door. Someone gave her a pair of goggles, which opened a new underwater world and increased her desire to learn more about the oceans. She finished high school at the age of 16, and though her parents were unable to help her financially, she began studying at Florida State University, paying her way by working in a laboratory. She earned her B.S. in marine biology and had graduated by the time she was 19 years old. It was during this time that she decided to study marine plants. Understanding the vegetation, she discovered, is the first step to understanding how the whole ecosystem works.

After earning her master's degree in 1955 from Duke University, she began work on her Ph.D., continuing her study of plant life in the Gulf of Mexico. During this time she married and her attention turned to having a family. In the next few years two children arrived, but she was able to continue working on her dissertation—"Phaeophyta of the Eastern Gulf of Mexico." When it was completed in 1966, it caused a sensation in the oceanographic community by giving the world its first glimpse at the marine plant species in the Gulf of Mexico. She has continued her effort to catalog the plant and animal species of the oceans to help people understand and protect this largely unexplored ecosystem.

Until this time, most of what was known about life in the oceans was learned from dead specimens. Inspired by Jane Goodall, who studied chimpanzees in their own habitat, Earle wanted to do the same for marine plants and animals. In 1970 she was selected to participate in Tektite, a project sponsored by the United States Navy, the Department of the Interior, and the National

Aeronautics and Space Administration (NASA). It allowed scientists to live in an enclosed laboratory 50 feet under water to conduct extended research projects. She and four other women lived and worked there for two weeks, and she treasured the chance to get to know some of the ocean creatures on their own terms.

The Tektite mission brought Earle new recognition and national attention. She used this opportunity to raise public awareness of the oceans and began speaking out against the damage being done to the oceans by development along coastal areas, pollution and dumping of chemicals, overharvesting of fish, and other environmental hazards. In addition to her ongoing research projects, she began writing books and producing films, always advocating greater care for the oceans of the world. She has also been ever-mindful of how little we actually know. Less than one-tenth of 1 percent of the ocean has been explored, and she emphasizes that it is difficult yet essential to protect marine species.

In continuing to push the boundaries of underwater exploration, Earle has been confronted with shortcomings in existing technology. In 1979 she made a record-breaking dive to 1,250 feet below surface in a special pressurized "Jim" suit near the Hawaiian island of Oahu. She was able to explore the depths for two and a half hours, untethered, yet this only fueled her desire to dive deeper. In the 1980s she and engineer Graham Hawkes started Deep Ocean Technologies, a company dedicated to developing tools for oceanic work and exploration. In 1984, using one of their innovative Deep Rover submersibles, Earle made another record-breaking dive, this time to 3,280 feet.

Earle continued through the 1990s to pursue her goal of helping people understand the role of oceans in maintaining the health of the planet. In 1990 she was appointed chief scientist of the National Oceanic and Atmospheric Administration (NOAA), where, among other duties, she was responsible for monitoring ocean health. As part of her work with NOAA she traveled to Kuwait and reported on the environmental damage caused when 11 million barrels of oil spilled into the Persian Gulf during the Gulf War. Frustrated with the bureaucracy facing a political appointee, she left the post in 1992 and continues to struggle with disappointment in a government that has no underwater equivalent of NASA, despite the fact that oceans cover 75 percent of the earth and that 90 percent of all known organisms live there. She continues to be a highly visible environmentalist—in 1998 she began leading a project called Sustainable Seas Expeditions, with a goal of documenting the wildlife in the 12 U.S. marine sanctuaries. These marine reserves, administered by NOAA, make up about 18,000 square miles of the coastal waters of the United States and are largely unexplored. Declines in numerous fish species, widespread coral loss, and other signs of increasing ecological damage make the Sustainable Seas project even more important. It creates baseline data for monitoring the future health of these reserves and for educating the public about protecting the diversity there.

Sylvia Earle has been married three times and raised four children. She is active in environmental organizations, including serving as a board member of the Center for Marine Conservation. She is in high demand as a lecturer, encouraging her audiences to think of the ocean as a cornerstone of the life-support system for the whole planet. She lives in Oakland, California.

Bibliography

Conley, Andrea, *Window on the Deep*, 1991; Earle, Sylvia A., *Dive! My Adventures in the Deep Frontier*, 1999; Earle, Sylvia A., *Sea Change: A Message of the Oceans*, 1995; Pangea Digital Pictures, *Oceanography/with Dr. Sylvia Earle* (video recording), 1995; White, Wallace, "Her Deepness," *New Yorker*, 1989.

Edge, Rosalie

(November 3, 1877–November 30, 1962)
Conservation Activist

Bird-watcher and environmental reformer Rosalie Edge was the first woman to become prominent in the American conservation movement. She established the Emergency Conservation Committee (ECC) in 1929 to investigate misdeeds of the National Audubon Society, of which she was a life member. She published numerous pamphlets describing environmental problems and recommending solutions and provided occasional testimony to the U.S. Congress about conservation issues. In addition to personally purchasing the land for Hawk Mountain Sanctuary in Pennsylvania, she fought successfully for the establishment of Olympic and Kings Canyon National Parks and for the expansion of Yosemite National Park. She lost a campaign in the 1930s to reform the U.S. Biological Survey, but her complaints were later addressed by the U.S. Fish and Wildlife Service during the 1960s and 1970s.

Mabel Rosalie Edge was born Mabel Rosalie Barrow in New York City on November 3, 1877. Cousin to author Charles Dickens and painter James McNeill Whistler, Edge was raised as a member of the privileged elite. She attended private schools but got into trouble for asking too many questions and annoying her teachers. She met her husband, British engineer Charles Noel Edge, on a trip to England and shortly after their marriage moved to the Orient, where he sold locomotives. They had two children and spent some time in England before returning to the United States. Edge became involved with the women's suffrage movement and served as treasurer for the New York Women's Suffrage Party. In 1915, when the family bought a vacation house in Rye, New York, Edge began noticing the area's birdlife.

Soon she joined the Audubon Society as a life member and came to know the zoologists at the American Museum of Natural History. One of them, Dr. Willard Van Name, was especially concerned about the dangers facing birds. They were hunted for sport and for their plumes, and although the Audubon Society had fought hard for federal laws to protect birds during the first two decades of the century, Van Name believed that the society was not doing enough to advocate the laws' strict enforcement. He wrote a pamphlet about the situation and mailed a copy to Edge, who was then vacationing in Paris. She read it and returned home in time for the Audubon Society's 1929 annual meeting. The board of directors and Audubon's president T. Gilbert Pearson were taken aback by Edge's forceful questions about how the Audubon Society was addressing the problems Van Name brought up. Instead of responding to her concerns, Pearson complained that she had spoiled the meeting. There was no time to see the movie that had been scheduled, lunch was getting cold, and the official photographer for the annual society photograph had been kept waiting too long.

Far from discouraging Edge, this rebuff fueled her drive to reform the Audubon Society. Edge formed the Emergency Conservation Committee with Van Name and journalist Irving Brant shortly after the annual meeting. From that year on, she and her small group of allies attended every Audubon Society annual meeting. The formerly subdued gatherings became lively happenings. Crowds attended, eager to witness the scandal Edge inevitably stirred. President Pearson felt the heat. The ECC publicized his collusion with sportsmen and the hunting industry and decried the commissions he earned from the generous contributions by these interests.

In 1932 Edge discovered that the Audubon Society was earning "rent" from its Paul Rainey Wildlife Sanctuary in Louisiana. The Audubon Society had been leasing trapping rights to local muskrat trappers and had earned some $100,000 in royalties from the almost 300,000 muskrat pelts obtained. Infuriated, Edge brought this up at the 1932 annual meeting. Pearson and the board justified it, saying that muskrats were eating the vegetation that the endangered blue goose depended upon, and that furthermore, the money raised was necessary for the depression-depleted coffers of the Society. Edge published a pamphlet on the situation, and after fighting in court to obtain Audubon's mailing list, she sent a pamphlet to every member of the National Audubon Society. Audubon Society membership plummeted from 8,400 to 3,400, but Pearson remained in place. Finally in 1934, Pearson was forced by his board to resign, and the new president, John Baker, put an end to the trapping.

This fight won, and confident that the Audubon Society was in better hands, Edge turned her focus to the federal government's conservation policies. She fought with the U.S. Biological Survey, which at that time appeased farmers and ranchers with an aggressive antipredator policy. She argued that predators were given unjustifiably bad press and that they actually helped farmers by preying on destructive rodents. One of her pamphlets deconstructed the popular myth that eagles routinely carried off human babies. Although the Biological Survey ignored her concerns at the time, when reformers took up the cause during the 1960s and 1970s, the later-reorganized U.S. Fish and Wildlife Service did cease its predator program.

Edge's concern for birds of prey led her to found a sanctuary for them along the Pennsylvania mountain ridge they follow on their southward migration. Edge visited the Kittatinny ridge one fall and found scores of hunters lying in wait for the hawks and shooting them for sport by the thousands. Edge believed that three-peaked Hawk Mountain, in the middle of the migration corridor, would serve as a valuable refuge, and she used $2,500 of her inheritance to purchase it in 1934. The ECC raised more funds to hire a caretaker and build hawk shelters at Hawk Mountain Sanctuary.

During the 1930s Edge joined with other conservationists to fight for the creation of two new national parks in the West: Olympic National Park and Kings Canyon National Park, established in 1938 and 1940, respectively. She was the principal proponent of the enlargement of Yosemite National Park, which resulted in the addition of 6,000 acres of old-growth sugar pine forest in 1937. Edge also joined the fight against the dam in Dinosaur National Monument's Echo Park in the 1950s.

Edge kept the ECC going as long as she lived. Since it was essentially a one-person organization during her last years, it folded after her death, but Hawk Mountain Sanctuary still exists and is administered by an association that she founded for that purpose. The National Audubon Society eventually expressed public appreciation for her persistent activism, treating her to a standing ovation at the 1962 annual meeting just a few weeks before her death. Edge died on November 30, 1962, in New York City.

Bibliography

Broun, Maurice, *Hawks Aloft: The Story of Hawk Mountain*, 1949; Fox, Stephen, *The American Conservation Movement: John Muir and His Legacy*, 1981; Graham, Frank, Jr., *Man's Dominion: The Story of Conservation in America*, 1971; "Hawk Mountain Sanctuary," http://www.hawkmountain.org/; Kaufman, Polly Welts, *National Parks and the Woman's Voice*, 1996; Taylor, Robert Lewis, "Oh, Hawk of Mercy," *New Yorker*, 1948.

Ehrenfeld, David

(January 15, 1938–)
Founder and Editor of *Conservation Biology*, Ecologist

David Ehrenfeld is a professor at Rutgers University who specializes in conservation ecology and its role in an increasingly technological world. He has gained attention for his leadership in the journal *Conservation Biology*, of which he was the founding editor, and through the books he has written. In his influential writings, he often focuses on society's obsession with power and technology and on how that obsession affects our relationship with the natural world.

David William Ehrenfeld was born January 15, 1938, in New York City to Irving and Anne (Shapiro) Ehrenfeld. He attended Harvard University, where he earned his bachelor's degree in history in 1959, and then attended Harvard Medical School, receiving his M.D. in 1963. At the University of Florida he did doctoral research on the orientation and navigation of sea turtles and received his Ph.D. in zoology in 1966. The next year, Ehrenfeld accepted an assistant professor position at Columbia University in New York and three years later became an associate professor of biology. In 1970, he and June Gardner, a plant ecologist, were married. They would later have four children: Kate, Jane, Jonathan, and Samuel. Ehrenfeld stayed at Columbia University until 1974, when he took a position as professor of biology at Rutgers University in New Brunswick, New Jersey.

In 1978, Ehrenfeld's book *The Arrogance of Humanism* came out. In it, he sets out to remind the world of its failures—and does so by commenting on soil depletion, technological mistakes such as dichlordiphenyltrichlor (DDT), plant and animal extinctions, and the concurrent rise in social violence. Rather than point a finger at capitalism or overpopulation or any of the other usual targets of environmentalists, Ehrenfeld blames the philosophy of humanism and its defining principle of supreme faith in human reason. He believes that humans are having a fatal love affair with technology and control and that problems arise when the only socially acceptable solution is to apply reason over emotion. It is an example of hubris to believe that all problems are soluble by people, he writes, and such intellectual arrogance simply reduces nature, thereby reducing conservation as well. *The Arrogance of Humanism* became a much-discussed and influential book that has gone into nine printings. The influence of his ideas is still felt more than 20 years later, with contemporary ecologists discussing the underlying assumptions of ecosystem management as to whether or not they imply a demonstration of human arrogance.

In 1987 Ehrenfeld founded and became editor of *Conservation Biology*, a scientific journal that deals with ecologically important studies and issues that contribute to the preservation of species and habitats. It is supported by the Society of Conservation Biology, which is committed to examining the scientific basis of conservation in order to counter environmental deterioration. Articles in the journal cover topics such as analyses of species declines, population modeling, discussions of vertebrates as indicator species, recommendations for management decisions and practices, the impacts of game ranching and poaching, and many others. *Conservation Biology*, now accepted as required reading for ecologists throughout the world, remains instrumental in the effort to move conservation biology to the forefront of the sciences. In the past decade it has become the most frequently cited journal in its field. Ehrenfeld also has compiled selected articles from the journal into a series of six volumes titled *Readings from Conservation Biology*, which cover a broad range of topics—illustrating the diversity of approaches to conservation decisions.

Ehrenfeld's book *Beginning Again: People and Nature in the New Millennium*, published in 1993, further builds on a foundation of ecology. He maintains that many of the problems in the relationship between humans and nature that have accompanied the end of the twentieth century lie in an obsession with control. The ability to manipulate the world is new and addictive, he says, and it overlooks complexity and diversity—the backbones of the natural world. This fixation with control has also led to a growing tendency toward overmanagement, which serves to disconnect those in power from those doing the groundwork—in other words, those who really understand what's going on. In searching for solutions for these disturbing trends, Ehrenfeld relies on the understanding gained by ecologists, who study the particularities of life in diverse systems. He discusses restoration ecology, saying that although species cannot be reconstructed, ecosystems can—as long as it is recognized that restoration is never as good as preserving the land in the first place. In the last chapter of the book, titled "Life in the New Millennium," Ehrenfeld emphasizes that the ultimate success in conservation will depend on a revision of everyday living—not just in increased efforts to save the world. When people who are not even making efforts at conservation are able to live in a way that is compatible with the existence of other native species in the region, then the destructive changing of nature will end. In other words, he concludes, conservation has to start at home.

Since 1989, Ehrenfeld has been a regular columnist for *Orion*, a magazine dedicated to characterizing humans' responsibilities to the environment, exploring the ethic of stewardship, and cultivating nature literacy. In his column, Ehrenfeld sometimes deals with biological issues, such as woodland restoration or biotechnology. But often his topics go beyond biology to explore things like the modern lifestyle and its disconnection from the outdoors or the problems that arise from the reliance on expert knowledge, which usually ignores cultural wisdom. A common theme for Ehrenfeld is that society's concept of progress leaves little room for traditional knowledge or for anyone with concerns for the future, and yet traditional wisdom can often hold brilliant solutions to problems that people face today.

In 1993 he stepped down as editor of *Conservation Biology* and now serves as a consulting editor. He continues his work at Rutgers University, where he teaches general ecology, field ecology, and conservation ecology. He makes his home in Highland Park, New Jersey.

BIBLIOGRAPHY

Ehrenfeld, David, *The Arrogance of Humanism*, 1978; Ehrenfeld, David, *Beginning Again: People and Nature in the New Millennium*, 1993; Norton, Bryan G., "Beginning Again: People and Nature in the New Millennium," *BioScience*, 1994; Stanley, T. R., "Ecosystem Management and the Arrogance of Humanism," *Conservation Biology*, 1995.

Ehrlich, Anne, and Paul Ehrlich

(November 17, 1933– ; May 29, 1932–)
Biological Researcher; Population Biologist

Anne and Paul Ehrlich are well-known spokespeople for the conservation and population control movements in the United States. They have cowritten ten books, all of which send a straightforward, hard-hitting message about the environmental dangers that face the planet and its living inhabi-

Paul and Anne Ehrlich (Courtesy of Paul and Anne Ehrlich)

tants. Paul Ehrlich—whose academic specialty is entomology and who cofounded the field of coevolution—emerged in popular media during the 1960s and 1970s when he issued dire predictions about population growth and the ensuing scarcity of the earth's resources. His bestselling *The Population Bomb*, published in 1968, warned of the devastation that would be suffered by a population grown too big. The Ehrlichs work out of Stanford University's Center for Conservation Biology, which they helped found in 1984.

Anne Fitzhugh Howland was born on November 17, 1933, in Des Moines, Iowa. As a child, she developed an interest in nature through outdoor activities, nature studies in school, and summer camps.

Paul Ralph Ehrlich was born on May 29, 1932, in Philadelphia and raised in Maplewood, New Jersey. As a child, he chased butterflies and caught and dissected frogs; as a teen, he was encouraged in a serious study of butterflies by a mentor at the American Museum of Natural History in Manhattan. Ehrlich studied at the University of Pennsylvania, graduating in 1953 with a B.S. in zoology.

The Ehrlichs met at the University of Kansas; Anne studied there from 1952 to 1955, and Paul earned his M.A. there in 1955 and his Ph.D. in 1957. They married and established a close collaborative relationship. Paul Ehrlich accepted a position in Stanford University's Department of Biological Sciences in 1959. Specializing in evolution, population dynamics, and ecology, he has studied butterflies, snake mites, birds, and coral reef fishes. With botanist PETER RAVEN, Paul Ehrlich discovered and described a phenomenon they called "coevolution," complex interactions between species that influence the way they evolve. Anne Ehrlich focused on raising their daughter, Lisa Marie, during the 1950s and early 1960s. She illustrated Paul Ehrlich's identification key *How to Know the Butterflies*, published in 1961, and joined Stanford's Department of Biological Sciences in 1962.

The Ehrlichs first became interested as teenagers in environmental problems caused by overpopulation. They both read *Our Plundered Planet*, by HENRY FAIRFIELD OSBORN JR., and Paul Ehrlich also read *Road to Survival*, by WILLIAM VOGT. The writers of both books warned that overpopulation and abuse of natural resources could lead to widespread famine and impoverishment. During a sabbatical in 1966, the Ehrlich family traveled to India and was horrified by the conditions in the poorest part of Delhi, where people were living on the street, barely surviving. This seemed to be a worst-case scenario for overpopulation, and Paul Ehrlich described it in *The Population Bomb*, which was published in 1968 and quickly rose to best-seller status.

An appearance on the *Tonight Show* shortly after *The Population Bomb* came out thrust Paul Ehrlich into the public spotlight. He announced publicly that he had had a vasectomy and founded the organization Zero Population Growth to promote smaller families in the United States. He argued for a "luxury tax" on such items as diapers and baby food and supported a proposal that the U.S. government not send food aid to countries such as India that had not taken any measures to curb population growth.

The Population Explosion, which the Ehrlichs cowrote in 1990, provides an update to *The Population Bomb*, detailing environmental problems created by overpopulation. It emphasizes that wealthy countries, even though their populations are not increasing at the same rate as poorer countries, cause more environmental problems because they consume natural resources and pollute on a much greater scale.

In addition to *The Population Explosion*, the Ehrlichs have cowritten nine other books. They write in a clear and compelling manner to make their books accessible to lay readers. This dedication to the education of nonscientists about science and conservation issues, as well as their vanguard research and compelling arguments for conservation, has made the couple prominent spokespeople for the environment. *Extinction: The Causes and Consequences of the Disappearance of Species* (1981), in addition to being an entertaining primer on evolution and extinction, describes the dangers that will result from the widespread loss of biodiversity currently taking place on planet earth. Their memorable metaphor for disappearing species on earth, rivets being popped out of an airplane one by one, has been adopted and widely cited by the biodiversity conservation movement. The central problem leading to loss of biodiversity and other environmental problems, in their analysis, is that the human enterprise—which includes the global population's size and growth, its consumption, and technologies—is highly unsustainable and is becoming more so as both the population and each person's consumption increase. The inequitable rich/poor gap makes the unsustainable situation even more unstable. The Ehrlichs describe the causes and effects of inequity in *The Stork and the Plow* (1995), cowritten with Gretchen Daily.

The Ehrlichs have been especially concerned about what they call the "brownlash," an attempt by anticonservationists to convince the public that environmentalists are overly pessimistic and exaggerate the dangers to the planet. In their books and articles, the brownlashers claim that human activity has negatively affected only a very small portion of the planet, that nature is incredibly resilient and capable of recovering from any destruction that does occur, and that technology and human innovation will prevail and solve serious environmental problems. The Ehrlichs' *Betrayal of Science and Reason: How Anti-Environmental Rhetoric Threatens Our Future* (1996) debunks outrageous statements one by one, tracing misunderstandings, miscalculations, and misleading statements.

Anne Ehrlich became senior research associate for the Department of Biology in 1975 and began teaching an environmental policy course for Stanford's Human Biology program in 1981. Together, the Ehrlichs helped found the Center for Conservation Biology (CCB) at Stanford in 1984; it has become an influential research and teaching center for the field of conservation biology and also studies such questions as population growth, environmental deterioration, and use of natural resources. Anne Ehrlich serves as its associate director and policy coordinator, and Paul Ehrlich is its president.

In addition to her writings and work at the Center for Conservation Biology, Anne Ehrlich has worked as an outside consultant to the Global 2000 Report of the White House Council on Environmental Quality (1977–1980), as an adviser for the Fate of the Earth Conferences in 1981 and 1984, and as a commissioner for the Greater London Area War Risk Study (GLAWARS; 1985–1986) of the Greater London City Council. She was elected a member of the American Academy of Arts in 1998 and holds two honorary degrees. She has served on the boards of directors of numerous nonprofit organizations, including the Sierra Club; the Pacific Institute for Studies in Environment, Development, and Security; the Rocky Mountain Biological Laboratory; Friends of the Earth; and the Ploughshares Fund. She received the 1985 Raymond B. Bragg Award for Distinguished Service and also in 1985 was honored by the American Humanists Association.

Paul Ehrlich, the Bing Professor of Population Studies at Stanford University, has received several honorary degrees, a MacArthur Prize Fellowship, the 1990 Crafoord Prize of the Royal Swedish Academy of Science (which he shared with entomologist E. O. WILSON), the John Muir Award of the Sierra Club, the 1987 Gold Medal Award of World Wildlife Fund International, the 1993 Volvo Environmental Prize, and the Blue Planet Prize in 1999.

Paul and Anne Ehrlich shared the 1994 United Nations Environmental Programme Sasakawa Environment Prize, the 1995 Heinz Award, the 1996 Distinguished Peace Leader Award from the Nuclear Age Peace Foundation, and, most recently, in 1998, the prestigious Tyler Award.

BIBLIOGRAPHY

"Center for Conservation Biology," http://www.stanford.edu/group/CCB/; Ehrlich, Paul, *A*

World of Wounds: Ecologists and the Human Dilemma, 1997; Ehrlich, Paul, and Anne Ehrlich, *Healing the Planet: Strategies for Resolving the Environmental Crisis*, 1991; "Heinz Awards," http://www.awards.heinz.org/ehrlich.html.

Eisner, Thomas

(June 25, 1929–)
Entomologist

Thomas Eisner's life work has been focused on insects, and he is especially known for his discoveries about how insects communicate with one another by secreting certain chemicals. His work has led him to cofound (along with his wife, Maria, and his colleague Jerrold Meinwald) several new entomological fields, including insect physiology, chemical ecology, comparative behavior, biocommunications, and defensive secretions. Eisner has been a leader in the movement to conserve endangered environments and promotes bioprospecting, the search for useful substances occurring naturally, to make nature more economically valuable.

Thomas Eisner was born on June 25, 1929, in Berlin, Germany. His family moved to Barcelona when Hitler ascended to power in 1933. Then the Spanish Civil War erupted, and after a brief stay in France and Argentina, his family emigrated to Uruguay, where he spent the remainder of his childhood. Eisner's parents believe their son learned to walk in order to chase insects in the backyard, and Eisner tells a story about how as a young boy in Barcelona, he was in a sandbox one day, intently observing some pill bugs. A tremendous explosion shook his neighborhood—one of the first acts of terrorism of the Spanish Civil War. But instead of frightening the boy, it annoyed him because the ensuing chaos interrupted his session with the bugs! As well as indulging his intense interest in insects, his parents instilled him with a love of music and an unusual olfactory ability. His father, a chemist, concocted perfumes and creams at home, and their scents opened his nose to a world of subtle smells.

When Eisner's family moved to New York in 1947, Eisner applied to Cornell University but was not admitted, mostly owing to his poor English. He went to Champlain College in Plattsburgh, New York, and after two years there, transferred to Harvard, where he earned a B.A. in 1951 and a Ph.D. in 1955. Eisner married Maria Lobell and became a naturalized U.S. citizen in 1952. He was offered a teaching position at Cornell University in 1957 and has been there ever since. Now the Jacob Gould Schurman Professor of Biology (since 1976), Eisner codirects the Cornell Institute for Research in Chemical Ecology and is a senior fellow at the Cornell Center for the Environment. He is one of the most popular lecturers on campus.

In the 1991 documentary film *The Bug Man of Ithaca*, Eisner is a convincing advocate for insects. He reveals one caterpillar's defensive strategy: it hides itself from predators by sewing itself a costume of pieces of the flower it eats. He shows how the green lacewing's larvae, which eat woolly alder aphids, pick the wool off each aphid before eating it and stick the wool onto their own bodies. Still other insects fit into a complex ecological web. As a stink bug stuck in a spider web dies and is devoured by the spider, its stench attracts other flies that in turn get stuck in the web. Another spider finds a moth in its web, tastes it, and if it does not like the moth, frees it. Certain caterpillars feed on poisonous plants, whose distasteful flavor remains in their bodies once they transform into moths. Spliced into this film are cuts from some of Hollywood's insect horror movies, which by the end of *Bug Man* seem crass compared to the brilliance of bugs as revealed by Eisner.

Eisner is best known for his research into the ways that insects defend themselves and communicate with one another using chemicals. One of Eisner's most famous discoveries was of the defense system of the bombardier beetle, which shoots hot poison from the tip of its abdomen at a rate of 500 shots per second. Eisner and Meinwald have discovered a chemical in millipedes that could be used as a nerve drug, a substance in fireflies that serves as a cardiac stimulant, and a cockroach repellent that occurs in an endangered mint.

Learning about the potentially useful chemicals manufactured by plants and animals led Eisner to develop a field he calls "chemical prospecting," which involves scouring natural habitats for potential medicines, pesticides, and chemicals of use to industry. Eisner sees four benefits from chemical prospecting: new jobs for local people, new opportunities for investment, more funds for conservation, and a stronger scientific infrastructure in biodiverse countries. This will help developing countries, especially in their quest for funds for conservation. Eisner helped convince the Merck & Company pharmaceutical giant to sign an agreement with the Costa Rican National Biodiversity Institute (INBio), whereby Merck would pay INBio half a million dollars per year in return for being provided with samples of what INBio researchers believe might be useful pharmaceutical substances. The agreement provides Merck with a steady stream of promising chemicals and helps maintain INBio.

Eisner has received many awards for his work in chemical ecology and conservation, including Harvard University's Centennial Medal, the 1990 Tyler Prize, the 1994 National Medal of Science, and the National Conservation Achievement Award from the National Wildlife Foundation in 1996. He sits on the boards of directors of Zero Population Growth and the National Audubon Society. Eisner resides in Ithaca, New York.

BIBLIOGRAPHY

Ackerman, Diane, *The Rarest of the Rare: Vanishing Animals, Timeless Worlds*, 1995; Eisner, Thomas, and Jerrold Meinwald, eds., *Chemical Ecology: The Chemistry of Biotic Interaction*, 1995; Eisner, Thomas, and Edward O. Wilson, eds., *The Insects: Readings from* Scientific American, 1997; Canadian Broadcasting Corporation, *The Bug Man of Ithaca*, (video recording) 1991; Wilson, E. O., *Naturalist*, 1994.

Emerson, Ralph Waldo

(May 25, 1803–April 27, 1882)
Writer

From the mid-nineteenth century on, Ralph Waldo Emerson has been cited by writers and naturalists as inspiration for their interest in the natural world. A leader of the American transcendentalist movement beginning in the mid-1830s, Ralph Waldo Emerson spoke and wrote about nature as a revealer of spiritual truths, declaring in his essay "Nature" that "every natural fact is a symbol of some spiritual fact." His work as philosopher, orator, theologian, and poet remains a prerequisite in the study of American literature, history, theology, and the natural sciences.

Ralph Waldo Emerson was born in Boston on May 25, 1803, when some areas of Boston were still half-rural. Like many naturalists, as a child he was drawn to open fields, ponds, and sylvan places. His father, Rev. William Emerson, died when Ralph Waldo was seven years old, and it was only through great resourcefulness that his mother succeeded in sending all six sons to Harvard College. Young Ralph Waldo was greatly encouraged by his aunt, Mary Moody Emerson, who enjoyed his poetry and provided him with literature from around the world. Emerson graduated at 18 years of age and took the pulpit of Boston's Second Unitarian Church at the age of 27. He continued to build on his studies and became increasingly convinced that deeper knowledge of the natural world would not only attest to the glory of God, but would also build one's moral foundation. He stated in a sermon called "Astronomy" that "religion will become purer and truer by the progress of science. This consideration ought to secure our interest in the book of nature."

By 1832, Emerson's curious mind led him away from the ministry and toward natural history. He went to Europe and visited the Jardin des Plantes at the Musée d'Histoire Naturelle in Paris, after which he wrote in his journal, "I am moved by strange sympathies, I say continually, I will be a naturalist." He referred back to this experience in his work continuously over the next decades. Emerson returned to the United States and began lecturing at the Natural History Society in Boston and so embarked on a career as a prophetic orator whose addresses were often published as essays and were disseminated by New England church ministers and literati. Rather than return to the pulpit, Emerson preferred to work as a free agent, organizing his own speaking engagements by renting a hall and paying himself with the proceeds.

Emerson's 1836 essay "Nature" presents his theory that the natural world is emblematic of larger truths: "1) Words are signs of natural facts. 2) Particular natural facts are symbols of particular spiritual facts. 3) Nature is the symbol of spirit." Criticism immediately followed its publication, as church officials recognized Emerson's ideas as an attack on the authority of organized Christianity. Emerson's "Nature" and the ensuing discussion in ecclesiastic circles led to a loosely organized movement in New England called transcendentalism. Transcendentalists referred to the book-length essay as "the Bible," and they

Ralph Waldo Emerson (courtesy of Anne Ronan Picture Library)

would gather at Emerson's Concord home for transcendentalist meetings. Transcendentalists held the Neoplatonic and Kantian notion of a parallel between the material world and the higher realms, and they regarded nature as a source of divine truth about the spirit and the human mind.

In 1841, while living in Concord, Massachusetts, Emerson took in a handy young writer, HENRY DAVID THOREAU, who in exchange for chores had a place to live and a visionary mentor in Emerson. Thoreau put Emerson's philosophy to the test by practicing it in the woods of Concord and northern New England and chronicling his experience in the American literary classic *Walden*. While Emerson and Thoreau did not agree about everything, they spoke and wrote on nature's behalf like none other before them. The magazine *The Dial*, which Emerson published with Margaret Fuller from 1840 to 1844, became an outlet for their transcendentalist writing. Emerson's productive relationship with Thoreau lasted until Thoreau's untimely death in 1862 of tuberculosis. The late twentieth century's ecology movement in the United States, nature writing, and the spread of Buddhist philosophy in the United States all claim some influence from the combined literary efforts of Emerson and Thoreau.

Some of Emerson's most quoted essays are "The American Scholar," which Oliver Wendell Holmes called the United States's "intellectual Declaration of Independence," and "The Poet," an essay pored over by WALT WHITMAN, Emily Dickinson, and a generation of American poets. Perhaps Emerson's most famous essay was "Self-Reliance," which reified the American characteristic trait of rugged individualism.

Critics were later to say that Emerson spent too much time doing desk work to be a true naturalist and that he preferred his creature comforts to the rugged lifestyles of Henry David Thoreau and JOHN MUIR. Late in Emerson's life he invited Muir to come to New England as his guest, advising that the solitude of the wilderness "is a sublime mistress, but an intolerable wife." But what Emerson may have lacked in woodsmanship and natural scientific knowledge, he made up for with his flair for the romantic, and his enthusiastic romanticism led many people into the woods and into the study of natural history. He also remained au courant in the natural science study of his day, reading extensively on developments in the field. As laudatory of natural scientific inquiry as he could be in his speeches and essays, he also admonished the direction of science when it became strictly empirical and denied the spiritual aspect.

Among the lasting works of Ralph Waldo Emerson, his essays and sermons are most widely known, but his journals (from a journal-writing practice that began when he was eight years old and continued until his death) especially chart his studies of the natural environment. Ultimately, Emerson was not a naturalist but a writer, public speaker, poet, and perhaps the first American philosopher. His work has had a lasting impact on American theology, environmental study, poetry, and nature writing.

At 79 years of age, on April 27, 1882, Emerson died from pneumonia that he had contracted walking in a spring rain without a coat or hat. He is buried in Sleepy Hollow Cemetery in Concord, Massachusetts, the town he called home for most of his life.

Bibliography

Brooks, Van Wyck, *The Life of Emerson*, 1932; Elder, John, ed., *American Nature Writers*, Vol. 1, 1996; Paul, Sherman, *Emerson: Angle of Vision: Man and Nature in the American Experience*, 1952; Richardson, Robert D., *Emerson: The Mind on Fire, A Biography*, 1995.

F

Fontenot, Willie

(November 26, 1942–)
Community Liaison Officer in the Louisiana Attorney General's Office

Willie Fontenot has worked for more than 20 years as an environmental specialist in the Citizens Access Unit of the Louisiana attorney general's office. He has helped organize over 500 citizens' groups throughout Louisiana and 30 other states to solve environmental problems faced by their communities. Fontenot also works with other state and national environmental organizations, including Clean Water Action, the Mississippi River Basin Alliance, and the Louisiana Environmental Action Network.

William Alexander Fontenot was born on November 26, 1942, in Opelousas, Louisiana, one of eight children. His early environmental education came from his father, who subscribed to several natural history magazines and was concerned about the effects of the campaign to eradicate fire ants in the Deep South in the 1950s by poisoning them with chlordane in ground-up corn cobs. Fontenot enrolled at the University of Southwest Louisiana (USL) in 1960 but left in 1961 to serve for four years in the United States Navy. Returning from military service, he reenrolled at USL and graduated with a B.A. in political science in 1969.

Fontenot moved to New Orleans in 1970. From 1970 to 1975, he worked in retail and business, owning an antique and traditional craft store with his wife from 1973 to 1975. When he learned about the area's severe environmental problems in 1970, he joined the local environmental movement. Lake Pontchartrain was fouled by pollutants from industry, dredging for clam shells, and sewage. Local developers were using the state highway department and the Corps of Engineers to push several projects that would have led to the draining and development of 300,000 acres of coastal wetlands around New Orleans. (Today most of these wetlands are preserved in Bayou Savage National Wildlife Refuge in New Orleans East and Jean Lafitte National Park south of the city.) Fontenot joined the Sierra Club in 1972, becoming chair of its New Orleans Group in 1973 and then chair of the Delta (Louisiana and Mississippi) chapter in 1974. He ran as the environmental candidate for the First Congressional District seat of Louisiana in 1974, losing the election but succeeding in drawing attention to the environment. He volunteered as vice president and director of community services at the Ecology Center of Louisiana, Inc., a clearinghouse where many local environmental groups held meetings and had offices. In 1975 Fontenot moved into full-time remunerated work for the environment, when he was hired as the executive director of the Louisiana Wildlife Federation, the state affiliate of the National Wildlife Federation. This position, as well as his prior volunteer work, provided him with knowledge about the workings of government regulatory agencies and effective strategies for working against pollution.

In 1978 Louisiana attorney general William J. Guste Jr. was encouraged by a staff attorney, Richard M. Troy, to hire Fontenot to help citizens understand how to deal with environmental problems and get better access to governmental agencies. Fontenot recalls that Attorney General Guste was aware of an inherent conflict of interest for government officials and agencies. The government has a constitutional obligation to protect resources and the environment, yet it is also charged with promoting economic development. When a regulatory agency approves any project or plan, the government employees involved generally become advocates for it. Explains Fontenot: "They become friends [with the project promoters] technocrat to technocrat, they have empathy. Then [when the project is environmentally destructive], citizens come in, upset and frustrated, and to the technocrats they seem unreasonable." Fontenot worked with Troy and the attorney general to create a job description for helping citizen organizations become involved in the regulatory process.

Fontenot was hired as an "investigator" in 1978, and although his job title has changed since then to "community liaison officer," his duties have not changed. Current attorney general Richard P. Ieyoub describes Fontenot's responsibilities as becoming "involved with communities, groups, people, and public officials who are facing environmental threats; and to help them organize, approach government, gain access to information, and interact with regulatory agencies effectively." In many states, when citizens try to contact the government and get information or complain about an environmental problem, they are shuffled around between offices and frequently do not find the person who can respond to their problem. In Louisiana they are sent straight to the one person charged with helping them, Willie Fontenot. (Only one other state, West Virginia, has a similar office.)

During his tenure, Fontenot has helped organize more than 500 citizens groups and has guided them in their struggle to solve environmental problems in their communities. He travels throughout the poor, rural parts of the state, meeting with members in living rooms and on front porches. Most groups are quite small when he is first contacted, just three to five people. Fontenot reminds such groups of Margaret Mead's observation: "Never doubt that a small group of thoughtful and committed individuals can change the world; indeed it is the only thing that ever has." Fontenot encourages the groups, teaches them research techniques and helps them find the resources they need, and accompanies them when they meet with officials and agencies.

Among the hundreds of successes of the groups he has helped, he recalls the work of the residents of rural, Gulf coast Vermilion Parish. In mid-1978, Vermilion Parish resident Gay Hanks called him and complained of trucks dumping toxic waste in 13 commercial dump sites in the parish and randomly in drainage ditches that farmers used to irrigate their rice and crawfish fields. She was worried about the health effects for local people; her 15-year-old daughter had just died of leukemia. With advice and encouragement from Fontenot, she and four others founded the Vermilion Association to Protect the Environment. They organized the neighbors of each of the 13 commercial dump sites to photograph the sites and keep logs of when trucks were coming in. They organized residents to turn out en masse to hearings and meetings with officials, and by the late 1980s, they had succeeded in having 12 of the 13 commercial dump sites shut down and had convinced the EPA to designate three of them as Superfund sites. They also fought off more than 20 more proposed dumps in the parish.

Another landmark citizens response that Fontenot cites is the Ascension Parish "Save Ourselves" organization, headed by local citizen Teresa Robert, which fought a hazardous waste incinerator proposed by IT Corporation in the courts. Save Ourselves based its suit on Article 9, Section 1 of the Louisiana Constitution, which states that "the important natural resources of the state including air and water and the healthful scenic and aesthetic qualities of the environment shall be protected and replenished. . . ." The battle began in 1980, and five years later, the Louisiana supreme court declared that this constitutional mandate superceded other state agency regulations. The ruling has been an important precedent for dozens of other suits in the state.

Fontenot serves on the boards of the Louisiana Environmental Action Network, an organization focusing on environmental justice and toxics; the national organization Clean Water Action, which campaigns for safe and clean water and for candidates who work for people and the environment; the regional Mississippi River Basin Alliance, which is concerned with locks on the northern part of the river, greenways, toxics, and the dead zone in the Gulf of Mexico; and the Labor Neighbor Project, which links organized labor with citizens living near industry. He resides in Baton Rouge with his wife, Mary Umland Fontenot. They have two grown children, Jacques Alexandre and Dona Frere.

Bibliography

Anderson, Bob, "Fontenot's a People Helper," *Baton Rouge Morning Advocate*, 1982; Ellis, William S., "The Mississippi River under Siege," *The National Geographic Special Edition on Water*, 1993; Schwab, Jim, *Deeper Shades of Green*, 1994.

Foreman, Dave

(October 18, 1946–)
Conservation Activist, Cofounder of Earth First!

Wilderness lover and conservationist Dave Foreman is best known—infamous to some—for the 10 years he spent as an ecowarrior. Working with a loosely knit but large group of individuals under the Earth First! banner, Foreman advocated radical measures to defend wilderness, ranging from sabotage to civil disobedience. He was arrested in 1989 and tried with the so-called Arizona Five on grounds of sabotaging nuclear plants in the Southwest. But this represents only part of Foreman's conservationist career, which also has included an eight-year stint on the staff of the Wilderness Society (TWS), serving on the board of the Sierra Club, and coordinating the Wildlands Project, which has designed a wilderness recovery plan for North America.

Dave Foreman was born on October 18, 1946, in Albuquerque, New Mexico, to Skip and Lorane Foreman. His father was an air force sergeant who moved his family frequently. As a child, Foreman sought solace in whatever natural areas were close to his family's current residence. He attended the University of New Mexico, earning a B.A. in history in 1968. As a college student, Foreman campaigned for Republican Barry Goldwater in his 1964 bid for the presidency. He also joined the conservative Young Americans for Freedom and chaired a group called Students for Victory in VietNam. After graduation, Foreman joined the Marines but was dishonorably discharged two months later. He had spent one of his two months in the brig for insubordination and for going absent without leave. Returning to New Mexico, Foreman taught on a Zuni Indian reservation and became a horseshoer in the northern part of the state before beginning his conservationist career.

Foreman joined the staff of the Wilderness Society in 1972 as its Southwest representative and went to Washington, D.C., in 1977 as a TWS lobbyist. There, he recalled in a 1985 *Mother Earth News* interview, he felt that TWS was being lobbied by the government instead of the other way around. Foreman worked with TWS on negotiations with the U.S. Forest Service about RARE II (Roadless Area Review Evaluation II), which proposed that 65 million of the last remaining 80 million acres of untouched wilderness be opened to mining and logging. TWS, he claimed, had presented convincing evidence and arguments for a greater degree of wilderness preservation in a polite and rational manner but nonetheless lost its case. And worse, TWS discouraged its members from trying to appeal RARE II because it was concerned that the meager conservation gains from RARE II would be lost in any litigation.

That experience sent an embittered Foreman back west in 1979. With four friends, Bart Koehler and Susan Morgan (both ex-TWS staff), HOWIE WOLKE (a onetime Friends of the Earth representative), and ex-yippie Mike Roselle, Foreman drove a Volkswagen van down to the Pinacate desert in Mexico. They mused on the ineffectiveness of mainstream conservation organizations, whose administrators seemed more eager to compromise with the government than to change government policies. During that trip, Earth First! was born. Inspired by EDWARD ABBEY's 1975 novel *The Monkeywrench Gang*, Earth First! was a loose confederation of ecoactivists who swore that they would make "No Compromise in Defense of Mother Earth." Although the group had no formal members, no constitution, no officers, no bylaws, and no nonprofit status, Foreman became its best-known spokesperson because he edited its newsletter.

Earth First!'s first public action took place at the 1981 spring equinox, when Foreman and some others draped a 300-foot black plastic streamer off the Glen Canyon dam to simulate a huge crack. (The Glen Canyon dam, on the Arizona-Utah border, was built as a result of a compromise that the Sierra Club under then-president DAVID BROWER had made in 1959. It agreed to allow the dam to be built if Dinosaur National Monument would be spared flooding.)

Earth First!ers meet at seasonal gatherings to define priorities and plan actions and through the 1980s succeeded in halting many environmentally destructive projects that the more mainstream conservationists could not stop with their traditional strategies. Earth First!'s successes usually involved large protests that drew media attention to the targeted environmental destruction. While Earth First!'s sabotage or civil disobedience delayed the loggers, miners, or road-builders, other groups with legal staffs were able to have court orders drawn up to halt the destruction. This was the case in a 1982 action at a Getty Oil road-building and gas-drilling operation, at a 1983 California old-growth redwood grove, and later in 1983 at a road-building project near the Siskiyou National Forest in Oregon. At the latter action, Foreman himself blocked a logging road and was seriously injured when he was dragged for 100 yards under a Plumley Construction Company pickup truck.

Foreman wrote *Ecodefense: A Field Guide to Monkeywrenching* in 1985, a manual for Earth First!ers. It was his authorship of this guide that put the Federal Bureau of Investigation on his trail. Foreman and four others (Mark Davis, Mark Baker, Peg Millet, and Ilse Asplund) were arrested in May 1989 and accused of sabotaging various nuclear installations in the Southwest. The main piece of evidence against Foreman was that his book was found on the dash of the car of the other defendants when they were caught torching an electric pole that carried power generated by the Palo Verde Nuclear Generating Station. The 1991 trial of the "Arizona Five" resulted in jail terms for all but Foreman, who pled guilty to a felony conspiracy charge in return for a five-year delay on his sentence and a reduction, at that time, of his felony to a misdemeanor. Following the trial, Foreman distanced himself from Earth First!, claiming that many Earth First!ers were straying from the main point of the group and allying themselves with socially progressive causes.

Foreman and his wife, nurse Nancy Morton, whom he married during the 1986 Earth First! Round River Rendezvous, continue their conservationist work, albeit with more mainstream environmental groups. Foreman works with the Sierra Club, of which he has been a member since 1973. He also currently heads the Wildlands Project, which is designing a plan for a recuperation of North America's rich wilderness and is working with environmental groups such as the Sierra Club to put this plan gradually into place. They live in Tucson, Arizona.

Bibliography

Berger, John J., "Tree Shakers," *Omni*, 1986; Bergman, B. J., "Wild at Heart," *Sierra*, 1998; Bookchin, Murray, and Dave Foreman, *Defending the Earth: A Dialogue between Murray Bookchin and Dave Foreman*, 1990; Foreman, Dave, *Confessions of an Eco-Warrior*, 1991; Looney, Douglass, "Protector or Provocateur?" *Sports Illustrated*, 1991; "The Ploughboy Interview: Dave Foreman: No Compromise in Defense of Mother Earth," *Mother Earth News*, 1985; Zakin, Susan, *Coyotes and Town Dogs: Earth First and the Environmental Movement*, 1993.

Fossey, Dian

(January 16, 1932–December 24, 1985)
Zoologist, Founder of Karisoke Research Center

Author of *Gorillas in the Mist* (1983), Dian Fossey was the world's foremost expert on the mountain gorillas of central Africa. Fossey established the Karisoke Research Center in Rwanda, where she spent more than 13 years studying the endangered species. Fossey gathered extensive scientific evidence about the gorillas' behavior, fought off poachers and other threats to their continued existence, and brought their plight to the attention of the world. Without Fossey's efforts, the mountain gorilla would almost certainly be extinct.

Dian Fossey was born on January 16, 1932, in San Francisco. After graduating from high school, she enrolled in the preveterinary medicine program at the University of California, Davis, and later transferred to San Jose State College. She received a bachelor's degree in occupational therapy from San Jose State College in 1954. She moved to Kentucky and worked as the director of the occupational therapy department at the Kosair Crippled Children's Hospital in Louisville. In 1963 she took out a loan to finance a safari trip to Africa, where she first saw a mountain gorilla and met the anthropologists Louis and Mary Leakey. In 1966 Louis Leakey went to Louisville to recruit Fossey for a long-term gorilla study project. Fossey was at first uncertain that she had the appropriate training, but Leakey assured her he was looking for someone who could bring a fresh approach, as well as great determination, to the project. In preparation for the trip, Fossey had her appendix removed prophylactically to demonstrate her determination, before realizing that Leakey had made the suggestion in jest.

Fossey began her study with Jane Goodall in Tanzania, where she learned something of Goodall's methods of studying chimpanzees. Fossey then went to the Republic of Congo (later Zaire) and began her research on the gorillas. On July 10, 1967, Fossey was taken into custody and kept for two weeks, during which time she saw many of her fellow captives tortured and killed and she herself was repeatedly raped. She managed to escape and was able to cross the border into Uganda. On September 24, 1967, she established the Karisoke Research Center in the Parc National des Vulcans in Rwanda. In 1970 she started doctoral studies at Cambridge University. She received her Ph.D. in zoology in 1976.

Fossey's work dramatically changed both the scientific and popular understanding of gorillas. Prior to Fossey's research, gorillas suffered from the "King Kong myth" and were seen as dangerous, hyperaggressive monsters. Fossey demonstrated that mountain gorillas are largely peaceful, family-oriented vegetarians. She described their personalities and family structure, identifying the individual gorillas she studied by name. She described the relations between gorillas and documented the devastation created by poachers. Adult gorillas will fight fiercely to protect their young, and poachers have to kill many adults to capture a single baby. One of the most famous cases involved an adult male gorilla she called "Digit," who was killed while protecting his family. CBS *Evening*

Dian Fossey (Yann Arthus-Bertrand/CORBIS)

News announced the death, and a fund was established in his name. Fossey's work was both scientifically rigorous and engaging for general readers and thus was effective in serving the interests of preservation.

Fossey fought passionately in defense of the gorillas, against poachers, farmers, and sometimes even game preserve officials. Rwanda is a poor nation with a rapidly growing population, and the small park surrounding the research station was seen as potential agricultural land. Poachers are a major threat to gorilla survival, because collecting gorillas for zoos and for trophies is a profitable business. Fossey had violent confrontations with armed poachers, and at one point kidnapped the child of a poacher whom she offered to exchange for a captured baby gorilla. Park rangers sometimes cooperated with poachers, so Fossey's relationship with wildlife officials was uncertain and often overtly hostile. She organized volunteer patrols to fight against poachers, because she believed that park officials were not active enough or completely trustworthy.

In 1980, suffering from physical exhaustion, Fossey took a leave of absence from the research center and accepted a three-year post at Cornell University. While there she finished work on *Gorillas in the Mist.* Fossey returned to Rwanda in 1983, heartened by the enthusiastic reception of the book. On December 27, 1985, her body was found in her camp in the Viruga Mountains in Rwanda. She had been hacked to death by a machete. A number of theories have been advanced about motive for the murder and the identity of the murderer, though no definitive answers have ever been found. In 1988, a movie starring Sigourney Weaver was made of *Gorillas in the Mist.* Today, the mountain gorillas are in a much less precarious position. Increased tourism in the park has been a boon to the local economy, so there is far greater local incentive to preserve Fossey's beloved gorillas. Once thought to have dwindled to as few as 250, the gorilla population is now believed to have risen to more than 500.

BIBLIOGRAPHY

Biermann, Carol, Louise Grinstein, and Rose Rose, *Women in the Biological Sciences*, 1997; Crouse, Debby, "Up Close with Gorillas," *International Wildlife*, 1988; Montgomery, Sy, *Walking with the Great Apes: Jane Goodall, Dian Fossey, Birutae Galdikas*, 1991.

Franklin, Jerry

(October 27, 1936–)
Forester, Plant Ecologist

After an intensive, 20-year study of old-growth forest, U.S. Forest Service plant ecologist and professor of forestry Jerry Franklin during the late 1980s developed an ecosystem management approach called New Forestry, which melded two goals that had traditionally been considered contradictory: maintaining the ecological health of old-growth forests and logging them. New Forestry emerged as a promising alternative when the northern spotted owl was listed as an endangered species in 1990, obligating the U.S. Forest Service to protect its old-growth forest habitat. In 1993, Franklin was invited by Pres. Bill Clinton to serve on the Ecological and Economic Assessments Roundtable at the White House Forest Conference and on the Executive Committee of the White House Ecosystem Management Assessment Team. He is credited with convincing President Clinton to significantly reduce logging of old-growth forests and set up a 9.5-million-acre system of old-growth forest and riparian reserves.

Jerry Forest Franklin was born on October 27, 1936, in Waldport, Oregon. He was raised in Camas, Washington, a small town on the Columbia River dominated by a large pulp mill. His father, a worker in the pulp plant, instilled a love of forests in his son. As a youth, Franklin camped with the Boy Scouts in nearby Gifford Pinchot National Forest and enjoyed long solitary hikes in the woods. He decided at an early age that he would become a forester. He earned his B.A. in forest management in 1959 from Oregon State University (OSU), the leading forestry school in the Northwest at that time. He earned an M.A. in forest management from OSU in 1961 and a Ph.D. in botany from Washington State University in 1966.

From 1959 until 1975, Franklin was employed as a research forester at the Pacific Northwest Research Station in Corvallis, Oregon. In 1969, he became deputy director of the Coniferous Forest Biome study for the United Nations' International Biological Program, a large-scale study of the earth's major ecosystems. Research by Franklin's team at the H. J. Andrews Experimental Forest, which is attached to and managed by the Pacific Northwest Research Station, was the first ever done on the old-growth coniferous forest ecosystem. Traditional forestry maintained that old-growth forests—those made up primarily of mature trees, many of them 200 or 300 feet high and several hundred years old—had reached their peak; they were no longer growing in height or girth. The approach of traditional forestry was to harvest these old giants and replace them with easy-to-harvest tree plantations. During the National Science Foundation–funded biome study and those that followed it during the 1970s and 1980s, Franklin and his associates were able to synthesize their discoveries and for the first time weave together a portrait of the complex ecological system of the old-growth forests of the Pacific Northwest.

William Dietrich's *The Final Forest* describes some of the many intricate ecological cycles that Franklin's team discovered: the clean, relatively silt-free streams and rivers of old-growth forests provide habitat for hatching insects and attract spawning salmon, who die upon laying their eggs. Rotting salmon carcasses fertilize trees. As insects die and fall into the streams, they feed recently hatched baby salmon. Biologist Chris Maser's observations revealed the relationship between tree voles and mycorrhizal fungi, which live on the root hairs of trees and help them extract nutrients from soil. Voles inhabit the upper reaches of old-growth trees but descend to dig the truffle-like fruit of mychorrhizal fungi. As voles move along the forest floor in search of truffles, they defecate, dispersing the fungi's spores.

Such elegant ecological relationships disappear when a forest is clear-cut. Humus, the nutrient-rich top layer of soil, erodes, making it harder for plants to grow; robbed of their habitat, animals leave; mudslides and flooding are more likely since vegetation does not capture and slow the movement of water over the land; and siltation fouls streams and rivers. During a 1986 sabbatical at Harvard Forest from Oregon State University—where he taught in the Departments of Forest Science and Botany from 1975 to 1992—Franklin learned that clear-cutting in dispersed patches can compromise conditions in the remaining forested areas, due to what came to be called "edge effects." In subsequent research with his student Jiquan Chen, he documented edge effects in old-growth forests as far as 800 feet away from cutover edges.

Franklin devised a system of techniques that he called New Forestry to meld the goals of ecological health and timber extraction. The New Forestry approach to ecosystem management was inspired by Franklin's knowledge of old-growth forest ecology and from his observations of the important role in ecological restoration played by small mammals inhabiting brush remaining at Mount St. Helens shortly after its 1980 volcanic eruption. New Forestry advocates "messy" logging, which allows some live trees and dead snags to remain standing and leaves fallen logs and other debris on the ground to anchor and nurture the soil and to provide habitat for wildlife. Seedlings of various useful tree species are planted, in an attempt to replicate the diversity of the forest. Cut areas are clustered together to minimize the need for logging roads.

New Forestry burst into public view when the northern spotted owl was listed as an endangered species in 1990. Eric Forsman, a biologist who had worked at H. J. Andrews at the same time that Franklin was forest officer there, studied the previously unknown but now famous northern spotted owl and determined that it was an indicator species for the health of old-growth forests in the Pacific Northwest. Indicator species are important because the health of their populations is indicative of the health of the ecosystem in general. Once the northern spotted owl was listed, the U.S. Forest Service was legally obligated to protect its habitat—large expanses of old-growth forests.

Pres. Bill Clinton convoked a Forest Conference in Portland, Oregon, in 1993 and assigned a team of scientists, including Franklin, to devise a plan to protect the spotted owl without completely crippling the timber industry. Franklin's New Forestry emerged as a promising approach, and Franklin quickly rose as its main spokesperson. He was appointed to the Executive Committee of the White House Ecosystem Management Assessment Team in 1993, which is credited for convincing Pres. Clinton to establish a system of old-growth forest reserves, expand stream buffers, and cut logging in national forests by 80 percent.

Although New Forestry was criticized early on by traditional foresters who felt that its guidelines were not specific enough and that it was too young to be proven to work and by environmentalists who feared that it might justify extensive cutting of old growth, Franklin and others have continued to refine the concept. Foresters experiment with New Forestry techniques at the H. J. Andrews Experimental Forest and in U.S. national forests. The faculty at the Department of Forestry Science at Oregon State University, considered to be the Vatican of forestry programs, has recruited specialists and become a national leader in the fields that Franklin pioneered: forest restoration and ecosystem management. With Kathryn Kohm, Franklin coedited *Creating a Forestry for the Twenty-first Century: The Science of Ecosystem Management* in 1996, to which more than 50 authors contributed chapters on aspects of New Forestry ranging from ecological processes and management systems to forest economics.

Franklin is the recipient of many honors and awards, including the Murie Award from the Wilderness Society in 1998, the 1996 William B. Greeley Award from the American Forests Association, and the Barrington Moore Award for Outstanding Achievement in Forest Research from the Society of American Foresters in 1986. Since 1986, Franklin has been a professor of ecosystem analysis at the University of Washington's College of Forest Resources.

Franklin lives with his wife, Phyllis, in Issquah, Washington, on the outskirts of Seattle.

Bibliography

Associated Press, "Experimental Forest Provides Scientists with Information," *Oregon News*, 1998; Dietrich, William, *The Final Forest: The Battle for the Last Great Trees of the Pacific Northwest*, 1992; "Dr. Jerry Franklin," http://depts.washington.edu/wrccrf/Franklin/; Durbin, Kathie, "Reformation in the Vatican of Sawlog Forestry: History Takes Oregon State for a Ride," *High Country News*, 1995; Kohm, Kathryn A., and Jerry F. Franklin, *Creating a Forestry for the Twenty-first Century: The Science of Ecosystem Management*, 1996; Luoma, Jon, *The Hidden Forest*, 1999.

Frome, Michael

(May 25, 1920–)
Environmental Journalist

Michael Frome chose to dedicate himself to environmental journalism at a time when there was very little information in the media about environmental issues. During the 1960s and 1970s he rose to become one of the nation's foremost conservation writers, known for his investigations of environmental crimes and his willingness to name perpetrators. He was fired in 1971 by *American Forests* and in 1974 by *Field & Stream* for resisting their censorship of his work; both of these firings received widespread attention in other national media. Frome has written 16 books about the environment and efforts to protect it, plus several outdoor travel guides.

Michael Frome was born on May 25, 1920, in New York City to William and Henrietta (Marks) Frome. His journalistic career began at Dewitt Clinton High School, where he wrote for the *Clinton News*. He attended City College of New York (now City College of the City University of New York) immediately after graduating from high school but left in 1941 to work as a copyboy for the *Washington Post* and International News Service, taking dictated stories by phone from reporters. He volunteered during World War II as a navigator for the U.S. Army Air Corps, flying transport missions all over the world, and then after the war returned to the *Washington Post* as a reporter. In spring 1946 he joined a United Nations relief mission to Czechoslovakia and Poland and wrote a series of front-page *Post* articles about those countries' postwar recovery efforts. From 1947 to 1957 he worked as editor of travel publications for the American Automobile Association (AAA) and began to freelance travel articles for such publications as *Holiday*, *Woman's Day*, *Parade*, *Changing Times*, the *New York Times*, the *Christian Science Monitor*, and the *Chicago Tribune*. One of his assignments at the AAA was to develop and promote a campaign for antibillboard legislation, which allowed him to coordinate with national conservation organizations. Through these contacts, as well as his research trips to national parks throughout the United States, Frome became committed to wilderness preservation. Knowing that the public would not support conservation without good information on the issue, he became concerned about the dearth of conservation news in the mainstream media.

In 1959, after writing several travel guides, he began to focus his writing on environmental issues. His first book in this vein was *Whose Woods These Are—The Story of the National Forests* (1962), a history of the U.S. Forest Service (USFS) and guide to national forests. That was followed up with a book he initially had conceived as a history and travel guide for Great Smoky Mountain National Park. Upon learning that the National Park Service was planning an ill-conceived highway and development scheme for that park that would have developed large areas of pristine forest, Frome was moved to assume the role of advocate. The resulting book, *Strangers in High Places*, published in 1966, argued convincingly against the plan. Park officials reacted by banning his book from park bookshops, but the proposed road was never built.

His reputation for good writing and strong advocacy established, Frome became a columnist for *American Forests* in 1966 and for *Field & Stream* in 1967. A freelance article for *Holiday* in 1966, "The Politics of Conservation," probed a question that bothered Frome: why politicians were not responding adequately to a growing public concern for wilderness preservation. This is a problem that Frome has returned to many times over during his career.

Although his columns were popular among readers of *American Forests* and *Forest & Stream*, Frome was fired from both publications in the early 1970s for resisting censorship. For *American Forests*, published by the American Forestry Association, whose board of directors was made up primarily of foresters associated with the timber industry and the USFS, Frome wrote critically about clear-cutting of national forests, decrying what he saw as an overemphasis on timber production by a government agency that was also charged with responsible stewardship of public lands. When William Towell, executive director of the American Forestry Association, wrote a memorandum in 1971 to the editor of *American Forests* demanding that Frome cease his criticism of the USFS and the timber industry and focus on other less controversial issues, Frome protested, and was fired. His dismissal, however, caused more trouble for the American Forestry Association than for Frome himself. Former interior secretary Stewart Udall and Jeff Stansbury wrote about his firing for the Los Angeles Times Syndicate: "How to Kill a Conservation Organization . . . Obituary for the American Forestry Association"; a colleague wrote a tribute in *Field & Stream* entitled "How the Clearcutters Tried to Gag Mike Frome"; and many of Frome's fans wrote letters of protest to the editor of *American Forests*.

Frome continued writing for the larger and more influential *Field & Stream*, but soon he faced opposition from publishers there as well. Although he did not write from the perspective of an outdoorsman, his focus on conservation was welcomed by hunters and anglers concerned about disappearing natural habitat. He wrote about the effects on wildlife of logging, dam building, wetland drainage, and other intrusive activities; named names of individuals and companies responsible for the damage; and gave suggestions for getting involved. As part of this effort to encourage activism, Frome coordinated a "Rate Your Candidate" feature in election years 1968, 1970, and 1972, in which congresspeople were rated according to their commitment and effectiveness on conservation issues. "Rate Your Candidate" upset those members of Congress who were given poor ratings, one of whom chaired the subcommittee that ruled on broadcasting relations. This was of especial concern to the Columbia Broadcasting System (CBS), the media conglomerate that owned *Field & Stream*, and the magazine discontinued the feature. By 1974, a new editor at *Field & Stream* was actively censoring Frome's work. Frome was told to write about conservation problems in general rather than describing specific crises and referring to specific culprits. Because he would not comply with this order, Frome was fired in late 1974. Conservationists reacted immediately. Individuals picketed CBS's Washington headquarters, and organization heads (as well as some members of Congress) wrote letters of protest to the chair of the board of CBS. *Time* featured the story in its November 4, 1974, issue.

Following this, Frome went on to write columns for *Defenders of Wildlife* for 18 years and for the *Los Angeles Times* and *Western Outdoors* as well. He has written 12 books since 1974, including *Battle for the Wilderness* (1974, revised in 1997), which recounts the long fight to pass the Wilderness Act of 1964 and describes the process for designation of wilderness. *Conscience of a Conservationist* (1989) is a collection of some of his best-known essays and articles, including exposés and conservation histories. *Regreening the National Parks* (1992) elaborates on a theme he has revisited frequently since 1962, the overdevelopment of national parks and other natural sites. Frome believes that rather than defining its primary mission as providing recreational opportunities, the National Park Service should acknowledge the importance of providing sanctuaries for wildlife. He advocates such changes to park visitation policy as determining the human carrying capacity of each park and limiting visitorship to that number, limiting automobile access to some parks, and deemphasizing the role of concessions. *Green Ink: An Introduction to Environmental Journalism* (1998) is both a memoir and a primer for good environmental writing.

In addition to practicing environmental journalism, Frome has taught courses on the topic. He has taught at the University of Vermont (1978), the Pinchot Institute for Conservation Studies in Milford, Pennsylvania (1981), the College of Forestry at the University of Idaho (1982–1986), the Sigurd Olson Environmental Institute at Northland College in Ashland, Wisconsin (1986–1987), and the Huxley College of Environmental Studies at Western Washington University in Bellingham, where he developed an environmental journalism program (1987–1995).

Frome has received many awards and recognitions, including the Thomas Wolfe Memorial Literary Award (1967); Trout Unlimited's Trout Conservationist of the Year (1972); a Mort Weisinger Award, presented by the American Society of Journalists and Authors for the best Magazine Article of the Year for his 1981 five-part series "The Ungreening of the National Parks" (published in *Travel Agent* and subsequently in *National Parks* Magazine and in the book *National Parks in Crisis*); and the National Parks and Conservation Association's Marjory Stoneman Douglas Award (1986). The University of Idaho offers the Michael Frome Scholarship for Excellence in Conservation Writing in his honor. He completed his doctorate in environmental studies through the Union Institute in 1993 and was named Outstanding Alumnus of the Year in 1999.

Frome lives with his wife, June Eastvold, a Lutheran pastor, in Bellingham, Washington. He has two adult children.

Bibliography

Frome, Michael, *Green Ink*, 1998; McNulty, Tim, "Respect for the Earth, Respect for Michael Frome," *Seattle Times*, 1996; "Michael Frome. Papers, 1959–1989," http://www.lib.uidaho.edu/special-collections/Manuscripts/mg174.htm.

Fuller, Buckminster

(July 12, 1895–July 1, 1983)
Architect, Inventor

Buckminster Fuller is most often remembered as an architect who designed buildings with an eye for the future. His lifelong goal was to develop a science of design that would solve the world's major problems, such as poverty and housing shortages, while conserving finite resources. Frustrated by the obvious discrepancy between the advanced state of available technology and its haphazard and mediocre applications to ordinary life, he continually and loudly advocated for holding technology to a higher standard. He was ahead of his time in supporting the use of renewable energy resources, which he always incorporated into his designs—claiming that our perceived energy crisis is actually a crisis of ignorance.

Richard Buckminster Fuller Jr. was born on July 12, 1895, in the Boston suburb of Milton, Massachusetts, the second of four children of Richard Buckminster and Caroline Wolcott (Andrews) Fuller. His father was a successful Boston tea and leather merchant who died when his son was 15. Buckminster was born cross-eyed and abnormally farsighted and for several years only saw large patterns and blurred colors. He got glasses when he was four and could suddenly see real details, yet his perception of patterns stayed with him and influenced the way he thought. He went to school at Milton Academy, where he excelled in math and science, and in 1913 he enrolled in Harvard University, where four previous generations of Fullers had received their education. However, he had long been frustrated with formal academics, often feeling that his teachers lacked adequate answers to his questions. By the middle of his first year at Harvard he had become impatient with school and ran off to New York City, where he squandered his tuition entertaining a girl he knew in the chorus line of a current theatrical show—and all her friends and coworkers. His spree resulted in his expulsion from Harvard. A week later his family arranged for him to work in a cotton mill in Quebec, which was not altogether unfortunate for him, as he learned about machines and mechanics and ended up an enthusiastic technician. Having supposedly demonstrated his reliability, he was readmitted to Harvard, only to have his disinclination for conventional education get the best of him; once again he was expelled for irresponsibility. At this point he went to work as a meat lugger at Armour and Company in New York.

In early 1917 he enlisted in the navy and was assigned to active duty, and a few months later he married Anne Hewlett. By the time the war ended, Fuller had gained important experience involving the mechanics of ships and their equipment; perhaps more relevant was his glimpse at the technology necessary to protect people from the hostile weather conditions found at sea. In 1919 he returned to work at Armour and Company as an assistant export manager; there he stayed until 1922. At that point, he went into business with his architect father-in-law, forming Stockade Company, which manufactured a new form of fibrous cinder blocks. That same year his first daughter, Alexandra, died at the age

Buckminster Fuller in front of his geodesic dome, ca. 1960 (Archive Photos)

of four—after suffering from a series of epidemic infections aggravated by poor housing conditions during the war. Fuller was shattered by her death and took years to emerge from a profound depression. Shortly after the birth of a second daughter, Allegra, in 1927, he lost his job at Stockade Company, which brought him near the point of suicide. But after a period of self-analysis, he emerged from his emotional crisis and resolved to renounce financial and personal gain and to use his talents to help humankind. He believed that worldwide needs, such as housing, could be met by taking advantage of all relevant scientific principles and technical designs, which could be implemented in such a way as to conserve resources and still be practical and cost-effective.

Fuller moved his family to Chicago and began the process of turning his meditations into realities. This led to the invention of several revolutionary structures, all of which adhered to his principle of doing more with less. Some of Fuller's other inventions include the Dymaxion house, whose name was coined by Fuller's publicist from "dynamism," "maximum," and "ions," all favorite words in Fuller's vocabulary. The Dymaxion house, a family dwelling that was to be mass-produced to cut down on energy expenditure, was to serve as a prototype for a worldwide housing industry. It incorporated a number of highly sophisticated labor-saving devices, such as compressed air and vacuum units that took care of dusting; laundry machines that washed, dried, and conveyed clothing to storage units; and revolving shelves that were rigged to move at the interruption of a light beam. Fuller intended to construct the houses using inexpensive, efficient materials, such as casein—a translucent sheeting made from vegetable refuse, which was to be used for the walls, windows, and ceilings. And, perhaps most remarkably considering that this was in the late 1920s, Fuller planned for the use of renewable energy (such as solar and wind-generated) that could be channeled directly into the house to work as electricity or air conditioning or could be stored in batteries for future use. The general public was perhaps not ready for these ideas, and the Dymaxion house never got beyond the stage of models. In 1932 he founded the Dymaxion Corporation to produce his inventions, including the Dymaxion car and a Dymaxion world map that showed all the continents on a flat surface without distortion. He invented a board game called the World Game, which utilized a large-scale Dymaxion Map for displaying world resources and allowed players to strategize resource-conserving solutions to global problems.

Fuller is most famous for his geodesic dome design, which demonstrates enormous strength with minimal quantities of materials. A dome can cover large areas of space without internal supports, since pressure applied to its surface is dissipated along the entire structure; and in fact it grows relatively lighter and stronger the larger it is. Geodesic domes are also extremely easy and quick to construct and use such a comparatively small amount of materials that they are dramatically more cost-effective than any other housing structure available. Fuller gained recognition for his design in 1953 when Ford Motor Company commissioned him to build a 93-foot cover for their Dearborn plant rotunda, and he was then increasingly in demand to build similar structures. He created them for the air force, the Union Tank Car Company, the Missouri Botanical Gardens, and an international trade fair in Afghanistan. Others began appearing in Tokyo, New Delhi, Bangkok, and Moscow.

Fuller was truly ahead of his time. Many of his inventions, considered impractical at the time, have gradually come into circulation over the years. Long before most people became concerned about energy conservation, he decried the inefficiency in most mechanical gadgetry—he once calculated that fully 96 percent of worldwide energy expenditure (most of which comes from nonrenewable sources) is wasted on friction, bad design, and general carelessness. His belief was that architects and engineers should be held to a higher level of responsibility on behalf of all humanity.

During the course of his career, he was awarded 25 U.S. patents, received 47 honorary doctorates, received dozens of awards in architecture and design, and circled the globe 57 times, lecturing to millions. On July 1, 1983, he died at his wife's deathbed in a hospital in Los Angeles.

Bibliography

"Buckminster Fuller Institute," http://www.bfi.org; Fuller, Buckminster, *Ideas and Integrities*, 1963; Fuller, R. Buckminster, and Robert Marks, *The Dymaxion World of Buckminster Fuller*, 1960; Kenner, Hugh, *Bucky: A Guided Tour of Buckminster Fuller*, 1973; McHale, John, *R. Buckminster Fuller*, 1962.

Fuller, Kathryn

(July 8,1946–)
President and Chief Executive Officer of World Wildlife Fund

Since 1989, Kathryn Fuller has served as president and chief executive officer of the World Wildlife Fund (WWF). She arrived to this post after leading that organization's Trade Record Analysis of Flora and Fauna in Commerce (TRAFFIC) program to monitor international trade in endangered species, directing public policy, and serving as its executive vice president. Previous to her work at WWF, Fuller helped create a Wildlife and Marine Resources Section at the U.S. Department of Justice.

Kathryn Fuller was born in New York City on July 8, 1946, the eldest of three children. Brought up in Westchester County just to the north of New York City, Fuller was from the very beginning interested in nature and the outdoors. Her mother, Carol Fuller, was an amateur naturalist dedicated to conservation and raised her children on books including accounts of naturalists and explorers, such as Jane Goodall, whose articles about chimpanzees in Tanzania were just being published for the first time in *National Geographic* magazine. Fuller attended Brown University, initially studying biology but switching to major in English after her biology classes proved to be premed rather than nature oriented. She graduated in 1968.

Fuller designed computer systems for Yale University's library and the American Chemical Society before taking a job in 1971 at Harvard's Museum of Comparative Zoology. In her free time, Fuller audited Harvard classes. One of her favorites was given by entomologist Edward O. Wilson, who rekindled her commitment to pursue a career in conservation. Fuller accepted an invitation from ecologists Richard and Runi Estes in 1972 to accompany them to Tanzania to study wildebeest migration. It was during this two-month trip that Fuller finally got to see firsthand the wilds that she had been daydreaming about since childhood and also that she realized how crucial conservation would be to the survival of such spectacular wildlands.

Kathryn Fuller (courtesy of Anne Ronan Picture Library)

Upon her return from Tanzania, Fuller enrolled in law school in Texas, hoping to become a wildlife lawyer. After graduating in 1976 from the University of Texas School of Law in Austin and clerking for a Texas U.S. District Court judge for one year, Fuller moved to Washington, D.C., and joined the Department of Justice. She began working for the Land and Natural Resource division of the Department of Justice in 1979 and helped set up the new Wildlife and Marine Resources Section there. She worked on cases pertaining to endangered species, fisheries, and the Lacey Act of 1900, which prohibits the importation or state-to-state transport of poached wildlife. During the early 1980s, while at the Department of Justice, Fuller enrolled in a master's degree program in marine, estuarine, and environmental studies at the University of Maryland. This, she thought, would allow her to further develop her scientific knowledge.

In 1982, when the youngest of her three children was born, Fuller left the Department of Justice to work as a private consultant in conservation. She was hired by the World Wildlife Fund to research all national and international laws that fell within the domain of the Lacey Act. Within a year she became an employee of WWF and directed WWF's TRAFFIC program, which monitors international wildlife trade and works to strengthen protection of threatened or endangered species. Fuller moved up the ranks of WWF, becoming executive vice president in 1987. When WWF president and chief executive officer (CEO) WILLIAM K. REILLY was named Environmental Protection Agency administrator by President Bush in 1989, Fuller was elected president and CEO of the organization.

WWF, perhaps the world's best-known conservation organization with over one million U.S. members, has sponsored more than 2,000 projects in some 116 countries. Its efforts are directed toward three global goals: "protecting endangered spaces, saving endangered species, and addressing global threats." Under Fuller's leadership, WWF has acknowledged that conservation efforts will not succeed unless human needs are addressed. One of her accomplishments during the mid-1980s was to help engineer many "debt-for-nature" swaps, through which countries rich in biodiversity but in debt to international financial institutions had portions of their debts forgiven in exchange for a commitment to conserve natural resources. This was seen as a win-win solution to the paralyzing external debts that many developing countries suffer. Current WWF conservation projects are designed with participation from local people, in order to address their needs and assure their full support. WWF also works on global threats to nature, such as climate change and the accumulation of toxic and persistent chemicals.

For her work in conservation, Fuller has received a United Nations Environmental Programme Global 500 Award, was given Brown University's 1990 William Rogers Outstanding Graduate Award, and has received several honorary degrees.

BIBLIOGRAPHY

Breton, Mary Joy, *Women Pioneers for the Environment*, 1998; Hitt, Greg, "Animal Passions," *Harpers Bazaar*, 1991; "Kathryn S. Fuller," *Current Biography Yearbook*, 1994; "World Wildlife Fund," http://www.worldwildlife.org/.

Futrell, J. William

(July 6, 1935–)
Attorney, President and Chief Executive Officer of Environmental Law Institute

J. William Futrell is president and chief executive officer of the Environmental Law Institute (ELI), a national, nonprofit research and educational institution that works to advance environmental protection by improving law, management, and policy. ELI publishes the environmental law journals *Environmental Law Reporter*, *The Environmental Forum*, and *National Wetlands Newsletter*, as well as a series of reports that summarize ELI's policy studies. More than 50,000 lawyers and other environmental professionals have participated in ELI's continuing education program, which includes introductory courses in environmental law and seminars on specialized topics. Through ELI, Futrell has promoted what he calls Sustainable Development Law: a synthesis of environmental protection measures and economic activities that merges concerns for the environment, economics, and equity.

John William Futrell was born on July 6, 1935, in Alexandria, Louisiana, to J. W. and Sarah Ruth (Hitesman) Futrell. Growing up in rural Louisiana, he was most influenced by the Methodist Church and the Boy Scouts, with whom he explored the Kisatchie National Forest woodlands that surrounded his home. In high school, he first read Albert Schweitzer's *Out of My Life and Thought* and was moved by its call for reverence of life. His church activities at college led to participation in sit-ins to desegregate church facilities in Louisiana, an experience that spurred a lifelong commitment to civil rights and that demonstrated how law and religion interact. He earned his B.A. in philosophy at Tulane University in 1957, was commissioned a second lieutenant in the United States Marine Corps, and received a Fulbright Scholarship to study recent German history and economics at the Free University of Berlin. This opportunity—which inspired his longstanding commitment to international work—led to his assignment to the Office of Naval Intelligence at the U.S. Mission in West Berlin. From 1960 to 1962, he served in East Asia, in turn commanding an artillery battery and a military police company and teaching a course on survival skills. These years taught him that communism posed a fundamental challenge to religious and democratic values.

Futrell earned his LL.B. from Columbia University's School of Law in 1965 and that same year became a law clerk for federal judge Edwin Hunter of the U.S. District Court, Western District of Louisiana, where he participated in desegregation cases for seven of the state's 64 parishes. He worked as a trial attorney for the New Orleans firm Lemle & Kelleher from 1966 to 1971, during which time he became interested in the new field of environmental law. Environmental law was just then emerging as a result of a broad social movement and citizen protests, which led to the passage of a dozen statutes and hundreds of regulations in a few short years. Futrell became one of the first environmental law professors in 1971, teaching at the universities of Alabama and Georgia from 1974 to 1979. Aside from his law practice and teaching, he organized Sierra Club groups in the southern United States and became southern regional vice president of the Sierra Club in 1971. Elected to the Sierra Club Board of Directors in 1971, he led campaigns for better coastal zone management and ocean protection, wilderness expansion, and urban environment. As Sierrra Club president in 1978, he worked to forge alliances with civil rights groups, an effort that led to the City Care Conference in Detroit in 1979 and the City Care movement.

Futrell was awarded a fellowship from the Wilson Center at the Smithsonian Institution for the academic year 1979–1980, during which he wrote and lectured on the relationship of administrative agency law and practice to nonprofit organizations and citizen participation. In 1980 he became president and chief executive officer of the Environmental Law Institute. ELI had been founded in 1969 by the Public Law Education Institute and the Conservation Foundation to train lawyers and other professionals to work in the new field of environmental law, and Futrell has developed and nurtured the organization as it has grown to become a major influence in the shaping of national environmental policy and law.

In addition to providing continuing environmental education to lawyers through its annual environmental law intensive and about 50 seminars and courses per year held throughout the country, ELI offers its educational services to corporate and public interest members as well. The Corporate Program, through which more than 90 major national and international companies have joined ELI, offers training in environmental law issues to the environmental staff members of these companies. ELI bills this program as a way in which "forward-thinking businesses and their environmental professionals" can promote "environmental leadership in the business community." ELI's Public Interest Program, established in 1995, opened a special membership category for public interest and citizen advocacy groups. Public interest members enrich ELI's seminars about specific environmental topics and issues, and ELI in turn helps assure that their viewpoints are heard by corporations and government agencies such as the Environmental Protection Agency and the National Oceanic and Atmospheric Administration. These "professional dialogues" convene a diverse group of experts to develop implementable solutions to pressing environmental problems.

ELI publishes three periodicals that have become major sources of environmental law, policy, and management information: the monthly *ELR—The Environmental Law Reporter®*; *The Environmental Forum®*, a bimonthly journal with articles on environmental management issues by lawyers, corporate executives, politicians, Environmental Protection Agency administrators, and citizen activists; and *National Wetlands Newsletter*, which specializes in wetlands regulation, policy, and science. Much of the material in these journals is published electronically on ELI's web site (http://www.eli.org). In addition, ELI publishes several research reports per year, concise summaries of the policy studies undertaken by EPI's staff. Research topics include air and water pollution, wetlands, economics, state environmental law, and toxic substances and hazardous waste.

Upon his arrival at ELI in 1980, Futrell initiated a series of international dialogues, focusing on the then Soviet Union and eastern bloc countries to encourage the nascent environmental groups that he calls "the edge of the wedge of freedom." ELI became the first western environmental organization to build environmental training programs for citizens of Eastern European countries, even before the fall of the Berlin Wall. Futrell has worked to forge an international network of environmental lawyers working on behalf of the planet. Approximately one-third of ELI's work is done outside of the United States. Projects emphasize capacity building for environmental groups and government agencies in India, Mexico, Ukraine, and Uganda, among other countries. Since 1993, Futrell has served as the North American vice chair of the International Union for the Conservation of Nature (IUCN) Commission on Environmental Law, working with the IUCN staff in Switzerland and with the six other vice chairs from the other regions to make the international network a reality.

Futrell has published many articles on topics ranging from environmental ethics to environmental law and constitutional rights, to hazardous waste management, to the National Environmental Policy Act. Responding to a mandate from the 1992 United Nations Earth Summit, Futrell and the ELI staff have worked to develop a blueprint for a new approach to environmental law. In the 1993 ELI book *Sustainable Environmental Law*, Futrell and the other contributors chart a shift from traditional environmental law—"a massive legal response aimed at curbing pollution and conserving resources"—to a new approach that would not eliminate the earlier approach but would "shift the focus to causes of impacts." Through this new approach, environmental concerns would be integrated into the development law of real estate, insurance, and other fields. The book is based on resource-to-recovery analyses of more than a dozen economic sectors, detailing all pollution control regulations for each stage of production (extraction, manufacture, distribution, disposal, and so on). The book concludes that the most important step in achieving effective sustainable environmental law will be "a reaffirmation of sustainability as a prime national goal."

Futrell lives in Arlington, Virginia, with his wife, Iva Macdonald Futrell. They have two adult children, Sarah and Daniel.

BIBLIOGRAPHY

Campbell-Mohn, Celia, ed., *Sustainable Environmental Law*, 1993; "Environmental Law Institute," http://www.eli.org; Futrell, J. William, "Environmental Ethics, Legal Ethics, and Codes of Professional Responsibility," *Loyola (Los Angeles) Law Review*, 1994; Futrell, J. William, "Love for the Land and Justice for Its People," Sierra Club History Series, Regional Oral History Office, The Bancroft Library, University of California, 1984; Futrell, J. William, "Sustainable Development Law," *The Environmental Forum*, 1994.

G

Geisel, Theodor (Dr. Seuss)

(March 2, 1904–September 24, 1991)
Children's Author and Illustrator

As a writer and illustrator of over 40 much-beloved children's books, Theodor Geisel (Dr. Seuss) has had a unique influence on four generations of Americans. In an era when children's books were dry and didactic, his whimsical and imaginative stories revived children's interest in reading—while also exposing such dangers as war, prejudice, or wasting natural resources. His classic book, *The Lorax* (1971), addresses the theme of conservation with the story of an ecosystem that becomes dismally polluted and deserted as a consequence of an individual's self-interest and greed. For many young children, *The Lorax* may be their introduction to environmental issues and may have a lasting influence on their thinking and sense of responsibility.

Theodor Seuss Geisel was born on March 2, 1904, in Springfield, Massachusetts, and grew up there with his older sister, Margaretha. His father, Theodor Robert Geisel, was the manager of a brewery and later, during Prohibition, served as a park superintendent. His mother, Henrietta Seuss Geisel, discovered young Ted's love of rhymes and fostered his awareness of the pleasure of words. In high school, his teachers found him bright, but not dedicated; and to fellow students he was funny and charming. In 1921 he entered Dartmouth College, planning to major in English. Before long he had made himself a fixture at the offices of the Dartmouth humor magazine, *Jack-O-Lantern*, and by his junior year he was editor in chief. As graduation approached, he faced questions from his father regarding his plans, and he jokingly wrote back that he would be attending Oxford University on a fellowship. His father, proudly believing it, had an announcement published in the newspaper. Upon graduating in 1925, Ted had to confess that there was no fellowship. His father insisted that the family must simply find a way to send Ted to England anyway, and that fall he was enrolled at Lincoln College in Oxford. He found his graduate studies there to be tedious and abandoned them after a year. The experience was not a total loss, however, as he met the woman he would eventually marry, a fellow American student named Helen Palmer.

In 1927, Geisel settled in New York, got a job as a writer and artist with *Judge* magazine, and married Helen. Geisel signed "Seuss" on his cartoons for the magazine, explaining to those who knew him that he was saving his full name for a "serious" novel he hoped to write someday. After a time he added "Dr." to the name to compensate for the doctorate he never earned at Oxford. Geisel also worked as a freelance cartoonist and essayist, and throughout the 1930s his creative ideas were seen on the pages of *Vanity Fair*, *Redbook*, and the *Saturday Evening Post*. Later, before his enlistment in the Army Signal Corps in 1943, Geisel wrote political cartoons, lashing out at the evils of fascism and racism that accompanied World War II.

His first children's book was written in 1936, about a boy who imagines a fantastic parade of scenes on the street as he

Theodor Geisel (Dr. Seuss) (courtesy of Anne Ronan Picture Library)

walks to school, instead of the drab horse and wagon that passes by. It was called *And to Think That I Saw It on Mulberry Street* and was rejected by 27 publishers as being too different from other children's books. It was finally published in 1937 when an old Dartmouth friend who was junior editor at Vanguard Press convinced his superiors to give Geisel's book a chance. This was the breakthrough that would shape his career.

Geisel went on to write over 40 children's books, all of them in his exuberant style—giddy illustrations and silly, rollicking verse. In the era of mundane Dick and Jane primers, his books were a breath of fresh air to children and parents alike. Geisel avoids preaching in his books, yet underlying the humor is often a moral emphasizing strong values important for readers of any age. In *The Lorax*, which Geisel listed as his favorite of all the books he wrote, he unabashedly voices his fears about the detrimental effects of greed and pollution on the environment. It tells the story of a creature called the Once-ler, who happens upon a beautiful place where Truffula Trees grow, Swomee-Swans sing, and Brown Bar-Ba-Loots frisk about in the shade. Filled with entrepreneurial ambition, the Once-ler proceeds to cut down the Truffula Trees and manufacture Thneeds, which resemble misshapen hairy pink underwear—and everyone starts buying them in great quantities. As the trees continue to fall, the Lorax appears, crying out against this destruction and calling the Once-ler "crazy with greed." The Once-ler replies that business is business, saying:

> Well, I have my rights, sir, and I'm telling you
> I intend to go on doing just what I do!
> And, for your information, you Lorax, I'm figgering
> on biggering, and biggering, and biggering, and BIGGERING,
> turning MORE Truffula Trees into Thneeds
> which everyone, EVERYONE, EVERYONE needs!

Soon the trees are gone, the sky is smothered with smog, and the water is brown with pollution. The Swomee-Swans and all the other creatures have long since left. The Lorax appears a final time, gives the Once-ler a mournful look, and departs—leaving behind a rock etched with the word "UNLESS." Years pass in this wasteland where only Grickle-grass grows in place of the Truffula Trees, and at the end of the book the Once-ler, now filled with remorse and worry over what he caused, explains the meaning of the Lorax's message: Unless someone starts to care, things are not going to get any better. But the story ends with hope: one Truffula Tree seed has been saved, and the Once-ler urges the reader to plant it and care for it.

Published in 1971, *The Lorax* was ahead of its time, and it took a decade before its popularity soared. But its influence was clear, to the point where it was seen as a threat to some in the timber industry. In 1989, in the logging town of Laytonville, California, a campaign was initiated to remove the book from the second-grade reading list after the child of a logging-equipment salesman read the book at school and then came home and questioned his father's motives. A media frenzy ensued, and eventually the school-board president closed the debate and left *The Lorax* on the shelves. *The Lorax* has proven to be an effective educational tool. Children who read the story can identify with it, giving them a taste of how costly an individual's selfish actions can be and cultivating a sense of commitment to protecting the environment. Like many books, this one can have a powerful and lasting influence on a child's thinking.

Though he and his wife never had children of their own, Geisel's love of children is apparent in his books. He won many awards, including Oscars, Emmys, and a Pulitzer. And 30 years after his graduation, Dartmouth bestowed an honorary doctorate on him, making "Dr." Seuss legitimate. In 1967, after 40 years of marriage, his wife, Helen, took her own life. Geisel married Audrey Stone the next year, and they lived in La Jolla, California, for over 20 years, until his death. Theodor Geisel died on September 24, 1991, in La Jolla.

Bibliography

Fensch, Thomas, ed., *Of Sneetches and Whos and the Good Dr. Seuss: Essays on the Writings and Life of Theodor Geisel*, 1997; Kanfer, Stefan, "The Doctor Beloved by All: Theodor Seuss Geisel," *Time*, 1991; Morgan, Judith, and Neil Morgan, *Dr. Seuss and Mr. Geisel: A Biography*, 1995; Zicht, Jennifer, "In Pursuit of the Lorax: Who's in Charge of the Last Truffula Seed?" *EPA Journal*, 1991.

Gelbspan, Ross

(June 1, 1939–)
Journalist, Writer

Ross Gelbspan, a journalist and writer, has been at the forefront of the movement to expose to public attention the disastrous consequences of global warming, which has propelled the climate into instability and exacerbates virtually every other environmental problem. Through his intensive investigations, he has helped launch a campaign to disclose the truth about issues surrounding the global addiction to oil and coal and the environmental damage this causes by trapping heat in the atmosphere. His book *The Heat Is On: The High Stakes Battle over Earth's Threatened Climate* (1997) definitively brings home the imminence of climate change brought about by combustion of fossil fuels and examines the campaign of deliberate confusion of the public by well-funded public relations professionals in the big coal and oil companies.

Ross Gelbspan was born on June 1, 1939, in Chicago, Illinois, and grew up there. He received his B.A. in political science and English from Kenyon College in Gambier, Ohio, in 1960. After a year of graduate study at the Johns Hopkins School of Advanced International Studies in Washington, D.C., he took a reporting job at the *Philadelphia Bulletin* and then switched to the *Washington Post* in the early 1960s. He later moved to New York City and worked for the *Village Voice* during the late 1960s and early 1970s. In 1971, he wrote a series on the Soviet underground for which he spent a month in the Soviet Union interviewing dissidents and human rights advocates. He began reporting on environmental affairs in 1972 when he covered the first United Nations environmental conference in Stockholm, Sweden. This was also where he met Anne Charlotte Brostrom, whom he married in 1973. At around this time Gelbspan accepted the position of national news editor for Scripps Howard News Service, based in New York. He also taught journalism as an adjunct professor at Columbia University.

In 1979 the *Boston Globe* hired Gelbspan as a senior editor. In his capacity as special projects editor, he conceived, directed, and edited a series of articles on job discrimination against African Americans in Boston-area corporations, universities, unions, newspapers, and state and city government. The series won a Pulitzer Prize in 1984. In 1992 he coauthored a front-page series for the *Globe* on the 1992 United Nations Conference on Environment and Development in Rio. In 1993 Gelbspan retired from the *Boston Globe* and started writing a novel, but he found himself getting more and more involved in the global warming issue instead. In 1995 he coauthored an article for the *Washington Post* on the link between the spread of infectious diseases and changes in climate. He found that although the article was well received, a number of readers wrote and assured him there was no proven link between fossil fuel combustion and global climate change. Their response made him wonder why there was such a controversy over the validity of global warming. He began to investigate the politics surrounding the issue more deeply and quickly discovered the main reason for the public confusion. Since the 1992 Earth Summit, a band of several high profile "greenhouse skeptic" scientists had helped to create a widespread public belief that the issue of climate change is filled with uncertainties and hardly cause for alarm. In 1995 Gelbspan wrote an article disclosing the fact that many of these skeptic scientists receive fuel industry funding. The article appeared on the cover of the December issue of *Harper's Magazine*. Gelbspan took the main ideas from this article, which was nominated for a National Magazine Award, and expanded them into a book called *The Heat Is On: The*

High Stakes Battle over Earth's Threatened Climate (1997). The book was published in paperback in 1998 as: *The Heat Is On: The Climate Crisis, the Cover-up, the Prescription.*

The Heat Is On begins with a scientific explanation of global warming and how it is related to human activities. Each year humans cause six billion tons of carbon to be released into the lower atmosphere through the burning of fossil fuels and deforestation. This carbon builds up and traps heat, causing some of the erratic and increasingly unstable weather patterns around the world. As Gelbspan points out, global temperature records already bear witness to climate change: 1997 just replaced 1995 as the hottest year ever recorded, the five hottest consecutive years on record began in 1991, and the 11 hottest years in recorded history have occurred since 1980. The planet is heating at a faster rate than at any time in the last 10,000 years. Other consequences of alterations in the atmosphere include altered drought and rainfall patterns, temperature extremes, and more severe storms. To leading climate scientists, the congregation of so many strange weather events worldwide is an early symptom of a rise in the average global temperature. Gelbspan affirms that scientific opinion is virtually unanimous in agreement that industrial activity is having an effect on the earth's atmosphere and that climate change is happening, yet the United States stands alone in questioning the reality of the situation.

Gelbspan devotes much of his book to examining the campaign of deception by coal and oil companies that keeps many people in this country confused and misinformed. He expands on the links between the greenhouse critics and the coal and oil industries that, together, constitute the biggest single industry in history. For example, in 1993 alone the American Petroleum Institute spent nearly the same amount on public relations as the entire expenditure budget combined of the nation's five major environmental groups that focus on climate change. Gelbspan also discusses the influence of the fossil fuel industries on congressional leaders who have fought to defund climate research and who have even been able to tie up international negotiations to reduce carbon emissions. Though there is no quick or easy solution to the problem, Gelbspan does have recommendations—for example, transferring government subsidies and tax incentives away from fossil fuels and into renewable energy. His book carries a powerful and influential message, and not surprisingly, it has also been the subject of numerous loud attacks by the fossil fuel lobby.

Since the book's publication, Gelbspan has done extensive traveling and speaking on global climate change. He has appeared on *ABC News Nightline* and the National Public Radio shows *All Things Considered* and *Talk of the Nation.* He was invited to the World Economic Forum in Davos, Switzerland, in 1998, where he addressed heads of state, government ministers, and leaders of multinational corporations. With Dr. Paul Epstein of Harvard Medical School and the Intergovernmental Panel on Climate Change (which the United Nations formed in 1988 to address the climate crisis), Gelbspan brought together a group of people in 1998 to work out a set of solution strategies. This group, which includes scientists, energy company presidents, economists, and other specialists, came up with a World Energy Modernization Plan that contains a set of prescriptions, including the elimination of national subsidies for fossil fuels and the use of revenues from a tax on international currency transactions to finance the development of alternative energy technologies for developing nations. Gelbspan continues to be involved in this group as they work to accelerate international climate negotiations. He and his wife, who is a nonprofit developer with the Women's Institute for Housing and Economic Development, make their home in Brookline, Massachusetts. They have two grown daughters, Thea and Joby.

Bibliography

Dunn, Seth, "The Heat Is On: The High Stakes Battle over Earth's Threatened Environment," *World Watch*, 1998; Gelbspan, Ross, "Global Warming: The Heat Is On," *Alternative Radio*, 1998; Gelbspan, Ross, "The Heat Is On," *Harper's Magazine*, 1995; Gelbspan, Ross, *The Heat Is On: The High Stakes Battle over Earth's Threatened Environment*, 1997; "The Heat Is Online," http://www.heatisonline.org; Simonsen, Kevin, "The Heat Is On: The High Stakes Battle over Earth's Threatened Environment," *Ecology Law Quarterly*, 1998.

Gibbs, Lois

(June 25, 1951–)
Activist, Founder of the Center for Health, Environment and Justice

As a mother of small children and a housewife in the subdivision of Love Canal in Niagara Falls, New York, Lois Gibbs's life was transformed when she and her neighbors discovered in 1978 that their neighborhood was built on a toxic waste dump and that levels of toxic chemicals in their homes and the local elementary school were high enough to cause birth defects, cancers, and a host of other health problems. Taking the helm of the Love Canal Homeowners Association, Gibbs worked to assure relocation and compensation for all of the neighbors who were affected. After relocating her family to Washington, D.C., Gibbs founded the Citizens Clearinghouse on Hazardous Waste in 1981 (now the Center for Health, Environment and Justice—CHEJ), an organization dedicated to helping community organizations facing similar toxic waste issues. Now commonly referred to as the "mother of Superfund" (the federal program to clean up toxic waste sites), Gibbs's work through CHEJ has inspired and aided more than 8,000 grassroots organizations to demand accountability from industrial polluters and the U.S. government.

Lois Marie Gibbs was born on June 25, 1951, in Buffalo, New York. Until 1978, she was a shy homemaker, married to chemical plant worker Harry Gibbs, and the mother of two children. In 1972, the Gibbses bought a home in the modest Love Canal subdivision of Niagara Falls, named after entrepreneur William Love,

Lois Marie Gibbs (courtesy of Anne Ronan Picture Library)

who dug the canal in the late 1800s to link the Upper and Lower Niagara Rivers. Halted by an economic depression, all that was left of the project was a 60-foot-wide by 3,000-foot-long pit that later became a dump site used by the Niagara Falls municipality, the United States Army, and the Hooker Chemical Corporation. When the pit was filled to capacity and covered with soil, Hooker sold the land to the local board of education for one dollar, on the condition that Hooker would be free of any future liability. A school and houses were built on the site in the 1950s, and except for strong fumes and strange substances surfacing in peoples' yards and in the school's playground, Love Canal's past was forgotten.

In 1978, after a series of articles on local hazardous waste problems by reporter Michael Brown was published in the *Niagara Gazette*, Gibbs began to wonder if her five-year-old son Michael's epilepsy and urinary and respiratory problems could be caused by the chemicals. Since the focus was initially on the elementary school built right in the middle of the former canal, she asked Michael's pediatrician to write a letter to the school board requesting that he be transferred to a school in a safer area of town. The school board refused. Gibbs realized that her son would continue to suffer unless she conquered her shyness and began to fight. Her first step was to write and circulate among her neighbors a petition calling for the closure of the elementary school. In addition to gathering signatures, Gibbs heard of horrifying health problems and of residents' fears that their homes would soon be worthless and they would have no way to escape the polluted neighborhood. In August 1978, Gibbs and two neighbors took the petition with 161 signatures to the state capital of Albany to present it to the State Health Commission. They were surprised by the commissioner's response: to close the school and to recommend that all pregnant women and children living closest to the school leave the area. Within a week after that meeting, President Jimmy Carter declared Love Canal a federal emergency area and allocated funds to relocate the 239 families living in the first two rings of homes around the school.

But that is where rapid government response ended, recounts Gibbs in her memoir, *Love Canal: My Story*. The next two years were a nightmare for the remaining neighbors living on the canal but farther away from the school. A "clean-up" effort involving channels that would drain toxic leachate out of the area was undertaken by the state health department, but residents were doubtful that it would solve the problem. In addition, as machinery dug into the toxic ground, more chemicals were released, and the area became even more dangerous to inhabit.

Gibbs became president of the Love Canal Homeowners Association (LCHA), a group residents formed so that they would have a say in the future of the area. She spent virtually all of her time spearheading a fight to get the government to purchase Love Canal homes at a fair price so that Love Canal residents could move away and regain their health. Gibbs worked without a salary full time at the LCHA office at the closed school, and her home became an after-hours meeting place for residents, officials, and reporters. Gibbs helped with the neighborhood health surveys, designed by scientist and ally of the homeowners Dr. Beverly Pagen. Gibbs became a practiced public speaker, testifying about Love Canal in Albany and for the U.S. Congress. She served as an advocate for residents, continually pressing government agencies for relocation funds. But it was not until the Environmental Protection Agency (EPA) released its study in May 1980 showing that residents exhibited chromosomal damage, and angry Love Canal residents took two EPA representatives hostage, that President Carter finally agreed to evacuate all Love Canal families who wished to leave. Carter first provided for temporary relocation and then in October 1980 allocated federal money to fund permanent relocations for all Love Canal residents wanting to leave. As a response to the pressure that Gibbs and her neighbors had put on the federal government, the Federal Superfund Program was established in 1980 to clean up toxic waste sites similar to Love Canal.

Gibbs relocated to Washington, D.C., with her two children. Thousands of Americans were contacting her to request her advice and assistance in dealing with toxic hazards in their own neighborhoods, and in 1981 she founded the Citizens Clearinghouse for Hazardous Wastes (now CHEJ) to respond. Staffed by scientists and organizers, CHEJ helps grassroots groups facing local environmental problems involving toxic waste, solid waste, air pollution, incinerators, medical and radioactive waste, pesticides, sewage, and industrial pollution. CHEJ produces books and information packets, responds to questions on environmental hazards, and produces both a monthly and quarterly magazine. It has served as an important informational and networking link for more than 8,000 communities throughout the country, with a commitment to helping grassroots groups be independent and self-sufficient. Additional accomplishments by CHEJ include successfully lobbying the government to provide communities near Superfund Sites with up to $50,000 in technical assistance grants; coordinating efforts to pass Right-to-Know laws that give people access to information about the chemicals stored, disposed of, and released near their homes; helping convince McDonalds to stop using styrofoam hamburger packages; and many more campaigns.

For her courage and effectiveness at Love Canal and her work at CHEJ, Gibbs received the Goldman Environmental Prize in 1991 and the Heinz Award in the Environment in 1998, was recognized by *Outside* magazine as one of the "Top Ten

Who Made A Difference," and was awarded an honorary Ph.D. from the State University of New York, Cortland College. She lives in Virginia with her husband and four children.

BIBLIOGRAPHY

Berger, Rose Marie, "What Sustains over the Long Haul?" *Sojourners*, 1997; Breton, Mary Joy, *Women Pioneers for the Environment*, 1998; "Center for Health, Environment and Justice," http://www.essential.org/cchw; Gibbs, Lois Marie, *Love Canal: My Story*, 1982; Gibbs, Lois Marie, and the Citizens Clearinghouse for Hazardous Waste, *Dying from Dioxin: A Citizens Guide to Reclaiming Our Health and Rebuilding Democracy*, 1995; Shribman, David, "Lois Gibbs: A Woman Transformed by a Cause," *New York Times*, 1981.

Gold, Lou

(March 5, 1938–)
Storyteller, Wilderness Advocate

Political scientist Lou Gold quit academia in 1973 and moved to Oregon where he fell in love with the old-growth forest of the Klamath-Siskiyou bioregion and decided to devote the rest of his life to celebrating and protecting wilderness. For more than a decade, starting in 1983, Gold spent every summer camped on top of Bald Mountain in the Siskiyou wilderness area of southwestern Oregon. He spent the fall and winter months touring the United States, lecturing about ancient forests and the dangers they faced. Gold's effectiveness in gaining fans for forests is obvious to U.S. Forest Service officials, who say they can trace Gold's lecture route by the postmarks on letters they receive decrying their permissive policies.

Lou Gold was born in Chicago, Illinois, in 1938. He led an urban childhood but enjoyed exploring the lagoon at Chicago's Garfield Park, his main contact with "wilderness." Gold studied at the Illinois Institute of Technology, receiving a B.S. degree in political science in 1959 and continuing on there to do graduate work until 1961. He left Illinois for Columbia University, where he studied for three more years, until becoming a popular professor of political science at Oberlin College (1964–1966) and at the University of Illinois (1967–1973).

By the early 1970s, Gold began feeling that his life was not whole and that his work was not as meaningful as he had hoped it would be. So in 1973, he quit academia and moved west to Takilma, Oregon, where some close friends lived. He became a carpenter and cabinetmaker, in the hopes that working with his hands would make him happier. In his free time, Gold explored the rugged, 700,000-acre Klamath-Siskiyou bioregion straddling southwestern Oregon and northwestern California. Gold and his neighbors knew how special the area was, and in recent years the incredible biodiversity of that region has been recognized by such organizations as the World Wildlife Fund, the International Union for the Conservation of Nature, and the Wildlands Project.

The area boasts the nation's highest concentration of Wild and Scenic Rivers and the largest remaining unprotected roadless forests.

During the late 1970s and early 1980s, the U.S. Forest Service was initiating construction of a road through the Siskiyou National Forest (one of five national forests in the Klamath-Siskiyou bioregion), to facilitate the logging that was to commence there. Local environmentalists protested the road building, and Gold decided to join them. He was arrested for the first time during a May 1983 road blockade, when he and seven others sat down in front of a bulldozer and refused to let it pass. Gold and the others were granted their freedom on the condition that they not enter national forest land for one year. But within a few days Gold returned. He declared that he was there in peace, in order to bear witness to the destruction of the forest and pray for its safety. Gold took off on a pack trip, climbing Bald Mountain, a 3,800-foot peak whose summit afforded a magnificent 360-degree view. He decided to camp under the luxuriant branches of a large tree near its summit. He constructed a Native American–inspired medicine wheel on the mountain's flat top and prayed there every day. He was delighted to find that within a few days of his arrival the animals were no longer afraid of him. His friends packed in food for him, and surprised hikers were delighted by his mission and often donated supplies for his stay. In his copious free time on top of Bald Mountain, Gold carved walking sticks to give to each of the hikers he met during the summer.

Despite his hopes that life in the wilderness would bring him the peace he sought, Gold was haunted by questions. As he wrote in his "Bald Mountain Vigil," "What sense does it make to ask a society that discards its old people to save its old trees? What sense does it make to ask a society that regularly abuses its children to preserve the forest for our great, great grandchildren? What sense does it make to ask a government, which is continually preparing for war, to maintain the peacefulness of the natural world?" He also caught his mind "concocting great dramas, holding arguments with the logging industry or delivering self-righteous lectures to the Josephine County Court." Eventually, through meditation and soul-searching, Gold realized that he did not know how to stop the road building and logging that was happening in the forests below Bald Mountain and that the best he could do was to pray from his mountaintop spot and to talk to anyone he could about the problem.

His original intention was to stay on Bald Mountain forever. However, the darkness, cold, and precipitation of fall convinced him that it would be better not to spend winters there. When he descended the mountain in September, he was immediately arrested for violating his promise not to enter national forest land during that year.

The next summer Gold again climbed Bald Mountain and within a few days met a group of 4-H campers there. Several of them helped Gold clean up the mountaintop (the Forest Service had demolished a lookout there, and each winter's snows and winds unearthed more rubble) and construct his medicine wheel. One of the most enthusiastic volunteers was the son of the local national forest supervisor. He brought his father up a few weeks later, and instead of evicting Gold for violating a rule that visitors can camp for no longer than two weeks in any spot, the supervisor arranged to haul down the hundreds of pounds of rubble that Gold and his 4-H volunteers had cleaned up and asked Gold to sign a volunteer agreement to maintain nearby trails. From that moment until he stopped summering on Bald Mountain in the mid-1990s, Gold served as a volunteer caretaker of the area.

Gold found that he was happy alone on Bald Mountain but that he was even happier when he was able to serve as storyteller and host to visitors. He then decided that he could use his formidable lecturing skills from his university days and put together a slide show and talk. He spent 15 falls and winters on a lecture circuit, talking to students, clubs, conservation organizations, any group interested in his message. His presentation received rave reviews, and the U.S. Forest Service claimed it knew where Gold had spoken last from the barrage of mail it received from that place.

Gold decided in 1999 to change careers and has embarked on a new path as a painter. He continues to live in the Klamath-Siskiyou bioregion and to work closely with the Siskiyou Regional Education Project, a wilderness protection and advocacy organization.

BIBLIOGRAPHY

Gold, Lou, "Bald Mountain Vigil," in Cass Adams, ed., *The Soul Unearthed*, 1996; Norman, Michael, "Lessons: A Former Professor Comes Down from a Mountain to Offer Insights about Nature," *New York Times*, 1988; "Siskiyou Regional Education Project," http://www.siskiyou.org; Watkins, T. H., "The Laughing Prophet of Bald Mountain," *Orion*, 1990.

Golten, Robert

(February 2, 1933–)
Attorney, Professor of Law

Attorney Robert Golten has dedicated his career to helping the disenfranchised, be they marginalized people or endangered species, and to training law students to do the same. He is known for litigating the first successful case based on the Endangered Species Act (ESA), for his effective mediation skills in controversies with several competing interests, and for founding the country's first environmental legal clinic, at the University of Colorado law school.

Robert Joseph Golten was born in Chicago on February 2, 1933. He muses that his beloved Chicago Cubs inspired him to work on behalf of the underdog and also believes that being Jewish might have influenced him to pursue a career dedicated to altruism. Golten attended the University of Michigan, graduating in 1954 with an A.B. cum laude and honors in economics. He went to Harvard University's law school, graduating in 1959.

Following two years in a conventional law practice, Golten was offered a job selling mutual funds to expatriate Americans in Europe. After thinking seriously about it—it would have been a glamorous and lucrative job—Golten decided against taking it and instead went to Washington, D.C., to work for the Justice Department. He moved to the Job Corps program of the Office of Economic Opportunity in 1965, serving as general counsel there. From 1970 to 1974, Golten was a trial attorney for the District of Columbia's Public Defender office. He began teaching at American University's law school in 1970 as well, specializing in psychiatry and law and professional responsibility. There he initiated what was to become the first of his legal clinics: students worked with him to defend mental patients who had been involuntarily committed to mental institutions. Golten immediately recognized the multiplier effect of working with students: many more clients could be served when the students got involved.

In 1974, Golten was recruited by the National Wildlife Federation (NWF) into the field of environmental law. He won the nation's first suit based on the Environmental Species Act, a 1975 case (*National Wildlife Federation* v. *Coleman*) in Mississippi that prevented a highway interchange from being built through the last remaining habitat of 40 endangered Mississippi sandhill cranes. Golten had overwhelming evidence that the ESA would be violated if the highway were built, but the real estate developers who owned the land at the proposed highway interchanges apparently enjoyed considerable political influence, and the judge (later impeached) ruled that a compromise could be forged for both "cranes and lanes." Golten achieved a reversal of this decision from the Fifth Circuit Court, and later recalled, "That put me on the map."

Golten was capable of more than feisty advocacy. In 1976 he was called to the I'on swamp in Francis Marion National Forest, one of the last sites where the very endangered and soon to be extinct Bachman's warbler had been seen. The U.S. Forest Service (USFS) was planning to allow timber cutting in the swamp, and local environmentalists wanted Golten to help them prevent it. Since he could foresee complications with a traditional lawsuit, Golten asked for permission to mediate and obtained a compromise with the USFS, whereby upland pines would be cut but not the lower hardwoods crucial to the swamp habitat. Golten's effectiveness as a mediator pleased him, because he felt that reason and dialogue were often better tactics for long-term success than hard-core litigation.

In 1978, Golten moved his family, wife Mary Margaret Addison and daughters Ryan and Lauren, to Boulder, Colorado, to set up the nation's first natural resource litigation clinic. The NWF paid Golten's salary, and the University of Colorado (CU) Law School provided an office, institutional support, and student interns. The litigation clinic allowed students—early on graduate students at Colorado State University's Natural Resources school as well as CU's law students, but later just law students—to research issues such as land use, timber management, mineral development, water resources, and fish and wildlife management. The most famous struggle in which the litigation clinic participated involved the proposed Two Forks dam. The Two Forks project, scheduled for construction near Deckers, Colorado, would have dammed the South Platte River, eliminating an ecologically important stretch of the riverine habitat and with it a highly prized trout fishery near the major urban area of Denver. WILLIAM K.

Reilly, then the administrator of the Environmental Protection Agency, finally rejected the dam in 1991 and instead adopted the environmentalists' alternative model for delivering water to Denver.

After leaving the Natural Resource Litigation Clinic in 1984, Golten worked for a private law firm in Boulder on various American Indian, natural resources, and land use issues. He also served as county attorney in Summit County, Colorado, during 1984 to 1987, acting director for the Native American Rights Fund in 1988, and special land use counsel for Jefferson County, Colorado, in 1989. In 1992 he started an American Indian law clinic at the University of Colorado, directing it until 1996.

Golten resides in Boulder with his second wife, Joan Brett. In addition to his private law practice, he teaches at the University of Denver (DU). Since 1998, he has directed an International Human Rights Advocacy Project at DU, which focuses particular attention on indigenous people worldwide in cases involving land and natural resource issues. He is on the board of directors of Global Response, which responds to imminent environmental threats worldwide with letter-writing campaigns and political and citizen action.

Bibliography

Bingham, Gail, *Resolving Environmental Disputes : A Decade of Experience*, 1986; Malmsbury, Todd, "Environmentalist Believes Political Power Remains," *Boulder Daily Camera*, 1981; Neuschaefer, Nan, "Environmental Law: New Field," *Boulder Daily Camera*, 1979.

Goodman, Paul

(September 9, 1911–August 3, 1972)
Writer, Social Critic

Paul Goodman was a social critic with multiple and varied targets; as an anarchist pacifist, he decried militarism, he criticized traditional schooling, and he attacked the government's bureaucracy and centralism. He approached the reconstruction of society with imagination and creativity, ever providing what he called "practical proposals" for change. It is mainly for his pioneering proposals in the area of urban planning—especially those put forth in the 1947 book *Communitas* that he cowrote with his brother Percival—that he is considered an important influence on the environmental movement.

Paul Goodman was born on September 9, 1911, in Greenwich Village, New York City, the youngest of Barnett and Augusta Goodman's three children. Before he was born, his father suffered a business failure and deserted the family. Goodman received an excellent education at a public magnet school and attended religious school as well. He studied at the College of the City of New York with philosopher Morris Raphael Cohen, from whom Goodman learned to question the status quo and think critically. He graduated with a B.A. in 1931 and then worked reading movie scripts at home while continuing to attend classes as an auditor at different colleges and universities in Manhattan, participating in discussions and writing papers. One of the papers he wrote for Columbia professor Richard McKeon was published in the *Journal of Philosophy* in 1934. During the mid-1930s, Goodman began publishing his prose, poetry, and criticism regularly in avant garde literary magazines.

When professor McKeon moved to the University of Chicago in the late 1930s, he invited Goodman to join him and begin work on a doctorate. Goodman did so but was expelled one year later for his sexual behavior—propositioning young men. Goodman, a bisexual who fathered three children during his two common-law marriages, encountered similar problems with the administrations of the Manumit School of Progressive Education in Pawling, New York, and Black Mountain College in North Carolina. Goodman eventually was awarded his Ph.D. from the University of Chicago (in 1954), and his doctoral thesis, *The Structure of Literature*, was published by the University of Chicago Press.

Goodman wrote prolifically during the 1940s; he published poetry, fiction, plays, literary criticism, and in 1947, *Communitas: Means of Livelihood and Ways of Life*, cowritten with his brother Percival. This book, published during the post–World War II construction boom, was revolutionary in that it was among the first American works to insist that cities did not have to be crowded and unhealthy, designed to facilitate industry and commerce, but instead could be planned with aesthetics, recreation, and quality of life in mind. The book analyzes how different city plans deal with transportation and quality of life for their residents. While earlier planners, Benton MacKaye and Frederick Law Olmsted Jr., for example, had hoped that the automobile would improve the lives of urban workers by allowing them to commute between their work in cities and their homes in garden-filled suburbs, the Goodmans recognized that dependence upon the automobile was a terrible tradeoff, because ensuing congestion and demolition of neighborhoods for highways diminished everyone's quality of life. The Goodmans provided suggestions for the renewal of New York City, such as banning cars from Manhattan, reducing pollution in the rivers, converting the island's riverfront into a swimmable beach, and building a "garden city" where people could work, recreate, and live comfortably within the city. Few read the book upon its first publication, but it was more widely read when it was reissued in 1960 with a cover quote from Lewis Mumford. The book's recommendations for New York City were taken up in the 1960s by a group of activist planners called Urban Underground. Although never implemented as Goodman had envisioned, Goodman's utopian ideas nevertheless served as inspiration that change was possible, according to Urban Underground member and environmental historian Robert Gottlieb.

Goodman's restless intellect shifted foci often throughout his life; his receptivity and flexibility served him as he explored a number of fields during the 1950s and 1960s. He underwent psychotherapy during the late 1940s and early 1950s, and he quickly became an expert on it, cowriting *Gestalt Therapy* with Dr. Fritz Perls and Ralph Hefferline in 1951 and working half-time as a lay psychotherapist at the New York Institute for Gestalt Therapy. His interest in psychology is reflected in the work he wrote for the Living Theatre of New York City, which was deeply psychological in nature. Facing rejection by mainstream publishers and critics during most of the 1950s and living in virtual poverty, Goodman's view became increasingly cynical. In 1960 when his book *Growing Up Absurd* was published by Random House, he shot into public view. *Growing Up Absurd* criticized society in the United States of the 1950s for its conformism, for its discouragement of creativity, honesty, and true community. The book resonated with youth of the day, and he became a popular speaker at college campuses—he spoke at over 100 during this period—as students erupted into rebellion with the Berkeley Free Speech movement and massive nationwide protests against the Vietnam War. Goodman sympathized with young people's disgust with "the system," drawing connections between society's excessive consumerism, developers' eagerness to turn rich land into parking lots, and the U.S. government's use of napalm in Vietnam to defoliate jungles in its attempts to uncover guerrillas. He had written pacifist reflections on the military draft during the 1940s, when he was draft age, and republished them during the 1960s. He urged young people not to aid and abet a violent state and to break the government's laws if they were in conflict with moral law. Goodman's son Matthew, a student at Cornell University, participated in the first public burning of draft cards at New York's Central Park in the spring of 1967.

In addition to editing the radical newspaper *Liberation* with activist and organizer David Dellinger from 1962 to 1970, Goodman wrote at least one book per year during the 1960s. *The Community of Scholars* (1962) and *Compulsory Mis-Education* (1964) criticized traditional universities, advocating instead student-centered institutions with a stronger sense of community. These are considered primers for the alternative education movement, inspiring the creation of many experimental free schools. His other books provided critiques of society in the United States for its various flaws, including militarism, centrism, rigid hierarchy, bureaucracy, and rigidity in terms of the sexual behaviors it permitted.

In addition to writing, Goodman taught during the 1960s as guest professor at the University of Wisconsin–Madison, the Experimental College of San Francisco State College, and the University of Hawaii–Honolulu and worked as a fellow at the Institute for Policy Studies in Washington, D.C.

Although initially excited by the potential of the 1960s youth movement to bring about major structural changes in the tradition of Martin Luther's reforming the church and Thomas Jefferson's constructing a new democratic nation, Goodman was eventually disappointed by the movement's stubborn, unrealistic goals. He faded out of the public eye when students rejected his critiques of their movement. Goodman never completely recovered from the tragic death in 1967 of his son Matthew, who fell to his death while picking huckleberries in the mountains near his father's New Hampshire farm, one week before having to report to an FBI office for his draft resistance.

Paul Goodman died of a heart attack on August 3, 1972, at his farm in Groveton, New Hampshire. He left two surviving daughters, Susan and Daisy.

BIBLIOGRAPHY

DeLeon, David, *Leaders from the 1960s: A Biographical Sourcebook of American Activism*, 1994; Gottlieb, Robert, *Forcing the Spring: The Transformation of the American Environmental Movement*, 1993; Parisi, Peter, ed., *Artist of the Actual: Essays on Paul Goodman*, 1986; Widmer, Kingsley, *Paul Goodman*, 1980.

Gore, Albert, Jr.

(March 31, 1948–)
Vice President of the United States

The publication in 1992 of Al Gore Jr.'s best-selling book, *Earth in the Balance: Ecology and the Human Spirit*, which explored the environmental problems that threaten the survival of life on earth, established Gore as one of the most environmentally aware politicians to occupy such a high-level public office. His subsequent election later that year as Pres. Bill Clinton's vice president brought considerable media attention to the tough environmental proposals that he listed in his book. Although Gore was able to persuade President Clinton to resist the antienvironmental backlash of the ultraconservative 104th Congress, elected in 1994, and to make some important strides in conservation, environmental health, and global climate change issues, the Gore-Clinton team still fell short of what environmentalists had hoped they would accomplish. Nonetheless, by 2000, when Gore was the Democratic nominee for president, he reiterated the goal stated in *Earth in the Balance* of making the rescue of the global environment the central organizing principle of society, and he was still seen by many environmentalists as their best hope for true reform on a national level.

Albert Arnold Gore Jr. was born March 31, 1948, in Washington, D.C., to Senator Albert Gore Sr. and Pauline (LaFon) Gore. Gore grew up in a political atmosphere in the nation's capital. He graduated cum laude from Harvard in 1969 and served a tour of duty in Vietnam as an army reporter. Gore married Mary Elizabeth ("Tipper") Aitcheson in 1970. When he returned from Vietnam, Gore began reporting for the Nashville *Tennessean* while attending divinity school at Vanderbilt University Graduate School of Religion. He later studied law at Vanderbilt, before giving it up to run for a seat in the U.S. House of Representatives in 1976. At the age of 30, Gore won this election and began what would be a long and successful political career.

During the eight years in which he served as a Democrat in the House of Representatives, Gore became increasingly interested and active in environ-

Albert Gore Jr. sits with President Bill Clinton at a global warming discussion, 1997. (Reuters/Mark Wilson/Archive Photos)

mental issues and was known for his depth of knowledge about nuclear weapons, in particular. He led the first congressional hearing on toxic waste and was the catalyst behind the passage of the 1980 Superfund bill, which set up a fund to clean up toxic waste sites. His environmental record was marred by his support for two Tennessee Valley Authority environmental fiascoes: the Tellico Dam, which was opposed by environmentalists because it was thought to threaten the endangered snail darter, and the Clinch River experimental nuclear "breeder" reactor. The former was built; the latter was not; Gore supported them both because of the jobs they would offer Tennesseeans. Gore left the House of Representatives in 1984, when he won a seat in the Senate, and served as a senator until winning the vice presidency in 1992.

Gore began writing *Earth in the Balance* in 1989, after his son, Albert III, was hit by a car and severely injured. Gore later explained that this trauma had prompted him to reevaluate serious life issues, and he began to consider the effect an environmentally unstable world would have on the future of his four children. The book was a resounding success both on the market and among the critics, because Gore effectively synthesized a variety of complex environmental issues and explained them in lay terms.

Earth in the Balance is divided into two sections. The first is a comprehensive discussion of the current global ecological crisis. The second, which garnered much media attention, presents a series of solutions to these problems. Gore drew from his experience in legislating environmental laws and policies to outline the more complex facets of global warming, acid rain, deforestation, and overpopulation. His understanding of these issues is indisputable, and his proposed market-based solutions are pragmatic, though controversial. Many conservative critics disparaged Gore's proposed solutions as being economically implausible.

The second section of *Earth in the Balance* received the most media attention during the 1992 presidential campaign, making environmental issues a major aspect of the Democratic platform. Gore proposed a unified global effort, with rich and poor nations working together, to stem the tide of population growth and environmental degradation inflicted by existing economic and political strategies. Such a plan would work under the umbrella of a United Nations Stewardship Council, which would monitor events and policies that are detrimental to the planet.

Gore's nomination as vice president earned the Democratic ticket the endorsement of many major environmental organizations, whose leaders were optimistic that Gore would be able to push Clinton to adopt a proenvironment agenda. Gore was able to convince Clinton to resist the antienvironmental offenses of the right-wing 104th Congress, elected in a conservative backlash in 1994; to negotiate the successful Kyoto Protocol for reducing greenhouse gas emissions; and to designate several new national parks and monuments in the West, including the spectacular 1.7-million-acre Grand Staircase–Escalante National Monument in Utah. However, Gore was unable to obtain many significant gains that he had reportedly lobbied hard for. For example, during the administration's first year in office, Gore tried to persuade the president to impose a new tax on energy consumption, but Clinton refused, afraid the middle class would revolt. Clinton also refused Gore's plea to raise fuel efficiency standards for automobiles, before the 104th Congress revoked the president's unilateral power to do so. Nor could Gore prevent the president from backtracking on issues of livestock grazing on public lands, mining reform, and the 1995 salvage-logging rider, which suspended all environmental regulations of logging in damaged forests, a piece of legislation that environmentalists universally considered disastrous.

Despite these setbacks, Gore continued to promote himself as an environmentalist during his 2000 campaign as Democratic nominee for president. *Earth in the Balance* was reissued in time for the campaign with a new, updated introduction; in interviews Gore reiterated his commitment to its ambitious goals.

When not in Washington, Gore and his family make their home in Tennessee, where Gore has been a livestock and tobacco farmer since the early 1970s.

Bibliography

Baden, John A., ed., *Environmental Gore: A Constructive Response to Earth in the Balance*, 1994; Burns, James MacGregor, and Georgia J. Sorenson, *Dead Center: Clinton-Gore Leadership and the Perils of Moderation*, 1999; Clinton, Bill, and Al Gore Jr., *Putting People First: How We Can All Change America*, 1992; Gore, Al, Jr., *Earth in the Balance*, 1992, 2000; Rauber, Paul, "The Great Green Hope," *Sierra*, 1997.

Gottlieb, Robert

(February 21, 1944–)
Professor, Urban Planner, Activist

Urban planner, activist, and professor of urban and environmental policy, Robert Gottlieb attempts to widen what is traditionally thought of as the environmental agenda in his writings, research projects, and college courses. Gottlieb believes strongly that there must be a link between research and action, and he has worked with community, environmental, and labor groups to try to bring about environmental change in such areas as water policy, food policy, transportation and land use policy, and industry production issues.

Robert Gottlieb was born on February 21, 1944, in Brooklyn, New York, in a household where radical politics was a mainstay. As a teenager Gottlieb was involved in this country's earliest protests against atomic testing and its radioactive fallout. Gottlieb attended Reed College in Portland, Oregon, and spent a year in Strasbourg, France, where he became involved with the International Situationists, a European-based group that prefigured many of the themes of the New Left, including its critique of a mass consumer and "spectacle" society. At Reed, Gottlieb majored in French literature, and upon graduating in 1965, he joined Students for a Democratic Society, a national political organization that was prominent in the New Left political movement responsible for many of the major youth and anti–Vietnam War actions of the 1960s.

Gottlieb returned to New York City after graduation and, while studying sociology at the New School for Social Research, founded and directed the Movement for a Democratic Society and cofounded several affiliated organizations, including the Urban Underground (or Planners for a Democratic Society). These groups, all founded between 1967 and 1969, represented an early marriage of radical politics and environmentalism. They translated the environmentally oriented utopian ideas of such radical thinkers as Murray Bookchin and Paul Goodman into action strategies, including, for example, limiting the access of cars into Manhattan.

In 1969, as several of the New Left groups began to implode, Gottlieb journeyed west. Many young New Yorkers were at this time moving to San Francisco, famous for its accommodation of those looking for an alternative lifestyle. When Gottlieb would mention to acquaintances en route that he was interested in Los Angeles, the response would be quizzical. Los Angeles was viewed as a conservative heartland, not a place of interest to a radical activist interested in environmental and urban planning issues. But Gottlieb saw Los Angeles as a challenge and opportunity in relation to his radical environmentalist perspective, given its role as the original city of sprawl and, at the same time, its emerging dynamic of new ethnically and culturally diverse communities connecting with the region's long history of experimentation that Carey McWilliams (who became an important influence on Gottlieb's own research and writing) had previously described.

One of Gottlieb's first activities in California was to cofound and manage the Midnight Special, a cooperatively owned bookstore in Venice (now located on the promenade in Santa Monica). After he left the Midnight Special in 1973, he began lecturing on politics and social movements, urban planning, and the history of the Los Angeles region at various local universities, including the California Institute of the Arts, University of California, Los Angeles (UCLA), Peoples College of Law, and California State University at Los Angeles. During this period he undertook his investigation of the history of the power structure, the political dynamics, and the land use issues of the Los Angeles area. This resulted in his first book—*Thinking Big: The Story of the* Los Angeles Times, *Its Publishers, and Their Influence on Southern California*—which was cowritten with Irene Wolt and published in 1977.

Since this first project, Gottlieb has continued to write prolifically on topics that reflect his wide range of interests, all of which are based on the common theme of the link of social movements and progressive politics with environmentalism. Since 1977, he has published eight more books, chapters in 12 books, and more than 100 articles in scholarly journals and magazines of general interest.

Far from remaining at his desk or inside the confines of an ivory tower, Gottlieb has been an active participant in the issues he studies. For example, in 1986 he served as faculty coordinator for a conference, "International Green Movements and the Prospects for a New Environmentalism," at UCLA that brought together academics and activists to explore opportunities for a more expansive environmental movement and its place among movements for change. Twelve years later, he hosted a conference at Occidental College, "Progressive L.A.: A Conference on Social Movements in Los Angeles—Uncovering Our Past and Envisioning Our Future," that helped launch the development of the Progressive Los Angeles Network (PLAN), an innovative marriage of progressive policy analysts and activists, including environmental activists. While serving from 1980 to 1987 as director of the Metropolitan Water District of Southern California, the largest water district in the country, Gottlieb challenged the dominant water industry perspective of ever-expanding water supplies imported from distant places to fuel the continuing expansion of the region. Gottlieb identified the need for a new water "ethic of place" and a "community value" rather than "commodity value" to water. He further elaborated these ideas in his two books on water policy, *A Life of Its Own: The Politics and Power of Water* (1988) and *Thirst for Growth: Water Agencies as Hidden Government in California* (1991), and in his many articles on this subject. Interested in issues of materials use and production and waste generation and disposal, and inspired by a student project he helped supervise, Gottlieb studied the struggles of such Los Angeles area grassroots groups as Mothers of East Los Angeles and Concerned Citizens of South Central against proposed waste incineration facilities in their communities. He documented these and other struggles, within the context of a history of waste generation and disposal policies, in his book *War on Waste* (1989), coauthored with his former student Louis Blumberg.

The rise of the antitoxics groups of the 1980s, which Gottlieb helped document, has been associated with the development of an environmental justice movement that challenged the range of environmental burdens, such as waste incinerators and other hazardous facilities, imposed on low-income neighborhoods and communities of color. Gottlieb's major study of environmentalism and the roots of the environmental justice movement, *Forcing the Spring* (1993), helped place in historical perspective and make more coherent this more recent development of a dynamic and broader-based environmental movement. From this work and through other research projects, Gottlieb became increasingly interested in identifying how the environmental agenda could be expanded to include daily life issues in communities and workplaces. Along these lines, Gottlieb began to explore issues of production and the workplace and industrial environmental policy through the Pollution Prevention Education and Research Center that he cofounded in 1991 and still directs today (as part of the Urban and Environmental Policy Institute in Los Angeles).

Most recently, Gottlieb has stretched his conception of what should be included in the environmental agenda even further, engaging in such diverse issues as cleaning clothes, janitorial work, and cultivation, production, and access to food. In 1992, while he was teaching at UCLA's Graduate School of Architecture and Urban Planning, his students became interested in food issues and persuaded him to supervise a research project on the topic. They were interested in the price and availability of food, its quality, and how a community's food system is experienced. The students' work became a new point of departure for Gottlieb, who saw an emerging community food security movement as directly addressing social, environmental, community health, and production issues. Through subsequent research projects and programs, Gottlieb helped develop new innovative policy and programmatic initiatives (such as the development of a farmers' market salad bar for low-income elementary schools and the passage in 1996 of the Community Food Projects federal legislation), as well as new movement groups (such as the Community Food Security Coalition). Gottlieb argues that these food issues resonate for people as a daily life concern and can embody an aspect of the new environmentalism that he has elaborated in his most recent book, *Environmentalism Unbound: Exploring New Pathways for Change*, to be published in 2000.

Gottlieb is currently the Henry R. Luce Professor of Urban Environmental Studies at Occidental College and director of the Urban and Environmental Policy Institute. He resides in Santa Monica.

Bibliography

Gottlieb, Robert, "Beyond NEPA and Earth Day: Reconstructing the Past and Envisioning a Future for Environmentalism," *Environmental History Review*, 1995; Gottlieb, Robert, "The Meaning of Place: Reimagining Community in a Changing West," in Hal K. Rothman, ed., *Reopening the American West*, 1998; Gottlieb, Robert, and Louis Blumberg, "Rethinking Place, Reinventing Nature: An Environmental Justice Perspective on the Public Lands," *George Wright Forum*, 1996; Gottlieb, Robert, and Andrea Brown, "Janitorial Cleaning Products: It's Not Just What Gets Used, But How It Gets Used That Counts," *Pollution Prevention Review*, 1998; Gottlieb, Robert, and Andrew Fisher, "'First Feed the Face': Environmental Justice and Community Food Security," *Antipode*, 1996.

Gould, Stephen Jay

(September 10, 1941–)
Paleontologist, Writer

Perhaps more than any other contemporary scientist in the United States, Stephen Jay Gould has presented the modes, implications, benefits, and shortcomings of science to a literate public. A highly productive scholar, Gould has defined and participated in crucial debates of the biological and geological sciences, particularly with regard to the theory of evolution, the interpretation of fossil evidence, and the meaning of diversity and change in biology. His forwarding of these debates has influenced our understanding of nature, our recognition of the human impact on the natural environment, and, ultimately, a significant portion of the environmentalist rhetoric. One of the leading evolutionary theoreticians today, Gould, with his nearly 20 books and hundreds of essays, reviews, and articles, is a popular and well-known writer and lecturer on scientific topics.

Stephen Jay Gould was born September 10, 1941, in New York City. When he was five years old, he was taken to the American Museum of Natural History by his father, a court stenographer with an avocational interest in natural history. Gould's fascination with paleontology developed throughout his childhood and teenage years, rivaling his passion for the New York Yankees. He completed his undergraduate degree in geology at Antioch College in 1963 and returned to New York for doctoral studies in paleontology at Columbia University, completing his studies in 1967.

In the course of his graduate studies, Gould became very aware of problems surrounding ecosystem disruption by humans. Working in Bermuda on his dissertation, which examined a remarkable Bermudian land snail named *Peocilozonites*, Gould watched the population of this snail become completely eliminated, not on purpose, but by accident as a result of biological control. In order to control *Otola*, an imported edible snail that had escaped from a garden and spread throughout the island as an agricultural pest, authorities introduced a "cannibal" snail from Florida, called *Euglandina*. Rather than controlling the *Otola* populations, *Euglandina* devastated the *Peocilozonites*. Returning in 1973, Gould was unable to find a single *Peocilozonite* on the island. This experience led Gould to argue for environmental protection, not from an ecological perspective but rather from an emotional humanistic point of view. Preaching for a humanist ecology to complement our scientific understanding for the need for biodiversity, Gould argues, in his 1993 collection of essays, *Eight Little Piggies*, that "we cannot win this battle to save species and environments without forging an emotional bond between ourselves and nature . . . for we will not fight to save what we do not love." Many of Gould's other writings also reflect his strong sense of environmental ethics.

In 1967, Gould became assistant professor of geology and assistant curator of invertebrate paleontology at the Museum of Comparative Zoology at Harvard University. In 1971, he became associate professor and associate curator; in 1973, he became professor of geology and zoology and curator of invertebrate paleontology at the Museum of Comparative Zoology and an adjunct member of the Department of the History of Science. During this time, Gould gained a reputation as one of Harvard's most visible and engaging instructors, offering courses in paleontology, biology, geology, and the history of science. In 1982, he was named Alexander Agassiz Professor of Zoology.

In 1974, Gould initiated a monthly series of essays for *Natural History*, under the title, "This View of Life." Writing on such topics as "Size and Shape," "Sizing Up Human Intelligence," and the "Race Problem," Gould established himself as a widely respected writer and is often praised for reviving the popular scientific essay. His critical genre of science writing represented a marked change from the traditional and traditionally dull, didactic style that had preceded his work. Some 25 years and almost 300 contributions later, Gould continues to write his monthly essays for *Natural History*. Gould's *Natural History* submissions have also helped forward his notions of an environmental ethics to a widespread audience.

Among Gould's more famous pursuits is certainly his critique of the central concepts of Darwinism and his notion of "punctuated equilibria." This theory, which he formulated with his colleague, Niles Eldredge, in 1972, postulates that species tend to mutate aimlessly, within the expected bounds of statistical variation, and that radical evolution is concentrated in relatively rapid events of speciation—within geologically short periods of 100,000 years or so. Gould's and Eldredge's punctuated equilibrium theory modifies the theory of evolution that had been accepted previously, which held that evolution took place gradually, as slow, continuous transformations of established lineages. While his theory of punctuated equilibrium has stimulated endless debates since the mid-1970s, Gould maintains that it fills a void in Darwin's theory. Darwin himself struggled to explain gaps in fossil records that could not be explained if evolution moved forward as the accretion of many small changes.

Gould's high visibility as a leading Harvard academic, his critical voice, and his enthusiasm for debate have prompted him to enter into scientific, cultural, and political controversies. His participation in the debate surrounding "creationist science" has made him a prominent detractor of this cause. He has openly opposed legislation to require its teaching alongside Darwinian evolution and has testified in several court cases concerned with this issue. As the chief evolutionist in the United States, he continues to stand out as a lightning rod for advocates of creationism, as made obvious by the endless Internet bulletin boards offering discussion threads on this or related topics.

The recipient of more than 40 honorary degrees, more than a dozen literary awards, and countless academic medals and awards, Gould's work is among the most highly respected in the United States of scholars in any field. Gould continues to teach at Harvard University. He lives in South Boston.

Bibliography

Gould, Stephen Jay, *Full House: The Spread of Excellence from Plato to Darwin*, 1996; Gould, Stephen Jay, *The Mismeasure of Man*, 1981; Gould, Stephen Jay, *The Panda's Thumb*, 1981; Gould, Stephen Jay, *Wonderful Life: The Burgess Shale and the Nature of History*, 1989.

Gray, Asa

(November 18, 1810–January 30, 1888)
Botanist

Few Americans, past or present, have moved their disciplines and the general understanding and perception of nature forward more substantially than nineteenth-century botanist Asa Gray. Gray was a pioneer and master in the field of plant geography and became the most persistent and effective advocate in the United States of Charles Darwin's theory of evolution. Gray's contributions in botany and to the creation of the Botanical Museum at Harvard University, to which he donated his immense collection of American flora, were instrumental in the development of the appreciation of nature in the United States.

Asa Gray was born November 18, 1810, in Sauquot, New York, the eldest of the eight children of Moses Gray, a farmer and tanner, and his wife, Roxana Howard. Gray attended the Clinton Grammar School, nine miles from his home, and transferred to Fairfield Academy, where he received his first lessons in natural science. Gray was drawn to botany at the age of 17, after reading an entry about the field in Brewster's *Edinburgh Encyclopedia.* He bought Amos Eaton's *Manual of Botany* that winter and immersed himself in it until spring, when he could put his newfound knowledge to practice.

Gray received his degree in medicine from the Fairfield Medical School in 1831, but already his heart was devoted to botany and he never did take up the practice of medicine. One of his earliest professional acquaintances was with Dr. John Torrey, a scientist whose early attempts to categorize plant species into broader plant families represented the first steps toward our modern classification system. Torrey quickly became Gray's friend and mentor. Gray spent 11 years teaching, traveling, collecting, and working as Torrey's assistant in his home in New York. Living in the Torreys' home, Gray benefited from the daily association with Torrey, while Mrs. Torrey worked to improve his taste, manners, and general culture. More important was Gray's immersion in the scientific community of New York. In December 1834, for the first time, Gray read two scientific papers before the New York Lyceum of Natural History. The following year, he moved permanently to New York to be close to Torrey. Gray's first major publication, *Elements of Botany,* was published in April 1836, and that summer he was appointed botanist of a projected government expedition to the South Seas. Delays and complications led to his resignation before the ship sailed. Instead, he became librarian and curator of the New York Lyceum of Natural History until 1838, when he accepted a professorship of botany at the University of Michigan. He spent the next year in Europe, purchasing books for the university and studying various specimens and hybridizations of American plants in European herbaria. The trip was successful and initiated lifelong friendships with several European botanists.

In 1842, after further expeditions to Virginia and the mountains of North Carolina, the first edition of Gray's *Botanical Text-Book* was published. He was offered the Fisher professorship of natural his-

Asa Gray (courtesy of Anne Ronan Picture Library)

tory at Harvard, and without ever having the time to assume his teaching duties at the University of Michigan, he accepted it. Already, Gray was the universally acknowledged leader in American botany. Gray's connections with European and North American botanists allowed Gray's herbarium at Harvard to thrive and serve as a general nucleus for the identification of newly discovered North American plants.

After a year-long engagement, in 1848 Gray married Jane Lathrop Loring, who survived him to edit his autobiography and letters. He traveled extensively in the ensuing years and created the Harvard Department of Botany, training many of the eminent botanists of the next generation. Gray held several distinguished positions in academic societies and was one of the founders of the National Academy of Sciences.

Gray was also a prolific writer. His frequent contributions to the *American Journal of Science* constitute an impressive and authoritative history of botany extending over half a century. He wrote less frequently for a variety of other periodicals and also published five important textbooks. In all, Gray produced more than 350 books, monographs, and shorter papers furthering the understanding and acceptance of descriptive botany in North America. Timing certainly played a role in Gray's success as a botanist. He lived at a time when exploration of the rich and diverse flora of North America was accelerating. The clarity and accuracy of Gray's prose made his writing accessible to the lay public as well as to scientists.

In September 1857, Charles Darwin wrote Gray a now-famous letter in which he outlined his theory of the evolution of species by means of natural selection and sent Gray an advance copy of *The Origin of Species*. The eminent American botanist championed Darwin's theory, and Darwin prized Gray as one of his most influential supporters and most searching critics. Much to the disappointment of many of Darwin's agnostic disciples, however, Gray continued to consider himself a theist. Gray believed that religion and Darwin's theory of natural selection were not in conflict. While it seemed that he had one foot in each camp, Gray anticipated later discoveries by Gregor Mendel and Hugo De Vries by suggesting that variations were not physical, but physiological, thereby allowing himself to maintain his faith in God as Creator and his belief in Darwin's theory of evolution.

In 1872, Gray retired from teaching, and in 1873, CHARLES SPRAGUE SARGENT succeeded him as director of the Botanical Garden. Gray donated his herbarium, containing some 200,000 specimens, and his library of 2,200 volumes to Harvard University upon the completion of a fireproof building designed for them. Gray continued his work and held onto his Fisher professorship until his death on January 30, 1888. He died at home, in Cambridge, Massachusetts, one month after a paralytic stroke. Appropriately, his last act was a letter to a fellow botanist, gently scolding him for coining a superfluous plant name.

BIBLIOGRAPHY

Dupree, A. Hunter, *Asa Gray: American Botanist, Friend of Darwin*, 1988; Gray, Jane Loring, ed., *Letters of Asa Gray*, 1893; Sargent, Charles Sprague, ed., *Scientific Papers of Asa Gray*, 1889.

Grinnell, George Bird

(September 20, 1849–April 11, 1938)
Cofounder of the Audubon Society, Cofounder of the Boone and Crockett Club, Editor and Publisher of *Forest and Stream*

George Bird Grinnell was an avid explorer of the American West and a prolific writer. The articles and editorials he wrote for his magazine, *Forest and Stream*, issued rallying calls for the conservation of big game animals and birds and more stringent protection of the national parks. His readers were urged to join the two major conservation-oriented organizations he helped found, the Audubon Society and the Boone and Crockett Club. Ethnologists laud his con-

George Bird Grinnell (Library of Congress)

tribution to the field of American Indian studies. He wrote ten books about Plains Indians and their folklore.

George Bird Grinnell was born in Brooklyn, New York, on September 20, 1849, to a well-off family. At the age of eight, he moved with his family to a house on the estate of the Audubon family in northern Manhattan. Madame Audubon, the widow of the artist and naturalist JOHN JAMES AUDUBON, ran a day school in her family's mansion in Audubon Park. While not at school, Grinnell and a band of friends who also lived on the estate had the run of the place, which extended all the way to the Hudson River. His interest in natural history and love for exploring developed at Audubon Park, along with a mischieviousness that marked his early years. At the age of 12, he and some other boys from Audubon Park were arrested for skinny-dipping in the river and were actually put in jail until their parents could pick them up. Grinnell was expelled from Yale University for climbing a lightning rod and painting his class's graduation year on the face of the college clock. He was readmitted after he passed his class exams and he earned a B.A. in 1870.

The summer after graduation, Grinnell left on the first of what would be 40 trips west during his lifetime. On this first trip, he accompanied a Yale fossil-collecting expedition to Kansas, Wyoming, Utah, and Nebraska. Upon his return, Grinnell moved to New York to help his father run the family dry goods business during the year, escaping for trips west to hunt, explore, and learn more about the Indians in the summer. In 1874, Grinnell reached a turning point in his life and career. He sold the family business and devoted himself wholly to science. Yale's Peabody Museum offered him employment as an osteology assistant, and he began working toward a Ph.D. in paleontology. During the summers, Grinnell continued to travel west. He served as naturalist for Col. George Custer on his Black Hills expedition in 1873, and made another trip west in 1875 to Yellowstone, writing an acclaimed account of the wildlife he observed there. While at Yale, Grinnell also began writing for *Forest and Stream*, a high-profile conservationist magazine with a readership of influential sportsmen. In 1880, after receiving his Ph.D., he bought majority rights to the magazine and became editor and publisher. Grinnell's frequent articles and editorials called attention to the environmental problems of the day, including diminishing game, depletion of forests, water pollution, and endangered watersheds.

Sportsmen formed 342 known conservation clubs throughout the country in response to *Forest and Stream*'s call for activism. When Grinnell became concerned in 1887 about the disappearance of birds that were valued for their plumes, he began publishing *Audubon Magazine*, devoted solely to bird protection. During the mere two years that he published *Audubon Magazine*, some 19 regional Audubon societies were founded with a total of about 50,000 members. The year 1900 marked the passage of the Lacey Act, for which Audubon members had lobbied heavily. It outlawed interstate commerce of protected birds killed by hunters, greatly decreasing the financial incentive for hunters interested in plume birds. That year, Grinnell resumed publishing a bird-oriented magazine, *Bird Lore*, which in 1904 became *Audubon*, and the scattered Audubon societies joined together under the banner of the National Audubon Society.

One of Grinnell's most influential readers was THEODORE ROOSEVELT. Roosevelt first went to meet Grinnell after *Forest and Stream* criticized his book *Hunting Trips of a Ranchman* (1886). They soon became friends, vowing to further the conservationist cause together. In 1887 they cofounded the Boone and Crockett Club, whose membership requirements included having killed at least one adult male of three separate species of big game. The first stated goal of the club was "to promote manly sport with the rifle," but the club's objectives were quickly distilled into the conservationist goals of protecting big game animals and their habitats. The club was particularly effective in its fight to maintain the wildness of Yellowstone, which had been declared a national park in 1872. The influential members of the Boone and Crockett Club lobbied Congress to prevent railway and concessions companies from cutting forests, killing animals, and even marring natural geysers in their attempts to transform the park into a huge, profitable resort. The club was also instrumental in the founding of Glacier National Park (1910), which Grinnell had first surveyed in 1885.

Roosevelt and Grinnell coedited the Boone and Crockett Club's publications during the 1890s. They put out three books together, *American Big Game Hunting* (1893), *Hunting in Many Lands* (1895), and *Trail and Campfire* (1897).

Grinnell's service to the conservation movement was vast: he helped establish the American Game Association in 1911, he served as president of ROBERT STERLING YARD's National Parks Association in 1925, he helped found the New York Zoological Society in 1895 and served as a trustee, and he was a member of the federal advisory board for the Migratory Bird Act.

In addition to his conservationist activities, Grinnell was an active advocate for the Plains Indians, with whom he had become familiar during his trips west. Presidents Cleveland and Roosevelt depended upon Grinnell to advise them on Indian affairs, and they both dispatched him to mediate land disagreements between Indians and White settlers. Grinnell developed many strong, long-lasting friendships with Indian chiefs, especially White Calf of the Blackfeet. Grinnell wrote ethnographic descriptions of the Pawnee, Blackfeet, and Cheyenne peoples and transcribed many folktales.

Grinnell married Elizabeth Kirby Curtis Williams in 1902, when he was 52 years old. Grinnell died in New York City on April 11, 1938, at the age of 88, after a long illness.

Bibliography
Cutright, Paul Russell, *Theodore Roosevelt: The Making of a Conservationist*, 1985; Diettert, Gerald A., *Grinnell's Glacier: George Bird Grinnell and Glacier National Park*, 1992; Evans, Robley, *George Bird Grinnell*, 1996; Parsons, Cynthia, *George Bird Grinnell, A Biographical Sketch*, 1992.

Grogan, Pete

(February 19, 1949–)
Recycling Expert

Pete Grogan cofounded the grassroots recycling organization Eco-Cycle in Boulder, Colorado, in 1976. He directed Eco-Cycle until 1987, fostering its growth as it became the largest nonprofit recycling organization on the continent. He then became an international recycling consultant, assisting numerous cities, states, and private companies in the development of their own recycling programs and assisting the paper and plastic industries with strategies for recovering recyclable commodities. At his current position of market development manager for Weyerhaeuser Recycling, Grogan works to increase paper recovery throughout North America.

Peter Laurence Grogan was born on February 19, 1949, in Newark, New Jersey. As a young child vacationing at the beach, he spent summer evenings with his family on the boardwalk, watching boats at sea that appeared to be on fire. One evening when he asked about this, his parents told him that they were actually barges that hauled garbage out to sea, burned it, and then dumped it into the ocean. The next day, swimming in the ocean, he noticed that it was full of tiny bits of incinerated garbage! This experience, at the age of five, was the root of Grogan's environmentalism, he now believes. Grogan's parents had learned to recycle during World War II and continued to do so after the war. The family diligently separated newspapers and magazines to give to a local ragman or the Salvation Army when they came through their neighborhood. Once Grogan left home, he was amazed at the waste generated by most households.

After spending some time exploring the United States, Grogan studied psychology at the University of Colorado–Boulder, graduating with a B.A. in 1974. One of Grogan's jobs after graduating was with a local program for juvenile offenders. Grogan wanted to experiment with sports therapy, but there were no funds available for sporting equipment. To raise the money, he and the teenagers collected recyclables from residents until they had raised a few thousand dollars. After he stopped the collections, people continued to call him to request pickups.

That was the inspiration for Eco-Cycle, a residential recyclable material collection program that he and friend Roy Young founded in 1976. For the first year, Grogan and Young volunteered their time; they coordinated truck rental, recruited community organizations to provide 40 volunteers to collect the recyclables in return for a donation of $500 to their organization, and sold the material to recycling processors. After a year, they approached the U.S. Environmental Protection Agency (EPA) for funding. Because of their success and their potential as a model for other communities, the EPA agreed to fund the program if it could also get financial support from the governments of the city of Boulder and Boulder County. With that seed money in place, Eco-Cycle was able to grow until it reached a point of financial self-sufficiency.

Grogan's background in psychology and his contacts with the university psychology department helped him encourage wide participation in Eco-Cycle. A block-leader program signed up volunteer community block leaders, who distributed Eco-Cycle's newsletters and posted signs to remind their neighbors of the monthly pickups. Prof. Stuart Cook of the Department of Psychology at the University of Colorado hypothesized that uniform recycling bins would increase participation levels, so he obtained research funds to buy bins and to pay graduate students for a study. Bins were distributed to 1,000 homes on specific blocks in different neighborhoods and indeed participation with the bins increased significantly. Today, 9,500 other cities provide residents with containers.

In 1987, when recycling was coming into vogue throughout the country, Grogan decided to leave Eco-Cycle and begin helping others to plan and implement recycling programs. He spent the next seven years as a consultant with R.W. Beck and Associates, which offers consulting services to government and industry throughout North America and Asia in recycling, composting, and solid waste planning; design of material recovery facilities; and other related areas. Grogan assisted numerous cities, states, and companies in the development of their own recycling programs: Miami, Honolulu, Orlando, Disneyworld, American Samoa, Guam, Denver International Airport, and many more. He also helped the paper and plastic industries increase their recovery of recyclable waste products.

Since 1994, Grogan has managed market development activities for Weyerhaeuser Recycling, one of the world's five largest paper recyclers, based in Tacoma, Washington. Weyerhaeuser Recycling buys recovered paper (Eco-Cycle has been one of its suppliers for 20 years) and repulps it into newsprint and corrugated containers. He works with legislative bodies and paper trade associations to promote the growth of paper recycling and recovery throughout the continent. The United States is currently approaching 50 percent paper recovery. Grogan is working to increase that percentage and to ensure that swings in the international market do not threaten the economic viability of recycling paper. Grogan also monitors Congress to ensure that it does not interfere with the growth of paper recycling. Congress has never passed legislation to promote recycling and, Grogan laments, does not recycle its own office paper!

Grogan is a popular speaker, appearing frequently at recycling conferences in the United States, Canada, Mexico, and Europe. He gave the keynote address at the White House Summit on Recycling in 1998. He was a member of the board of directors of the National Recycling Coalition from 1982 to 1987 and again from 1989 to 1994, serving as president from 1992 to 1993. Between 1993 and 1995, he was a member of the board of directors of Responsibility Inc., a Tijuana, Mexico, organization that builds schools and medical clinics to serve the recycling scavengers who live and work in the dumps of Mexico.

Bibliography

Booth, Michael, "Recycling Guru Laments Colorado's Waste," *Denver Post*, 1999; Grogan, Peter, "Emerald City Looks to Tomorrow," *Bio-Cycle*, 1998; Grogan, Peter, "Recyclables like Bananas Need Buyers," *Bio-Cycle*, 1997.

Grossman, Richard L.

(1943–)
Activist, Director of Environmentalists for Full Employment, Cofounder of Program on Corporations, Law and Democracy

Richard L. Grossman helped bring about the passage of the first antinuclear ballot initiative, in California in 1974. As director of Environmentalists for Full Employment (EFFE) from 1976 to 1985, he initiated cooperation among environmental and labor groups. Today he focuses on stimulating debate among environmental and social activists that challenges the authority of corporations.

Born in Brooklyn, New York, in 1943, Richard L. Grossman was a member of a family that, early in his life, instilled in him an appreciation for justice and democracy. His family, he says, was always very clear about what was right and what wrong. He graduated with a B.A. from Columbia University in 1965 and joined the Peace Corps, where he was a teacher in a small town on the Filipino island of Leyte. He was in the Philippines from 1965 to 1967, a series of years he found very instructive. Living there—in a former American colony—during the Vietnam War, he gained a new perspective on the United States and the rest of the world. He remembers the people of the Philippines as being very patient with the "young arrogant Americans."

Upon returning to the United States in 1967, Grossman participated as an adult education teacher in various War on Poverty Program efforts throughout the country. Then, from 1971 to 1972, he farmed in the Hudson Valley of New York, wrote a novel that was not published, and wrote and published several short stories. In 1973, Grossman took a position in California as a writer and production worker for the San Francisco *Phoenix*, a biweekly newspaper. He also cofounded and published a literary magazine entitled *Gallimaufry* and was a fellow with the CORO Foundation for Public Affairs during these years.

In 1974, he became involved with the antinuclear protest movement in California. He was an organizer for California Proposition 15, the first antinuclear initiative on a state ballot, challenging government officials and utility providers to actually engage in a meaningful debate on the subject of nuclear power and attempting to force them outside of their "no nukes, no jobs" mantra. The ballot initiative was successfully passed into law. Owing to his efforts in organizing, Grossman was offered the position of director of Environmentalists For Full Employment in Washington, D.C., in 1976. He accepted the offer and left California. He acted as director of EFFE from 1976 to 1985, helping to build a foundation of unity and cooperation between labor and environmental organizations. He saw a large part of his responsibility as being to provide support for local unions interested in protesting potential nuclear power generator projects in their neighborhoods. The American Federation of Labor–Congress of Industrial Organizations (AFL-CIO) was, at that time, a staunch supporter of nuclear power projects, and local unions that wanted to protest their construction could not do so without facing recriminations from the national labor organization. Grossman created a coalition of antinuclear unions that broke the public unity of unions, allowing many local unions to protest nuclear projects. During his time as director of EFFE, Grossman also organized a giant labor/environmental conference in Pittsburgh that attracted thousands of workers and environmentalists.

After leaving EFFE in 1985, Grossman became a guest fellow at the Institute for Policy Studies in Washington, D.C., and was executive director of Greenpeace USA for a short time during 1985 and 1986. In 1987, he cofounded the Stop the Poisoning (STP) schools project at the Highlander Center in Tennessee. These schools, which concentrated on ending corporate poisoning, were modeled after the citizenship schools that the Highlander Center ran for the civil rights movement. Grossman helped to organize an STP committee composed of people from all over the South who lived in at-risk communities. The committee met once or twice a month. Participants came to strategize and compare notes, and they eventually became quite skilled at stopping particular corporate assaults on the health and well-being of their communities. However, Grossman saw that while these dump-by-dump and chemical plant–by–chemical plant victories were very important, there simply was not enough time in the world to combat each of these problems one at a time. He saw people organizing, educating, and passing new laws, yet somehow the destruction continued. He began wondering about the causes behind the fact that corporations have such a large degree of decision-making power in our society, while the best that ordinary people can normally do is react to the decisions and try to make things, as he says, "less bad."

In 1992, Grossman left the STP schools project to concentrate his efforts on exploring and challenging the existing power structure that so effectively recognizes the authority of corporations while completely disregarding the power of people in the United States. To this end, he cofounded the Program on Corporations, Law and Democracy (POCLAD) in 1993. POCLAD's official mission is "to instigate democratic conversations and actions that contest the authority of corporations to define our cultures, govern our nations, and plunder the earth." POCLAD works to replace the illegitimate power of corporate institutions with democracy. Corporate power, according to Grossman, is illegitimate simply because it is not constitutional and has usurped the authority of a sovereign people. Corporations make decisions that have a fundamental impact on our lives. They decide things like how our health, food, and energy services operate. These are decisions that should, he says, be made instead by elected officials.

According to Grossman, in order to reclaim these decision-making powers, Americans need to stop treating organizations as if they were entities subject to the Bill of Rights (which they are; legally a corporation has the right to free speech, while an employee on company property does not) and instead redefine corporations as public instruments subordinate to the people. Corporations should no longer be treated as private contracts but should be treated as limited, subordinate institutions. Grossman provides a simple example of what this means practically, in an interview published in *Corporate Crime Reporter*. He says we should tell Exxon,

> Your job is to find oil. Dig it up. Refine it. Get it where it needs to go. Your job is not to write environmental laws. Your job is not to write the worker laws. Your job is not to educate people on how to use energy. Your job is not to tell us the future. Your job is not to influence how we think. Your job is not to lobby. Your job is not to get involved in any elections. Your job is not to give any charitable contributions. Your job is not to endow chairs in universities. Your job is to get the oil and bring it back. Your job is not to be a cultural, political institution.

Grossman has written hundreds of newspaper and magazine articles, and he has written and coauthored several books, including *Fear at Work, Job Blackmail, Labor and the Environment*, and a frequently reprinted pamphlet entitled *Taking Care of Business: Citizenship and the Charter of Incorporation.* He lives with his family in New Hampshire.

Bibliography

"Corporate Crime Reporter—Interview With Richard Grossman," http://www.ratical.org/corporations/CCRivRG1197.txt; Grossman, Richard L., "After Seattle . . . The WTO, The US Constitution and Self Government," *By What Authority*, http://www.poclad.org/bwa/fall99.htm, 1999; Kazis, Richard, and Richard L. Grossman, *Fear at Work: Job Blackmail. Labor and the Environment*, 1982; "POCLAD," http://www.poclad.org/resources.html.

Gussow, Joan Dye

(October 4, 1928–)
Nutritionist, Professor

Troubled by trends in agriculture and food processing, Joan Dye Gussow has sought to educate people on the complexities of the food system in the United States and its impact on human nutrition and the environment. She is an emerita professor of nutrition and education at Columbia University Teachers College and has written about consumer food choices, agricultural systems, and prospects for the future. With the growing popularity of convenience foods—which are often highly processed and heavily packaged—the diets of most people are becoming radically disconnected from the raw materials produced by farmers, and Gussow has made a career of trying to reverse this trend. She is a dedicated gardener and champions a greater reliance on fresh, locally grown foods instead of products of the multinational food industry.

Joan Dye was born on October 4, 1928, in Alhambra, California, the daughter of Chester and Joyce (Fisher) Dye. She attended Pomona College in Pomona, California, receiving her B.A. in zoology in 1950. She then moved to New York City, where she took a job as a researcher for Time, Incorporated. In 1956 she left that job and became a free-lance writer for Street and Smith Publications, and on October 21 of that year she married Alan M. Gussow, an artist. She and Alan had two sons, Adam and Seth. During the mid-1960s, she worked as an editorial and research assistant at Yeshiva University and then held a similar position at Albert Einstein College of Medicine, from 1966 until 1969. In 1970 she began working as an instructor in nutrition for Columbia University Teachers College. She received her master's in community nutrition education from Columbia University in 1974, where she also received her doctorate in nutrition education in 1975. At that point she became assistant professor of nutrition and education and chairperson of the nutrition program.

Gussow's original area of specialization was nutrition, and she has always worked at generating resistance to nutrition miseducation in the United States. But she also knows that the realities of agriculture and food production are invisible to a great majority of food consumers and that in order to make the right nutrition decisions, people need to have a better understanding of where the food comes from. Hoping to reach a wider audience, she began writing magazine articles and books. In 1978 her book, *The Feeding Web: Issues in Nutritional Ecology*, was published, which dealt with some of the problems with the existing agricultural systems and how they relate to consumer food choices.

Current industrial agricultural patterns are shaped by the desire to maximize profit, and Gussow's writings explain how this trend unleashes destructive forces on farming's resource base, threatening long-term sustainability. Of primary importance is the need to encourage more localized food production in order to lower the energy costs of transport, processing, and distribution. In an article she wrote for *Food Monitor* in 1985, Gussow points out other problems with a globalized food supply—such as contamination disasters in which tainted food gets dispersed over great distances. She illustrates this with an example of a polychlorinated biphenyl (PCB) spill at a packing plant in Billings, Montana, in 1979 in which PCBs got into meat meal used for animal food. By the time the problem was discovered, contamination had spread to 18 states and British Columbia, and 1.2 million chickens, 5,300 hogs, 30,000 turkeys, 800,000 pounds of animal feed, and 74,000 bakery items had to be destroyed. Despite an apparently elaborate system of safeguards, this was by no means an isolated incident; and Gussow reminds readers of the vulnerability inherent in such large distribution systems.

Gussow continued to examine problems with the food industry in *Chicken Little, Tomato Sauce and Agriculture: Who Will Produce Tomorrow's Food*

(1991), where she tackles the issues of biotechnology, consumer preference, and alternatives to current unsustainable agriculture patterns. Part of the book's title comes from Frederik Pohl's 1952 science fiction classic *The Space Merchants*, which features a product of misguided technology called "chicken little"—a headless, featherless, wingless lump of flesh fed by tubes, created to feed a growing global population. Gussow comments on real technological "progress" in the food industries and questions what drives it. She reveals that the number of food items available in a typical supermarket has increased from 800 around 1930 to over 30,000 today, with an average of 15,000 new products introduced a year. Consumers are bombarded with choices, and more and more are leaning toward food selections that are instant, artificial, disposable, individually portioned, and overpackaged. Gussow rightfully condemns this current trend as intolerably wasteful. When it comes to making food choices, she recommends keeping it simple and advises consumers not to eat anything their ancestors would not recognize. She also points out that although the food industry has become a daunting force that may seem impervious to change, consumer concern really is capable of creating pressure on the food system. In her fight to help reform some of the problems in large-scale industrial agriculture, Gussow has come to believe that food is an indication of other problems and that the way to engage people in helping to sustain the processes that supply their food is to get them to care about the environment as a whole.

As a nutritionist, Gussow advises people to eat whole foods and to eat lower on the food chain; as an environmentalist her advice would be the same, and she would also advocate buying food locally. Proving that it is possible to live year-round off a home garden, Gussow buys very little from grocery stores and relies mainly on what she grows. She retired from Columbia Teachers College in 1994 but continues to teach seminars as an emerita professor. She also serves as a member of the National Organic Standards Board, as she has done since 1996. She tends the garden that she and her late husband designed for their home on the west bank of the Hudson River in Piermont, New York.

Bibliography

Fraser, Laura, "Homegrown Harvest," *Health*, 1997; Gussow, Joan Dye, *Chicken Little, Tomato Sauce and Agriculture*, 1991; Gussow, Joan Dye, *The Feeding Web: Issues in Nutritional Ecology*, 1978; Gussow, Joan Dye, "PCBs for Breakfast and Other Problems with a Food System Gone Awry," *Food Monitor*, 1982; Liebman, Bonnie, "A Chicken Little in Our Future?" *Nutrition Action Healthletter*, 1991.

Guthrie, Woody

(July 14, 1912–October 3, 1967)
Singer, Songwriter

A refugee of the dust bowl of the 1930s, hobo, and champion of the working man, Woody Guthrie wrote more than 1,000 folksongs, including such classics as "This Land Is Your Land" and "Pastures of Plenty" that Americans still sing today. His songs instilled generations of Americans with a deep pride in the natural beauty of the land and the humble people on it. As an advocate for marginalized populations, be they migrant workers or lonesome hobos of the Depression, his work was a source of inspiration for many underdog causes, including the antinuclear movement, farmworker solidarity, and regional save-the-river campaigns all over the North American continent.

Woodrow Wilson Guthrie was born on July 14, 1912, in Okemah, Oklahoma, to Nora Belle (Tanner) and Charlie Guthrie, a real estate agent and district clerk. Guthrie's childhood was wracked with natural and unnatural disasters. When he was six, his 14-year-old sister Clara's dress caught on fire while she was doing household chores, and she died. Another sister, Mary Josephine, was born in 1922, just as Charlie Guthrie's real estate business died out and he lost reelection as district clerk. His mother began to show signs at this time of the disease that would eventually claim Woody as well, Huntington's chorea.

After Charlie Guthrie committed his ailing wife to an asylum and moved to Pampa, Texas, with his youngest children, 15-year-old Woody began living outdoors and found that he enjoyed it. One family took him in and taught him some music. Guthrie learned to play jaw harp and the harmonica, beat rhythm with carved animal bones, and began singing his way out of the hard times that had befallen his family. The year 1929 brought the stock market crash, the beginning of the Great Depression, a worsening drought in Texas and Oklahoma, and the death of Guthrie's mother.

Guthrie married Mary Jennings in 1933. While awaiting the birth of their first child in Pampa, Texas, in 1935, they witnessed a devastating dust storm. The winds blew across the Great Plains carrying hundreds of thousands of tons of Oklahoma and Texas Panhandle topsoil into the sky. Farms in South Dakota, Colorado, Iowa, Arkansas, Missouri, Nebraska, Kansas, Oklahoma, and Texas—and the life savings of the farmers as well—were blown away. Guthrie recorded the storm's horror in his song "Dust Storm Disaster."

> We saw outside our window
> Where wheat fields they had grown,
> Was now a rippling ocean
> Of dust the wind had blown.
>
> We loaded our jalopies
> And piled our families in,
> We rattled down the highway
> To never come back again.©

This dust storm and the many others Guthrie observed yielded him more than a song. The dust bowl—the result of a protracted drought that parched fields left fallow owing to the reduced demand for wheat during the Great Depression—was a formative political experience for Guthrie. What made the situation worse

Woody Guthrie (Frank Driggs Collection/Archive Photos)

for small farmers was the mechanization of agriculture, which pushed small undercapitalized farmers off the land, because only large, highly capitalized agribusiness could afford modern combines and harvesters. The drought exacerbated bad circumstances for the small farmer in the Wheat Belt, and many undertook a great migration westward.

Guthrie began hopping freight trains, to Oklahoma, Arkansas, and eventually California. The songs he wrote during his wanderings, to his surprise, had a profound effect upon his listeners. In 1937 Guthrie went to California and, with his cousin Jack Guthrie, created "The Oklahoma (Jack) and Woody Show" on KFVD in Los Angeles. Woody's popularity among the migrants in California grew as he sang their tales. For KFVD, he covered migrant camp conditions in northern California. Guthrie became more politicized after seeing the camps, and by 1939 his work on behalf of the Communist Party led to a rift with the radio station.

On New Year's Day 1940, Woody Guthrie set out for New York City, hitchhiking from Pampa, Texas. On the trip he heard the entire country singing along with Kate Smith's "God Bless America," and he wrote a retort he called "God Blessed America for Me," which he later renamed "This Land Is Your Land." The song has come to stand as an anthem for the environment, celebrating America's spectacular landscape and emphasizing that the beauty and bounty of the United States belong to the people of the land. Two stanzas were deleted from Guthrie's first recording of it:

As I went walking, I saw a sign there,
And on the sign it said "No
Trespassing."
But on the other side it didn't say
nothing,
That side was made for you and me.
In the shadow of the steeple I saw my
people,
By the relief office I saw my people—
As they stood hungry, I stood there
asking
Is this land made for you and me.©

In March 1940, Woody sang "This Land Is Your Land" at a benefit for West Coast migrant workers. At this performance he met folklorist Alan Lomax and Pete Seeger, who befriended him and collaborated with him on many later recordings. Lomax recorded Guthrie for the Library of Congress, which was followed by recordings for Victor Records and radio appearances. Lomax recommended Guthrie for a part in the documentary film to promote the Bonneville Power Administration and the Coulee Dam project. Guthrie spent a month at the project's site, writing 26 songs that are compiled in the Columbia River Collection, including "Roll On Columbia":

Green Douglas firs where the waters cut
through
Down her wild mountains and canyons
she flew,
Canadian Northwest to the ocean so blue,
Roll on, Columbia, Roll on.©

1941 was an important year for Guthrie. He joined the Almanac Singers with Pete Seeger, Lee Hays, Millard Lampell, and others and published his memoir *Bound for Glory*, which received rave reviews. That same year, after eight years of marriage, most of which Guthrie spent on the road, Mary Guthrie gave her husband an ultimatum: that he choose to live at home with his family or leave them permanently. He chose the latter and soon fell in love with Marjorie Greenblatt Mazia, a dancer he met in New York. They married in 1945 and had four children. World War II interrupted his career, but Guthrie joined the merchant marine with song mate Cisco Houston, and together they chronicled in song this phase of U.S. history.

After the war ended, Guthrie reunited with Pete Seeger and returned to recording songs for the labor movement. However, once communist witch hunts began in the spring of 1947, with Pres. Harry Truman's Executive Order 9835 to track down disloyal Americans, Guthrie's career faltered. The unions that he and Seeger had hoped to collaborate with had blacklisted them as probable communists and members of subversive organizations, and his concerts began to be canceled unexpectedly. By 1949, Guthrie had begun losing his concentration, drinking alcohol more heavily, and disappearing more frequently. In 1952 he was diagnosed with Huntington's chorea, an incurable disease of the nervous system. He divorced Marjorie in 1952 and married Anneke Van Kirk Marshall, with whom he lived briefly in a community of blacklisted entertainers in Topanga Canyon, California. He spent the last 15 years of his life in and out of hospitals, until his death in Creedmore State Hospital in Queens, New York, on October 3, 1967.

Bibliography

Guthrie, Woody, *Bound for Glory*, 1941, 1995; Guthrie, Woody, *Pastures of Plenty*, ed. Dave Marsh and Harold Leventhal, 1990; Guthrie, Woody, *Seeds of Man: An Experience Lived and Dreamed*, 1995; Klein, Joe, *Woody Guthrie: A Life*, 1980, revised, 1999; Santelli, Robert, and Emily Davidson, eds. *Hard Travelin': The Life and Legacy of Woody Guthrie*, 1999; Yates, Janelle, *Guthrie: American Balladeer*, 1995

Gutiérrez, Juana Beatriz

(1932–)

Cofounder and President of Mothers of East Los Angeles

Juana Beatriz Gutiérrez was instrumental in founding Las Madres del Este Los Angeles–Santa Isabel (the Mothers of East Los Angeles), or MELA, a nonprofit group of women activists who have agitated and organized to improve conditions in their neighborhood since 1985. She is now president of this organization, which came into existence to protest the construction of a state prison in East Los Angeles. Since that time, MELA has been active in protesting many other environmentally and socially unsound projects in East Los Angeles and elsewhere. Currently, it operates the Mothers of East Los Angeles Water Conservation Program.

Born in the north-central region of Mexico in the town of Sombrete, Zacatecas, in 1932, Juana Beatriz Gutiérrez grew up in Ciudad Juárez near El Paso, Texas. She moved to the United States in 1956 and married Ricardo Gutiérrez, who was then a marine. He later became a warehouse shift manager in Los Angeles, California. Gutiérrez and her husband have lived in the Boyle Heights section of East Los Angeles for the past 40 years, where they have raised 9 children.

Gutiérrez became concerned about the safety of her children in 1979 when drug dealers and gang members began assembling on a regular basis in the park across the street from her house. In response to their presence, she organized a neighborhood watch program and became president of the local Parent Teachers Association. Also, she and her husband began organizing after-school activities in the park for the children of the neighborhood. Thus began her role as a leader in organizing effective community responses to threats to the health, safety, and happiness of her neighborhood.

In 1985, California state officials announced plans to construct a $100 million state prison on Santa Fe Avenue in East Los Angeles, prompting Gutiérrez, with the help of California assemblywoman Gloria Molina, to organize a group to oppose the construction of the prison. Upon the advice of a local priest, Father Moretta from Resurrection Church, they named the group Mothers of East Los Angeles. Gutiérrez and several other women from her neighborhood, including Aurora Castillo, collected 1,500 signatures on a petition against the building of the prison. They gave this petition to Molina to take to the State Assembly in Sacramento. Gutiérrez also organized community demonstrations protesting the prison and participated in the Earth Day 1990 festival in Exposition Park, distributing educational materials and T-shirts and raising funds through the sale of tamales. MELA became adept at attracting media attention, especially utilizing protest marches to draw regional, and even national, focus to their cause. Their campaign ended successfully in 1992, with the passing of a law that prohibited the building of any state prison in East Los Angeles.

The Mothers of East Los Angeles next acted in response to the proposed construction of a hazardous waste incinerator in the industrial town of Vernon, just one mile from Gutiérrez's home. This incinerator would have further degraded the already overburdened air of East Los Angeles by burning up to 22,500 tons of commercial hazardous waste per day. In 1982, MELA led 500 protesters to a Department of Health Services meeting and demanded that an environmental impact review be conducted. MELA also threatened to sue the city of Vernon as well as California Thermal Treatment Systems, the company that planned to build and operate the incinerator. In response to the rising tide of opposition to the incinerator, Tom Bradley, who was then mayor of Los Angeles, eventually joined Las Madres in denouncing the project, and it was defeated.

In the years since these successes, Las Madres has continued combating the construction of toxic facilities and other health-threatening projects in East Los Angeles and elsewhere. MELA was a part of the coalition organized to protest the diverting of an oil pipeline away from the wealthy, politically connected cities of Santa Monica and Pacific Palisades and through the low-income communities of East Los Angeles. The group also joined the protest against the construction of a cyanide hexavalent chromium treatment plant next to Huntington Park High School, fought against the spraying of insecticide, and marched against a toxic waste incinerator planned for the Latino farming community of Kettleman City.

Gutiérrez's most recent project with MELA has been the Mothers of East Los Angeles Water Conservation Program. This program, launched in 1992, has focused on replacing old water-guzzling toilets and shower heads to conserve water and has also served to invigorate the neighborhood's economy. The primary objective of this program is to get Angelenos (residents of a naturally water-poor city) to trade in their old 7-gallon toilets for new 3.5-gallon toilets provided by the Los Angeles Department of Water and Power (DWP). DWP has calculated that 48 million gallons of water would be conserved daily if all Los Angeles residents were equipped with these Ultra Low Flush Toilets. Residents can obtain these low-flush toilets simply by showing up at MELA's headquarters at the Gutiérrez house with identification and their DWP bill. They are given an Ultra Low Flush Toilet, as well as instructions on how to install it. When they bring back their old toilet to be recycled, they get a low-flow showerhead. This program has led to the installation of 50,000 low-flush toilets since 1992. The water conservation program employs 25 full-time and three part-time staff members. Its profits are channeled into community projects: for instance, paying high school students to go door to door encouraging mothers to have their children fully immunized and tested for lead poisoning and hiring teams of students to go out in the morning before school to sweep streets and paint over graffiti.

Gutiérrez, along with MELA, has become a political force to be reckoned with in East Los Angeles, not to mention in the rest of California. She sees her neighborhood group's emergence as such a force on toxic-waste and other issues as natural. She sees the environment not as something "out there," but rather as the entire physical and social setting where people live and work every day. The poverty and violence in inner city Los Angeles, then, are just as much a part of her environment as the toxic lead levels and air pollution; all problems she faces in attempting to help her neighborhood stay healthy and safe. She is quoted in Steve Lerner's *Eco-pioneers* as saying that her organization is "not economically rich—but culturally wealthy; not politically powerful—but community conscious; not mainstream educated—but armed with knowledge, commitment, and determination that only a mother can possess; The Mothers of East Los Angeles–Santa Isabel is an organization striving to protect the health and safety of East Los Angeles Neighborhoods."

Gutiérrez's work has been recognized through numerous honors and awards. She was named 1988 Woman of the Year by the California State Legislature. She received the Pope John Paul II Benemereti Award in 1992 and the Mexican Mother of the Year Award in 1993.

BIBLIOGRAPHY

Bullard, Robert, ed., *Unequal Protection: Environmental Justice and Communities of Color*, 1994; Lerner, Steve, *Eco-pioneers: Practical Visionaries Solving Today's Environmental Problems*, 1997; Schwab, Jim, *Deeper Shades of Green*, 1994.

H

Hair, Jay

(November 30, 1945–)
President of the National Wildlife Federation

Jay Hair was head of the National Wildlife Federation (NWF), one of the most prominent and important conservation organizations in the United States, for the 15 years from 1981 to 1996. It was largely during Hair's term as president that this organization began to assume the role of influence that it occupies today. Since leaving the National Wildlife Federation, Hair has acted as a consultant and as executive vice president of business development for a company called GreaterGood.com, which helps garner support for nonprofit organizations by utilizing Internet e-commerce.

Jay Dee Hair was born on November 30, 1945, in Miami, Florida. When he was three months old his parents divorced, and Hair moved with his mother, Ruth, and his two siblings, to Leota, Indiana, where Ruth's family maintained a farm. Hair lived on the farm for the first 11 years of his life, catching fish in the farm ponds and learning to hunt with the help of his uncles. In 1957, Hair's mother remarried, and the family moved to Landenberg, Pennsylvania. Hair's stepfather, William Johnson, was a DuPont chemical engineer, and he encouraged Hair to excel academically. Hair attended Avon Grove High School in West Grove, Pennsylvania, where he was president of his senior class. After graduating from high school in 1963, Hair entered Clemson University in South Carolina. He earned a B.S. degree in biology four years later, in 1967. He remained at Clemson for two more years to complete an M.S. degree in biology, as well.

Upon receiving his M.S. degree in 1969, Hair moved to Edmonton, Canada, to work as a graduate research fellow in the Department of Zoology at the University of Alberta. He remained in this post for one year. In 1970, he was commissioned as a second lieutenant in the United States Army, and he left the United States for Vietnam, where he served as a public health specialist. After one year of service, Hair returned to the University of Alberta and completed his Ph.D. in zoology in 1975. In the meantime, Hair had accepted a position as assistant professor of wildlife biology in the Department of Entomology, Fisheries and Wildlife at Clemson University in 1973, a position he occupied until 1977. During his time as assistant professor at Clemson, Hair also worked as a research and management consultant to the South Carolina Wildlife and Marine Resources Department from 1976 to 1977. Hair left Clemson in 1977 for North Carolina State University in Raleigh, where he served as administrator of the fisheries and wildlife sciences division, as well as associate professor of zoology and forestry. He worked in Raleigh until 1981, holding a second position as well, this one with the U.S. Department of the Interior as special assistant from 1978 to 1980, in which he helped to develop national fish and wildlife policy.

The National Wildlife Federation is one of the largest, most influential private, nonprofit conservation-education organizations in the United States. Founded in 1936, its mission is "to educate and inspire" individuals to conserve wildlife and other natural resources, in an effort to promote an environmentally sustainable future. Hair had been heavily involved with the NWF after attending his first meeting of the organization in South Carolina in 1974. He became president of the South Carolina Wildlife Federation in 1976, the year in which that group was named the NWF's outstanding affiliate of the year. And, in 1981, after learning that the NWF was conducting a search for a new president to replace Thomas Kimball, who had served since 1960, Hair applied for the position. He was elected president of the NWF at the age of 35. Hair acted as the NWF president for the next 15 years, moving the organization away from its conservative stance on many environmental issues and into the environmental mainstream. Hair revitalized the sportsman image of NWF, creating a larger, more active, more diverse, and more influential organization. During his tenure, membership grew from 4.2 million in 1981 to nearly 6 million in 1994, and the NWF annual operating budget more than doubled, reaching $90 million in 1990.

As president of NWF, Hair played a significant role in many national environmental issues. In one of his first publicly visible actions as president of the NWF, he criticized the actions of James Watt, President Reagan's secretary of the interior. Watt was a strong supporter of development, and during his time as secretary he caused an uproar in the environmental community by advocating development on public lands and by supporting oil drilling in wildlife preserves. Hair, along with other well-known environmentalists, called for Watts's resignation, an outcry that made front-page news across the Unites States. Hair was also the first head of a major environmental organization to arrive on the scene in Prince William Sound when the *Exxon Valdez* spilled its 11 million gallons of oil in 1989. He used the opportunity to speak with the national press about the dangers of drilling and transporting oil in ecologically sensitive locations. The NWF and the Natural Resources Defense Council successfully sued Exxon, requiring the company to establish a fund for the continuation of cleanup and to provide assistance to those Alaskans who suffered financially from the spill.

Despite his criticism of Exxon, Hair believed that business and industry could play an important role in the environmental movement. He created the Corporate Conservation Council, whose members include such companies as DuPont, Monsanto, ARCO, Ciba Geigy, and others, each of which pays a $10,000 membership fee and is invited to seminars and on special excursions. Many corporate representatives have also been invited to sit on NWF's board of directors. In one well-known case, described in MARK DOWIE's *Losing Ground*, this has been controversial for the organization. The chief executive officer of Waste Management, Inc., known for its violations of toxic release regulations, not only sat on the board but was the recipient of Hair's personal intervention on its behalf. Hair helped Waste Management persuade the Environmental Protection Administration (EPA) director under President Bush, WILLIAM K. REILLY, to override EPA policies and issue permits to two stalled Waste Management projects. Congress investigated, and Hair eventually admitted that his involvement had been a mistake.

Hair left the NWF in 1996 to act as an international environmental policy consultant. That same year, he was commissioned by the International Finance Corporation (IFC), the private sector arm of the World Bank, to review the IFC's handling of the environmental and social mitigation aspects of the Pangue hydroelectric project on the Bio Bio River in Chile, which the IFC helped to finance. The "Hair report," as it has come to be known, contains strong criticisms of the practices followed by the IFC in the development of this controversial $367 million hydroelectric dam. Since 1998, Hair has been executive vice president of business development for a company called GreaterGood.com. This company, founded in 1998, operates "shopping villages" on the Internet, which link to the home pages of nonprofit groups. The nonprofits use these "villages" to sell merchandise, operating under the assumption that having their merchandise grouped in one location on the Internet will lead to an increase in membership and support for the nonprofit organizations.

Hair has received many awards for his environmental efforts, including the Edward J. Cleary Award from the National Academy of Environmental Engineers in 1989 and the National Park Foundation Theodore and Conrad Wirth Environmental Award in 1990. He currently lives with his second wife, Leah, and his two daughters.

BIBLIOGRAPHY

Allen, Thomas B., *Guardian of the Wild*, 1987; Dowie, Mark, *Losing Ground*, 1995; Kinch, John A., *Newsmakers*, 1994; Mosher, Lawrence, "Washington's Green Giants," *The Amicus Journal*, 1989.

Hamilton, Alice

(February 27, 1869–September 22, 1970)
Physician, Social Reformer

In a time when occupational health and safety were hardly considered, Alice Hamilton became an active advocate against the use of industrial toxins and hazardous working conditions. A lifelong proponent of social justice and pacifism, Hamilton is most widely recognized for her endeavors to detect and combat the medical problems caused by industrialization. Through her tireless efforts, toxic substances in the lead, mining, painting, pottery, and rayon industries were exposed, and legislation was passed to protect workers.

The second of the five children of Montgomery Hamilton, a wholesale grocer, and Gertrude Pond, Alice Hamilton was born in New York City on February 27, 1869. She grew up in a secure material and financial environment on her grandmother's estate in Fort Wayne, Indiana, after her father's business failure. Taught only by her parents and tutors, Hamilton received little formal training

Alice Hamilton (R) with Marion Edward Park (M) and Mrs. Louis Slade at Bryn Mawr's commencement, June 1936 (Bettmann/Corbis)

except in languages until 1886, when she entered Miss Porter's School in Farmington, Connecticut. Hamilton's mother always encouraged her children to follow their minds and inclinations.

Hamilton's decision to pursue a career in medicine was influenced by two main factors: first, it was one of the few professional fields open to women in the latter half of the nineteenth century, and second, as she stated in her autobiography, "as a doctor I could go anywhere I pleased . . . and be quite sure that I could be of use anywhere." Hamilton started in medicine at the University of Michigan in March 1892, after studying science at the Fort Wayne College of Medicine. She received her M.D. the following year and interned in Minneapolis and Boston. She returned to the University of Michigan in 1895 to work in the bacteriology laboratory of F. G. Novy but left that fall to travel to Germany with her sister Edith, a noted Greek scholar. In Germany, Hamilton studied bacteriology and pathology at the universities of Leipzig and Munich. Returning to the United States, she trained for a year at Johns Hopkins Medical School before taking a job in 1897 as a professor of pathology at the Woman's Medical School at Northwestern University, a position that she held until the Woman's Medical School closed in 1902. Hamilton accepted a position as a bacteriologist at the Memorial Institute for Infectious Diseases, but before starting there, she studied briefly at the Pasteur Institute in Paris.

When Hamilton returned to Chicago in the autumn, she found that the city had been struck by a typhoid epidemic. Hamilton gained acclaim for a paper presented to the Chicago Medical Society that suggested that flies were the agents in spreading the disease. While her theory proved to be wrong—it was later discovered that the epidemic had stemmed from a break in the local pumping station that had allowed sewage to escape into the water pipes—Hamilton and her work were gaining respect.

Prior to the closing of the Woman's Medical School, Hamilton became a resident at Hull House, a settlement designed by its founder JANE ADDAMS to give care and counsel to Chicago's poor and sick. At Hull House, Hamilton witnessed firsthand the disease, disability, and premature death common to workers in certain industries. Upon reading Sir Thomas Oliver's 1902 book *Dangerous Trades*, Hamilton began her lifelong mission to oppose and treat the excesses of industrialization. She quickly learned that occupational safety laws, workers' compensation laws, and effective factory-inspection systems in the United States lagged far behind those in European countries such as Germany and England and began to use her growing reputation to publicize these problems.

In 1908, Hamilton was named to the Illinois Commission of Occupational Diseases by Gov. Charles S. Deneen. The commission's preliminary investigation revealed the need for a larger study that, in 1910, Hamilton directed. A year later, Hamilton accepted an appointment as special investigator for the U.S. Bureau of Labor. Her work required that she make field investigations of mines, mills, and smelters. Her first statistical study—on lead, the most widely used industrial poison—documented the alarmingly high mortality and morbidity rates of workers exposed to the poison. She later produced similar studies on aniline dyes, picric acid, arsenic, carbon monoxide, and other industrial poisons. During World War I, Hamilton investigated the high explosives industry and revealed that nitrous fumes were the cause of a great number of supposedly natural deaths. In spite of trying not to sensationalize her findings, Hamilton became widely known as a crusader for public health and an advocate of other social causes such as woman suffrage, birth control, a federal child labor law, state health insurance, and workers' compensation.

Hamilton was also a pacifist and accompanied Jane Addams to the International Congress of Women at The Hague as well as on a mission to the war capitals to present the women's peace proposals. In 1919, she investigated the famine in Germany and became involved in the Quaker famine relief effort. That same year she was named the first female faculty member of Harvard University as an assistant professor of industrial medicine. Hamilton suffered much discrimination as the university's first female professor. She was denied access to the Harvard Club and to participation in graduation ceremonies. Nevertheless her classic textbook, *Industrial Poisons in the United States* (1925), established her as one of the world's leading authorities on the subject. During this time she was also successful in persuading the surgeon general to investigate the dangerous effects of tetraethyl lead and radium.

While at Harvard, Hamilton maintained her international contacts. She served two terms on the Health Committee of the League of Nations (1924–1930) and became an ardent proponent of the league. In 1924, she was the only woman delegate on the League of Nations Health Commission to the USSR. Concluding that "there was more industrial hygiene in Russia than industry," Hamilton admired certain aspects of the Bolshevik system but deplored its suppression of free speech. Of Nazi Germany, which she visited in 1933, Hamilton was far more critical; from a very early stage, she felt that the United States should oppose Hitler.

Hamilton retired from Harvard in 1935 and moved to Hadlyme, Connecticut. She remained publicly active, however, and in 1935 accepted a position as consultant to the Division of Labor Standards in the U.S. Department of Labor. In 1937 and 1938, Hamilton conducted her last field investigation, a survey of the viscose rayon industry. She successfully demonstrated that rayon processes involved a high level of toxicity, a finding that resulted in Pennsylvania's first compensation law for occupational diseases.

Hamilton, who never married, published her autobiography in 1943. Between 1944 and 1949, she served as president of the National Consumers' League. In her later years, Hamilton continued to be politically active and vocal, advancing causes of social justice and pacifism; she withdrew her opposition to the equal rights amendment in 1952 and in 1963 called for an end to U.S. military involvement in Vietnam. Hamilton died of a stroke in her home in Hadlyme, Connecticut, on September 22, 1970; she was 101.

BIBLIOGRAPHY

Grant, Madeline P., *Alice Hamilton: Pioneer Doctor in Industrial Medicine*, 1967; Hamilton, Alice, *Exploring the Dangerous Trades: The Autobiography of Alice Hamilton, M.D.*, 1943; Sicherman, Barbara, *Alice Hamilton: A Life in Letters*, 1984; Slaight, Wilma Ruth, "Alice Hamilton: First Lady of Industrial Medicine," Ph.D. dissertation, Case Western Reserve University, 1974; Young, Angela Nugent, "Interpreting the Dangerous Trades: Workers' Health in America and the Career of Alice Hamilton, 1910–1935," Ph.D. dissertation, Brown University, 1982.

Hardin, Garrett

(April 21, 1915–)
Biologist, Human Ecologist

Biologist and human ecologist Garrett Hardin is known for his bold assertions about difficult issues facing humanity. His famous 1968 essay, "The Tragedy of the Commons," warns that unless there are strict regulations in place, people will destroy the land, water, and air that they share. His 1974 article for *BioScience*, "Living on a Lifeboat," proclaimed that it was irresponsible for wealthy nations to share resources with poor countries unless the poor countries had population control policies in place and the wealthy countries donated expertise and technology for improvement along with the material aid. Controversial as they may be, Hardin's opinions have inspired new fields of studies and fruitful debates among scholars.

Garrett Hardin was born on April 21, 1915, in Dallas, Texas. His father worked in the offices of the Illinois Central Railroad and was transferred every few years. Hardin grew up in five different midwestern cities but spent every summer on his grandfather's Missouri farm. He contracted polio at the age of four, which weakened and shortened his right leg. In his Chicago high school he excelled in both writing and drama and was awarded scholarships to a drama school as well as to the University of Chicago. Realizing that his handicap would relegate him to a limited number of roles, Hardin opted for the University of Chicago, where he studied under ecologist W. A. Allee, one of the only scholars at the time who taught that unimpeded population growth would be a major challenge to life on earth. Hardin majored in zoology and graduated in 1936. He continued studying at Stanford University, earning a Ph.D. in 1941 in biology. He stayed at Stanford on a fellowship from the Carnegie Institute in Washington to study algae and its potential as a food source, but in 1946 he stepped down, because he had come to believe that increasing the production of food would only aggravate population problems. He was hired at the University of California–Santa Barbara as assistant professor of bacteriology in 1946 and has been there ever since. He was named full professor of biology in 1957 and in 1960 began teaching human ecology there.

Hardin began to diversify in 1949 with his popular textbook *Biology: Its Human Implication.* He wrote the scripts for and starred in teaching films and during the 1950s became interested in genetics and evolution. He wrote *Nature and Man's Fate* in 1959, which continues to be a useful introduction to evolution, the history of evolutionary theory, and its moral and social implications. In 1960, Hardin designed a course in human ecology. It dealt with controversial issues such as evolution and population growth. At that time abortion was illegal and socially taboo, but Hardin included it in his course. He became an advocate of freeing women from what he termed "compulsory pregnancy" and lectured throughout the United States on the need to legalize abortion. Abortion did not become legal until the *Roe* vs. *Wade* ruling in 1973, but in 1963 Pres. Lyndon B. Johnson signed the nation's first birth control bill, and in 1964 the first federally funded family planning program was established in Corpus Christi, Texas.

Hardin's "The Tragedy of the Commons" was published in *Science* in December 1968 and provided a philosophical and historical context for the population concerns that were popularized by PAUL EHRLICH'S *The Population Bomb*, also published in 1968. "The Tragedy of the Commons" challenges the thesis of economist Adam Smith, who believed that decisions about individual behavior made by rational human beings would always lead, in the long run, to the greater good for all. Instead, Hardin expanded upon the argument of William Forster Lloyd, a nineteenth-century Oxford philosopher who described what happened when pastures held in common (or "commons") were open to all herdsmen. Each would try to gain the greatest individual benefit, running as many animals as possible. But eventually the pasture would be depleted, and no one would be able to use it. Hardin's conclusion was that "freedom in a commons brings ruin to all" and that the only way to avoid disaster was to establish a policy of "mutual coercion, mutually agreed upon," which would regulate behavior. "The Tragedy of the Commons" was reprinted in dozens of anthologies and was made part of the curriculum for courses in biological and social sciences.

For Hardin, limiting population growth was basic to avoiding degradation of the commons. More people leads to more resource use, which leads to the speedy depletion of those resources. He advocated worldwide population growth policies, with limits to the number of children one could have. Bearing as many children as they wish brings satisfaction to parents, but the presence of too many children leads a society to economic and environmental disaster. "Freedom to breed will bring ruin to all," he told a biographer for the 1974 edition of *Contemporary Biography.* During the 1970s Hardin wrote about pollution disposal as an additional problem of the commons. With increases in population and its ensuing production, more and more waste is generated, yet there is no growth in areas that can be used for waste disposal.

Hardin wrote another seminal piece for *BioScience* in 1974, "Living on a Lifeboat." Here he wrote that countries were like lifeboats, each having only a limited supply of the essentials for survival. Poor countries' lifeboats were much more crowded and supplies were shorter. Wealthier nations had more food and water and fewer citizens but should refrain from sharing with the poor unless the poor nation had a strict population control policy in place. Wealthy donors, in his opinion, should also share technology and expertise to help the poorer countries become self-sufficient. Hardin favors restrictions on immigration into the United States for similar reasons: emigration serves as a pressure valve for poor nations, and this release of pressure only delays the poor countries' solutions to their problems.

Hardin's analyses have been controversial, earning him stars in some camps and attacks in others. Respondents to the lifeboat article point out that the world did not start out with wealthy and poor nations; rather, many wealthy nations gained their wealth by pilfering resources from poor nations; also, they say, cooperation among nations is potentially necessary for the survival of all nations. The authors of *Environment*'s 30th anniversary revisitation of "The Tragedy of the Commons" cite many examples in which communities have been able to regulate their own use of natural resources more successfully than when an outside regulator such as the government intervenes. Yet despite critiques of the essay, Hardin's point that commons are delicate and must be protected from overuse and the resulting devastation was seminal for much recent work on issues of population control, conservation of resources, and sustainable development.

Hardin and his wife, Jane Coe Swanson, whom he married in 1941, live in Santa Barbara, California. They have four children, Hyla, Peter, Sharon, and David.

BIBLIOGRAPHY

Burger, Joanna, and Michael Gochfeld, "The Tragedy of the Commons 30 Years Later," *Environment*, 1998; Hardin, Garrett, "Extensions of 'The Tragedy of the Commons,'" *Science*, 1998; Hardin, Garrett, *Living within Limits : Ecol-*

ogy, Economics, and Population Taboos, 1993; Hardin, Garrett, *The Ostrich Factor: Our Population Myopia*, 1999; Ostrom, Elinor, Joanna Burger, Christopher B. Field, Richard B. Norgaard, and David Policansky, "Revisiting the Commons: Local Lessons, Global Challenges," *Science*, 1999; Spencer, Cathy, "Garret Hardin Interview," *Omni*, 1992.

Harrelson, Woody

(July 23, 1961–)
Actor, Cofounder of Oasis Preserve International

Woody Harrelson was first known to millions of American television viewers as Woody Boyd, an endearing, dimwitted bartender on the very popular situation comedy series of the 1980s and 1990s, *Cheers*. Nominated five times for an Emmy for this role, Harrelson won the award in 1989. Harrelson has since appeared and starred in many major Hollywood films, including *The People vs. Larry Flint*, *Indecent Proposal*, and *Natural Born Killers*. Harrelson is also a well-known celebrity activist who has participated in many public acts of civil disobedience to protect the environment. His activism has included protesting the logging of ancient redwood forests, advocating for the legalization of hemp, and withholding federal income taxes. He is also the cofounder of Oasis Preserve International, an international nonprofit organization to protect sensitive rain forests in Latin America.

Woodrow Tracy Harrelson, the second of three boys, was born in Midland, Texas, on July 23, 1961. His father, Charles Harrelson, abandoned his family when Harrelson was seven years old. He was a gambler who was soon convicted of murder and sent to prison. Harrelson's mother, Diane, a legal secretary, divorced Harrelson's father and raised her three sons on her own with great financial difficulty. As a young child, Harrelson was prescribed the drug Ritalin for hyperactivity and violence. When his difficulties continued, Harrelson was given a scholarship to Briarwood, a private school in Houston for children with learning disabilities. When Harrelson was 12, his family moved to Lebanon, Ohio, where he performed in his first theatrical production. Harrelson's father was released from prison in 1978 during Harrelson's senior year at Lebanon High School but was arrested a year later, charged once again with murder, and sentenced to two life terms. He remains in prison today.

Despite his difficulties, Harrelson was an intelligent student and maintained good grades in school. In 1979, Harrelson won a scholarship to Hanover College in Indiana. A theology major, Harrelson began college with very conservative religious and political views and voted for Ronald Reagan in 1980. During his junior year of college, however, Harrelson began to pursue his interest in the theater as well as to experiment with drugs and alcohol. He appeared in several college theatrical performances before graduating with a B.A. in theater arts in 1983. Harrelson moved to New York City after graduating from Hanover, and his first break came in 1984 when he was hired as an understudy in the Broadway play *Biloxi Blues* by Neil Simon. Shortly thereafter he was cast with a small part in the Hollywood film, *Wildcats*. While filming in Hollywood, Harrelson decided to audition for a role on a popular television series, even though he felt television was a step down from the theater. Offered the part and given 24 hours to decide, Harrelson said yes and became Woody Boyd on *Cheers*.

Harrelson's success gave him the opportunity to play many small parts in feature films, such as *Casualties of War*, *Cool Blue*, *L.A. Story*, and *Ted and Venus*. His first major movie role was in *Doc Hollywood* in 1991, with Michael J. Fox. Harrelson continued to perform in theatrical productions, as well, including *The Boys Next Door*, *The Zoo Story*, and *Brooklyn Laundry*. Harrelson's first starring role in a Hollywood movie was in 1992 with *White Men Can't Jump*. Shortly thereafter he was cast in a leading role in the movie *Indecent Exposure*.

After *Cheers* went off the air, Harrelson began to focus on his film career. His most controversial film to date was Oliver Stone's *Natural Born Killers* in 1994. Due to its theme of violence, the film received a great deal of media attention. Harrelson went on to star in another controversial film, *The People vs. Larry Flint*, a biographical drama about the publisher of the pornography magazine *Hustler*, for which Harrelson received an Oscar nomination. Harrelson continued on to leading roles in *Welcome to Sarajevo* and *The Thin Red Line*.

In his personal life, Harrelson has struggled with a self-admitted addiction to fame, power, violence, and sex. In 1990, he came to the realization that he was unhappy and decided to embark on a spiritual search. Traveling to Peru, India, and Africa, Harrelson became a strict vegetarian and began practicing holistic healing and yoga. In the early 1990s, Harrelson began to speak out publicly on issues of environmental preservation, hemp legalization, racism, and sexism, and he voiced his opposition to the Persian Gulf War. His first well-known act of civil disobedience was in 1995 when Harrelson withheld money from the Internal Revenue Service and wrote them a letter explaining that he could not support the government's disregard for the environment. A very outspoken hemp advocate, Harrelson was arrested in Kentucky in August 1996 for planting four seeds of industrial hemp. Harrelson says that industrial hemp is a very important source of ecologically friendly fiber and that the distinction must be made between industrial hemp and marijuana. A couple of months later, in November 1996, Harrelson was arrested on the Golden Gate Bridge in San Francisco along with eight others protesting the logging of ancient redwoods in northern California; he was fined and sentenced to community service. In August 1997 Harrelson posted bail for a young man who was arrested for growing marijuana; he was a medical marijuana proponent and amateur scientist experimenting with different plants for the treatment of cancer. Harrelson donates his time and energy to many environmental organizations, including American Oceans Campaign, Rainforest Action Network, and Surfrider Foundation. Harrelson is also the cofounder of

Oasis Preserve International, a nonprofit organization dedicated to protecting ecosystems through the establishment of preserves and conservation programs in Latin America.

Woody Harrelson married his long-time partner and second wife, Laura Louie, in 1998. Both Harrelson and his wife are yoga instructors, and they are the owners of an oxygen bar called O2. Louie is co-founder of Yoganics food products. They have two daughters, Deni and Zoe. Harrelson and his family reside in both Costa Rica and Malibu, California.

Bibliography

"Crusader Woody Supports the Legalization of Hemp," *People Weekly*, 1996; "Oasis Preserve International," http://www.oasispi.org; Schick, Elizabeth, ed., "Woody Harrelson," *Current Biography Yearbook*, 1997; Wood, Campbell, "Woody Harrelson: Acting on His Convictions," *E*, 1997.

Harry, Debra

(July 9, 1957–)

Executive Director of the Indigenous Peoples Council on Biocolonialism

Debra Harry's career has focused on rebuilding indigenous communities after the destructive impacts of colonization and protecting the human and collective rights of indigenous peoples. Most recently, her work has centered on protecting the genetic resources of indigenous peoples from exploitation by corporate, scientific, and government interests. She is an expert in nonprofit management and encourages communities to initiate their own programs and generate culturally appropriate solutions from within. As a recipient of a three-year Kellogg Foundation leadership fellowship in 1994, she studied the field of human genetic research and its implications for indigenous peoples. Currently she is the executive director of the Indigenous Peoples Council on Biocolonialism (IPCB), which she founded in 1998. She also serves on the Board of Directors of the Council for Responsible Genetics, based in Cambridge, Massachusetts.

Debra Harry was born in Reno, Nevada, on July 9, 1957, to Floyd Harry and Charlotte Davis. A Northern Paiute, she grew up on the Pyramid Lake Reservation and graduated from Fernley High School. At age 21 she became an activist and community organizer against the MX land-based missile system that was proposed for Nevada and Utah and also opposed plans for uranium mining in Nevada.

Throughout her twenties Harry continued to work with Native American rights and community development efforts, specializing in nonprofit management. In 1981, Harry become a board member of the Tribal Sovereignty Program, which later became the Seventh Generation Fund, a national foundation established to provide philanthropic support to Native American communities. In

1985, she served as program officer at the Seventh Generation Fund. She is a faculty member of the Fund Raising School at Indiana University, in Indianapolis. In the early 1990s she served as the capital campaign manager for the Institute of American Indian Arts (IAIA) in Santa Fe, New Mexico. She was responsible for major gift development, volunteer management, and campaign management. The campaign raised $4.9 million, providing the necessary funds to support the IAIA's museum for student and alumni work. The museum, housed in a renovated historic building, is located in downtown Santa Fe.

Harry received her M.S. degree in community economic development from New Hampshire College in 1994. That same year she also received a Kellogg Foundation Leadership Fellowship and studied the implications of human genetic research for indigenous peoples. In 1998, she founded the IPCB to assist indigenous peoples in the protection of their genetic resources, indigenous knowledge, and cultural and human rights from the negative effects of biotechnology. For the past ten years she has been an activist, researcher, and frequent speaker on the topics of human rights, genetic research, race relations, and community development.

Harry considers herself to be a continuing link in a struggle for indigenous peoples' survival that began centuries ago. In an interview, Harry said, "Since the first contact with western civilization, all of our attention and energies have been devoted to responding to external assaults on our sovereignty and human rights. Native people have never been passive—we can't afford to be because we have responsibilities to insure the wellbeing of our future generations. They have a right to live free from oppression and external control." The problems facing indigenous people today include forced economic dependency, erosion of sovereignty, appropriation of natural resources, diminished territories, racism, and now, biopiracy. Harry's interest in biodiversity runs parallel to her work on human rights. They are both ethical issues in which the dominant society repeatedly refuses to recognize the indigenous peoples' right to self determination.

Much of the earth's genetic and cultural diversity is found within indigenous territories. Unfortunately, this diversity is viewed as a commodity by many governments and corporations. The immense resources of technology-rich countries are being pooled in worldwide public and private collaborations in order to carry out genetic research. These efforts, combined with increased technological capabilities for genetic sequencing, are fueling a worldwide effort to collect genetic samples from plants that produce foods or medicines, from animals, and from diverse human populations. The genetic resources that indigenous societies have nurtured, and that in turn have nurtured their lives for centuries, are at risk of genetic theft. The patents and profits from products that come from appropriated indigenous biological resources and knowledge do not benefit the tribes, but rather go to shareholders and those who can afford to buy the products. Greed, racism, and oppression are at the heart of the matter because the human and collective rights of indigenous people are often violated in the quest to own and control the world's genetic resources. The biodiversity that exists on indigenous land is not there by chance. A thorough knowledge of the land, species, and natural cycles creates a respect for the ecosystem that allows indigenous people to live on their land without exploiting it for profit. Yet, the rights of indigenous people to use, control, and protect the resources within their territories are under constant pressure from a dominant society built upon extraction and commodification of natural resources.

The field of human genetic engineering has potential to further exploit indigenous people as genetic material is collected, studied, and commercialized by scientific, corporate, and government interests. Unique genetic patterns can be patented under U.S. patent law, turning life and human traits into property. Harry finds this to be unethical for a variety of reasons, the most basic being the idea that the human genetic code can be privatized for profit. Other disturbing aspects are that blood samples have been taken from indigenous people without fully informed consent and that the genetic material is then made available to other researchers for secondary uses unknown to the original donors. Indigenous people cannot control who has access to the samples or how those samples are manipulated or commercialized. "We must set limits on what is acceptable." Harry says. "Patenting life is not acceptable, and many people would fundamentally agree with that. Human genetic materials are passed down from our ancestors, and belong to our future generations. It is not a commodity to bought, sold, owned, or manipulated."

Harry founded the IPCB in order to "assist indigenous peoples in the protection of their genetic resources, indigenous knowledge, and cultural and human rights from the negative effects of biotechnology . . . [and provide] educational and technical support to indigenous peoples in the protection of their biological resources, cultural integrity, knowledge, and collective rights." IPCB projects include a genetic research database that tracks genetic research affecting indigenous peoples as well as community education, technical assistance, and intervention programs. To prevent exploitation of indigenous peoples, the IPCB provides technical expertise to tribes that are affected by genetic research.

Harry is a board member of the Council for Responsible Genetics in Cambridge, Massachusetts. The council insists that the public have an understanding of the true nature and intent of genetic research, that the research fall within ethical boundaries, and that scientific endeavors be socially responsible. Efforts to permanently alter the genetic structure of living things are unethical. Genetically modified organisms should not be treated as commodities without recognizing their huge potential for ecological harm. The council has prepared a Genetic Bill of Rights that states that all people have a right to say how and when technology should be used in society and should not be subjected to a genetically modified environment against their will.

Although Harry's work takes her on frequent travel around the United States and internationally, she maintains her residence in Nixon, Nevada.

BIBLIOGRAPHY

Harry, Debra, "Globalization and Indigenous Peoples," *National Network of Grantmakers Newsletter*, 2000; Harry, Debra, and Frank

Dukepoo, *Indians, Genes and Genetics: What Indians Should Know about the New Biotechnology,* 1998; "Indigenous Peoples Council on Biocolonialism," http://www.ipcb.org/; "Race to Crack the Code, an Interview with Debra Harry," *Midwest Soarring Wings,* 1999.

Harvey, Dorothy Webster

(October 25, 1915–)
Environmental Activist, Writer

Writer, activist, and environmental researcher Dorothy Harvey spent ten years working against environmental exploitation in Utah. She was primarily concerned with the development of water resources in that arid state, especially the massive Central Utah Project (CUP), which was designed to take water from high in the Uinta Mountain range southward into the Uinta Basin. By diverting water from the streams to the west along a 100-mile aqueduct, the project would have left the streams parched and the range without an adequate water supply. Through fundraising efforts she mounted a major public education campaign in Utah about the growing concerns of biologists and recreation users regarding the CUP and its potentially negative impact upon Utah's wilderness areas.

Dorothy Davis Alexandria Webster was born on October 25, 1915, in Galt, Ontario, Canada. She was one of two children born to Charles Webster, an engineer, and Maude Davis, a schoolteacher. She survived the depression and worked at various jobs for a few years after graduating from high school. In 1937, thanks to family members who supported her quest for education, she was accepted at the University of Michigan and focused on American studies, but she never finished her degree. It was there that she met and eventually married William R. Harvey, an engineering student.

After World War II, she followed her husband to jobs in Maryland and Maine until the couple settled in Madison, Wisconsin, in 1952. An ordained Episcopalian priest, her husband was assigned to St. James Church in Manitowoc, Wisconsin, and kept that position for the next 30 years. In the meantime, Dorothy Harvey bore four children, Kathleen, Patricia, Pamela, and Mark; took a position as a school secretary; and wrote an advertising column for the *Manitowoc Herald* under the pseudonym "Sadie Snooper."

During the 1960s, the Harvey family made annual treks to the Rocky Mountain states, visiting various national parks. It was during these trips that park ranger–led campfire programs introduced her to natural resources management concepts. She became actively involved in the nation's wilderness designation activities, seeing her mission as a devout Christian to protect God's creation.

Harvey's early environmental activism was centered on wilderness designation in Wisconsin. She volunteered for the

Sierra Club to lead a series of studies of the surrounding national forest and roadless areas. But seeing the wilderness in the state of Wisconsin as relatively limited and knowing that there was much more wilderness in the West, very vulnerable to development, Harvey felt "called" to pursue environmental work in the western region of the country.

In 1971, working under Prof. Ross Smith of the University of Nevada, Harvey participated in wilderness studies of the Ruby Mountains within the Humboldt National Forest in Nevada. During the following year, 1972, she led a group of volunteers studying areas in the Payette National Forest of Idaho that had been subjected to excessive logging. Here she was exposed to the politics of resource management within the national forest. Her group became disillusioned as it faced the corporate power of the Boise Cascade Company and a Forest Service administration unwilling to ensure environmental compliance; eventually it disbanded, and she returned to Wisconsin.

During 1973 and 1974, she participated in several Sierra Club–sponsored studies of the Uinta Range of northeast Utah. Often working alone, she hiked extensively, encountering a variety of wildlife species and landscapes. While on the north slope of the Wasatch Forest she became acquainted with Jim Kimbel, a wildlife biologist for the U.S. Forest Service, who told her about a project to design logging proposals that would create meadows for elk habitat. Her interest in the project resulted in statewide efforts to solicit funding for it. Working closely with Kimbel and other biologists, "Mrs. Elk," as she became known, developed a close relationship with National Forest and various university biologists. These relationships would prove important later in her efforts to combat the CUP.

In 1975 Kimbel suggested she investigate the impact of phosphate mining in the Caribou National Forest in Idaho. She discovered that a third of the forest had been completely denuded of vegetation as a result of the digging of 150-foot canyons to extract the ore, which was not sold but was rather stored by the company in hopes that it could earn higher profits in the future. She wrote an extensive report on the impact of this activity on the habitat of local wildlife, including the migratory patterns of elk. Unfortunately the Forest Service's hands were tied because of provisions in the 1872 Hard Rock Mining Law, which permitted mining on the public lands without environmental regulations or standards. Despite a state government that sought to exploit rather than to protect the environment, her publications of these abuses increased public awareness and led to curtailment of further developments.

Having accepted an investigative assignment from *High Country News* in 1976 and 1977, Harvey began to write about oil shale development within the Uinta Basin of Utah. Fearing the loss of a potential wilderness area, Harvey teamed up with Peter Hoving, a biochemist, and led canoe trips on the White River where a dam for the water supply was being proposed. These trips increased awareness of the fragility of the ecosystem and the potentially disastrous impact of a dam upon local flora and fauna. No dam would be built, and eventually oil shale extraction from the basin was determined to be economically unfeasible.

In 1977, a government biologist introduced Harvey to the CUP on a tour of dams currently under construction. The CUP was under development by the U.S. Bureau of Reclamation and the Central Utah Water Conservancy District (CUWCD) in the central and east-central part of Utah. The project provided Utah with the opportunity to use a sizable portion of its allotted share of Colorado River water. This water would be provided to meet the municipal and industrial requirements of the most highly developed part of the state along the Wasatch Front, where population growth and industrial development were continuing at a rapid rate.

It was during the 1977 dam tour that Harvey gained her recognition of the value of outdoor recreation along the south slope of the Uinta basin, inspiring her to organize opposition to the CUP. She felt the Bureau of Reclamation and the state were failing to meet the mandates of the National Environmental Policy Act (NEPA), were basically ignoring the values of outdoor recreation and wilderness, and were willing to sacrifice the rivers for economic benefit. In 1977, she helped to form an anti-CUP grassroots organization, Citizens for a Responsible CUP, whose name changed the following year to Utah Water Resources Council and then again to Intermountain Water Alliance in 1979. The group proved instrumental in informing the public about the state and federal government's plans for water development in Utah and their potential impact upon the environment. It sought to educate the public on the natural value of stream flows and the wildlife facing destruction as a result of the CUP. Unfortunately, Harvey and her cohorts faced a public that maintained a "use it or lose it" mentality and harbored the impression that the CUP would "make the desert bloom." She left the state of Utah in the mid-1980s and although she never saw results, her efforts laid the groundwork as the public slowly began to see the rivers as possessing value beyond economics. Other groups spawned in the 1990s, such as the Utah River Council, would continue the efforts she began and would ultimately prove successful in saving portions of the Uinta and Bear Rivers from the developers of the CUP.

Harvey currently resides in Fargo, North Dakota, and at the age of 84 is working on a book regarding the CUP and the efforts of grassroots environmentalists in confronting government agencies responsible for resource management on public lands. She continues to support the concept of sustainable water management so that the integrity of the natural systems is maintained and intact ecosystems are preserved. Her son, Mark, is an associate professor of history at North Dakota State University, emphasizing environmental history and the wilderness movement.

BIBLIOGRAPHY

"Central Utah Water Conservancy District," http://www.cuwcd.com; Harvey, Mark T., *A Symbol of Wilderness: Echo Park and the American Conservation Movement*, 1994; Reisner, Marc P., *Cadillac Desert: The American West and Its Disappearing Water*, 1987; "U.S. Bureau of Reclamation–Central Utah Project Overview," http://dataweb.usbr.gov/html/cupoverview.html; "Utah Rivers Council," http://www.wasatch.com/~urc/.

Hawken, Paul

(February 8, 1946–)
Entrepreneur, Writer

Paul Hawken is an entrepreneur who writes and works in the fields of industrial ecology and sustainability, where he seeks to marry environmentalism with economic and industrial development.

Paul Gerard Hawken was born on February 8, 1946, in San Mateo, California. Involved in the civil rights movement as a young man, he served as press coordinator for the 1965 march on Montgomery, Alabama. He worked in Selma, Alabama, with Martin Luther King's staff, registering the press, giving interviews on national radio, and acting as a marshal for the final march. Later that same year he worked as a staff photographer for the Congress on Racial Equality in New Orleans, where he was kidnapped by the Ku Klux Klan but escaped thanks to intervention by the Federal Bureau of Investigation. The photographs he took during this time were published around the world.

In 1966, he moved to Boston and created the Erewhon Trading Company, the first company in the United States to focus on selling organically produced food. Erewhon introduced many of the health food items that have since become staples: locally produced organic fruits and vegetables, soy sauce, and local spring water, for example. Erewhon was also the first company to produce organically grown rice in the United States. By 1973, Erewhon revenues had risen to $50,000 a day. Erewhon had its own manufacturing facilities and contracted with farmers in 37 states on 56,000 acres of farmland. Erewhon owned two stone mills and two rail cars and leased three warehouses. In 1974, Hawken sold Erewhon, beginning his pattern of founding, successfully leading, and then walking away from businesses.

In 1979, he cofounded the Smith and Hawken mail order gardening supply company with a partner, David Smith. Smith and Hawken originally began as a nonprofit offshoot of Ecology Action, specializing in hand tools used in a type of gardening known as French-intensive/biodynamic gardening. Eventually,

Paul Hawken (courtesy of Anne Ronan Picture Library)

the company branched off into other horticultural areas and began importing tools made by a 200-year-old foundry in Lancashire, England. Together, the partners turned Smith and Hawken into a $12 million a year business with four retail store locations and 600,000 yearly catalog customers. More important to Hawken than the bottom line, however, was the philosophy behind the business. Hawken is a noted business theorist whose views on running a business have been widely read and are vastly influential. He expressed in an interview with the magazine *Mother Earth Review* in March 1985, that "a business has to resonate with your sense of who you are, and what you do in the world." All of the other "business things" such as inventory and accounts and budgets and profit should come secondary to the *reason* that business is being done. He urged readers not to "confuse the vehicle with the destination" and said that doing business is nothing more than a way to accomplish something and should never be looked upon as an accomplishment in and of itself.

It was this sense of purpose that led him to leave Smith and Hawken in 1991. In his 1993 book, *The Ecology of Commerce*, he explained his reasons for leaving. "The recycled toner cartridges, the sustainably harvested woods, the replanted trees, the soy based inks were all well and good, but basically we were in the junk mail business. All the recycling in the world would not change the fact that doing business in the latter part of the twentieth century is an energy intensive endeavor that gulps down resources." Even though the business he was running operated in a manner that was 99.9 percent more ecologically sensitive than almost every other business operating in the United States, he was still unable to justify carrying out business in a way that generated so much waste. So, he stopped.

Today, Hawken concerns himself with confronting the problems of global commercial and biological sustainability. He works with the U.S. and international branches of the Natural Step, an international nonprofit organization whose goal is to change the industrial paradigm in which business is conducted. The Natural Step was originally founded in Sweden in 1989 by Karl-Henrik Robèrt. An oncologist, Robèrt observed firsthand the connections between human illness and the presence of toxins. He became concerned with the fact that much of the environmental debate focused on "downstream issues" rather than on the causes of the problems themselves. The Natural Step was brought to the United States by Hawken in 1995. As an organization, it focuses on building an ecologically and economically sustainable society, by offering businesses and individuals a scientifically based framework that allows them to operate in harmony with earth's cyclical processes.

Hawken finds something to be fundamentally wrong with an economic system that makes it cheaper to destroy resources than to sustain them, and he believes that large-scale changes need to be enacted to avoid the continued physical degradation of earth and the consequences that go along with that degradation. He suggests that we create a market where people and businesses are rewarded for internalizing costs, rather than externalizing them. To do this we should, over the course of 20 years, phase out the entire current tax system and replace it with one that would no longer tax those things that we encourage: jobs, profits, income. Instead, we would tax through the utilization of what he calls "green fees." Green fees would adjust the prices of commodities to reflect the actual cost of the commodity being purchased. Gasoline, he says, actually costs between four and a half and seven dollars per gallon when you take into consideration environmental costs such as air pollution and acid rain. This, according to Hawken, is the price we should pay at the pump.

Hawken is currently involved with several business organizations, including the Global Business Network (a private consulting network of 100 members), Interface, Inc. (he is a part of a 12-member group of consultants responsible for turning RAY ANDERSON's Interface into the world's leading industrial ecology company), and Datafusion (a data mining, warehousing, and knowledge synthesis company), among others. His work with nonprofit organizations is varied; currently he serves on the boards of the Natural Step International, Urban Ecology, the Aldo Leopold Leadership Program, the Materials Efficiency Project, the Center for New American Dream, and Friends of the Earth. He has also received numerous awards for his work on the intersection of environmentalism and industry. He was named "Best of a Generation" by *Esquire* in 1984, was given the Environmental Stewardship Award by the Council on Economic Priorities in 1991, and was named "One of 100 Visionaries Who Could Change Our Lives" by the *Utne Reader* in 1995.

Hawken currently resides with his family in Sausalito, California.

BIBLIOGRAPHY

Frons, Marc, ed., "A Garden-Tool Maker Who's Cultivating Slow Growth," *Business Week*, 1986; Hawken, Paul, *The Ecology of Commerce*, 1993; Hawken, Paul, *Growing a Business*, 1987; Hawken, Paul, *Natural Capitalism*, 1999; Katz, Donald R., "Guru of the New Economy," *Esquire*, 1984; "The Natural Step/US," http://www.naturalstep.org/.

Hayes, Denis

(August 29, 1944–)
Earth Day Organizer

Denis Hayes became famous in 1970 when Sen. GAYLORD NELSON recruited him to organize the first Earth Day. The participation of some 20 million people in Earth Day's varied activities brought the environment to the forefront of the mainstream public. Twenty years later, Hayes became the international chair of the second major Earth Day celebration. The intent of Earth Day II was to remind the public that all of the original environmental problems that Earth Day I addressed were still in existence and that individuals did have the power to respond to these problems by becoming more conscious of the environmental impact of their lifestyle. When he has not been organizing Earth Days, Hayes has worked with government and nonprofit agencies to promote alternative, nonpolluting sources of energy and solutions to the environmental problems that plague the planet today.

Denis Allen Hayes was born in Wisconsin Rapids, Wisconsin, on August 29, 1944, and raised in the small mill town of Camas, Washington. An early impetus to environmental work was the putrid sulfur smell emanating from the town's paper mills. Hayes left Camas for a three-year hitchhiking trip around the world and then entered Stanford University, where he became student body president. Hayes attended Harvard Law School, but once he was invited in 1970 by Wisconsin senator Gaylord Nelson to organize Earth Day, he quit Harvard and moved to Washington, D.C. Thanks to Hayes's energetic publicity and organizing campaign, 20 million schoolchildren, college students, church members, and others participated in the first Earth Day. This show of public concern for the environment helped push Congress into passing the string of environmental legislation that came before it during the 1970s, including the Clean Air Act, the Water Quality Act, the Environmental Protection Act, and the Endangered Species Act.

After Earth Day, Hayes spent a year as a visiting scholar at the Smithsonian Institution (1971–1972) and directed the Illinois State Energy Office in 1974 and 1975. He joined the Worldwatch Institute's staff in 1975, where he researched alternative energy sources, especially solar energy. He wrote *Rays of Hope* in 1977, which described how the United States could kick its fossil fuels habit and convert to solar energy. Next, Pres. Jimmy Carter named Hayes director of the Solar Energy Research Institute (SERI) in Golden, Colorado. This position lasted only two years, since the Reagan administration took funds away from government agencies such as SERI that were researching alternative energy sources. Hayes moved to San Francisco in 1981 and returned to law school. He practiced corporate law for a short time, then returned to his environmental work.

The year 1990 was a banner year for Hayes. It was the 20th anniversary of Earth Day, and Hayes was the international chair of a celebration by 200 million participants in 3,600 U.S. cities and 141 countries. Hayes saw Earth Day II not as a cure-all for the world's serious environmental problems but rather as an effective means to communicate with the public. Perhaps owing to increased environmental awareness, people made changes in their lifestyle that year. In an *E Magazine* interview in 1995, Hayes cites these statistics: in 1990, 3,000 new curbside recycling programs were initiated, and there was a 100 percent increase in recycling of postconsumer waste that year. Also in 1990, Hayes founded Green Seal, an organization that independently certifies environmentally friendly products. The program helps consumers support companies doing their part to reduce pollution, and it can help "green" manufacturers reach a conscientious market.

In addition to serving as chair of the board of Green Seal and the Coalition for Environmentally Responsive Economics, which helps corporations improve their environmental practices, Hayes currently serves as president of Seattle's Bullit Foundation. The Bullit

Foundation is dedicated to defending the plants, animals, water, and land of the Northwest with its $100 million endowment. One of its main projects currently is Eco-Trust, a financial institution that funds businesses that offer environmentally friendly employment to people in areas where extractive industries such as timber cutting have collapsed. Hayes resides in Seattle, Washington.

BIBLIOGRAPHY

Cahn, Robert, and Patricia Cahn, "Did Earth Day Change the World?" *Environment*, 1990; DeLeon, David, *Leaders from the 1960s*, 1994; Motavalli, Jim, "Cutting through the Fog: Denis Hayes on Earth Day Strategy," *E Magazine*, 1995.

Hayes, Randy

(July 11, 1950–)
Founder and President of Rain Forest Action Network

Rainforest Action Network (RAN) and its founder and director Randy Hayes were largely responsible for bringing the problem of rain forest deforestation to the attention of the American public during the 1980s and 1990s. RAN's partnerships with grassroots rain forest conservation groups in 60 countries around the world, its effective publicity campaigns in the United States, and its ability to mobilize its 30,000 members make it an effective, visible force against rain forest destruction.

Randall Hayes was born on July 11, 1950, in East Liverpool, Ohio, to Ace and Beverly Hayes. Exploring the forests of West Virginia and the swamps of Florida inspired an early interest in nature and in keeping it wild. His father's entrepreneurial spirit and his mother's focus and kindness contributed to Hayes's career choice and success. He earned a B.A. in psychology and sociology from Bowling Green State University in Ohio in 1973 and went on for graduate work in environmental planning at San Francisco State University in 1983. His master's thesis was the film "The Four Corners, A National Sacrifice Area?" that documented the tragic effects of uranium and coal mining on Hopi and Navajo Indian lands in the Four Corners region of the American Southwest. It won the Best Student Documentary award from the Academy of Motion Picture Arts and Sciences in 1983 and served as his introduction to the environmental problems faced by native peoples.

During a 1984 trip to Costa Rica, Hayes was shocked by the rain forest destruction there. He visited a pristine rain forest area where a new road had recently been built. The blood-red clay swath through the forest left an indelible impression on him. After vowing to do something to combat worldwide rain forest destruction, Hayes returned to the United States and in 1985 organized the first international rain forest strategy conference in Sausalito, California. It was attended by about 70 participants from 12 countries, representing several environmental organizations and groups of indigenous peoples. The conference revealed that rain forest destruction was the result of many converging conditions in less-developed countries: official government disrespect for the rights of indigenous people to live in their native rain forests; land ownership inequities that drive landless peasants to slash and burn virgin forests, the only lands left for colonization; and road construction and dam building projects financed by international financial institutions such as the World Bank and the International Monetary Fund. Out of this founding conference, the Rainforest Action Network was born. It existed as a working group of the Earth Island Institute, which nurtured it until 1987, when RAN was large enough to sustain itself.

RAN's first goal was to bring the rain forest issue to public attention. One of its early projects was a well-publicized boycott of Burger King, because that fast-food chain bought beef from Latin America where ranchers graze their cattle on pastures cleared from rain forests. After the boycott cut Burger King's 1987 profits by 12 percent, Burger King officials announced in 1988 that the company would no longer buy its beef in Latin America.

Its effectiveness in mobilizing boycotts established, RAN also came to be known for the creativity and innovation of its direct action protests. Hayes was present at the birth announcement of Earth First!: the draping of a black plastic 300-foot "crack" down the side of the Glen Canyon Dam in March 1981. Inspired by the drama of that event, Hayes has organized RAN volunteers with technical climbing skills to undertake many similar actions. Their first was a 1986 climb of the World Bank building in Washington, D.C., during a meeting of World Bank finance ministers. Activists hung a two-story banner that declared "World Bank Destroys Tropical Rainforests" from the front of the building, and police were not able to cut it down for two days. Hayes later told the *San Francisco Chronicle*, "That image went around the world." The banner revealed what before then had been a little known fact: that World Bank–funded development projects frequently resulted in severe environmental degradation. Public outcry put pressure on the World Bank to develop stricter environmental criteria for its projects, and vocal RAN members helped influence the World Bank's 1994 withdrawal of funding from the environmentally destructive Altamira dam project in Brazil's Amazon Basin.

Other targets of RAN's campaigns have included Mitsubishi, which logged southeast Asian and Canadian rain forests for its wood products division, and Canada's largest logging company, MacMillan Bloedel. Pressure resulted in MacMillan Bloedel's halting all clear-cutting operations in 1998. On another campaign, Stone Container Corporation was denied permission to log Honduras's Mosquito Coast and build a chipping mill in Costa Rica after RAN and local activists exerted significant pressure on the company and local authorities.

RAN has grown to include about 30,000 members working through some 150 local affiliates (Rainforest Action Groups or RAGs). Recognizing the potential of electronic networks, RAN early on began to use the computer network EcoNet to maintain close contact with grassroots environmental groups and indigenous people's organizations in 60 rain forest–rich countries. Alerting its membership of problems, RAN coordinates national publicity and letter-writing campaigns, nonviolent direct-action protests, and boycotts of the culprits. In many cases, RAN collaborates not only with local grassroots groups but also with environmental giants such as the Sierra Club, Greenpeace, and the Natural Resource Defense Council.

Currently, Hayes is focusing on a national campaign to reduce consumption of wood and wood-based paper by 75 percent in the next decade. He resides in San Francisco.

Bibliography

Buchanan, Rob, "Looking for Rainforest Heroes," *Outside*, 1992; Freistadt, Margo, "The Man behind the Rainforest Action Network," *San Francisco Chronicle*, 1989; Hayes, Randy, "Activism: You Make the Difference," *Lessons of the Rainforest*, 1990; Hayes, Randy, "Lessons of the Rain Forest," *Call to Action: Handbook for Ecology, Peace and Justice*, 1990; "Rainforest Action Network," http://www.ran.org.

Hays, Samuel P.

(April 5, 1921–)
Environmental Historian

Fondly referred to by his colleagues as "the grandfather of environmental history," Samuel Hays has spent a lifetime as a leader and respected writer of the environmental history of the United States. While his writings cover such diverse matters as environment and urbanization, environmental regulation, forest issues, air quality, the role of values, and postwar environmental politics, the consistent thread though all of Hays's work is his deep commitment to environmental issues and to his belief in the importance of history to understanding the environmental impulse, environmental policy, and environmental controversy.

Samuel Pfrimmer Hays was born April 5, 1921, in Corydon, Indiana, the son of Clay Blaine, a small-town lawyer and dairy farmer, and Clara Ridley (Pfrimmer) Hays. His interest in the environment emerged during his boyhood. He acquired some knowledge of conservation from experience with the U.S. Soil Conservation Service on his family's farm and more from a Civilian Conservation Corps camp at a neighboring state forest. Hays also spent two months at an American Friends Service Committee work camp on a demonstration farm in eastern Tennessee that was part of the program of the Tennessee Valley Authority. These experiences were augmented between 1943 and 1946, when he worked with the Oregon and California Revested Lands Administration on public forests in western Oregon. His work for the so-called O&C ranged from tree planting and fighting forest fires to cooking and tool maintenance. During this two-and-a-half-year stint, Hays became interested in the agency itself and was made "project education" leader. His new task was to inform members of the camp about the O&C, its history, and its program.

Hays completed his B.A. in psychology at Swarthmore College in 1948 and then moved on to Harvard, where he earned an M.A. in 1949 and a Ph.D. in history in 1953. At Harvard, Hays was able to integrate his interest in conservation with his focus on history. His dissertation dealt with the Progressive movement and conservation. Harvard University Press published the revised dissertation in 1959 as *Conservation and the Gospel of Efficiency*. This work became one of the seminal contributions to the field of environmental history and was reprinted by the University of Pittsburgh Press in 1999. It is as relevant today as it was when it was first published.

Applying to agricultural schools, Hays hoped to teach conservation history but had no success securing employment in his then-obscure field. Instead, he opted for the equally new, but more popular, fields of political, urban, and social history, teaching briefly at the University of Illinois and for seven years at the University of Iowa before accepting the position as chair of the Department of History at the University of Pittsburgh in 1960. During the 12-year hiatus between the publication of *Conservation and the Gospel of Efficiency* and Hays's return to the study of environmental issues, he published widely in the fields of urban political and social history.

Hays's return to historical work on the environment after 1970 coincided with a period of great environmental enthusiasm. The celebration of Earth Day in April 1970, the 1969 passage of the National Environmental Policy Act (NEPA), and the founding of the Environmental Protection Agency in 1971 suggested that the human relationship with the environment was becoming an important—and hitherto relatively unexplored—avenue for historical inquiry. For Hays, it was a return to an area of scholarly and personal interest, but one that was driven by new interests. Through the 1970s to the present, Hays has written prolifically in journals and book chapters on conservation history, environmental politics, and the city of Pittsburgh. Not least of these interests was the opportunity to engage in mutual scholarly interests with his wife, Barbara Darrow, a biologist whom he married in 1948. The two collaborated on the highly acclaimed *Beauty, Health and Permanence: Environmental Politics in the United States, 1955–1985* in 1987. At the heart of his work is the relationship between conservationism and politics, as evidenced in his books and his 1992 article in *Environmental History Review* (now *Environmental History*), "Environmental Political Culture and Environmental Political Development: An Analysis of Legislative Voting, 1971–1989."

In addition to his books and essays, Hays is an archivist of primary urban and industrial resource materials. His work in collecting archival material for the University of Pittsburgh has earned the History Department its prominence in the field. Hays was responsible for amassing a tangible chronicle of the industrial growth and development of Pittsburgh and the surrounding region. Under his leadership, this collection, known as the Archives of Industrial Society, grew to incorporate original source documents and served as a basis for several dozen doctoral dissertations. The scholarly result was brought together in a volume called *City at the Point: Essays in the Social History of Pittsburgh*, which he edited and which was published by the University of Pittsburgh Press in 1990.

In 1982, a subdivision of the collection was created, the Environmental Archives. It included both Pennsylvania and national environmental records. The Environmental Archives now houses on a continuing basis records of the Pennsylvania chapter of the Sierra Club, the Western Pennsylvania Conservancy, and the Pennsylvania Environmental Council. National records include those of Environmentalists for Full Employment, Environmental Action, and the Environmental Action Foundation, one of the sponsor organizations for the first Earth Day in 1970 and no longer in existence. Over the years Hays and his wife have gathered a massive number of documents pertaining to environmental affairs since 1970, featuring citizen organizations from all 50 states as well as national organizations. These are now housed in the Environmental Archives and constitute its largest collection, totaling over 800 linear feet of documents. It includes a major collection of British environmental documents that Hays gathered beginning with his stint as Harmsworth Professor of History at Oxford in 1982–1983, to which a collection from an English environmental historian was added.

Aside from his academic work, Hays has also been active in environmental affairs, writing countless letters and articles to clarify environmental issues. He has been involved with the Sierra Club on a variety of issues, especially those pertaining to eastern state and national forests. He donated his father's 360-acre dairy farm to the Harrison County, Indiana, Park Board and the Indiana Nature Conservancy, to create the Hayswood Nature Reserve in 1976 and the Indian Creek Woods in 1986. And in 1994 he created an endowment through the Community Endowment for Southern Indiana to support upkeep of Indian Creek Woods. It is perhaps not irrelevant to his personal philosophy of life that in 40 years of teaching at three major universities he has never held a parking permit, always walking daily from his home to his office.

In 1982, Hays received the Theodore C. Blegen Award from the Forest History Society. In 1991, he received the Pennsylvania Governor's Award for Excellence in the Humanities, and in 1993 he was awarded the Historical Society of Western Pennsylvania History Makers Award. More recently, in 1997, the American Society for Environmental History awarded him its first Career Achievement Award for Lifetime Contributions to Environmental History. In 1999 he received a distinguished service award from the Organization of American Historians.

Hays and his wife live in Pittsburgh. They have four adult children.

Bibliography

Hays, Samuel P., *American Political History as Social Analysis: Essays by Samuel P. Hays*, 1980; Hays, Samuel P., *Explorations in Environmental History: Essays by Samuel P. Hays*, 1998; Hays, Samuel P., *Response to Industrialization, 1885–1914*, 1957 (revised, 1995).

Henderson, Hazel

(March 27, 1933–)
Economist, Futurist

Hazel Henderson began her environmental career in New York City by forcing her local television stations to broadcast daily air quality reports, a practice soon adopted by television networks and newspapers nationwide. In the early 1970s, she shifted her focus to economics, and since then she has written numerous articles and books criticizing such traditional economic concepts as the Gross National Product (GNP). She believes that economics must become a multidisciplinary field if it is to successfully create economically and environmentally sustainable policies. She advocates the creation of a sustainable future through the cooperation of local and international grassroots organizations and business enterprises to manage the planet's resources, establishing a "win-win" economy.

Hazel Henderson was born on March 27, 1933, to Kenneth and Dorothy Mustard in Bristol, England. She dropped out of school at the age of 16, and in 1956 she emigrated to the United States, settling in New York City. She became a naturalized citizen in 1962. Henderson became an activist in 1964 when she began a letter-writing campaign in response to the terrible air quality in New York. She wrote letters to the presidents of the major broadcasting networks requesting that an air pollution index be included in the daily weather report. She also sent letters to New York's elected officials and to Newton Minow, chairman of the Federal Communications Commission. Several weeks later, Henderson received a telephone call from an American Broadcasting Companies (ABC) vice president who supported the idea, and not long after, WABC-TV in New York City began broadcasting a daily air pollution index. Other local television stations and newspapers followed suit. Soon, air pollution information was being made available by television stations and newspapers nationwide.

As a follow-up to her success, Henderson and two friends created an organization called Citizens for Clean Air (CCA), in order to inform New York City residents about the health hazards that go along with air pollution. With the help of an advertising agency that was willing to donate time and resources, CCA launched an educational campaign that quickly gained 25,000 members. In recognition of these efforts, Henderson was named Citizen of the Year by the New York Medical Society in 1967.

In the early 1970s, Henderson began to move away from activism and started developing theories about how to bring about major changes in society. In a 1995 interview with David Kupfer in *Whole Earth Review*, Henderson said, "I always knew I was unemployable, so I invented my own job and have been self-employed for 25 years." She calls the job she invented "social innovator." Others have called it "futurist," "alternative political economist," and "visionary." In the early 1970s, Henderson began educating herself on the principles of economics. She also began publishing articles in such periodicals as the *Harvard Business Review*, the *Columbia Journal of World Business*, and *Science*. In 1972, she helped to create

Hazel Henderson (courtesy of Anne Ronan Picture Library)

the Princeton Center for Alternative Futures, Inc., an organization for which she continues to act as a director.

In 1974, when Congress created the Office of Technology Assessment (OTA), Henderson was appointed to its advisory council as a citizens' advocate. OTA's purpose was to determine the social, environmental, economic, and political consequences of emerging technologies. In *Women Pioneers for the Environment*, Henderson is quoted as saying that her time on the OTA advisory council, as well as her service on panels of the National Science Foundation and the National Academy of Engineering between the years of 1974 and 1980, served as her "Ph.D. course." In 1975, she served on Pres. JIMMY CARTER's election task force on economics, recommending that the President's Council of Economic Advisors be expanded into a multidisciplinary Council of Social Science Advisors.

Henderson published her first book, *Creative Alternative Futures: The End of Economics*, in 1978. In this book, as well as in her second, *The Politics of the Solar Age: Alternatives to Economics* (1981), Henderson makes the case that the focus of economics is much too narrow. She criticizes the use of the GNP as a true indicator of the economic and social health of a country, since the GNP overlooks nearly half of the production occurring in the world—production by unpaid workers, mostly women. The GNP also fails to classify natural resources as assets, actually treating the clean-up of pollution as additional production rather than as an expense. "Trying to run an economy using only economic indicators like the GNP is rather like trying to fly a Boeing 747 with a single oil pressure gauge," she says. "What we need to do is fill out the instrument panel." To this end she has helped to create the Country Future Indicators (CFI) index, which measures how well a country is investing in its own people. People, Henderson believes, are the true resource base of a country. She also believes that it is time to send the economists back to school to take courses in "ecology, cultural anthropology, social psychology, thermodynamics and every other discipline concerned with human development," as she stated in her *Whole Earth Review* interview. These are the types of subjects in which we need our macropolicy makers to be educated, she believes, if we are to ask them to create environmentally and economically sustainable policies.

In Henderson's third book, *Paradigms in Progress* (1991), she outlines the shift from the industrial age into the solar age, which she believes is already occurring. She describes the fundamental differences in the philosophies underlying economics and environmentalism, stressing the fact that economics can only view parts of the environmental problems that must be solved and can, therefore, only provide partial solutions. She prescribes the abandonment of our historical anthropocentric world-view and the embrace of biocentrism with its wider, more encompassing perspective. In order to develop sustainable, ecologically viable forms of productivity, she points out, we need to gain an appreciation for, and an understanding of, the system of interdependent species that makes up the planet upon which we live.

Her fourth book, *Building a Win-Win World: Life beyond Global Economic Warfare* (1996), provides an alternative to what Henderson refers to as today's global economic warfare based on capitalism, nationalism, and overconsumption. Henderson depicts a possible future that would include sustainable development achieved through decentralized enterprise, energy efficiency, recycling, and cooperation. She argues that the United Nations, currently too patriarchal, should be restructured to fill a role as global networker and broker of partnerships between local grassroots groups.

Henderson writes a syndicated newspaper column that appears in 27 different languages in over 400 newspapers worldwide, none of them in the United States. She travels the world as a lecturer and consultant, and she has taught courses at various academic institutions, including Schumacher College in the UK, Harvard University, Cornell University, and Brown and Dartmouth Colleges. She is a member of the board of directors for the Worldwatch Institute and the Council for Economic Priorities. And she is a member of the World Future Society and the World Future Studies Federation. In 1995, Henderson was named one of *Utne Reader*'s 120 visionaries who could change our lives. She lives on a small island off the coast of Florida and has a grown daughter, Alexandra.

BIBLIOGRAPHY

Black, Sherry Salway, "Paradigm in Progress: Life beyond Economics," *Christianity and Crisis*, 1993; Breton, Mary Joy, *Women Pioneers for the Environment*, 1998; Kelly, Kevin, ed., "Interview, Hazel Henderson," *Whole Earth Review*, 1988; Kupfer, David, "To Stitch the World Back Together Again," *Whole Earth Review*, 1995.

Hermach, Tim

(1945–)
Founder and Executive Director of Native Forest Council

Tim Hermach believes that like "big tobacco," the big timber corporations that strip-mine the nation's forests have lied to the people, used bribery and extortion to influence politicians, and mounted billion-dollar public relations campaigns to mislead the public, and he acts accordingly. He founded the Native Forest Council, which has taken unprecedented risks to assure the preservation of the nation's remaining native forests as living life-support systems for future generations. The group was the first to advocate a total logging ban on public lands, calling their position Zero Cut; and though they were dismissed at the outset as politically unrealistic, several powerful environmental groups such as the Sierra Club and the Oregon Natural Resources Council have now adopted Zero Cut for public lands. Passionate, outspoken, and uncompromising, Hermach has widened the focus of political debate over logging and forest management practices and has inspired many other progressive environmental advocates with his can-do arguments.

Tim Hermach was born in 1945 and grew up in Eugene, Oregon, in a family of four boys. He and his brothers spent a lot of time in the woods, often on publicly owned national forests and Bureau of Land Management (BLM) land, playing in the clear streams, drinking the water, and chasing enormous salmon. As a young boy he worked in the local fields and orchards and in his father's small businesses. He dropped out of college in the 1960s after attending for a few years and then toured Europe for a year. By the time he was 25 he had already sold encyclopedias, vacuum cleaners, and cars and had worked for a phone company and for the Department of Defense. He eventually ended up in apparel manufacturing, a business in which he was successful for many years. He returned to college in 1983, attending the University of Oregon and graduating with a business degree in 1986.

In 1985, Hermach learned that the United States had already logged over 90 percent of its one billion acres of native forests—public and private. In stark contrast, only 14 percent of the Amazon rain forest had been logged, although that area was receiving much anxious attention at the time. By then the forests Hermach had played in as a child had been stripped bare from ridgetop to ridgetop, and the streambeds had filled with mud. There was little left besides monocropped tree plantations. Most of the remaining small percentage of the country's native forests existed on public lands, but they were subject to logging at taxpayer expense and with no accounting for the cost and losses to the public. To do something about this, Hermach became heavily involved with the Sierra Club. But over the next few years, he grew disillusioned with the large national group, realizing that its main strategy for environmental conflict resolution was compromise. As with much of what Hermach calls the environmental establishment, the Sierra Club's highest priority seemed to be political access and protection of its wealthy donors, and Hermach wanted nothing to do with it.

In 1988 Hermach founded his own organization, the Native Forest Council, with the purpose of disclosing the truth about national forests and the politics that drive their management. From the start, the group began taking controversial stands and maintained that the United States could no longer afford to chip away at the 5 percent of native forests that remain unlogged. Arguing that with so little left uncut in the United States, clearly too much compromise had already occurred, Hermach offered a simple solution he calls Zero Cut: to pass national legislation outlawing all logging of native and old-growth forests on publicly owned federal land. At the outset, many dismissed his stance as quixotic and too radical, and his persistent condemnation of repeated or unwarranted compromise gained him no allies among Washington, D.C.–based environmental organizations. Undaunted, he gathered input from more than 200 grassroots organizations, and in 1989 he drafted the Native Forest Protection Act, which called for an immediate halt to old-growth forest logging and a ban on the export of unfinished wood products (logs, chips, and pulp). The act mandated ecological restoration and included a provision to ensure government accountability.

When Hermach took his proposed legislation to Washington, D.C., he recruited the support of 14 Democrats in Congress, though they all withdrew their backing after being approached by several national environmental groups expressing opposition. For the next few years, Hermach persistently stuck with the campaign, attracting media attention, distributing information, and raising public awareness of the issue. Eventually editorials appeared in several major newspapers, such as the *Washington Post* and the *New York Times*, questioning and criticizing national forest management policy. Growing public concern moved the U.S. Forest Service to adopt new management philosophies that promised to be ecologically sound, but changes were only cosmetic. In 1994, Hermach realized that his original plan to protect native and old-growth forest while allowing logging to continue on planted tree farms placed too much trust in the Forest Service, which had essentially become a servant of the timber industry. Hence, he altered his proposal. His new policy would end public land logging altogether. It met with predictable opposition for all the usual reasons: that it was politically unrealistic, that there would be a loss of jobs, that it would disrupt the economy of logging communities, and that aging, crowded stands of trees needed to be thinned. But the simplicity and explicitness of the Zero Cut proposal, backed up by Hermach's moral and economic arguments that the abuse of the nation's forests had gone on too long, appealed to many and reinvigorated the grassroots community that had been fighting losing battles against the timber industry.

The issue gained more and more attention, and although there was still no major environmental organization willing to back Zero Cut, the public became increasingly critical of governmental forest management practices. Hermach never broke his stride; he continued campaigning for Zero Cut for the next few years—attending conferences, educating the media, and constantly networking by phone and fax. By 1995 he had scored endorsements from the Oregon Natural Resources Council, Greenpeace, the New

York City and Long Island chapters of the Audubon Society, and Washington's Inland Empire Public Lands Council. And in 1996, after an exhausting battle in which the Sierra Club's management threw many obstacles up and tried to prevent its membership from voting in favor of it, the club officially voted to support Zero Cut.

Finally, in October 1997, nearly ten years after Hermach had introduced the notion of Zero Cut, Cynthia McKinney (D-GA) and Jim Leach (R-IA) introduced a bipartisan proposal called the National Forest Protection and Restoration Act (NFPRA) that would phase out all commercial logging within two years. Doing so would protect the country's forest heritage, cut corporate welfare, and save taxpayer money; the bill also included a provision to assist logging communities with economic recovery and diversification. Though the bill has not yet passed, its introduction was a victory in itself, dramatically improving the parameters of the political debate and proving that there was broad support of the Zero Cut idea.

The Native Forest Council continues to build strong coalitions for uncompromised and honest economic, social, and environmental solutions. It serves as a powerful information clearinghouse for the media and those within the forest movement, and its *Forest Voice* newsletter is read by activists all over the country. Hermach continues his work as executive director of the group and remains committed to fighting for Zero Cut and for the total protection of 650 million acres of federally owned public land, rivers, and streams. He lives in Eugene, Oregon.

Bibliography

Broydo, Leora, "Mutiny at the Sierra Club," *Mother Jones*, 1998; Hermach, Tim, "True Colors," *Forest Voice*, 1997; Hermach, Tim, "What Is the Value of Our Wilderness?" *Seattle Times*, 1998; Mazza, Patrick, "Sierra Club Supports Zero Cut," *E Magazine*, 1996; Sonner, Scott, "Activist Finally Sees Logging Ban Proposed," *Columbian*, 1997.

Hill, Julia Butterfly

(February 18, 1974–)
Forest Activist

Going beyond what most environmental activists would do to save a tree, Julia Butterfly Hill spent two years in a 200-foot-tall old-growth redwood in northern California, in protest of the notoriously destructive logging practices of Pacific Lumber, a division of the Houston, Texas–based Maxxam Corporation. She lived on a six-by-eight-foot tarp-covered platform, where she endured high winds and damp cold in the winter and searing heat in the summer. Before coming to an agreement with Pacific Lumber (which owns the property where her tree stands) and leaving her post, she withstood a barrage of scare tactics from the timber company, such as flying a helicopter right above her head, burning slash

Julia Butterfly Hill (AFP/Corbis)

piles all around her tree, and cutting off her supply crew for over a week. Hill and some of her supporters have created the Circle of Life Foundation in affiliation with the nonprofit Trees Foundation, so that they can continue restoration and conservation efforts in forest ecosystems.

Julia Hill was born on February 18, 1974, in Mount Vernon, Missouri, to Dale and Kathy Hill. Her father was an itinerant preacher, and many of Hill's early memories revolve around religion. She and her two brothers were raised nomadically, traveling from town to town, living in a small camping trailer. Her parents lived by putting their faith before their own concerns, and Julia grew up learning to follow her own convictions above everything else. By the time she had reached high school age, her family had settled in Jonesboro, Arkansas, and she began to attend school for the first time.

In the summer of 1996, while working as a bartender in Fayetteville, Arkansas, she was in a nearly fatal car accident. It took ten months of intensive therapy for her to recover her short-term memory and motor skills. The experience changed her, and she promised herself never to take anything for granted again. Determined to travel and discover her purpose in life, she went on a trip with some friends in the summer of 1997 to the West Coast. Hill was awed by the giant redwoods in northern California and resolved to come back and fight to save the remaining stands of these trees from being clear-cut. She returned to Arkansas, sold all but a few of her belongings, bought camping gear, and headed back west in mid-November 1997. When she reached Arcata, California, she got in touch with Earth First! activists who were rallied at their nearby base camp—the center of activity during their campaigns to protest the logging of old-growth trees. Hill was eager to join in their civil-disobedience activities, so when an opportunity arose to participate in a tree sit, she enthusiastically volunteered.

The tree, named Luna by the activists who built the tree-sitting platform on the night of a full moon, was a 1,000-year-old redwood, stretching 200 feet tall at the top of a windy ridgeline. Located on Pacific Lumber property, it was slated to be one of the millions of trees to be cut and sold by the timber company. Pacific Lumber, once a family-owned operation that upheld a policy of long-term sustainable logging, was acquired in the mid-1980s by Maxxam Corporation, headed by chief executive officer Charles Hurwitz. In an effort to pay off the taxpayer-funded $1.6 billion bailout of his bankrupt savings and loan, Hurwitz used high-interest junk bonds to finance his acquisition of Pacific Lumber, which owned rights to a large tract of ancient redwoods in northern California. To restructure his debt, Hurwitz doubled and in some areas tripled the rate of logging, destroying huge tracts of forest, sparking lawsuits, and provoking outraged environmental protests. In December 1996, seven homes in Stafford, California, were completely destroyed by a mudslide caused by Pacific Lumber's clear-cut on the steep slope above them, which had left no vegetation to absorb the winter rains.

Hill, who adopted the "forest name" Butterfly, used safety ropes and a harness to climb Luna for her first tree sit, which lasted six days. On December 10, 1997, she and another activist returned for another tree sit, assuming they would stay up for a few weeks. By this time, Pacific Lumber was ready to begin logging the area and had started intimidation tactics—sending a professional climber up after them and cutting down trees all around them. In early January, the activist who had accompanied Hill on the tree sit went back down. She was now on her own, and for the next few months, she endured almost constant harassment from the logging company. They flew their twin-propeller helicopter right over her head, nearly knocking her out of the tree. She managed to catch the act on videotape and sent it to the Federal Aviation Administration, which threatened to revoke the helicopter operator's license if that ever happened again. Pacific Lumber then began posting security guards around the clock under the tree, cutting off Hill's supply crew. The siege lasted ten days, until the guards were pulled out owing to the severe winter storms on the ridge. The El Niño storms of that year were unusually fierce, and Hill on one occasion was battered by a 16-hour storm with winds up to 70 miles per hour and driving sleet. Later, when Pacific Lumber had cut most of the trees around her, they began burning the slash—filling the air with suffocating smoke for six straight days.

Hill accomplished more than she had originally planned when she climbed into the tree—she became an international celebrity. By the end of her two-year stay, she had a cellular phone, a pager, a radio, a walkie-talkie, a video camera, a tape recorder, and a solar-powered battery charger, all of which she used to network with the press and keep in touch with her ground crew. She did countless radio, newspaper, and magazine interviews, and even had television crews climb up the tree to film her. Using the media attention as an opportunity for outreach, Hill spoke out and wrote editorials about her protest. She became a spokesperson—getting the message out to the public about issues such as the Headwaters Forest Agreement, in which the California and U.S. governments agreed to purchase an old-growth grove in Pacific Lumber's Headwaters Forest for $480 million and turn it into a preserve. Hill and her supporters decry the plan since it leaves to the logging company tens of thousands of acres of prime old-growth habitat. She even had the clout to address the floor of the U.S. Senate by speakerphone about the issue.

Along with some of her supporters, Hill created the Circle of Life Foundation, fiscally sponsored by the nonprofit Trees Foundation, to promote the sustainability and preservation of forests. In December 1999, Hill and Pacific Lumber came to an agreement in which she and the Circle of Life Foundation would pay the company $50,000 in return for a promise not to cut Luna and to reserve a 200-foot buffer zone around the tree. Pacific Lumber pledged to donate the money to Humboldt University to be used for forestry research. For her part, Hill agreed not to repeat her tree sit and to give the company 48-hours notice before visiting Luna. On December 18, 1999, she climbed down and touched the ground for the first time in two years. Hill's decision to end her tree sit was publicly criticized by other direct-action forest activists in the area, who said they wished that when Hill decided to leave her post, she had offered them the opportunity to continue the tree sit in Luna.

Hill tours the country to publicize the plight of the redwoods. She makes her home in Humboldt County, California, where she continues to work with the Circle of Life Foundation to save old-growth forests.

Bibliography

Fortgang, Erika, "The Girl in the Tree," *Rolling Stone*, 1999; Hill, Julia Butterfly, *The Legacy of Luna*, 2000; Hornblower, Margot, "An Ecowarrior Who Calls Herself Butterfly Has Set a Tree-squatting Record," *Time*, 1998; "The Luna Tree Sit," http://www.lunatree.org.

Hoagland, Edward

(December 21, 1932–)
Writer

Edward Hoagland is an essayist, nature writer, novelist, and short story writer, best known for his free-flowing, digressive essays that focus on the interactions between humans and nature. He has written more than 100 of these essays and has published them in book form in seven collections.

Edward Morley Hoagland was born on December 21, 1932, in New York City to Warren Eugene and Helen Morley Hoagland. His father was a financial attorney who, over the course of his career, was employed by Standard Oil and the Defense and State Departments of the U.S. government. As a child, Hoagland had a severe stutter, which led him to seek entertainment and companionship in the woods near his home (his family had moved out of New York City to New Canaan, Connecticut, when he was eight years old) rather than with other children. This speech impediment continued as he grew older and had a profound influence on his writing. In the introduction to the 1993 editions of *The Courage of Turtles* and *Walking the Dead Diamond River*, two essay collections that share the same introduction, Hoagland writes, "I wasn't writing primarily for money or even fame, but simply to speak. A bad stutter had rendered me functionally mute for years, and this [writing] at last was my breakthrough."

As a youth, Hoagland was intrigued by the wild. He rode horseback for a summer in Wyoming, fought forest fires in the Santa Ana Mountains in southern California, and joined the Ringling Bros. and Barnum and Bailey Circus, where he cared for the big cats. Hoagland's first novel, *Cat Man* (1955), was based on his experiences working for the circus. Hoagland entered Harvard University in 1950, graduating in 1954 with an A.B. degree. During his time at Harvard, Hoagland studied with many prominent writers, including John Berryman and Thornton Wilder. Hoagland served in the United States Army for two years from 1955 to 1957, working as a laboratory technician in an army morgue in Phoenixville, Pennsylvania. After his discharge in 1957, Hoagland returned to New York City, where he lived on the Lower East Side and "strode the streets hungering for experience, convinced that every mile I walked the better writer I'd be," as he wrote in the introduction to the 1993 editions of *The Courage of Turtles* and *Walking the Dead Diamond River*.

Hoagland's father was in no way supportive of Hoagland's career choice. His father, in fact, wrote Houghton Mifflin in an attempt to stop the publication of *Cat Man* on the grounds that it was obscene. After his failure to convince Hoagland to pursue another career, or at least to use a pen name, Hoagland's father effectively wrote his son out of his will. This opposition from his family provided Hoagland with a strong desire to succeed as a writer, and he began publishing numerous pieces in periodicals such as *Paris Review*, the *Transatlantic Review*, and the *New American Review*. As a young writer, Hoagland published three novels—*Cat Man* (1955), *The Circle Home* (1960), and *The Peacock's Tail* (1965)—each of which received varying degrees of critical acclaim and attracted little popular success.

In the summer of 1966, after the breakup of Hoagland's first marriage, he traveled to the wilderness of northwestern British Columbia, where he hiked hundreds of miles on backcountry trails and interviewed about 80 people. They were homesteaders, Native Americans, trappers, and prospectors, and he was struck by their gaiety and consistency, which had grown out of their "long labor of opening a new country to settlement," as he states in the resultant book's introduction. The book, *Notes from the Century Before: A Journal from British Columbia* (1969), asks the simple question, "How shall we live?"; it is considered by many to be the turning point in Hoagland's career. In *Notes from the Century Before*, Hoagland searches for experiences in a wild country before that country disappears. It was his first foray into nonfiction writing, and it was extremely well received by both reviewers and the general public.

Around the time *Notes from the Century Before* was published, Hoagland began to focus on writing essays, a genre to which he found that he was perfectly suited. In the foreword to *City Tales* (1986), he wrote, "I found at the end of the 1960s that what I wanted to do most was tell my own story. . . . I discovered that the easiest way to do this was by writing directly to the reader without filtering myself through the artifices of fiction." He began publishing essays regularly in journals such as the *Village Voice*, *Harper's*, and the *New American Review*. The subjects of his essays have varied widely, ranging from the Golden Rule to EDWARD ABBEY and horseback riding with his daughter in Wyoming. Hoagland has published seven collections of his essays. His most recent collection, *Tigers and Ice*, published in 1999, contains accounts of trips to India and Antarctica. The essays, like the majority of Hoagland's, mix astute observations of the natural world with thoughtful contemplation of his own life.

Hoagland's nature essays nearly always focus on the interactions between humans and nature. To him, there is no wilderness if there is no one there to observe it. In *American Nature Writers*, David W. Teague refers to a digression on "the American brand of walking" in one of Hoagland's best-known essays, "Walking the Dead Diamond River," which is collected in the 1973 book of the same name. In this digression Hoagland examines the manner in which modern Americans traverse a landscape. Hoagland explores people's relationships with the natural world, providing suggestions on how they might be improved. In this case, he believes that people should learn to travel more gracefully in nature. Teague writes, "In Hoagland's nature essays, then, wild things provide one of his subjects, but people invariably provide the other."

Although many of his essays focus on the natural world, Hoagland is not solely a wilderness writer. He admits to loving the city and is unique among nature writers in that his attention is not held by wilderness alone, but also is focused on humans and the cities they create. He has published essays with such titles as "New York Blues," "City Walking," and "City Rat." And, on the opening page of *Notes from the Century Before*, he writes that the only relief equal to leaving the city is the relief of coming home to it. Hoagland also continues to write fiction. In 1992, he published *The Final Fate of Alligators*, a collection of previously published short stories. Hoagland has written a total of 17 books and has edited six more. In 1999, he was guest editor for Robert Atwan's *Best American Essays* series. He has received many awards for his literary accomplishments, including an O. Henry Award in 1971, a New York State Council on the Arts Award in 1972, and a National Magazine award in 1989. Today, he lives in a country house in Vermont. He is twice divorced and has a daughter, Molly, from his second marriage.

Bibliography

Hoagland, Edward, *Notes from the Century Before: A Journal from British Columbia*, 1969; Hoagland, Edward, *The Courage of Turtles*, 1993; Moritz, Charles, ed., *Current Biography Yearbook*, 1982; Teague, David W., "Edward Hoagland," *American Nature Writers*, John Elder, ed., 1996.

Hornaday, William Temple

(December 1, 1854–May 7, 1937)
Director of the New York Zoological Park, Wildlife Preservation Activist

William Temple Hornaday was an influential wildlife preservationist active early in the twentieth century. Through his post as director of the New York Zoological Park and member of the influential Boone and Crockett Club, Hornaday worked to save birds, bison, and seals from extinction. Horna-

William Temple Hornaday (Library of Congress)

day also founded the Permanent Wild Life Fund (PWLF), which he controlled exclusively. Because of his uncompromising positions and his tendency toward racism, modern environmentalists have allowed Hornaday to fall into obscurity.

William Temple Hornaday was born on December 1, 1854, on his parents' Indiana farm near the town of Plainfield. As a child, his family moved to Iowa, where he roamed woods and marshes, hunting small game with his rifle. While studying zoology at Iowa State Agricultural College, he decided to become a naturalist and taught himself taxidermy. At the end of his sophomore year, he left Iowa for Ward's National Science Establishment in Rochester, New York, a supplier of mounted specimens for museums, where he perfected his taxidermy skills. Ward's sent him on a series of collecting expeditions throughout the 1870s. In 1882, Hornaday became chief taxidermist at the U.S. National Museum in Washington, D.C. He proposed that the National Museum include exhibits with live animals, and soon this idea evolved into a plan for a National Zoological Park.

Hornaday was appointed superintendent of the emerging National Zoo in 1888, but in 1890, after fighting with its board of directors about the zoo's layout, Hornaday quit his job and moved to Buffalo, New York, to work in real estate. He continued writing about taxidermy and wildlife and was offered a job directing the New York Zoological Park (also known as the Bronx Zoo) in 1896. From this influential position, Hornaday led many early-twentieth-century wildlife preservation battles. He became known for his all-or-nothing positions, for the urgency of his battles, and for his acid pen. In his numerous articles and books, Hornaday used xenophobia and racism to argue his views. During different struggles Hornaday vilified Africans, Jews, and recent immigrants from Italy and Germany. Perhaps for this reason, modern environmentalists do not honor him as they do other, less controversial turn-of-the-century conservationists, such as THEODORE ROOSEVELT, GIFFORD PINCHOT, and GEORGE BIRD GRINNELL.

One of Hornaday's early battles was against the decline of bird populations in the United States. Both for sport and market, hunters at the turn of the century were decimating birds valued for their plumes and their flesh. Innovations in technology and transportation were giving hunters advantages they had not enjoyed previously: for example, train routes and roads that penetrated formerly inaccessible wilderness, decoys and highly trained bird dogs, automatic rifles, and new telephone lines that made it easy for hunters to tell their friends where large flocks of birds were roosting. Hornaday by this time had totally renounced hunting and did not tolerate the activity, especially when practiced as an economic activity rather than for sport. He targeted "alien hunters"—mostly Italian immigrants—and successfully lobbied for the Armstrong Fire-Arms Law of 1905, which made it illegal for aliens to carry guns in public places. Hornaday also went after gun manufacturers for their increasingly deadly weapons, but this industry responded with its own proposals for smaller bag limits, shorter hunting seasons, and a new organization called the American Game Protective Association (AGPA). Although Hornaday and the AGPA both fought for the passage of the 1913 Migratory Bird Act, which assigned the Biological Survey the task of regulating the hunting of all migratory waterfowl, Hornaday abhorred the thought of working with the arms industry. When AGPA director John Burnham was nominated for membership in the Boone and Crockett Club, an exclusive organization of big game hunters and conservationists, Hornaday lobbied hard to exclude him.

Hornaday convinced the Boone and Crockett Club to organize the American Bison Society in 1905. This society facilitated the transfer of 15 bison from the New York Zoological Society to the federal government, which used them to produce a new, protected herd at the Wichita Mountains Wildlife Refuge in Oklahoma. This successful program, which Hornaday had first proposed when he was superintendent of the National Zoo, prevented the American bison from going extinct.

Another of Hornaday's causes was the protection of the northern fur seal, whose waterproof fur coat was highly valued. After the 1868 Alaska Purchase, the U.S. government leased hunting privileges on the Pribilof Islands, where the seals bred. The companies that signed contracts with the government were prohibited from killing female seals, but other parties were allowed to hunt the seals from boats at least three miles away from the islands, and these hunters shot the seals indiscriminately. By 1909, the seals' numbers had fallen to about 150,000, and their extinction seemed imminent. Hornaday called his contacts in Congress and gave a fiery, fact-filled speech to the Senate Committee on the Conservation of Natural Resources. The Committee asked Secretary of Commerce and Labor Charles Nagel to cease the hunting contracts. Instead of halting the killing, however, Nagel had the Bureau of Fisheries carry out the 1910 hunt, killing 12,000 seals, of which 8,000 were cubs. Nagel then tried to suppress Hornaday's outraged response. Hornaday published a booklet of their correspondence, including Nagel's threats of suppression, and shocked congressmen passed a law that put a five-year stop on the seal hunt. Hornaday also successfully urged the government to sign a treaty with Great Britain, Russia, and Japan in 1911 to end their seal hunts.

Hornaday remained at the New York Zoological Gardens until his retirement in 1926. He founded the Permanent Wild Life Fund in 1913, convincing Henry Ford, Andrew Carnegie, and George Eastman to contribute to a $100,000 endowment. He made all the decisions at the PWLF, using it as a vehicle for the preservationist lobbying that he could not do with funds from the Zoo. Hornaday continued as chief administrator of the PWLF until his death on May 7, 1937, in Stamford, Connecticut. He and his wife, Josephine Chamberlain, left one daughter, Helen.

Bibliography

Fox, Stephen, *The American Conservation Movement: John Muir and His Legacy*, 1986; Graham, Frank, Jr., *Man's Dominion: The Story of Conservation in America*, 1971; Hornaday, William Temple, *Wild Life Conservation in Theory and Practice*, 1914; Stroud, Richard, *National Leaders of American Conservation*, 1985.

House, Donna

Ethnobotanist

Donna House, Dineh (Navajo) and Oneida, integrates traditional knowledge and practices with western science in an attempt to mitigate the impact of human activity on the earth. In all of her multifaceted work—which has included protecting rare and endangered species on Native land, designing landscapes for museums that celebrate Native American relationships with the land, helping develop capacity in local indigenous organizations, and working with students at Northern New Mexico Community College (NNMCC) to design a solar car—she strives to maintain the philosophy, values, and ethics of her Native heritage.

Donna Elizabeth House, who sees herself as a reflection of her ancestors, all the elements that make up the earth, the millennia of her culture, and the ideas of Native American philosopher and historian Vine Deloria Jr., prefers to present herself in the Dineh way, by first introducing her ancestors. One pair of her maternal great grandparents was forced on the 1864 Long Walk to Fort Sumner, during which thousands of Dineh people perished as they were forced to march to a prison at Fort Sumner in southern New Mexico. The other pair hid in the mountains to escape the forced march. Her maternal grandparents were Carrie and Henry Taliman, of the Dineh Towering House and Honeycomb House People clans, respectively. Her paternal grandparents met at an Indian boarding school in Phoenix; her grandfather had been abducted by the school's recruiters, then kept at the school for a decade—much longer than usual—because he was a quick learner. His Dineh name was unpronounceable in English, so a girl from the Oneida Turtle Clan, Elizabeth House, gave him her last name and called him Albert. Later, they married and settled on the edge of the Dineh reservation. Donna is the daughter of Carolyn Taliman House (Dineh Towering House) and John House (Oneida Turtle Clan). She was raised in her Dineh maternal Towering House clan, accompanying her grandparents and other members of the community as they gathered plants for food and medicine; she was included in songs, stories, and dances about the plants and animals populating their homeland. House came to respect the traditional Dineh way of being and seeing the world. She accepts and puts into practice the Dineh view of all beings on earth—plants, animals, and elements such as water and stone—as people, meriting respect just as one would pay another person.

When House entered the University of Utah, she tried to fulfill the requirements of her biology courses without violating Dineh values and ethics. When she was assigned the task of preparing an insect collection, House carefully sought out insects that had died of other causes. This innovative attitude—meeting the requirements of science while respecting traditional indigenous knowledge and practice—has allowed House to straddle two worlds and to serve them both. While studying at the University of Utah, House organized and secured funds for a four-day conference there in 1983, during which Utah state and Native government agencies explored how they could collaborate on environmental issues that concerned them both. The meeting spawned a Utah Indian Task Force to address environmental issues.

Upon graduating with a B.A. in environmental science in 1984, House accepted a position as coordinator and botanist for the Nature Conservancy's Navajo Natural Heritage Program, the first U.S. program to inventory rare and endangered species on Indian land. She directed the inventory, reviewed environmental impacts of development projects, and worked with communities to develop environmental protection strategies. House then went on to work as a botanical consultant from 1987 until early 1989, surveying federally listed threatened and endangered plant species on Native lands and collecting data on economic and cultural use of threatened flora in the U.S. Southwest and in Mexico.

House returned to the Nature Conservancy in 1989, serving first as western regional tribal lands protection planner and then in 1991 as western regional tribal lands program director. This program works to protect habitat of endangered and threatened species by combining traditional Native knowledge, ethics, values, and practices with scientific information about conservation. In addition to creating a network of indigenous organizations, scientists, and federal and state land management decision makers, House conducted plant surveys, taught Native communities about environmental protection laws, and helped them develop programs to protect and conserve biological diversity on their lands. She collaborated with a team of Utah-based botanists to produce *Utah: Threatened, Endangered, and Sensitive Plant Field Guide* in 1991, a book that has been useful to Native caretakers of these plants throughout the region. House's method for working with local Native communities is described in a 1992 *Nature Conservancy* article: "She makes her pitch at tribal meetings, weaving biological facts and cultural concerns, then waits for someone to come forward and take up the message." House has recruited such Native conservationists as Tohono O'odham (Papago) Jefford Francisco, a volunteer caretaker of Kearney's blue-star, once considered by federal officials the rarest plant in Arizona. Dineh Jane Nez looks after Navajo sedge, an endangered broomlike grass that grows mostly on the sides of cliffs. In the *Nature Conservancy* article, Nez articulates a conservation ethic that House believes is an important aspect of Native culture, that to her, the sedge represents "home. . . . It's part of you. And if [the plant] should disappear a part of you goes, too." House feels that the relationships that Native people have developed over thousands of years with their land are why Indian reservations boast the most pristine landscapes and cleanest water in the country. She believes that protecting the culture, songs, languages, prayers, traditions, and traditional lifeways of Native people is absolutely necessary for the protection of biological diversity.

In 1994, House began to consult and independently research the environmental issues and policies of indigenous peoples. House currently is working as an ethnobotanical consultant for the Smithsonian Institution's National Museum of the American Indian project. This museum, which will occupy the last remaining site on the Washington, D.C., mall, between the Capitol and the National Air and Space Museum, is slated to open in 2002. She and other museum designers have met with representatives of indigenous peoples throughout the hemisphere to gather ideas for such aspects of the museum as landscape, architecture, and exhibits. By incorporating their ideas, the museum will allow Native peoples themselves to help decide which cultural information and artifacts to share. The landscaping surrounding the museum, for which House is key designer, will be an essential aspect of the design of the museum as a whole. House told the *Los Angeles Times*, "When people come onto the land, they are going to be greeted by water and boulders—ancient boulders that are the ancient people of the Earth." What is outside—part of the Native landscape—will also be inside; the museum cafeteria will serve a wholly Native cuisine, and baskets displayed inside will be woven with grasses grown in the landscape outside.

In addition to this hemispheric-level work, House is very active locally. She volunteers at NNMCC in Española, New Mexico, assisting a group of students—mostly Hispanos and Native Americans, immersed in the region's low-rider culture—who are working to design a flawless solar-powered car. Neither House nor the students had training in engineering, yet they recruited investors in the project, researched mechanics and solar energy, networked with engineering students and other with specific fields of expertise, and built El Águila. This car was an entrant in the 1999 Sunrayce, a 1,425-mile race from Washington, D.C., to Disney World's Epcot Center, sponsored by General Motors. Although El Águila suffered a solar array battery problem and was disqualified in the first round, House's students were proud to be one of only three community college teams among a field of competitors that included prestigious private and technical universities. House was the only adviser who was a woman of color. A by-product of this project is that the participants have begun to promote the use of solar and other alternative energy sources in their community. They have done outreach at local schools and have convinced Habitat for Humanity to install solar panels on its houses. House sees this project as one that embodies her view of the world as a giant ecosystem, with each member contributing with his or her special gift.

House is a member of the U.S. Fish and Wildlife Service Recovery Team for Endangered Plant Species in Region 2. She is a past member of the American Association for the Advancement of Science's advisory board on the Collaboration for Equity: Women in Science Initiative. She chaired the Utah Endangered and Sensitive Plant Interagency Committee, has been a board member of Tohono O'odham Soil and Water District in Sell, Arizona, and has served on the board of trustees for the Wheelwright Museum of the American Indian. She was activist-in-residence at the A. E. Havens Center for Study of Social Structure and Social Change at the University of Wisconsin, Madison, in 1991 and was the Richard Thompson Memorial Lecturer (an honor dedicated to the perpetuation of moral concerns and humanistic values) at Iowa State University in 1996.

House lives on a farm on the banks of the Río Grande, in northern New Mexico.

BIBLIOGRAPHY

Haithman, Diane, "A Legacy Reclaimed," *Los Angeles Times*, 2000; Oldfield, Margery L., and Janis Alcorn, eds., *Biodiversity: Culture, Conservation, and Ecodevelopment*, 1991; Stolzenburg, William, "Sacred Peaks, Common Grounds," *Nature Conservancy*, 1992; Young, Wendy, "Navajo Woman Is 'Key Designer' for New Washington Museum," *Navajo Times*, 1999.

Huerta, Dolores

(April 10, 1930–)
Cofounder of United Farm Workers

A cofounder of the United Farm Workers (UFW) in the 1960s, Dolores Huerta has played numerous pivotal roles in organizing and representing California's farmworkers, many of them migrants, who are generally underpaid and exposed to toxic agrochemicals and who endure unhealthy living conditions. From 1965 to 1970, a UFW table grape and wine boycott brought California's powerful agribusiness interests to the negotiating table, where Huerta helped mediate the first collective bargaining agreement for farmworkers. The contract objective was to improve not only the workers' wages and job security but also community and workplace environments, including sanitation and reduction of exposure to agricultural insecticides. In the years since, Huerta has directed the UFW's political and lobbying efforts, with the use of pesticides still at the forefront of battles waged between the union and growers.

Dolores Huerta was born Dolores Fernández in the small mining town of Dawson, New Mexico, on April 10, 1930. Her father was a coal miner, field-worker, union activist, and state assemblyman. Her parents divorced when she was five, and her mother moved to the agricultural community of Stockton, California. Huerta's

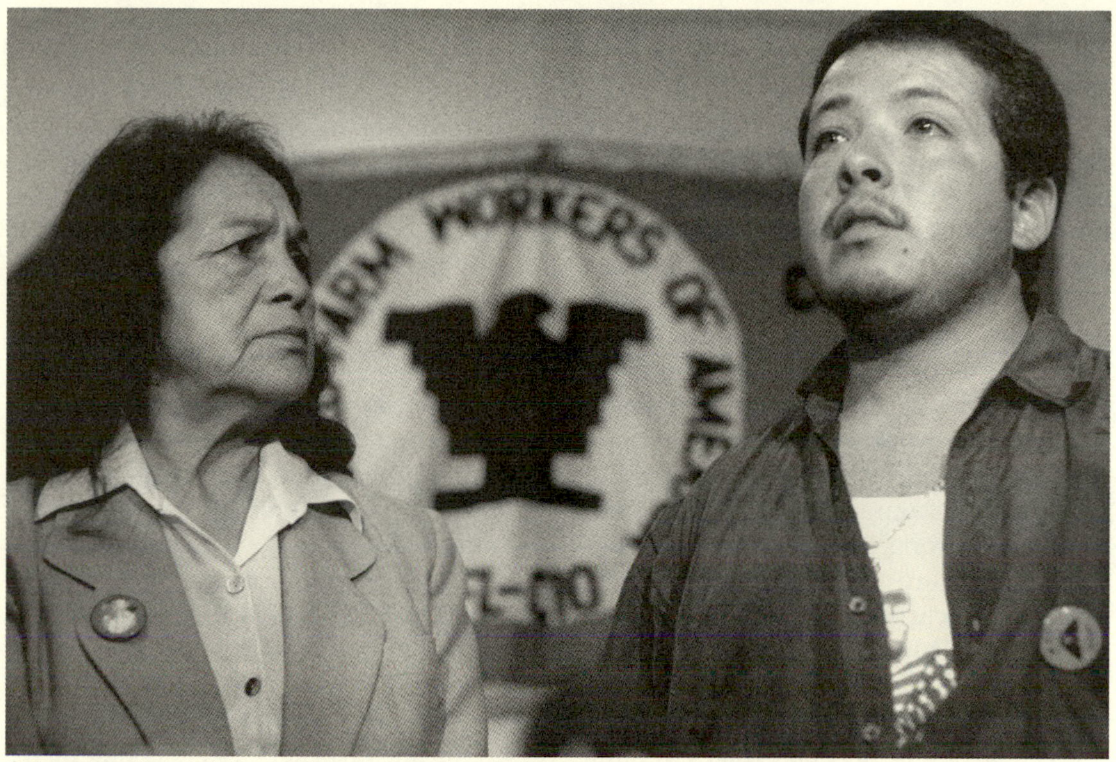

Dolores Huerta translates for Horacio Ortiz, an allegedly wrongly fired farmworker, during a news conference, 31 May 2000. (AP Photo/ Jakub Mosur)

mother supported three children as a waitress, cannery worker, and cook before buying and operating a small hotel and restaurant, where she frequently took in migrant farmworker families. That charity inspired Huerta, as did her mother's activities with women's groups providing assistance to the needy. Although she had infrequent contact with her father—who supplemented his miner's wages with migrant labor during beet harvests—Huerta was nevertheless profoundly proud of and influenced by his indignation over poor working conditions and his contributions as secretary-treasurer of a branch of the Congress of Industrial Organizations (CIO).

Huerta married a Stockton High School classmate in 1950 and gave birth to two daughters. The marriage did not last. As a single mother, Huerta worked and attended Stockton College, where she earned a teaching degree. Yet teaching frustrated her: "I couldn't stand seeing kids come to class hungry and needing shoes. I thought I could do more by organizing farm workers than by trying to teach their hungry children," she told the San Diego *Union-Tribune* in 1995.

In 1955, she met Fred Ross, an organizer with the Community Service Organization (CSO). Although initially reluctant, Huerta was persuaded to join the CSO because of its Los Angeles successes in building health clinics and registering voters. She became a founding member of the Stockton branch, and as a CSO lobbyist, she helped push through legislation that included requiring businesses to provide pensions to legal immigrants. She worked closely with CÉSAR CHÁVEZ, who became the organization's general director in 1958. By 1962, Chávez had left the CSO and founded the National Farm Workers Association (NFWA). Huerta joined him two years later, and together they chose grape pickers as their first group to organize. There had never before been a systematic effort to improve the labor and wage conditions of field-workers, who were considered too powerless to extract meaningful concessions from growers. Huerta and Chávez confronted a labor force suffering from severe underemployment, low wages, and chronic ill health, owing to pesticide poisoning and such unsanitary living conditions as open sewers and polluted drinking water.

Despite California Department of Food and Agriculture (CDFA) regulations, which extended beyond crop residue and sought to protect birds, mammals, and state waters, any concerns about worker exposure were discredited by all relevant California regulatory bodies. The CDFA did require growers to warn workers of dangers when fields had been sprayed, but it did not restrict growers from sending laborers into contaminated fields. Chemical burns, eye injuries, lung disease, systemic poisoning, cancer, and birth defects were among the deadly hazards affecting farmworkers exposed to pesticides. Most state safety efforts required that workers, rather than growers, carry the entire responsibility for their own welfare. The NFWA, which by 1966 had merged with another farmworkers union to become the United Farm Workers Organizing Committee (UFWOC), initiated lawsuits against individual growers and took an unflinching stance against pesticides.

While consumers might have little interest in the welfare of farmworkers, Huerta and Chávez understood that the purity of their food would be of great concern. In addition to mobilizing a strike against grape growers in the San Joaquin Valley, which eventually extended to the Coachella Valley of southern California, they called for a national boycott of table grapes and wine. Revealing information about pesticide dangers to the public and encouraging safety codes and regulations proved important leverage. Right-to-know laws, essential to the environmental battle, were rare in the 1960s. With health professionals, including volunteer nurse MARION MOSES, Huerta and Chávez began to amass a huge database on health and safety issues for both workers and consumers. Injuries to workers from sulphur, dichlordiphenyltrichlor (DDT), parathion, and other pesticides were drastically underreported, but so were the dangers to the public of pesticides leaking into the food chain. The UFWOC took an active stance in favor of regulations and against a policy of secrecy about the poisons employed by agroindustry.

In 1966, Huerta negotiated the UFWOC's first contract with Schenly Wine Company. She continued to direct the boycott until sales of grapes fell so drastically that on July 29, 1970, Delano-area growers capitulated en masse. The contracts, again negotiated by Huerta, provided a myriad of first-time benefits for grape workers. Union membership swelled to an all-time high of 50,000 members. In 1973, the newly renamed United Farm Workers (UFW) of America was chartered as an independent affiliate of the American Federation of Labor–Congress of Industrial Organizations (AFL-CIO).

Successive UFW strikes against lettuce producers and other agroindustries resulted in the 1975 passage of the Agricultural Labor Relations Act by the California legislature, which guaranteed the rights of California farmworkers to organize and hold elections for union representation. Huerta led the way for further major legislation, such as the removal of citizenship requirements for public assistance, disability, and unemployment benefits for farmworkers and the enactment for Aid to Dependent Children.

In a frequently cited quotation of 1970, Chávez described Huerta as "totally fearless, both mentally and physically." Through the births of nine more children, and more than 22 arrests, Huerta never rested in the fight for La Causa (the farmworker cause). In 1988, she was attacked by a police officer as she was demonstrating in front of the St. Francis Hotel in San Francisco. The incident left her with six broken bones and a ruptured spleen. As a result, the city agreed to change crowd control procedures, eliminating the presence of Special Weapons and Tactics (SWAT) teams at demonstrations. She still works long hours for the union she cofounded and in ongoing battles for social and environmental justice. When Gov. Pete Wilson pushed through legislation in 1996 for the extension of the use of methyl bromide, a deadly, odorless gas used as a pesticide on strawberries, Huerta, at age 66, spearheaded the UFW campaign to ban its use.

In addition to fighting for farmworkers' causes, Huerta has campaigned to get more women elected to public office in California. She has sat on the boards of the Feminist Majority, Latinas for Choice, the California Labor Federation, and National Farm Workers Service Center, among others. With Chávez, she helped create the Farm Workers Credit Union and the Juan De La Cruz Farm Worker Pension Fund. She has served as vice president of the Coalition of Labor Union Women and the California AFL-CIO.

In 1993, the year of Chávez's death, Huerta received the American Civil Liberties Union Roger Baldwin Medal of Liberty Award and the Eugene Debs Outstanding American Award and was inducted into the National Women's Hall of Fame.

Bibliography

Ferris, Susan, and Richard Sandoval, *The Fight in the Fields: César Chávez and the Farmworkers Movement*, 1997; Genasci, Lisa, "Dolores Huerta: She, Too, Founded UFW, but Still Toils in Chávez' Shadow," *San Diego Union Tribune*, 1995; Griswold del Castillo, Richard, and Richard Garcia, *César Chávez, A Triumph of the Spirit*, 1995; Pulido, Laura, *Environment and Economic Justice, Two Chicano Struggles in the Southwest*, 1996; Telgen, Diana, Jim Kamp, and Eva M. Neito, *Notable Hispanic American Women*, 1993; United Farm Workers, http://www.ufw.org.

I

Ickes, Harold

(March 15, 1874–February 3, 1952)
Secretary of the Interior

Harold Ickes was secretary of the interior between 1933 and 1945 under Presidents Franklin D. Roosevelt and Harry Truman. During his term, the longest ever for a secretary of the interior, the Civilian Conservation Corps (CCC) was mobilized to build campgrounds, trails, roads, and phone lines in national parks; the Soil Conservation Service combated the erosion problem that had caused the Midwest dust bowl; and huge public works projects began taking shape. Although the nation was challenged by the Great Depression and World War II, Ickes nevertheless managed to keep conservation as one of the prioritized items on President Roosevelt's list. He was a thinker ahead of his time, warning the nation that if the environment were allowed to deteriorate, huge public expenditures would be necessary to remedy the resulting problems.

Harold Ickes was born on March 15, 1874, in Frankstown Township, Pennsylvania. His father was an abusive alcoholic, and his mother died when he was 16 years old. Ickes was sent to Chicago to live with his aunt. He studied at the University of Chicago, teaching night school and scrimping on luxuries such as new clothes and using public transportation to make ends meet. He graduated with a B.A. in 1897 but did not participate in the graduation ceremony because he was embarrassed by his shabby clothes. He learned how to write and how Chicago's notorious political machine worked while reporting for the *Chicago Record*. He sided with the poor but was also intrigued by the political bosses. Once he decided to devote himself to political reform, he realized that he would be much better prepared with a law degree. So he studied for one at the University of Chicago, graduating with honors in 1907. Dedicating himself to political reform became more possible after his 1911 marriage to Anna Wilmarth Thompson, whose family's money allowed him to practice law less and politics more.

Ickes was an enthusiastic supporter of Pres. Theodore Roosevelt, but began to court reformist Democrats after Roosevelt's death. He managed the National Progressive League, which supported Franklin D. Roosevelt during the 1932 election. Roosevelt surprised the country by appointing Ickes, who had very little previous experience with the federal government, as secretary of the interior.

Immediately Ickes began to distinguish himself as an effective administrator. As secretary of the interior, he had to oversee the National Park Service, the Bureau of Indian Affairs, the Bureau of Mines, government buildings in Washington, D.C., the General Land Office, the Geological Survey, and more. Ickes made conservationist allies in these offices, depending upon their recommendations for his decisions. He concurred with his first National Park Service director, Horace Albright, who had been with the National Park Service since its 1916 inception, that national parks should be left in as natural a state as possible. His decisive leadership prevented a later National Park Service director from build-

Secretary of the Interior Harold Ickes signed the first constitution and by-laws to be issued under the Indian Reorganization Act and handed the document to delegates of the Confederated Tribes of the Flathead Indian Reservation in Montana. Looking over a war club are Chief Victor (Bear Track) Vandenberg, Chief Martin (Three Eagles) Charlo, and Ickes. (Bettmann/Corbis)

ing the infamous Skyline Drive along the crest of Great Smoky National Park. He worked closely with Bureau of Indian Affairs forestry director ROBERT MARSHALL to protect wilderness on Indian reservations from roads and development. Just weeks before the United States entered World War II, Ickes refused to allow the army to use Henry Lake, Utah, as a firing range because the soldiers' activity would threaten the endangered trumpeter swans.

But Ickes was not purely preservationist. Roosevelt's pet project was the Civilian Conservation Corps, which put over two and a half million young men to work in national forests and national parks and on Army Corps of Engineers construction projects. Under Ickes's supervision, the CCC built campgrounds, trails, roads, and phone lines in national parks. This was done primarily in order to provide jobs for young unemployed men, but also to prepare national parks for more visitors, who would contribute to the economic development of nearby towns. Ickes also oversaw another of Roosevelt's New Deal programs, the Public Works Administration. The projects that Ickes managed included the construction of Hoover Dam, the water system for Denver, Colorado, and the electrification of the Pennsylvania Railroad between Washington, D.C., and New York City.

Ickes combined his concerns for conservation and development when he founded the Soil Conservation Service and appointed soil scientist Hugh Hammond Bennett to head it. Unrestricted livestock grazing on public lands was a main culprit in the Dust Bowl. Ickes was able to address the problem thanks to a bill introduced by Colorado congressman Edward Taylor, which gave the secretary of the interior the power to establish grazing districts and sell grazing permits. This measure decreased erosion and has allowed for some recovery on federal lands.

Ickes's primary administrative goal while secretary of the interior was to consolidate all national resources within the Department of the Interior. He felt that conservation of national forests—which were managed by the Department of Agriculture—would be more successful and efficient if they were managed by the same department that took care of other natural resources, such as the national parks. His struggle, culminating unsuccessfully in 1939, earned him enemies in the U.S. Forest Service and Department of Agriculture.

Ickes never fully recovered from his failure to unite the natural resources within the Department of the Interior. After that setback and further difficulties that arose after Franklin D. Roosevelt died and left a less-conservationist Harry Truman to take over, Ickes resigned from his post.

Ickes's personality is highlighted whenever his name comes up in environmental histories. He was a scrappy, cantankerous man, who trusted no one and was the frequent victim of a paranoid imagination. His caricature-like personality amused his colleagues, however. Franklin D. Roosevelt nicknamed him "Donald Duck," and Robert Marshall recalled being more entertained than disturbed once while Ickes was berating him for some official misbehavior. Despite his crusty personality, however, Ickes was generous and liberal in his thoughts and actions. In the pre–civil rights years of the 1930s, Ickes hired African Americans for important positions in the Department of the Interior and desegregated the Department cafeteria. He believed that Indians had a right to their own culture, was in favor of their being allowed to own land communally, and was against Indian children's being taken from their families in order to attend boarding schools.

In 1945, after his resignation, Ickes moved with his second wife, Jane Dahlman, and their two small children to a farm in Maryland. Ickes continued to write for such national publications as the *New York Post* and the *New Republic*, but according to Dahlman, his lively spirit withered after he left office. He died on February 3, 1952, in Washington, D.C.

Bibliography

Glover, James M., *A Wilderness Original: The Life of Bob Marshall*, 1986; Ickes, Harold, *The Autobiography of a Curmudgeon*, 1943; Strong, Douglas H., *Dreamers & Defenders: American Conservationists*, 1988; Watkins, T. L., *Righteous Pilgrim: The Life and Times of Harold L. Ickes, 1874–1952*, 1990.

Ingram, Helen

(July 12, 1937–)
Director of the Udall Center for Studies in Public Policy, Political Scientist

Known for her public policy research, particularly in relation to resource-use issues, professor and researcher Helen Ingram has made significant contributions to the understanding of federal, state, and local policy design regarding water conservation and American Indian water rights. She has also focused closely on environmental problems along the U.S.-Mexico border region, including issues of water allocation and economic development. Ingram directed the Udall Center for Studies in Public Policy, an organization dedicated to finding new approaches to serious policy issues in the southwest, including American Indian rights, environmental conflict resolution, and U.S.-Mexico border dynamics. She also taught in the political science department at the University of Arizona in Tucson and more recently at the University of California, Irvine.

Helen Hill Ingram was born on July 12, 1937, in Denver, Colorado, to Oliver Weldon Hill, a salesman, and Hazel Margaret (Wickard) Hill, a teacher. She graduated from Oberlin College in 1959, with a B.A. in government. In 1963 she began working as assistant professor of political science at the University of New Mexico in Albuquerque, where she stayed until 1969—meanwhile earning her Ph.D. from Columbia University in 1967. She then served on the political science staff at the National Water Commission in Arlington, Virginia, for two years, before taking a position of associate professor at the University of Arizona in Tucson. During the next few years she also worked at the Institute of Government Research, first as associate director, then by 1974 as director, a post she held until 1977. In that year, Ingram accepted a two-year senior fellowship at the Washington, D.C.–based Resources for the Future. In 1979, she became professor of political science at the University of Arizona and continued to teach there until 1996.

From 1981 to 1983, Ingram served as a consultant for the Office of Technology Assessment in Washington, D.C. She had become interested in water resources in arid environments and in how water often acts as an indicator of how wealth, political power, and economic development are distributed. Early in her research and writing she mainly explored patterns of national politics and the mechanisms by which water resource projects were authorized and funded at the federal level. But her interests began shifting to state and local water policy and its impact on conservation and American Indian water rights, and she especially began to focus on the U.S.-Mexico border region. In 1988, Ingram became director of the Udall Center for Studies in Public Policy, established in 1987 through the University of Arizona. The center sponsors interdisciplinary research that informs policy making and specializes in issues involving the environment and natural resources, American Indian governance, the U.S.-Mexico border, and economic development. Up until this point, there had been virtually no binational research on border issues, and an uncomfortable relationship between the two countries added to the

Helen Ingram (courtesy of Anne Ronan Picture Library)

lack of understanding. Part of Ingram's work at the Udall Center involved attempting to ameliorate problems in the border region, such as natural resource exploitation and environmental degradation, by rethinking strategies for policy research and implementation.

Her work at the center included organizing and attending symposia for in-depth explorations of binational public policy. For example, in October 1991, Ingram attended a symposium in Guaymas, Sonora, Mexico, to help formulate a research agenda for socioeconomic and environmental issues on the U.S.-Mexico border. One of the studies later undertaken by the Udall Center analyzed the politics along the border, using the twin cities of Nogales to assess strategies for binational cooperation in managing water resources. Ingram cowrote a book summarizing the study and advocating new approaches to managing water resources for multiple use in this arid region. In the book, called *Divided Waters: Bridging the U.S.-Mexico Border* (1995), Ingram and her coauthors argue that current water management practices are unresponsive to the increasing urbanization and political dynamics in this border community. In order to deal proactively with the issue, they write, management institutions need to address a broader range of environmental issues and be open to working jointly with all levels of binational agencies. The book is a useful source for understanding environmental policy, regional development, and water management and illustrates the capacities of new environmental institutions formed by the North American Free Trade Agreement.

Among her many publications on water and natural resource policy, some of the more recent include *Policy Design for Democracy* (1997), *Public Policy for Democracy* (1993), *Water and Poverty* (1987), and *Saving Water in a Desert City* (1985). She has also contributed numerous articles to scholarly journals. Ingram left the University of Arizona in 1995 and took a position as professor of political science at the University of California at Irvine, where she continues her research and writing and teaches undergraduate and graduate courses and seminars in public administration, formation of public policy, and environmental policy.

Bibliography

Ingram, Helen, and F. Lee Brown, *Water and Poverty*, 1987; Ingram, Helen, Nancy K. Laney, and David M. Gillilan, *Divided Waters: Bridging the U.S.-Mexico Border*, 1995; Mumme, Stephen P., "Divided Waters: Bridging the U.S.-Mexico Border," *American Political Science Review*, 1996; Schneider, Anne Larason, and Helen Ingram, *Policy Design for Democracy*, 1997.

AREA DE CONSERVACION GUANACASTE
COSTA RICA
J

Jackson, Henry

(May 31, 1912–September 1, 1983)
U.S. Senator from Washington

A member of Congress for more than 40 years, Sen. Henry "Scoop" Jackson helped author a number of important pieces of environmental legislation, most notably the landmark National Environmental Policy Act of 1969 (NEPA). He also helped create the Big Cypress National Preserve in Florida, Washington's North Cascades National Park, and California's Redwood National Park. He chaired the Senate Committee on Interior and Insular Affairs and its successor, the Committee on Energy and Natural Resources, from 1963 to 1980 and played a leading role in conservation and energy legislation.

Henry Martin Jackson was born in Everett, Washington, on May 31, 1912, son of Norwegian immigrants. Although he earned a reputation for hard work as a newspaper carrier, his nickname "Scoop" had nothing to do with journalism. One of his sisters gave him the name because she thought he resembled a comic strip character known for getting others to do his work for him. Jackson graduated from Everett High School in 1930 and from the University of Washington's law school in 1935. In 1938 he ran a successful campaign for prosecuting attorney of Snohomish County. As prosecutor he gained attention for tough treatment of gamblers and bootleggers, and in 1940 he parlayed this attention into his election to the U.S. House of Representatives from Washington's Second District. During his time in the House, Jackson concentrated on military affairs and became an expert on nuclear energy. He enlisted in the military and served in World War II until he was recalled to Congress by Pres. FRANKLIN D. ROOSEVELT. Jackson accompanied the army during the liberation of Buchenwald, an experience that shaped his lifelong fight for Israel's security and the rights of Jews in the Soviet Union. While in the House, he became an opponent of the House Un-American Activities Committee and helped craft public land use policy, vital to his Western constituents.

In 1952, a year in which Republicans dominated Congress and Eisenhower was president, Jackson was one of the relatively few new Democrats elected to the Senate. He served on the Armed Services Committee and came to be known as "the Senator from Boeing" and "the hawk's hawk" because of his unwavering support for the Washington defense contractor and funding requests of the Pentagon. On other issues he was called "the liberal's liberal," backing labor rights, public education, and other social welfare programs. Jackson was an early critic of Sen. Joseph McCarthy and served on the committee that conducted the Army-McCarthy hearings. Jackson also authored the legislation that led to statehood for Alaska and Hawaii.

As a freshman senator, Jackson was appointed to the Interior Committee, where he had jurisdiction over a number of environmental issues important to his constituents, including mining, hydroelectric power development, and national parks. As he rose through the ranks of Senate seniority, Jackson was

Henry Jackson (Corbis)

dubbed, along with Sen. Warren Magnuson, one of the "Gold Dust Twins" of Washington State, able to bring home a wealth of federal dollars. Jackson pushed through a number of hydroelectric projects and was also active in several fights to preserve wilderness areas. His role in the creation of North Cascades National Park exemplifies his approach to conservation. The park in Washington was important to Jackson because he had hiked there as a boy and knew its beauty firsthand. In 1963 President Kennedy appointed a team to study the possibility of creating a park in the northern Cascades, and the team recommended a 700,000-acre site. A group of powerful interests—the Seattle power company, truckers, lumber companies, and those with mining claims—opposed the park. Jackson negotiated a compromise in which a smaller, 500,000-acre park was created. Jackson's version left a corridor in which it was still theoretically possible to mine and log, appeasing economic interests, but which made those activities nearly impossible because of the configuration of the terrain. Jackson was, first and foremost, a politician, master of the science of compromise and deal making.

Jackson's most important environmental legislation was the National Environmental Policy Act, drafted by consultant to the Senate Committee on Interior and Insular Affairs LYNTON CALDWELL, and often called the nation's environmental Magna Carta. In 1959, Sen. James Murray of Montana had introduced a bill called the Resources and Conservation Act. Although it was never passed, the bill contained several provisions later adopted by NEPA, including a declaration of conservation policy and the creation of an advisory council. Various advances in environmental policy were proposed throughout the 1960s, but it was not until 1969 that enough national momentum existed to enact sweeping legislation. In 1969, Jackson introduced S. 1075. The bill was passed out of the Interior Committee on July 9 and passed without further debate or amendment the following day. Meanwhile, similar legislation, introduced by Congressman JOHN DINGELL of Michigan, was pending in the House but was held up in procedural delays. In December 1969, a House-Senate Conference Committee meeting resolved differences between the two bills, and on January 1, 1970, President Nixon signed NEPA into law. The bill stated broad principles of the federal government's role as environmental steward, including the responsibility to "assure for all Americans safe, healthful, productive, and aesthetically and culturally pleasing surroundings" and to "enhance the quality of renewable resources and approach the maximum attainable recycling of depletable resources." One important practical result of NEPA was the creation of the Council on Environmental Quality (CEQ). Modeled after the Council of Economic Advisors, the CEQ was located within the Executive Office of the President and charged with submitting an annual report on the state of the nation's environments. A second provision required all federal agencies to submit an environmental impact statement (EIS) before proceeding with any action "significantly affecting the quality of the human environment." This requirement proved a potent weapon in the hands of environmentalists, who were able to use the courts to stop possibly destructive actions. By 1975, 119 injunctions had been issued on the basis of agencies' failing to complete the appropriate EIS. NEPA and its EIS requirements continue to direct federal policy today, and in fact some credit the act for initiating the concept of environmental assessment policy worldwide.

Jackson's environmental legacy includes a number of other legislative achievements. He was a key supporter of the Wilderness Act of 1964. He was instrumental in the establishment of the Land and Water Conservation Fund, the National Wilderness Preservation System, the Nationwide System of Trails, and the System of Wild and Scenic Rivers. He also helped enact the Youth Conservation Corps, the Surface Mining Control and Reclamation Act, and the Alaska National Interest Lands Conservation Act.

Henry Jackson died on September 1, 1983, at his home in Everett, Washington, while still serving in the Senate. In his remarks at the funeral, Robert M. Humphrey said Jackson "was an environmentalist long before it became fashionable." The Henry M. Jackson Foundation was established in 1983, guided by Jackson's interests and values. The foundation makes grants in four principal areas: international affairs, public service, human rights, and the environment. In 1987 North Cascades National Park was officially dedicated to his memory.

BIBLIOGRAPHY

Canter, Larry, and Ray Clark, *Environmental Policy and NEPA*, 1997; "Henry M. Jackson Foundation," www.hmjackson.org; Larsen, Richard, and William Prochnau, *A Certain Democrat*, 1972; Ognibene, Peter, *Scoop: The Life and Politics of Henry M. Jackson*, 1975.

Jackson, Wes

(1936–)
Plant Geneticist, Founder and Director of the Land Institute

Wes Jackson, a plant geneticist, is the founder and director of the Land Institute in Salina, Kansas, a research facility that is attempting to use modern crop-breeding techniques to produce high-yielding perennial grains. Jackson's new approach to agriculture, known as Natural Systems Agriculture (NSA), mimics the workings of natural ecosystems. A MacArthur Fellow and Pew Scholar, Jackson is recognized internationally as a leader in the field of sustainable agriculture.

Sharon Wesley Jackson was born in 1936 on a farm in the Kansas River Valley near Topeka, Kansas. He earned a B.A. in biology from Kansas Wesleyan in 1958 and an M.A. in botany from the University of Kansas in 1960. In 1967, he received his Ph.D. in genetics from North Carolina State University. Jackson taught biology at Kansas Wesleyan before moving to Sacramento, California, to accept a job at California State University. There he created the environmental studies program and served as the department chair until 1976, when he left his tenured position. Jackson and his wife, Dana (since divorced), moved back to Kansas to pursue their interest in sustainable agriculture and alternative energy. The Jacksons established the Land Institute on 370 acres located 28 miles south of Salina along the Smoky Hill River. Concerned about soil erosion and depletion, Jackson began to focus on developing a new method of agriculture that would be less detrimental to the soil. His Natural Systems Agriculture, outlined in his 1980 book *New Roots for Agriculture*, is patterned after the natural ecosystem of the prairie. Jackson proposed an agricultural system based on perennial grains, reversing 10,000 years of plant breeding focused on high-yielding annuals. Unlike conventional crops, the grains would be planted in polyculture plots containing a mix of plant species. Pests and diseases that run rampant in monoculture fields planted with one plant species would be held in check by the variety of grains growing in close proximity. The fields would not require tilling, thus reducing a major cause of soil erosion. Modern crop-breeding techniques would be used to increase the yield of such native perennial crops as wild rye, Eastern gramagrass, and Illinois bundleflower.

In 1991, the Land Institute set aside 150 acres for a ten-year research and demonstration project; the Sunshine Farm Research Program was established to explore ways to reduce the ecological costs of farming. Sunshine Farm produces livestock and conventional crops without the use of fossil fuels, irrigation, or chemicals. The project uses alternative sources of energy, including wind turbines, draft animals, and tractors powered with soy oil. Innovative management techniques are used to reduce the need for tillage, irrigation, and pest control. The annual Farm Field Day draws farmers from around the area to see demonstrations of sustainable farming techniques. Area residents also participate in the annual Prairie Festival, a celebration of the aesthetics of place featuring guest speakers, nature walks, and musical performances.

Jackson collaborated with his friend Wendell Berry on a collection of essays, *Meeting the Expectations of the Land: Essays in Sustainable Agriculture and Stewardship* (it was coedited with Bruce Colman and published in 1984). Jackson was strongly influenced by Berry's idea of stewardship evolving out of an attachment to place. Jackson explored this idea further in *Becoming Native to This Place* (1994) and *Rooted in the Land: Essays on Community and Place*, edited by William Vitek (1996), pointing out the need to integrate agriculture, ecology, and the rural economy to create what he calls a "coherent community."

Jackson is known primarily for his research on perennial grains. However, under his leadership, the Land Institute has initiated programs in other areas, particularly environmental education. Each year, the institute serves as a training center for ten graduate interns who work on the farm, attend classroom lectures and discussions, and conduct field research. Jackson's interest in the role of agriculture in the community led to the establishment of a Rural Community Studies Center in Matfield Green, a village in the Flint Hills about 100 miles southeast of Salina. Educators at the center, founded in 1999, work with local school districts to introduce ecology into the curriculum, with the goal of developing a place-based education. Scientists and historians will assist rural communities in balancing economic, social, and environmental factors so that they can become self-sustaining, retaining their young people as well as their topsoil, clean air, and water. In addition to his books, Jackson has written many scientific papers, which have appeared in such publications as *Cereal Chemistry* and *Ecology* as well as the Institute publication *The Land Report*. In 1990, he was named a Pew Conservation Scholar by the Pew Charitable Trusts. Jackson received a MacArthur "genius grant" in 1992 in recognition of his innovative ideas, far-reaching vision, and groundbreaking work. Under his direction, the Land Institute has grown to include a staff of ten, including plant breeders, an ecologist, and a plant pathologist and has attracted the attention of researchers around the world. Nevertheless, Jackson has struggled to gain credibility for his studies; harvests of the perennial crops have been disappointing, with low yields and other problems. Jackson has also been criticized for failing to address practical issues such as how to harvest multicrop fields and market unfamiliar grains to consumers.

By the mid-1990s, the inability to produce quick, conclusive results led to a drop in funding from the foundations that support the Land Institute. In response, the institute began to develop joint programs with various academic institutes in an effort to build credibility and to bring in funding. One such partnership is the "plant materials centers" project created in conjunction with Kansas State University. Together they plan to set up ten centers in various locations throughout the United States where ecologists, plant breeders, biotechnologists, and environmental historians will apply NSA to various climates and local conditions. Despite the inconclusive results of its own research, the Land Institute has greatly influenced other researchers in the field of sustainable agriculture. Promising work on perennial grains has been done at Texas A & M and Washington State University and by Jackson's daughter, Laura, a biology professor at the University of Northern Iowa.

Jackson's vision extends 25 years into the future, with a long-range plan he estimates will cost between five and seven million dollars a year. Jackson acknowledges that reversing 10,000 years of domesticated agriculture is a monumental task that will require many years' work and the support of major academic institutions, agribusiness firms, and governmental agencies. Jackson's efforts to produce perennial grains must also face the challenge of growing public concern about genetically engineered foods. Jackson believes it is possible to produce genetically improved grains without compromising the integrity of plant genomes, which he has likened to miniature ecosystems.

Jackson lives in Salina, Kansas, and takes pride in the fact that his grandchildren are the sixth generation of his family to live in Kansas.

Bibliography

Jackson, Wes, *Becoming Native to This Place*, 1994; Jackson, Wes, *New Roots for Agriculture*, 1980; "The Land Institute," www.landinstitute.org; Martin, Sam, "Defending Food: A Talk with Dr. Wes Jackson," *Mother Earth News*, 2000; Sanders, Scott Russell, "Lessons from the Land Institute," *Audubon*, 1999; Soule, Judith D., and Jon K. Piper, *Farming in Nature's Image: An Ecological Approach to Agriculture*, 1992.

Janzen, Daniel

(January 18, 1939–)
Tropical Ecologist

Biologist Daniel Janzen is as well known for his discoveries about the intricate ecology of tropical ecosystems as he is for his innovations in tropical forest conservation. A pioneer in the field of coevolution (evolutionary change by two species in response to each other), Janzen developed a field course in biology for the Organization of Tropical Studies (OTS) that is still taught today and that has helped train most North American tropical biologists currently active. Janzen's collaboration with Costa Rican scientists and government officials in establishing the Area de Conservación Guanacaste (a network of protected areas in northwestern Costa Rica) and in founding the Costa Rican National Biodiversity Institute (INBio) have made him an expert in how to make conservation succeed and attempt to pay its own way in ecologically rich but economically poor tropical countries.

Daniel Hunt Janzen Jr. was born on January 18, 1939, in Milwaukee, Wisconsin, son of an artistic mother and a father who directed the U.S. Fish and Wildlife Service. He was raised in central Minnesota, where he spent as much time as possible hunting and fishing, alone or with his father. Janzen told a *Rolling Stone* reporter that unbeknownst to him at the time, "the hunting and trapping I did as a kid—figuring out where the deer is going to be when and why—was experimental biology." Janzen began collecting

Daniel Janzen (courtesy of Anne Ronan Picture Library)

specimens as a child. He assembled a vast collection of Minnesota butterflies during preadolescence and went on his first international butterfly-collection expedition (to Mexico) at the age of 15. Janzen married his first wife while at college and supported her, their child, and his studies by trapping fur-bearing animals and hunting meat for their table. He graduated with a degree in entomology and botany in 1961 and rented a U-Haul to tote his butterfly and moth collection to the University of California at Berkeley (UCB) to study insect ecology and biological control.

In 1963, while at UCB, Janzen attended a National Science Foundation–supported course in Costa Rica. He was astounded at the richness and diversity of the remaining tropical forests of that region, far more impressive than the few remnants in Mexico, where he was doing his thesis research. While waiting to take his first job at the University of Kansas in 1965, he returned to Costa Rica and designed a course for the Organization of Tropical Studies in tropical biology for graduate students. Most North American tropical biologists currently working throughout the Americas have attended this eight-week course. Students attend a series of miniseminars in varied ecosystems throughout Costa Rica and then design their own research project. Janzen taught the course until the early 1970s, and since then he has hosted the classes as they visit his base in the Area de Conservación Guanacaste (ACG; http://acguanacaste.ac.cr).

In 1965 Janzen made a discovery about the relationship that has evolved between ants and acacia trees that revolutionized tropical ecology. During a study trip in Mexico, Janzen observed as an ant patrolling the branch of an acacia tree stung and immediately repelled a larger beetle that had alighted on the tree. Janzen discovered that when he removed the ants that lived in the enlarged thorns of the acacia, insects and vertebrates defoliated and eventually killed the tree. The tree's leaves, he told *Omni* interviewer Bill Moseley in 1993, "were like lettuce; there are no chemical defenses." The ant had evolved to provide defensive services for the acacia tree, and in return, the acacia provided room for the ants to live inside the swollen thorns and food in the form of protein-rich bodies on their leaftips and a sweet nectar oozing from glands at the leaf bases. Janzen learned that this relationship had been described a century earlier by English mining engineer Thomas Belt after a visit to Nicaragua, but until Janzen described it in his landmark 1966 paper, "Coevolution of Mutualism between Ants and Acacias in Central America," for *Evolution*, the concept of coevolution had been unfamiliar to ecologists.

In 1975, Janzen established a research base in Parque Nacional Santa Rosa, the largest remaining area of dry tropical forest in Mesoamerica. He and his biologist wife, Winnie Hallwachs, had become so concerned about the severe threats to wildlands in Costa Rica by 1985, however, that they abandoned many of their field biology studies and focused instead on conservation and restoration of dry tropical forest. Janzen devised a plan to help the forest regenerate on abandoned pastures and low-grade cropland surrounding the national park. The seeds for the forest trees are distributed mainly by the wind and by animal consumers. As agoutis, birds, bats, and monkeys travel through the forest munching and carrying fruits, they defecate or bury the seeds in good places for seedling survival. Janzen realized that if the degraded pastures surrounding the park were purchased and added to the park, and if the anthropogenic fires were stopped, the forest would regenerate. Santa Rosa was expanded and incorporated into a network of protected areas in Guanacaste Province that became the ACG, an area large enough to contain the entire dry forest ecosystem and the neighboring rain forests to which the dry forest organisms migrate seasonally and to absorb a wide variety of planned light human uses.

Although Janzen is not by nature a public person or an extrovert, he and Hallwachs became the main international spokespersons and fundraisers for the project. Their reputation grew as they were featured in such magazines as *Smithsonian* and *Rolling Stone.* People became convinced of their vision for the ACG and forest restoration, and Janzen's fame as a wildman—he was reportedly so fascinated by nature that he once allowed a warble-fly larva that had burrowed into his leg to mature and emerge just so he could watch the process—encouraged generous contributions. Most of the necessary funds were raised, and the ACG now has an endowment fund for most of its administration costs.

Janzen and Hallwachs insist that wildlands need to generate enough money to pay their own administrative costs and provide economic benefits for residents of surrounding areas, the nation, and the rest of the world. This has made Janzen and Hallwachs proponents of biodiversity prospecting—the search for compounds in nature that can be of use to industry: pesticides, foods, medicines, fragrances, and the like—and any other kind of nondamaging uses of wild biodiversity. Janzen joined with several Costa Rican scientists to found INBio (http://www.inbio.ac.cr) in 1989 to conduct an inventory of the nation's biodiversity. One by-product of the inventory was INBio's ability to facilitate bioprospecting in Costa Rica. Through its path-breaking contracts with Merck & Company, Inc., Givaudan-Roure, and other companies interested in the compounds discovered through bioprospecting, INBio contributes to the costs of conservation and provides jobs to Costa Ricans trained as paraecologists, parataxonomists, and parabioprospectors. Janzen and Hallwachs specialize in training rural inhabitants to carry out complex inventory and field research programs.

The responsibility for the success of Janzen's and Hallwachs's projects can by no means be attributed solely to them, but they deserve credit for their creative thinking, their insistence that conservation will not succeed unless the humans responsible for it recognize its benefits for them, their tenacity for seeing these projects through, and their willingness to serve as spokespersons abroad. Janzen has received numerous awards and recognitions for his work, including the 1984 Crafoord Prize (considered an equivalent of the Nobel Prize for ecology), a 1989 MacArthur grant, and the 1997 Kyoto prize in basic science. Janzen and Hallwachs have donated all of these monetary awards to conservation projects in the ACG.

Janzen and Hallwachs split their time between the University of Pennsylvania and their cinder-block lab in Sector Santa Rosa of the ACG.

BIBLIOGRAPHY

"Daniel Janzen," http://janzen.sas.upenn.edu; Gallagher, Winifred, "Recall of the Wild," *Rolling Stone*, 1988; Janzen, Daniel, "Gardenification of Wildland Nature and the Human Footprint," *Science*, 1998; Janzen, D. H., "How to Grow a Wildland: The Gardenification of Nature," *Nature and Human Society: The Quest for a Sustainable World*, Peter Raven and T. Williams, eds., 1999; Janzen, D. H., W. Hallwachs, J. Jiménez, and R. Gámez, "The Role of the Parataxonomists, Inventory Managers, and Taxonomists in Costa Rica's National Biodiversity Inventory," *Biodiversity Prospecting*, W. V. Reid et al, eds., 1993; Janzen, D. H., ed., *Costa Rican Natural History*, 1983.

Johnson, Glenn S.

(March 24, 1962–)
Sociologist

Focusing on issues of environmental justice, and specifically on the impact on communities of color of transportation, urban sprawl, and land use policy, Glenn S. Johnson is an assistant professor of sociology at Clark Atlanta University and a research associate at its Environmental Justice Resource Center (EJRC). Johnson has coedited two recent books in the environmental justice field, *Just Transportation* (1997) and *Sprawl City* (2000), and has cocompiled the *Environmental Justice Curriculum Resource Guidebook*, a collection of environmental justice–oriented curricula used by professors at universities throughout the country.

Glenn Steve Johnson was born on March 24, 1962, in Memphis, Tennessee. He attended the University of Tennessee, earning two B.A. degrees (in academic psychology in 1986 and in sociology in 1987), an M.A. in 1991, and a Ph.D. in 1996, both in sociology. For his M.A., he wrote a thesis entitled "What Are Cities Doing with Their Garbage: A Case Study on the Decision-making Process of Solid Waste Disposal of Knoxville, Tennessee." His Ph.D. dissertation further explored the problem of solid waste disposal; it was entitled "Garbage Disposal, A Case Study of the Impact of Environmental Racism on Landfills: The North Hollywood Dump in Memphis, Tennessee." These projects were inspired by the work of Johnson's mentor, sociology professor Robert Bullard, whose research on racial discrimination by solid waste companies led to the first environmentally oriented lawsuit filed under the 1964 Civil Rights Act. Bullard spent the academic year 1987–1988 teaching at the University of Tennessee, during which he and Johnson met, and Johnson became Bullard's research assistant for Bullard's landmark 1990 book, *Dumping in Dixie: Race, Class and Environmental Quality.*

As a graduate student, Johnson was already becoming an active participant in the emerging subfield of environmental sociology that analyzed the environmental justice movement, hosting symposia and presenting papers at numerous sociology conferences starting in 1990. He taught introductory courses in sociology and social problems at the University of Tennessee and classes on race and ethnicity at Knoxville College and worked for a Knoxville-based youth program during 1990 and 1991. Once he had been awarded his Ph.D., Johnson was offered an assistant professorship in sociology at Clark Atlanta University and a position as research associate at the Environmental Justice Resource Center, which Bullard founded in 1994 and still directs.

The EJRC at Clark Atlanta University is a comprehensive, university-based research center whose mission includes education, training, research, and the storage, retrieval, and dissemination of information. EJRC serves as a national clearinghouse, repository, and archive of the largest collection of environmental justice materials in the world (books, reports, monographs, proceedings, photographs, slides, videos, audiotapes, and so on). Its three major research areas include environmental justice, transportation equity, and suburban sprawl/smart growth. EJRC's four major objectives include (1) increasing the quality and quantity of environmental, health, landuse, economic development, transportation equity, and smart growth information available to community-based organizations; (2) linking via the Internet the various community-based organizations, networks, and minority leaders working on environmental justice, transportation equity, smart growth, and health/sustainable communities; (3) expanding the focus, stakeholder interaction, and agenda sharing among community-based and national groups; and (4) assisting community stakeholders in implementing environmental, health, energy, transportation equity, sustainability, and smart growth benchmarks they have set for themselves.

One of Johnson's initial projects at the EJRC was to compile the *Environmental Justice Curriculum Resource Guidebook* (with collaborators Robert Bullard and Chad Johnson). This book catalogs the variety of environmental justice–oriented courses offered at colleges and universities throughout the country, among them courses from the disciplines of sociology, urban planning, law, health, natural resources, and geography. It was first published in 1997, with an update in 2000.

In addition to coediting two special environmental justice issues of the *Race, Gender, and Class* journal of the Race, Gender, and Class Section of the American Sociological Society, Part I in 1997 and Part II in 1998, Johnson has coedited two scholarly books. *Just Transportation: Dismantling Race and Class Barriers to Mobility* (1997), coedited with Robert Bullard, reveals that the wealthy, educated classes receive disproportionately greater transportation benefits than people of color and others at the lower end of the socioeconomic spectrum. In many large cities throughout the United States, new highways connecting new, predominantly White, middle- and upper-class suburbs, take precedence over improving mass transit systems that serve the mostly poor residents of color who live in inner cities and rely on public transportation. Not only is this an issue of equity, it is also one of environmental health. More highways mean more use of automobiles, which results in more pollution, and the poor are generally more seriously afflicted by asthma and other conditions exacerbated by pollution. Contributors to *Just Transportation* detail how transportation problems have affected communities throughout the country, from Harlem to New Orleans, to southern California, to Atlanta and how activists have responded.

Johnson, together with Bullard and EJRC geographical information system specialist Angel Torres, next addressed the issue of urban sprawl with *Sprawl City: Race, Politics, and Planning in Atlanta* (2000). Focusing on sprawl from the dual perspectives of environmental justice and civil rights, this book examines how racial and class divisions are heightened through random, unplanned suburban growth. African Americans, in particular, are prevented from escaping increasingly unbearable inner cities by such obstacles as restrictive zoning practices, inadequate public transportation, and discrimination by real estate brokers, banks, and mortgage companies. The eight suburban counties surrounding Atlanta, which are growing as fast as 6.6 percent per year, provide an illustrative showcase for the environmental and equity problems caused by such rapid growth. The region has become economically and racially polarized; traffic is thick on the highways yet public transportation in the suburbs is sparse; air quality is so bad that the Environmental Protection Agency has classified Atlanta as a "nonattainment area"; each day some 50 acres of forest are leveled and converted into new development. *Sprawl City* contributors are local experts in the fields of law, urban planning, economy, education, and health care.

Johnson is a member of numerous scholarly societies, including the American Sociological Association, Association of Black Sociologists, Southern Sociological Society, Society for International Development, and the Society for the Study of Social Problems. He resides in Stone Mountain, Georgia.

BIBLIOGRAPHY

Bullard, Robert D., Glenn S. Johnson, and Angel O. Torres, "Atlanta Megasprawl," *Forum for Applied Research and Public Policy*, 1999; Bullard, Robert D., and Glenn S. Johnson, eds., *Just Transportation: Dismantling Race and Class Barriers to Mobility*, 1997; Bullard, Robert D., Glenn S. Johnson, and Angel O. Torres, eds., *Sprawl City: Race, Politics, and Planning in Atlanta*, 2000.

Johnson, Hazel

(January 25, 1935–)
Founder and Executive Director of People for Community Recovery

Hazel Johnson has been called the mother of the environmental justice movement for her three decades of activism to clean up the highly contaminated "toxic doughnut" surrounding her home in the Altgeld Gardens housing development of southeast Chicago. After her husband died suddenly in 1969 of lung cancer and Johnson discovered that many of her neighbors also suffered from cancer, birth deformities, premature deaths, rashes, eye irritation, and respiratory illnesses, Johnson began lobbying for change. She founded People for Community Recovery (PCR) in 1978 and has pushed city, state, and federal authorities to study and correct the area's environmental contamination and resulting health problems. Although by no means have all of the problems of Altgeld Gardens been solved, Johnson and other PCR activists have enjoyed some success: The neighborhood has been hooked up to the municipal water supply because its wells were contaminated; the group successfully demonstrated against the expansion of one landfill and a moratorium on further landfills has been called; and PCR worked with Greenpeace to prevent the construction of an incinerator nearby. For her courage, persistence, and accomplishments, Johnson was presented with the President's Conservation and Challenge Award for Communication and Education in 1992.

Hazel Washington Johnson was born in New Orleans on January 25, 1935; of her parents' four children, she was the only one who survived. Even as a young girl she was opinionated; she shocked her family by declaring that churchgoers were hypocritical. By the time she was 12, her parents had died, and she moved to Los Angeles to live with an aunt for five years. She returned to New Orleans, where she married John Johnson. In the course of their marriage she had seven children, two of whom, Cheryl and Valerie, help in her work now. In 1956 the Johnsons moved to Chicago, and in 1962 they moved into a new complex in the southeast part of the city, Altgeld Gardens, owned by the Chicago Housing Authority. Unbeknownst to her at the time she moved there, the 17-block development, whose residents were all African Americans, had been constructed on top of the old Pullman Railroad dump, in an area of the city that had been used for more than a century as the city's industrial dumping grounds.

In 1969 her husband began to suffer extreme back pain, and within 10 weeks he was dead, a victim of what doctors said was lung cancer. Dumbfounded, Johnson began to listen as her neighbors spoke of cancers, rashes, miscarriages, and hacking coughs in children. After Johnson saw a television report on the high incidence of cancer in South Chicago, she began to suspect that it might be caused by the area's severe contamination. Altgeld Gardens was encircled by what Johnson eventually dubbed "a toxic doughnut": a steel mill, sewage treatment plants, landfills, abandoned factories, chemical companies, and incinerators. Johnson attended a meeting sponsored by the Illinois Environmental Protection Agency (IEPA) on problems in the area, and after she spoke up about the problems she had noticed, she was sent 12 copies of a health evaluation questionnaire to take to some of her neighbors. She made more than 1,000 copies and organized concerned friends and neighbors to canvass the entire neighborhood. The canvassers found that approximately 90 percent of those they surveyed suffered from symptoms that could have been caused by air pollution. However, the IEPA refused to consider the data because it had not been gathered in a sufficiently professional manner. The city of Chicago then conducted a study that showed that African Americans living on Chicago's south side suffered high rates of cancer, but the city did not compare that rate with those of African Americans living elsewhere nor did it look at possible causes of the cancer.

Johnson founded People for Community Recovery in 1978, with an initial goal of lobbying the Chicago Housing Authority for tenant rights. PCR worked to assure that no apartment would be rented unless it was lead free and to have asbestos removed from the buildings as well. In the face of all of the health problems that were surfacing, and the likelihood that they were caused by air and groundwater pollution, PCR soon began to work on larger environmental issues. Because no satisfactory health study had been completed, PCR and the University of Illinois School of Public Health collaborated in 1993 on another survey that had two particularly significant findings: Fifty-one percent of pregnancies involved some abnormalities, and 66 percent of those who experienced symptoms while living at Altgeld Gardens reported relief from them upon leaving the area.

Although the local, state, and federal governments' environmental and health agencies have not come through for the residents of Altgeld Gardens to draw definitive connections between the environmental hazards and residents' health problems, and then to clean up the area, PCR activists were successful in drawing media attention to the problems and in solving some problems by lobbying politicians and through grassroots protest politics. One of residents' main worries was that their water supply—well water—was contaminated. PCR lobbied for and eventually obtained a hookup for Maryland Manor to the municipal water line. Johnson and other PCR activists, along with Greenpeace, organized effective protests to stop the expansion of a local waste management landfill and to prevent the construction of a chemical waste incinerator nearby. She has been arrested twice during these protests. In these struggles, PCR's predominantly African American activists were able to form bonds with predominantly White groups against their common enemies: polluters and the government agencies that do not enforce laws against environmental contamination. Despite the fact that environmental racism has damned 60 percent of African Americans and Latinos to live in areas contaminated by toxic waste (according to the 1987 United Church of Christ Commission on Racial Justice report, *Toxic Wastes and Race in the United States)* and that a study by the *National Law Journal* in 1992 showed that from 1985 to 1991 the [EPA] fined violators of the Clean Water Act from 500 to 1,000 percent more if they polluted White communities than if they polluted communities of color, Johnson understands the necessity of collaboration. She told Josh Getlin of the *Los Angeles Times*, "We all breathe the same air, and everybody has to live on this planet. So if the air is lousy where I live, sooner or later it's gonna get to your home too."

In 1992 Johnson was one of 13 African American delegates to the United Nations Conference on the Environment in Rio de Janeiro. In that same year she received the President's Environmental & Conservation Medal given by then President Bush. In 1995 she was asked to join the EPA's Common Sense Initiative (CSI) which, teamed with polluting industries, was looking for opportunities to write regulation that would emphasize prevention rather than cleanup. In 1997, Johnson helped arrange one of the many "toxic tours" she has conducted since the 1970s; she arranged for 20 members of the Automobile Manufacturing Subcommittee of the CSI to see nine sites around southeast Chicago from an environmental justice perspective. The CSI was disbanded in December 1998, though not without some success in an attempt to balance the burden of reporting required of industry, on the one hand, and the community's right to know, on the other. Johnson sat on the National Environmental Justice Advisory Council that oversaw the siting of landfills and incinerators.

PCR continues to operate out of a storefront in Altgeld Gardens. In addition to its continued work for environmental justice, it offers the 10,000, primarily African American residents of Altgeld Gardens whatever help it can: job training, human immunodeficiency virus (HIV) prevention, advocacy in the courts. Johnson still resides in Altgeld Gardens.

Bibliography

Ayres, William, Jean Ann Hunt, and Therese Quinn, *Teaching for Social Justice*, 1998; "Environmental Justice Resource Center: Unsung Heroes and Sheroes on the Front Line for Environmental Justice," http://www.ejrc.cau.edu/(s)heros.html; Ervin, Mike, "The Toxic Doughnut," *The Progressive*, 1992; Getlin, Josh, "Fighting Her Good Fight," *Los Angeles Times*, 1993; Greenpeace, *Rush to Burn* (video documentary), 1989; Miller, Stuart, "Green at the Grassroots: Women Form the Frontlines of Environmental Activism," *E Magazine*, 1997; Taylor, Dorceta E., "Environmental Justice: The Birth of a Movement," *Dollars & Sense*, 1996.

Johnson, Lady Bird (Claudia Alta)

(December 22, 1912–)
First Lady, Conservation Activist

Wife of Pres. Lyndon Baines Johnson, Lady Bird Johnson pushed for a variety of reforms as part of her First Lady's campaign for "beautification." Her efforts resulted in the 1965 Highway Beautification Act, which controlled the proliferation of billboards and the spread of roadside junkyards. In 1982 she founded the National Wildflower Research Center, now known as the Lady Bird Johnson Wildflower Center, to "educate people about the environmental necessity, economic value, and natural beauty of native plants."

Claudia Alta Taylor was born on December 22, 1912, in Karnack, Texas, a small town in the eastern part of the state. Taylor had an affluent upbringing; her mother was from a wealthy Alabama family, and her father was a successful businessman. She acquired the nickname "Lady Bird" from a family cook, who said she was as pretty as the ladybird beetle common in the area. Johnson's mother died when Claudia was five, and she was raised thereafter by her father and her mother's sister. As a child Taylor spent a great deal of time alone in the outdoors and later attributed her interest in conservation to a love of the land and natural beauty acquired in childhood. After graduating from high school at age 15, Taylor spent two years boarding at St. Mary's School for Girls in Dallas, before enrolling at the University of Texas. In 1933 she earned a B.A. with honors, which she followed in 1934 with a degree in journalism, also received with honors.

She met Lyndon Johnson in August 1934, and they were married later that year. Her husband's political career advanced swiftly, with his election to the House of Representatives in 1937, and Mrs. Johnson became closely involved in his political life. While he was serving in the Pacific during 1942, she ran his Washington, D.C., office and learned firsthand the nuts and bolts of American politics. In 1943 the Johnsons bought radio station KTBC in Austin, Texas, the beginning of what would be extensive media holdings, managed in large part by Lady Bird. Lyndon Johnson was elected to the Senate in 1948, and after two unsuccessful campaigns for the Democratic nomination to the presidency, he became Pres. John Kennedy's vice president in 1960.

When President Kennedy was assassinated in 1963, and Lyndon Johnson as-

Claudia Alta (Lady Bird) Johnson and Secretary of the Interior Stewart Udall hike along the lost mine trail in the Chisos Mountains, accompanied by Douglas Evans, Park Service guide and park naturalist, plus reporters and photographers. (Bettmann/Corbis)

sumed the presidency, Lady Bird Johnson inherited the office of First Lady from the enormously popular Jackie Kennedy. Mrs. Kennedy had used the attention she attracted to push for a well-regarded, historically attuned renovation of the White House, and in the weeks following the assassination, Lady Bird began to feel a responsibility to make a contribution of her own. In the summer of 1964 she made a tour of national parks and Indian reservations in the West. STEWART UDALL, the secretary of the interior, accompanied Mrs. Johnson on this tour, and their discussions and shared appreciation of the beauty of the West contributed to her growing commitment to conservation and beautification. Udall said later that Lady Bird played a pivotal role in galvanizing the president's commitment to preserving natural beauty, a commitment evident in the formation in 1964 of the Task Force on Natural Beauty. When the Task Force made its recommendations in November of that year, after Johnson was elected to another term as president, it focused on preservation of landscape and open space, highway beautification, and creation of and rehabilitation of the nation's urban parks. All of these became important parts of Lady Bird Johnson's agenda: the beautification of America. In Lewis Gould's *Lady Bird Johnson and the Environment*, she is quoted as saying, "'Getting on the subject of beautification is like picking up a tangled skein of wool—all the threads are interwoven, recreation and pollution and mental health, and the crime rate, and rapid transit, and highway beautification, and the war on poverty, and parks—national, state, and local.'" The push for the preservation of national beauty was an integral part of the Johnsons' Great Society vision, an ambitious effort to improve life for all Americans.

One of her first initiatives was the beautification of Washington, D.C., which she hoped could serve as an example for the rest of the nation. A Committee for a More Beautiful Capital was established in early 1965. She spearheaded efforts to clean up and organize neighborhoods, create and renovate parks, plant flowers in public places, and refurbish the Mall, historic sites, and monuments. Mrs. Johnson used her background in journalism to attract favorable publicity for her campaign, which was focused both on the tourist's Washington—monuments, parks, museums—and inner city Washington, which suffered from all the problems of poverty and racism. She was involved in projects that included the designation of Pennsylvania Avenue as a national historic site and Project Pride, which assisted residents of the Shaw neighborhood, one of Washington's poorest, with urban renewal and education programs. Project Pride involved 250 neighborhood children.

Mrs. Johnson faced many obstacles in completing her ambitious slate of projects, including a constant search for private funds and a Congress reluctant to spend money on what was often perceived as frivolous, "feminine" interest in beauty. Congressional opposition was even more intense in the campaign to pass the Highway Beautification Act, the achievement for which Lady Bird Johnson is best known. The billboard lobby fought the act vigorously and found a receptive audience in Congress. Many members of Congress traditionally waged their campaign battles on roadside billboards. Both Johnsons lobbied hard for passage of the act throughout 1965, and with some compromises to the wishes of the Outdoor Advertising Association of America, the bill was signed into law in October of that year. Highway beautification continued to be a focus for Mrs. Johnson throughout the rest of her term as First Lady, along with an increased involvement in creating new national parks. She was particularly active in campaigns to protect the Grand Canyon and establish a national park to preserve California redwoods. Some of the conservationist legislation passed in the final year of Johnson's term in office included setting up the National Wild and Scenic Rivers System and the creation of North Cascades and Redwoods National Parks.

After they left the White House, the Johnsons returned to Texas, where Mrs. Johnson continued her conservation work at the state and local level. She presided over Texas's highway beautification projects into the 1980s. She was active in Austin conservation efforts, including beautifying the riverfront and developing a system of parks and open space. Throughout her career she was particularly drawn to projects involving plant life, and in 1982 she gave $125,000 and 60 acres of land east of Austin to establish the National Wildflower Research Center, which has worked to preserve and promote the conservation of native wildflowers. Lyndon Johnson died in 1973. Lady Bird Johnson resides in Texas and has two daughters and seven grandchildren. She serves on the board of the National Geographic Society as a trustee emeritus.

Bibliography

Gould, Lewis, *Lady Bird Johnson and the Environment*, 1988; Johnson, Claudia, *A White House Diary*, 1970; "Lady Bird Johnson Wildflower Center," http://www.wildflower.org; "National First Ladies Library," http://www.firstladies.org.

Johnson, Robert Underwood

(January 12, 1853–October 14, 1937)
Editor

Editor of the influential national literary magazine *Century*, Robert Underwood Johnson collaborated closely with pioneer conservationist John Muir. Johnson helped Muir revive his literary career in 1889 and worked closely with him on establishing Yosemite National Park and fighting the Hetch Hetchy dam. Without the boost and friendship of Johnson, Muir might not be known as the wilderness preservation icon he is today.

Robert Underwood Johnson was born in Washington, D.C., on January 12, 1853. He was raised in Centreville, Indiana, and attended Earlham College in Richmond, Indiana. Upon graduation in 1871, Johnson moved to Chicago and worked as a clerk at the textbook publisher Charles Scribner's Sons. Two years later, Scribner's transferred him to its New York City office, where he began working as an editor at *Scribner's Monthly*, which later became the famed *Century*.

Century was the nation's most influential magazine of the late 1800s. It published new work by the nation's top novelists and took reformist editorial positions on such issues of the day as the improvement of tenement housing and corporate control of politics. Its readership consisted of 200,000 well-educated, affluent subscribers, people who wielded influence in their communities. Johnson prepared popular series for the magazine, most notably a collection of memoirs of Civil War combatants from both sides. Somewhat of an elitist, Johnson felt that by working for *Century*, he was contributing to a growing sophistication of tastes and morality of the century-old United States. Johnson rose through the ranks of the magazine, becoming associate editor under editor Richard Gilder in 1881.

During the 1870s, *Century* had published several articles by John Muir. Muir stopped writing in the 1880s once he became occupied with the day-to-day life on his Martinez, California, farm. Johnson wrote to Muir a few times during the early 1880s, urging him to take up the pen again, but Muir demurred. In 1889, Johnson traveled to California to research a historical series on the California pioneers. He contacted Muir, they met in person, and a mutual affinity bloomed almost instantly.

Stephen Fox, in his *The American Conservation Movement: John Muir and*

Robert Underwood Johnson (right) and several notable scientists help Auguste Piccard celebrate his forty-ninth birthday. (Bettmann/Corbis)

His Legacy, claims that each man had an ulterior motive for their friendship. Johnson's was to pump Muir for material for *Century*. Muir's was to recruit Johnson to work for the preservation of the Yosemite Valley he so loved. As soon as they met, Muir proposed a hiking trip through Yosemite. Johnson, who was not an outdoorsman yet, was curious to visit Yosemite and agreed. They spent several days in early summer trekking through remote areas that left Johnson breathless with admiration. Muir told Johnson of his worries that Yosemite, owned by the state of California, would be ruined by the unregulated grazing and tree cutting that was already degrading the area. Johnson said he thought the two should work to give Yosemite inviolable status as a national park, just like Yellowstone had gained in 1872. Muir agreed to write two Yosemite *Century* articles if Johnson would propose the idea to his contacts in Washington, D.C.

In March 1890, California congressman William Vandever introduced a bill to convert some 288 square miles of Yosemite into a national park, but Muir felt that 1,500 square miles deserved protection. While Muir was writing his *Century* articles, Johnson testified in favor of the larger park size at the House of Representatives' Commission on Public Lands. Muir's articles appeared in the August and September issues of *Century;* a bill to establish the 1,500-square-mile Yosemite National Park was passed by Congress on September 30, 1890, and signed into law by the president the next day.

Pleased by their success, Muir nonetheless worried about the Yosemite's safety. At this time there was no legislation and no national park service to protect national parks from the "utilitarians," as Muir called stockmen, lumbermen, and those who believed in the right to use public lands for economic gain. Again Johnson was to propose a prophetic idea: that they found a "defense association" for Yosemite and Yellowstone. Muir, who was not comfortable in most social situations, was intimidated by the thought of establishing a club. But by 1892, a group of California wilderness proponents formed the Sierra Club to keep the utilitarians out and facilitate the wilderness lovers' exploration of the park. John Muir was persuaded to serve as the Sierra Club's first president; he led annual wilderness trips and enjoyed the opportunity to tell his stories to a receptive crowd.

Johnson persistently urged Muir to continue writing throughout the 1890s. *Century* published Muir's first book, *The Mountains of California*, in 1894, and Muir's famous story about an adventure with his valiant dog, Stickeen. But after Johnson edited down "Stickeen" more than Muir wished, held one article for three years, and rejected another, Muir jumped ship and signed on with Boston publisher Houghton Mifflin. Johnson's and Muir's friendship survived this painful professional split, however, and they collaborated several years later on the fight against the Hetch Hetchy dam.

Hetch Hetchy was a narrow, deep valley carved by the Tuolumne River in northwestern Yosemite. Muir and Johnson agreed that this was one of the most beautiful, wild areas of the park. But to city officials of San Francisco, it seemed an ideal place to build a dam to supply San Francisco with needed water and hydroelectric power. There were several attempts in the first few years of the century to obtain federal permission to dam Hetch Hetchy, but they all failed, due to Interior Secretary Ethan Hitchcock's refusal to permit development in national parks. Once Hitchcock was replaced by an ally of the utilitarian Forest Service director GIFFORD PINCHOT, however, the Hetch Hetchy dam project rolled forward. Muir and Johnson led a national protest against Hetch Hetchy for seven years. They railed against it with their most energetic prose, in articles for *Century*, of which Johnson became editor in 1909; newspaper editorials; and privately published pamphlets. Johnson saw the struggle as one between spirituality and commercialism; he and Muir both referred to the Hetch Hetchy valley as a creation of God that it would be sacrilegious to destroy. Yet the utilitarians prevailed, claiming that if beauty were the criterion, the new reservoir created by the dam would be beautiful too. In December 1913, Pres. Woodrow Wilson signed a bill to grant the Hetch Hetchy valley to San Francisco's water utility for dam construction.

Both Muir and Johnson suffered deep personal setbacks after their defeat. Muir died one year later. Johnson left *Century* during the Hetch Hetchy fight, owing to conflicts with the magazine's business management department. He dedicated the remainder of his career to poetry. During World War I, he organized The American Poets' Ambulances in Italy, a fund for ambulances and field hospitals in Italy. Following the war he was appointed U.S. ambassador to Italy by Pres. Wilson. Although Johnson no longer remained on the forefront of wilderness conservation struggles, he retained his membership in the American Forestry Association, until he resigned in protest when the association came under the control of the lumber industry in the mid-1920s.

Johnson died on October 14, 1937, in New York City. He was married to Katherine McMahon and left two children, Owen and Agness.

BIBLIOGRAPHY

Fox, Stephen, *The American Conservation Movement: John Muir and His Legacy*, 1981; Nash, Roderick, *Wilderness and the American Mind*, rev. ed., 1973; Stroud, Richard, *National Leaders of American Conservation*, 1985.

Jontz, Jim

(December 18, 1951–)

U.S. Representative from Indiana, Executive Director of American Lands Alliance, Forest Activist

During his three terms in Congress as a representative from Indiana, Jim Jontz was most noted for his opposition to logging in old-growth forests in the Pacific Northwest. First elected to Congress in 1986, Jontz was at the center of the controversy surrounding protection for the spotted owl. He led resistance to resumption of logging after it was halted by concern for the owl and fought for tougher regulation on timber sales in national forests. Defeated in his bid for reelection in 1992, Jontz went on to lead several environmental lobbying and activist groups, including Audubon's Endangered Species Coalition and American Lands Alliance.

James Jontz was born on December 18, 1951, in Indianapolis, Indiana. He graduated from Indianapolis's North Central High in 1970 and from Indiana University in Bloomington in 1973. He was elected to the Indiana House of Representatives in 1974, as a Democrat in a traditionally Republican district. Jontz served in the House until 1984 and earned a reputation for hard work and support for working people. Jontz was called a "full-time legislator" in a state where legislators are expected to work only part of the year. While in the House he was particularly active in oversight of the public utilities; he sponsored legislation that encouraged conservation and worked to prevent the state power company from passing costs of a failed nuclear plant back to consumers. Jontz also actively advocated reducing medical costs for the poor and elderly and fought for legislation benefiting women. In 1984 he was elected to the Indiana Senate, again surprising observers by winning in a Republican stronghold.

Jontz left the Indiana Senate in 1986, when he was elected as U.S. representative from the 5th District. Jontz earned a reputation as a critic of monoculture practices, a dangerous stance for someone from a farm state, but he gained the trust of farmers by voicing their issues. Jontz demonstrated that it was possible to simultaneously support the environment and farmers. He argued against the overreliance on pesticides because of the danger to farmers, whose health and land were most directly impacted by exposure to pesticides and their runoff. Jontz was instrumental in passing an amendment to the 1990 Farm Bill that protected farmers' relief benefits from counting against their ability to qualify for other assistance programs such as food stamps. In addition to his work on agricultural issues, Jontz also played a key role in fighting the nuclear power industry by blocking a 1992 provision that would have allowed the industry to pass cleanup costs on to taxpayers. He also worked to protect Indiana's Salt Creek Corridor, a key wetlands habitat.

Jontz's time in Congress was defined by his work to preserve forests. He was especially concerned about private timber sales on public land. The U.S. Forest Service spent millions of dollars every year building roads and marking boundaries so that private logging firms could cut lumber on Forest Service land. Jontz argued that this amounted to a public subsidy of the timber industry, at a habitual loss to taxpayers. He introduced legislation to cut below-cost timber sales and federal outlays for lumber-harvesting support. Jontz received national attention for his efforts to stop logging in old-growth forests in the Northwest, particularly during the 1989–1990 battle over the spotted owl. In 1989, a federal court issued an injunction that blocked cutting on Forest Service land while considering whether cutting would interfere with efforts to protect the endangered owl. Mark Hatfield of Oregon and Brock Adams of Washington sponsored an amendment to an Interior Department appropriations bill that would have lifted the court injunction, and Jontz urged his colleagues to defeat the amendment. The debate was fierce and sharply polarized, and Jontz became a lightning rod for anger from timber interests. In April 1990, a protest was held in Hoquiam, Washington, at which angry loggers burned Jontz in effigy. Jontz was targeted for ouster by the timber industry and in

1992 was defeated in his bid for reelection by Steve Buyer.

After leaving the House, Jontz unsuccessfully challenged Republican Sen. Richard Lugar in the 1994 Senate race. In 1995, he served as director of the National Audubon Society's Endangered Species Coalition, where he again lobbied Congress in support of the Endangered Species Act. In November 1995 he became the executive director of the Western Ancient Forest Campaign (WAFC), a coalition of activist groups fighting for wilderness preservation. WAFC was active in the campaign against the North American Free Trade Agreement (NAFTA), which was criticized for promoting free trade at the expense of the environment. WAFC also campaigned in support of environmentalists in developing countries, including a 1997 campaign to block the expansion of NAFTA to South America. Jontz and WAFC were active in the 1995 Sugar Loaf timber protests in Oregon, as well as the 1999 World Trade Organization protests in Seattle.

In 1999 WAFC changed its name to American Lands Alliance. Jim Jontz still serves as executive director. He lives in Portland, Oregon.

BIBLIOGRAPHY

"American Lands," www.americanlands.org; Dustin, Thomas, "Jontz's Commitment Paying Off Again," *South Bend Tribune*, 1996; Lancaster, John, "Debate on Timber Harvesting Is Major Environmental Battle," *Washington Post*, 1989.

Jukofsky, Diane

(February 11, 1953–)
Journalist, Director of the Conservation Media Center

As worthy as their causes may be, environmentalist organizations will enjoy meager success unless they are able to effectively publicize the problems they are trying to solve and the solutions they propose. Journalist Diane Jukofsky has devoted her career to helping environmental organizations spread the word and, since 1989, has focused exclusively on Latin America's conservation groups through the Rainforest Alliance's Conservation Media Center.

Jukofsky was born in Saint Louis, Missouri, on February 11, 1953. She spent her childhood summers at her grandparents' Missouri farm, tending to farm chores alongside her grandmother, Elizabeth Zahorsky Cushing, who had been a founding member of most of the national conservation organizations and was unusually well educated for a girl growing up at the beginning of the twentieth century. At her grandmother's side, Jukofsky learned the names of flowers and constellations and fought her very first environmental battle: a struggle against a dam project on the Meremec River. These experiences are what Jukofsky credits as inspiration for her career.

Jukofsky attended her grandmother's alma mater, Mount Holyoke College, and graduated with a degree in English in 1975. She immediately went to work for the National Wildlife Federation, where she was the assistant director of public information. It was there that she discovered she could combine her academic training with her passion for conservation. Later Jukofsky worked on public relations matters for several nonprofits, including Twin Cities Public Television, the Minnesota Conservation Federation, Congressman Sam Gejdenson of Connecticut, and Scientists' Institute for Public Information.

During the 1980s, the steep loss of biodiversity in tropical rain forests began to alarm the scientists with whom Jukofsky worked. As the disturbing evidence of the disaster continued to roll in, Jukofsky made a personal commitment to devote herself to rain forest conservation. Along with her husband, Chris Wille, their friend Daniel Katz, and several others, she helped found the New York Rainforest Alliance in 1987. Originally it was conceived as a small, local group devoted to rain forest conservation, but it has become an international organization with a multimillion-dollar budget. The Rainforest Alliance distinguishes itself from other rain forest conservation groups through its innovative and economically sound programs to convert rain forest destroyers into allies.

Jukofsky and Wille moved to Costa Rica in 1989 in order to establish the Rainforest Alliance's Conservation Media Center so as to be able to report to conservationists from the site of rain forest destruction. Her Latin America–related projects have varied from coordinating public information campaigns for large-scale conservation projects so that vital scientific information may reach the people who can make it succeed, to producing a reference book on world rain forests, to publishing the bimonthly, bilingual periodical *Eco-Exchange* (*Ambien-Tema* in Spanish), which covers controversial environmental issues and innovative conservation projects.

Jukofsky's public relations trainings for nonprofit conservation organizations have been especially successful. During the workshops, the staff of a participating organization learns how to communicate more effectively with the press about their work. The final project includes a field trip with journalists to the area the group is trying to conserve. Jukofsky recounts one such expedition in Guatemala:

> It was the first time the Guatemalan journalists had seen a quetzal in the wild (their country's national symbol). When I saw the tears in the eyes of the Guatemalan reporter who watched an emerald green quetzal disappear into the mists of the Sierra de las Minas cloudforest, I knew she would forever understand the passions of those who are fighting to save the reserve. We also showed them a logging operation in the biosphere reserve (which was illegal, but the owner had managed to get permits). The Guatemalan reporters (with a leading daily newspaper and an influential weekly newsmagazine) wrote articles about the operation, which caused quite a stir. As a result, the government rescinded the logging permits. The power of the press! (personal communication, 1998)

Jukofsky has been a full-time employee of the Rainforest Alliance since 1990. She resides in San José, Costa Rica, with her husband Chris Wille.

Bibliography

Braus, Judy, ed. *Wow! A Biodiversity Primer*, 1994; "Conservation Media Center," http://www.rainforestalliance.org/programs/cmc/index.html; "Rainforest Alliance Staff Biographies," http://www.rainforest-alliance.com/about/biographies.html.

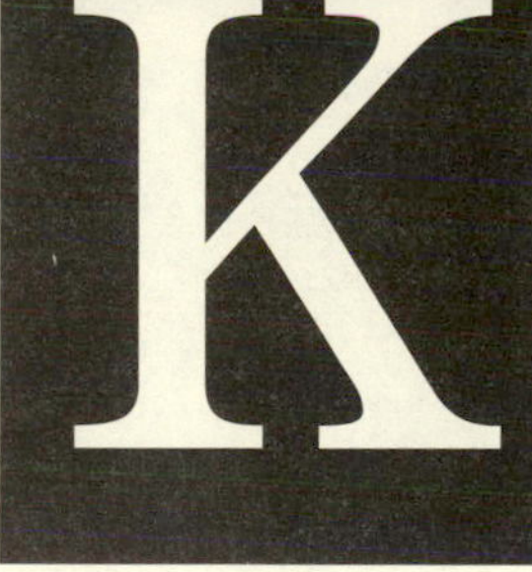
K

Kane, Hal

(October 9, 1966–)
Writer, Strategic Planner for Nonprofit Organizations

Hal Kane has worked for several environmental nonprofit organizations, including the think tanks Worldwatch Institute and Redefining Progress and the advocacy organization Pacific Environment & Resources Center (PERC). His work has focused on the links between development, the environment, population growth, and consumption. His primary area of interest has been how personal choices and lifestyle affect the environment and the economy. He is a prolific writer, with over ten books to his name that have been translated into 31 foreign languages.

Hal Moss Kane was born in Ann Arbor, Michigan, on October 9, 1966. His father, Gordon Kane, a professor of particle physics at the University of Michigan, subscribed to journals that covered issues of population and the environment. Kane attributes his interest in these topics to his exposure to them as a budding reader. Through his mother, Lois Kane, who wove and wrote, Kane learned how useful art and words could be for effective communication.

Kane attended the University of Michigan (UM) in Ann Arbor. He intended to study political science but was distressed by the verbose and poorly written texts he was assigned. Knowing that skillful writing would be important in any career choice he made, Kane decided that the best training would be to read good writing and then to learn to write well. So he majored in English language and literature and graduated in 1988. As an undergraduate, Kane organized an annual symposium on venture capital at UM, in which small companies that needed finance could meet potential investors. Kane's experience helping the companies formulate business plans came to be useful as he helped nonprofit groups devise viable financial strategies. During his years at UM Kane also worked for Senators Gary Hart (Colorado) and Don Riegle (Michigan). There, he gained an insider's view of politics and also made the decision not to pursue a career as a staff member for elected officials.

Following his graduation from UM, Kane worked for several environmental organizations, including the Geneva-based Centre for Our Common Future, an organization funded by the United Nations' World Commission on Environment and Development. Kane considers himself lucky to have had the opportunity, at that early point in his career, to draft several speeches for its director, Norwegian prime minister Gro Harlem Brundtland. Kane also wrote and consulted for other nonprofit organizations, including the National Congress of American Indians, Island Press, Partners for Livable Places, and Global Tomorrow Coalition. The U.S. Citizen's Network on the United Nations Conference on the Environment and Development (UNCED) asked Kane to write a book to prepare the American public for the big 1992 conference in Rio. With their support, Kane authored his 1991 *Time for Change: A New Approach to Environment and Development*, which outlines the interconnectedness of economics, the environment, and social issues.

Kane earned an M.A. in international relations and economics from Johns Hopkins University (JHU) in 1992. During the Washington-based time of his studies (he also studied at JHU's Bologna, Italy, campus) he worked as an environmental management fellow for the U.S. Environmental Protection Agency, researching the connections between international trade and the environment.

Following his graduation from JHU, Kane moved on to Worldwatch Institute (WI), where as a research associate he studied such issues as the sustainability of the world's fisheries, environmental economics, international trade, resource scarcity, hunger, refugees, migration, and population growth and wrote articles for *World Watch*, WI's bimonthly magazine. He cowrote two of WI's *State of the World* books, contributing chapters on industry and the environment for the 1995 edition and on refugees and migrants for the 1996 edition. Kane cofounded a new annual WI series, the *Vital Signs* books, cowriting the editions from 1992 to 1996. This series, which focuses on specific, influential world trends, has proven to be important, selling more than 50,000 copies per year and being translated into 18 languages. Kane also collaborated on a 1994 book with WI director LESTER BROWN, *Full House: Reassessing the World's Population Carrying Capacity.*

In 1997, Kane moved to the San Francisco–based nonprofit think tank Redefining Progress, which states its mission as opening up the debate on the "true meaning of progress, one that goes beyond economic growth to incorporate environmental sustainability and social equity." At Redefining Progress, Kane directed the National Indicators Program, which studies more accurate ways to measure a country's well-being than the common Gross Domestic Product measure. The Genuine Progress Indicator (GPI), for example, begins with economic output but adds and subtracts other values, including environmental improvement or decline, crime, divorce, unpaid work in the home, commute time, car accidents, and income inequality. In 1998 and 1999, as a senior fellow for Redefining Progress, Kane wrote *Triumph of the Mundane: The Unseen Trends That Shape Our Lives and Environment*, which describes the way of life in the United States today through statistics on the rise of such indicators as the number of people who live alone, the average distance we travel everyday, the frequency with which we move to new homes and new jobs, and many other personal indicators. The book argues that the way lifestyles have changed over the years has dramatically affected the health of our environment and ourselves and has done so more than politics or most of the other factors that we hear about in the news.

Since 1998, Kane has served as executive director of the Pacific Environment & Resources Center. PERC is a nonprofit environmental organization that supplies information and funding to nonprofit organizations in countries around the Pacific Rim and has fostered citizen environment movements in countries such as Russia and China, whose governments have traditionally been closed to public participation. PERC offers its members the opportunity to visit sites of interest in the United States where potential environmental problems are approached with innovative solutions: well-designed mines, sustainably managed logging sites, and responsible industry, for example. PERC also monitors the work of bilateral export-credit agencies and international financial institutions and alerts the rest of the world to environmental problems along the Pacific Rim that otherwise would have gone unnoticed.

Kane has become a popular public speaker, traveling throughout the United States and in foreign countries to give public speeches, to appear on television and radio stations, and to participate in professional conferences. He resides in San Francisco, California.

Bibliography

Brown, Lester, and Hal Kane, *Full House: Reassessing the World's Population Carrying Capacity*, 1994; Brown, Lester, et al., *Vital Signs*, 1992, 1993, 1994, 1995, 1996; Kane, Hal, *Time for Change: A New Approach to Environment and Development*, 1991; Kane, Hal, *Triumph of the Mundane: The Unseen Trends That Shape Our Lives and Environment*, 2000; "Pacific Environment & Resources Center," http://www.pacenv.org; "Redefining Progress," http://www.rprogress.org/; "Worldwatch Institute," http://www.worldwatch.org.

Katz, Daniel

(November 7, 1961–)
Executive Director of the Rainforest Alliance

In 1986, Daniel Katz cofounded the Rainforest Alliance. This organization, which Katz has directed ever since, promotes economically viable and socially desirable alternatives to the destruction of rain forests, through education, social and natural science research, and the establishment of cooperative partnerships with businesses, governments, and local peoples.

Daniel Roger Katz was born on November 7, 1961, in Cincinnati, Ohio. Raised in the city, one of his more memorable childhood exposures to nature was seeing Henri Rousseau's primitivist paintings in a children's art book, which even at his young age made an impression. Katz was an energetic organizer and founded a string of clubs in high school and college. He entered Ohio State University in 1979 and studied abroad, in 1981 at the National Chengchi University in Taiwan and from 1983 to 1984 at the Central China University of Science and Technology in Wuhan, People's Republic of China. Katz graduated from Ohio State with a double degree in Chinese and political science in 1984.

In 1985, Katz moved to New York City and did China-related work for a law firm for one year. He attended a small workshop on rain forests in 1986; he found himself fascinated with them and very troubled about the massive deforestation that was not receiving any media attention at the time. With four others who had attended the workshop, Katz created the New York Rainforest Alliance. He quit his job and devoted ten months of full-time work to organize "Tropical Forests, Interdependence and Responsibility," the largest international conference on rain forest conservation that had ever been held. Drawing such well-known figures in rain forest ecology as Thomas Lovejoy, Peter Raven, and Ghillean Prance, the 1987 conference featured 51 speakers and was attended by some 700 people. The international press covered the event, and Katz and his partners felt that they had successfully challenged scientists and governments to give rain forest conservation the attention it deserved.

The New York Rainforest Alliance continued to offer lectures and workshops to educate the public about deforestation issues. Katz felt that it was important to begin working in rain forests themselves as well. In 1988 the Alliance organized a workshop on the tropical timber industry. This led to the project eventually named SmartWood, the world's first forestry certification program. SmartWood, which has been directed by RICHARD DONOVAN since 1992, works with timber companies to assess their impact on the environment and awards a seal of approval to those companies that manage forests in an environmentally and economically sustainable manner. As of January 2000, some 100 managed forests on more than three million acres spanning four continents had been certified.

Following the success of SmartWood, the Rainforest Alliance developed the Better Banana Project, a program that certifies banana producers that comply with strict social, economic, and environmental criteria. This project, initiated in 1990, was eventually renamed ECO-O.K. and under the leadership of CHRIS WILLE, headquartered in Costa Rica, was given a big push in 1994 when banana giant Chiquita Brands vowed to certify all of its farms. In 1995, ECO-O.K. was awarded the prestigious Peter F. Drucker award for nonprofit innovation, and in the latter half of the 1990s it gradually expanded to include certifications of other tropical agricultural crops including coffee, citrus, sugarcane, and cacao. Wille has drawn together affiliate certification groups in several Latin American countries. Although each affiliate of the Conservation Agriculture Network uses slightly different criteria, overall they subscribe to the three pillars of sustainable agriculture: community well-being, environmental protection, and economic vitality.

The Conservation Media Center, also headquartered in Costa Rica, and directed by DIANE JUKOFSKY, teaches grassroots environmental organizations how to run effective public information campaigns and networks them with its bimonthly, bilingual newsletter *Eco-Exchange* (*AmbienTema* in Spanish). Rainforest Alliance sponsors various research projects, including the Amazon Rivers Program, through which one of the world's leading icthyologists, Michael Goulding, led local researchers on a nine-year (1990 to 1999) study of the consequences of deforestation in the Amazon basin. The Periwinkle Project, later renamed the Natural Resources and Rights program, looked at how indigenous people were being impacted by the extraction from rain forests of medicinal plants including the Madagascar periwinkle, used to treat leukemia and Hodgkin's disease. Director Charles Zerner wrote *People, Plants and Justice* (2000) and organized a conference to discuss how local communities could most democratically manage their own plant resources. In addition to these long-term programs, the Rainforest Alliance also offers the two-year Kleinhans research fellowships to scientists studying rain forest conservation issues.

The Rainforest Alliance links groups of conservation-minded schoolchildren or donors in the United States with specific rain forest–based grassroots organizations abroad through its educational web site, and the Rainforest Alliance also provides Catalyst Grants of under $3,000 to grassroots projects that are not otherwise eligible for government or international funding because of their size or their geographical location. The Rainforest Alliance had given approximately 70 of these grants, worldwide, by early 2000.

Because the organization often looked at market-based conservation solutions, Katz decided to study for an executive M.B.A., which he earned in 1996 from New York University. Katz resides in New York City with his wife, Maggie Lear, and their young son, Griffin. He has served on a number of boards for conservation and international affairs organizations. He was awarded a Kellogg Foundation National Fellowship from 1989 to 1992.

Bibliography

Katz, Daniel, and Miles Chapin, *Tales from the Jungle: A Rainforest Reader*, 1995; "Rainforest Alliance," http://rainforest-alliance.org/; Reiss, Bob, *The Road to Extrema*, 1992.

Kaufman, Hugh

(January 14, 1943–)
Hazardous Waste Investigator

Employed by the Environmental Protection Agency (EPA) since its inception in 1971, Hugh Kaufman is best known for his role as a whistleblower, calling attention to EPA's failures and political corruption. Kaufman was a central player in the scandal at EPA in the mid-1980s, which resulted in the ouster of EPA head Anne Gorsuch Burford and the conviction of assistant administrator Rita Lavelle on charges of lying to Congress. Kaufman also worked on drafting two important pieces of legislation, the Superfund Law and the Resource Conservation and Recovery Act.

Hugh Kaufman was born on January 14, 1943, in Washington, D.C. He was raised in Arlington, Virginia, in one of the few Jewish families in the community. His experience as an outsider growing up contributed to the development of a fighting spirit. He once said his tough-kid approach was to "hit first and ask questions later." Kaufman attended George Washington University, from which he earned a B.S. in 1965 and an M.S. in 1967, both in engineering. After serving in the United States Air Force for four years, Kaufman joined the new Environmental Protection Agency in 1971. Employed first in the noise program, Kaufman was active in creating rules to control noise around highways and airports and helped turn the agency against unlimited landing rights for the Concorde. In 1974 he moved to the hazardous waste division and began on-site investigations for the EPA. The EPA was a new agency, and Kaufman played an important role in developing its toxic waste programs. He compiled reports about hundreds of cases of land, air, and water contamination by hazardous waste, which helped convince Congress to pass the 1976 Resource Conservation and Recovery Act.

The late 1970s saw a dramatic rise in public awareness of the problem of hazardous waste, focused most dramatically on New York's Love Canal case, which made national headlines in 1978. Kaufman helped investigate the case, before Steffen Plehn, deputy assistant administrator for solid waste, recalled Kaufman out of the field and back to headquarters. The agency was afraid of the magnitude of the nation's toxic waste problem and its associated costs, and Plehn ordered Kaufman to stop visiting sites and pressing for clean-up efforts. Kaufman did not comply silently with this order. He and fellow EPA employee WILLIAM SANJOUR gave information to the press and to congressional staff and members, on and off the record, and eventually testified before Congress. Both employees put their jobs at risk and were the targets of internal EPA investigations. Their testimony caused a storm of publicity that helped generate the momentum to pass the landmark 1980 Superfund legislation.

Despite his reputation as a whistleblower, Kaufman managed to keep his job even after President Reagan's election, which brought strong deregulation pressure to bear on the EPA. During this time Kaufman was assistant to the director of the Hazardous Waste Site Control Division; the division was part of the Superfund program, which was led by assistant administrator Rita Lavelle. By March 1982, Kaufman was again in the national news. He appeared before several congressional committees and, on April 24, 1982, on CBS's *60 Minutes*. Kaufman accused the EPA, under the direction of Anne Gorsuch Burford, of failing to meet its legal obligations to enforce environmental regulations. The agency was not spending Superfund allocations to investigate polluters and was cutting illegal sweetheart deals with industry. Kaufman cited a memo from Rita Lavelle to White House counsel Robert J. Perry, in which she accused Perry of "alienating" President Reagan's most important constituency: business. The immediate result of Kaufman's television appearance was an investigation of his activities by the EPA, in which they tapped his phone and photographed Kaufman going into a hotel room with a woman who turned out to be his wife. Congress began an investigation of the EPA, during the course of which Burford invoked executive privilege to keep incriminating documents out of the hands of Congress and EPA critics. In a showdown with Congress, the Reagan administration eventually backed down; President Reagan fired Lavelle and forced Burford's resignation. Lavelle was tried and convicted of perjury and served four months of a six-month sentence. President Reagan appointed a new EPA director, William Ruckelshaus, who had been the agency's first administrator. Ruckelshaus restored order to EPA and returned focus to its enforcement programs.

Kaufman continued to work for the EPA and to serve as an insider advocate for citizens' groups. He was active in the successful fight against a nuclear waste dump proposed for the tiny farming town of Nora, Nebraska, in the late 1980s. In 1993 Kaufman again played the role of whistleblower, this time testifying to Congress about the EPA's issuing a questionable waste-disposal permit to a hazardous waste incinerator next door to an elementary school in East Liverpool,

Ohio, operated by Waste Technologies Industry, which had close ties to the Clinton administration. In 1994 Kaufman appeared on Michael Moore's television show *TV Nation,* in a segment called "Sludge Train," about Merco Joint Venture, a reputed mafia project to spread wastewater sludge on a ranch near Sierra Blanca, Texas. Merco successfully sued Kaufman for libel, but an appeals court overturned the verdict and found that Kaufman had not libeled Merco Joint Venture or its venture partners. Kaufman was also a vocal critic of the EPA's plan to clean up the Shattuck Chemical Company waste dump in Denver. Kaufman argued that the plan was not aggressive enough and would eventually lead to groundwater contamination. As a result of his work, the EPA agreed to move the waste to a licensed radioactive waste disposal facility.

Kaufman still works for the Environmental Protection Agency in Washington, D.C.; he is principal investigator for the Solid Waste and Emergency Response. Kaufman resides in Washington, D.C.

BIBLIOGRAPHY

Johnson, Jeffrey, "Hugh Kaufman: EPA Whistle-Blower," *Sierra,* 1988; Johnson, Roberta Ann, "Bureaucratic Whistleblowing and Policy Change," *Western Political Quarterly,* 1990; Mitchell, Greg, *Truth and Consequences,* 1981; Morson, Berny, "The Hero of Shattuck," *Rocky Mountain News,* 1999.

Kellert, Stephen

(October 10, 1944–)
Social Ecologist

A professor of social ecology, Stephen Kellert has spent decades studying the relationship between humans and nature. His research projects have included studies of basic perceptions relating to the conservation of biological diversity, and he has explored and eloquently written about a concept set forth by EDWARD O. WILSON called "biophilia"—the idea that humans possess a deep and biologically based urge to connect with the natural world. Kellert argues that because of this innate affinity for the complexities of nature, natural diversity must be protected for its capacity to enrich and enlarge the human experience. His message takes the rationalization for conservation beyond narrow utilitarian or material needs and makes a new case for preservation of diversity and natural spaces.

Stephen Robert Kellert was born on October 10, 1944, in New Haven, Connecticut. He spent the first five years of his life there, and then his family made a series of moves, mostly within Connecticut. In 1966 he received his bachelor's degree from Cornell University with a major in social psychology and a minor in biology. He earned a Ph.D. from Yale University in 1972 and began teaching and researching there that same year. In 1981, he and Priscilla Whiteman, a teacher, were married. They have two daughters, Emily and Libby. Kellert has continued teaching and conducting research as a professor of social ecology at Yale.

The basic human relationship to nature became the essence of Kellert's work. In the late 1970s and early 1980s he studied human perceptions of animals, including wolves, invertebrates, marine mammals, bears, and diverse endangered species. He also explored the nature-related perspectives of such different groups of people as hunters, birders, farmers, and the general public distinguished by age, gender, socioeconomic status, and place of residence. Through his studies, Kellert began to see recurring patterns of thought in people's attitudes toward wildlife and conservation, and it became more and more apparent to him that this reflected a basic tendency in humans to affiliate with nature and diversity.

As these ideas took shape, he became acquainted with the work of Edward O. Wilson, an entomologist at Harvard University. In 1984, Wilson published a book called *Biophilia*, which sought to provide an understanding of how human tendencies to connect with life and natural processes might be the expression of a biological need. Wilson defined *biophilia* as the "innate tendency to focus on life and life-like processes." Kellert and Wilson worked together on this concept when they coedited *The Biophilia Hypothesis*, which was published in 1993. This book brought together 20 scientists from various disciplines, each examining and refining the idea of biophilia in different contexts. The various perspectives—psychological, aesthetic, cultural, ethical, and political—help to frame a broad discussion of biophilia that ultimately goes farther than the standard ecological or economic arguments for the preservation of other species and ecosystems. There are many examples in the book that reinforce the notion that biophilia, and its opposite, biophobia, have evolutionary adaptations. For instance, humans show strong aversions to snakes and spiders with little or no negative reinforcement—while more dangerous modern items such as guns, knives, or electric wires rarely initiate such phobias. Also, people have a common desire to live amid green vegetation and flowers and trees, or near water. Another essay examines the development of myth and the use of natural symbols and how people often ascribe human qualities to animals. As a whole, these essays advance reasons for conservation that are intangible, yet still important: the biophilia concept means that estrangement from the natural world through continued levels of biological destruction could lead to psychological and spiritual impoverishment.

Kellert expanded on this new dimension of conservation in his next book, *The Value of Life: Biological Diversity and Human Society* (1996). Drawing on 20 years of research, he outlines a framework of biologically based values that humans exhibit, such as an aesthetic attraction to nature, a scientific inclination to understand it, a humanistic affection for animals, and others, and then examines differences among varying demographic and cultural groups. In discussing how vital living diversity is, he notes that more Americans visit zoos during an average year than attend all the professional baseball, basketball, and football games together. Visits to national parks and wildlife areas are increasing dramatically, and ecotourism is now the fastest-growing segment of the travel industry. Also, growing numbers of

Americans seek wildlife-related diversions such as birding, whale watching, or hunting and fishing. However, in comparing Americans' perceptions of nature with people in other nations, Kellert found that most Americans have a limited understanding of biological processes, and he decries the fact that environmental education receives far too little support. His argument, which illustrates the importance of biological diversity to human psychological health, pleads for greater environmental literacy—both directly, through schools; and indirectly, through mass media.

Building further on these concepts, Kellert wrote another book called *Kinship to Mastery: Biophilia in Human Evolution and Development* (1997). He continues to address concerns over education, since our affinity to nature, though biologically based, is also shaped by mediating influences of learning, culture, and experience. Kellert contends that biophilia is a "weak tendency" and that without positive reinforcement through education and social support it will not develop or thrive. In this book he makes some broad suggestions on how to slow the damage being done to the natural world. These include habitat conservation, integrating nature into people's daily lives, and respecting different cultural attitudes in respecting wildlife. Ultimately the richness of connections with nature will foster a greater quality of life.

Kellert has received much recognition for his work and many awards. In 1983 he received the Special Achievement Award from the National Wildlife Federation, and in 1985 he received the Best Publication of the Year Award from the International Foundation for Environmental Conservation. He received a Distinguished Individual Achievement Award from the Society for Conservation Biology in 1990, and in 1997 the National Wildlife Federation presented Kellert with its National Conservation Achievement award, recognizing his contributions toward an understanding of how people perceive wildlife and wild places. He continues to serve on International Union for the Conservation of Nature Species Survival Commission groups, which he has done since 1992. Kellert has authored more than 100 publications and is currently working on a book called *Ordinary Nature: Exploring and Designing Nature in Everyday Life.* He is also directing a large-scale watershed ecosystem study to examine the complex mental and physical dependency of people on healthy natural systems. He continues to teach social ecology at Yale University and lives in New Haven, Connecticut.

BIBLIOGRAPHY

Johnson, Andrew, "The Biophilia Hypothesis," *BioScience*, 1994; Kellert, Stephen R., *Kinship to Mastery: Biophilia in Human Evolution and Development*, 1997; Kellert, Stephen R., *The Value of Life: Biological Diversity and Human Society*, 1996; Kellert, Stephen R., and Edward O. Wilson, eds., *The Biophilia Hypothesis*, 1993.

Kendall, Henry

(December 12, 1926–February 15, 1999)
Physicist, Founding Member of the Union of Concerned Scientists

Winner of the 1990 Nobel Prize for Physics for his work in particle physics, Henry Kendall became a vocal critic of nuclear power and the nuclear arms race. Kendall was one of the founding members of the Union of Concerned Scientists (UCS), formed in 1969 to unite scientists to take action against dangerous uses of science and technology. He was particularly influential in the debate over President Reagan's proposed Strategic Defense Initiative (SDI, or "Star Wars") and in calling attention to global warming.

Henry Kendall was born on December 12, 1926, to a wealthy Boston "Brahmin" family. The Kendalls owned the Kendall Company, a health supplies company best known for its Curad line. As a boy, Kendall was more interested in sports than science and was an expert sailor and diver. Upon graduation from high school, he entered the Merchant Marine Academy, where he was studying when the United States dropped the first atomic bomb in 1945. The event awakened Kendall's interest in science, and he entered Amherst College, from which he graduated in 1950 with a B.S. in mathematics.

After graduation, Kendall returned to his interest in the outdoors and operated a diving and salvage operation. He also wrote books about shallow-water diving and underwater photography. Eventually he decided to pursue a career in science, and he earned a Ph.D. in nuclear physics from the Massachusetts Institute of Technology (MIT) in 1955. From 1954 to 1956, Kendall was a National Science Foundation Fellow at Brookhaven National Laboratory and MIT. In 1956 he joined the faculty at Stanford University, where he met Jerome Friedman and Richard Taylor, with whom he would eventually share the Nobel Prize. At the time Stanford had one of the largest particle accelerators in the world and by 1960 had plans to build a new accelerator, two miles in length. Kendall returned to MIT in 1961 and in 1967 began a collaboration with Friedman and Taylor that led to important advances in the understanding of the fundamental nature of matter. Using Stanford's accelerator, they fired electron beams at protons and neutrons and discovered unexpected patterns of deflection. These patterns demonstrated the existence of quarks, up to then only a theoretical proposition. This was the first experimental evidence of subatomic particles. The men's work thus was an important step in the development of the current standard model of the elements of matter.

As a member of the scientific elite, Kendall was in an ideal position to see the ways in which the U.S. military influenced the development and use of science and technology. He served as a consultant to the Department of Defense from 1960 to 1971 and became disturbed by what he saw: the harnessing of the power of science for increasingly destructive ends. At the height of the Vietnam War, Kendall signed the 1968 MIT Faculty Statement that led to the founding of the Union of Concerned Scientists. The statement was a call to scientists and engineers to assume responsibility for the uses of science. The 50 senior faculty

Henry Kendall airing his views on Three Mile Island Nuclear Plant as Robert Pollard, the Union of Concerned Scientists' nuclear safety engineer, looks on, 14 May 1980 (Bettmann/Corbis)

who signed proposed, in addition to several other basic ideas, "to initiate a critical and continuing examination of governmental policy in areas where science and technology are of actual or potential significance" and "to devise means for turning research applications away from the present emphasis on military technology toward the solution of pressing environmental and social problems." In early 1969, the group formed the UCS. Kendall was a key figure in the organization from the beginning and served as chairman of the Board of Directors from 1973 until his death.

UCS has been an influential voice in public policy in several areas. One of its initial projects was to call into question the safety of the nation's nuclear arsenal and power plants. In 1977 the UCS issued "Scientists' Declaration on the Nuclear Arms Race," which documented the destructive effects of the Cold War arms buildup. Armed with technical expertise and accepted scientific research, UCS has high credibility with journalists and public officials and used that credibility throughout the 1970s and 1980s to help slow the arms race and close dangerous nuclear plants. Kendall was particularly visible during the fight over the Strategic Defense Initiative, which UCS argued was not only a dangerous escalation of the arms race but also technologically unfeasible. In

1985 UCS issued "Appeal by American Scientists to Ban Space Weapons." SDI eventually died in development. UCS has also been important in publicizing and gaining credibility for theories of global warming. In 1990 it released "Appeal by American Scientists to Prevent Global Warming." In 1992 Kendall authored UCS's "World Scientists' Warning to Humanity," which brought into focus "critical stress" on the environment on a number of fronts, including atmosphere, water, oceans, soil, forests, and living species.

Henry Kendall died on February 15, 1999, while assisting a team from the National Geographic Society in its efforts to map underwater caves in Florida. The cause of death was gastrointestinal hemorrhaging, unrelated to diving.

BIBLIOGRAPHY

Chandler, David, "He Still Brown-Bags It," *Boston Globe*, 1990; "Professor Henry Kendall," *London Times*, 1999; "Union of Concerned Scientists," http://www.ucsusa.org.

Kennedy, Danny

(February 24, 1971–)
Cofounder and Director of Project Underground

Danny Kennedy cofounded and serves as director of Project Underground, an organization that works in solidarity with Native Americans and with communities worldwide that face environmental destruction due to mining by large multinational oil and mineral companies.

Daniel Ian Kennedy was born on February 24, 1971, in Los Angeles, California. He spent his childhood in New South Wales, Australia, and at an early age became an amateur naturalist. At the age of 12, in 1983, he worked with the Australian Conservation Foundation against the Franklin Dam, which was slated to be built in Tasmania. He canvassed for a Labor Party promise to stop the dam, and for the first time ever, a candidate for prime minister was swept into office by the green vote. Throughout his teens, Kennedy continued his environmental activism, especially on forest issues. He was sent as a youth representative to the Montreal Protocol negotiations in London in 1990, and then he coordinated the Australian student activist group Action for Solidarity, Equality, Environment, Development (ASEED), which organized young people to lobby government delegations at the United Nations Conference on Environment and Development in Río de Janeiro in 1992. Officially, he attended the conference as a journalist for the Australian Broadcasting Corporation, but he was fired after he was arrested in a civil disobedience action to protest the failure of the Summit.

Kennedy received his B.Sc. with honors in resource management from Macquarie University, New South Wales, in 1994. After helping to set up a green bookshop in Sydney, Kennedy went to work for Greenpeace in 1993, in Papua

New Guinea, to document the work of a local community in its struggle against oil drilling.

In 1996, after returning to the United States, Kennedy collaborated with Aaron Freeman, Pratap Chatterjee, and Steve Kretzmann to found Project Underground. They, along with environmentalists worldwide, had been horrified by the November 1995 execution in Nigeria of Nobel Prize–winning author Ken Saro-Wiwa and eight other Ogoni activists. The "crime" of the Ogoni Nine was effective organizing against Royal Dutch Shell's environmentally devastating oil drilling on their land, which had formerly been the most fertile agricultural land in West Africa. Project Underground hoped to help prevent such barbarity elsewhere, by working in solidarity with communities threatened by environmental damage and human rights abuses from resource extraction operations. The mission of Project Underground was to provide information and technical assistance to grassroots activist groups, network them with others facing the same companies or types of mining and drilling practices, and, when it was deemed necessary, mount legal battles, boycotts, and stockholder meeting protests in the home countries of the companies. By 2000, Project Underground had grown from its shoestring beginnings to have a full-time staff of six and an annual budget of $500,000.

The first struggle in which Project Underground became involved was in Dominica, an island in the eastern Caribbean, where Australian mining company Broken Hill Proprietary Company, Limited (BHP) proposed a large open pit copper mine. The Dominican Conservation Association approached Project Underground for information, and in August 1996, one month after it was founded, Project Underground sent an Australian ecologist to evaluate the proposal. He found that the mouth of the Dominica mine would occupy approximately 10 percent of the island's land mass and that it would use the same waste dumping practice, known as riverine tailings disposal, as was used at the company's Ok Tedi mine in Papua New Guinea, where hundreds of square kilometers of river and rain forest had been destroyed. After a massive publicity effort by Dominican activists, one in five Dominicans signed petitions against the mine, but BHP pressured the Dominican government heavily to approve its proposal. Project Underground activists attended BHP's annual meeting in Australia in September 1996 to reveal to shareholders the level of dissension about the project in Dominica, and an embarrassed BHP abandoned the mine proposal within six months.

In solidarity with the Movement for the Survival of the Ogoni People (MOSOP), Project Underground has organized mass protests against Shell Oil. Monthly pickets of Shell gas stations were organized in the San Francisco Bay area, continuing for two years, and expanding to about a dozen other cities throughout North America. More significantly, 11 cities and counties passed ordinances prohibiting local governments from buying from those doing business in Nigeria. Project Underground's Steve Kretzmann helped organize the first international meeting of Ogoni leaders and their international supporters after the execution, and in March 1997 he traveled undercover into Ogoniland to take the first independent water samples, proving the claims of Saro-Wiwa of environmental contamination and documenting the existence of "Shell Police"—a paramilitary security squad operating within the Ogoni community to intimidate further protest. Together with the Rainforest Action Network, Project Underground published the first of its series of "Independent Annual Reports" on the activities of Royal Dutch Shell in Nigeria and the Peruvian Amazon. It was released in London during Shell's annual meeting in May 1997. In part due to this pressure and to Project Underground's threat of a boycott should the company continue its environmental and human rights abuses in Ogoniland and Peru, Shell has not pumped oil in Ogoniland since 1993, and it abandoned the Camisea oil-drilling project that it had been considering in Peru.

Project Underground has also worked with the U'Wa people of the Colombian Amazon, who oppose an oil-drilling project planned by U.S.-based Occidental Petroleum. Their spiritual beliefs consider oil to be the blood of Mother Earth, and they find it sacrilegious to extract it. In addition, they fear that the presence of Occidental in their remote homeland would attract the Colombian military and the paramilitary guerrilla group Fuerzas Armadas Revolucionarias de Colombia. In 1997, three North American indigenous rights activists, including Terry Freitas, a former Project Underground staff member, were murdered while on an investigative and solidarity mission in U'Wa territory. Project Underground, with its allies and the U'Wa people, has forced Royal Dutch Shell, an original partner in this project, to back out. The united groups have also organized Occidental shareholders to express their opposition to the project to the company's management and have sued for mediation of the U'Wa peoples' case in Colombian courts by the Organization of American States' Interamerican Commission on Human Rights. To date, Occidental has renounced its claim to 75 percent of its Colombian concession, but it is still pursuing drilling on the remaining quarter of its claim. The U'Wa people say that they will defend their land with their lives; they have threatened to commit mass suicide if Occidental goes on with the project.

Project Underground not only opposes oil drilling and other extractive activities for the human rights and environmental abuses that accompany them but also questions the value of industries themselves. At least 80 percent of the gold now mined, for example, is used for jewelry and other decorative purposes. This includes gold from Freeport McMoRan's Grasberg mine in the Irian Jaya region, the largest gold mine in the world, which has ruined the water and land of the mine's neighboring indigenous Amungme people; from Newmont Corporation, the second-largest gold producer in the world, which contaminates the land and water around each of its sites with massive quantities of cyanide; and from South African gold mines, in which one worker is killed and another dozen are seriously injured for each ton of gold extracted from its mines. Gold prices are dropping at the same time that gold is growing more scarce and lower-grade ores are being exploited at higher costs (with higher quantities of cyanide and the other toxic chemicals necessary for leaching out the gold). Project Underground works with communities who question whether the use that gold is given today merits such sacrifice of human rights and environmental health. Similarly, Project Underground brings up global warming issues when discussing the exploration of petroleum extraction. Its No New Exploration campaign calls for an end to oil exploration not only because of what it does to local people and their ecosystems but also because "from a climate change perspective, we cannot even afford to burn the reserves we know we have, so why spend billions every year looking for more?"

Project Underground publishes a biweekly newsletter, *Drillbits and Tailings*, and has published several reports about the extractive activities it has investigated worldwide; they are available through its web site (www.moles.org).

Kennedy is a board member of the Pacific Environment and Resource Center and a charter member of the International Forum on Globalization. He lives in Berkeley, California.

BIBLIOGRAPHY

Kennedy, Danny, "Kutubu: A Case-Study of Unsustainable Development," *Resources, Nations, and Indigenous Peoples: Case Studies from the Asia-Pacific*, John Connell and Richard Howitt, eds., 1996; Kennedy, Danny, "Trade and Environment: An NGO Perspective," *Environmental Outlook*, Ben Boer, Robert Fowler, and Neil Gunningham, eds., 1996; Kennedy, Danny, Pratap Chatterjee, and Roger Moody, *Risky Business, An Independent Annual Report on PT Freeport Indonesia*, 1998; "Project Underground," http://www.moles.org.

Kennedy, Robert F., Jr.

(January 17, 1954–)
Attorney

Robert F. Kennedy Jr., nephew of Pres. John F. Kennedy and son of Atty. Gen. Robert F. Kennedy, is a senior attorney for the Natural Resources Defense Council (NRDC), chief prosecuting attorney for the Hudson Riverkeeper, and director of the environmental law clinic at Pace University Law School in White Plains, New York. He has led Hudson Riverkeeper in more than 200 legal victories over polluters and is best known for negotiating an agreement to protect watersheds surrounding the 19-reservoir water supply of New York City.

Robert Francis Kennedy Jr. was born on January 17, 1954, in Washington, D.C., the third of Robert F. and Ethel Skakel Kennedy's 11 children. He was raised at Hickory Hill, an antebellum estate in McLean, Virginia, across the Potomac from Washington, where his uncle, John F. Kennedy, served as U.S. president from 1961 until his assassination in 1963. Kennedy's devout Catholic parents instilled in their children a profound appreciation for democracy and commitment to social justice. The Kennedy children were frequently reminded that their country had been very good to their family, and they often heard St. Paul's dictum, "To whom much is given, much is expected."

Although Kennedy took seriously the family's debt to society, his early, primary interest was in nature and animals. As a child, he kept a large collection of amphibians, reptiles, and rodents gathered from the streams and forests of Hickory Hill, and when he was 11 years old, his father gave him a red-tailed hawk to train. This awakened what would be a lifelong fascination with raptors. After his father was assassinated in June 1968, Kennedy was sent to board at Millbrook School in upstate New York; the school offered ornithology and natural history curricula and an extracurricular falconry club. Kennedy learned to tame and fly hawks and to trap, band, and release wild raptors. At this point in his life, he hoped to become a veterinarian.

Kennedy attended Harvard College, earning his A.B. in history and literature in 1976. His senior thesis was an analysis of the work of the progressive Alabama federal judge Frank M. Johnson Jr., who desegregated buses and all public facilities, including schools, of Montgomery, Alabama; abolished state poll taxes; ordered that Ku Klux Klan membership lists be published; and reformed the prison system. Kennedy's *Judge Frank M. Johnson, Jr.: A Biography* was published in 1978 by Putnam. Kennedy went on to earn a J.D. degree from the University of Virginia Law School in 1982 and spent one year at the Manhattan district attorney's office prosecuting misdemeanor cases. In 1984 he began work for the Natural Resources Defense Council and the Hudson River Fishermen's Association (HRFA). This organization had been founded in 1966 by Hudson River fishermen angry about the river's contamination and the ensuing decline of the fish population. By the time Kennedy began working with the group, it had a full-time director and "riverkeeper," John Cronin, who patrolled the river to follow up on reports of contamination by the fishermen.

Kennedy's first cases for the HRFA, between 1986 and 1987, were in the town of Newburgh, about 40 miles up the Hudson from Manhattan. Newburgh was run by corrupt politicians who allowed private industry and municipal utilities to poison the Quaissaic tributary and the Hudson River itself. Cronin had been tipped off by local businessman Joe Augustine about the situation, and he, Kennedy, and volunteers from the HRFA collected samples of effluents and stream water that showed undeniably that the Quassaic was being illegally contaminated by 20 separate entities. The New York Department of Environmental Conservation joined with the Fishermen's Association to bring legal action against the polluters. Every polluter—ranging from Mobil Oil to the town water treatment plant to a local textile factory—settled out of court with the plaintiffs, and a well-endowed fund was set up for the cleanup of the Quassaic.

Working on this case proved to be a turning point for Kennedy. Previously, he had not known how to integrate the Kennedy family commitment to social justice with his true love for wildlife and nature. But after learning how environmental contamination affected all of the inhabitants of the Hudson—nonhuman animals and humans alike—he understood that environmental law melded both passions. The success at Newburgh coupled with the challenge of cleaning up the rest of the 315-mile Hudson River convinced Kennedy to accept a position as chief prosecuting attorney for the HRFA (now called the Hudson Riverkeeper). During the Newburgh case, Kennedy had been taking night classes in environmental law at Pace University, and upon his graduation in 1987, Hudson Riverkeeper and Pace agreed to establish a jointly run environmental litigation clinic for Pace University law students. Under supervision by Kennedy and Pace law professor Karl S. Coplan, ten second- and third-year law students spend a year representing Hudson Riverkeeper in federal and state courts and administrative proceedings. Not only have these students gained valuable real-world experience for future careers in environmental law, their work has resulted in ground-breaking decisions in federal courts on decisions under the Clean Water Act and the Resource Conservation and Recovery Act.

Hudson Riverkeeper has been involved in hundreds of legal actions involving the cleanup of the Hudson and Long Island Sound and has forced polluters to pay billions in penalties and remediations. Kennedy is best known for the agreement he forged in 1997 to protect the source of New York City's drinking water. He convinced Republican governor George Pataki—against whom he had campaigned in 1994—to protect the declining watersheds around the 19 upstate reservoirs that provide New York City's drinking water. New York City saved the $8 billion that it would have had to spend on a water filtration plant, and in return, paid watershed protection fees to the municipalities surrounding the reservoirs. Kennedy and Cronin coauthored a 1997 book that described Hudson Riverkeeper and its work, *The Riverkeepers*.

As the accomplishments and the reputation of the Hudson Riverkeeper grew, other water keeper organizations emerged around the country. Currently there are more than 40 similar organizations throughout the United States and Canada. Kennedy serves as national coordinator for the umbrella group, Water Keeper Alliance. In 1999, Kennedy and two partners launched Keeper Springs Mountain Spring Water, a new brand of bottled water, whose after-tax proceeds will all be channeled to the Water Keeper Alliance.

In addition to his work for Hudson Riverkeeper and Water Keeper Alliance, Kennedy is a senior attorney for the NRDC, where he has played a leading role in a number of international environmental campaigns to preserve endangered ecosystems and the rights of indigenous peoples. These include successful battles to stop such activities as the damming of James Bay in northern Quebec; clear-cut logging in British Columbia's Clayoquot Sound; the Mitsubishi Salt Plant in Laguna San Ignacio, Mexico; the Clifton Cay Development on New Providence Island in the Bahamas; and the Upper Dam on the Bio Bio River in Chile.

Kennedy still flies hawks and holds vigil on Schunemunk Ridge overlooking the Hudson River during their annual fall migration, to trap, band, and release them as part of an effort to keep tabs on the population. He has a state permit to operate a wildlife rehabilitation center and takes in injured animals, primarily raptors. He has served as president of the New York State Falconers Association.

Kennedy lives with his wife, Mary Richardson, and their three children outside of New York City. He has two older children from a previous marriage.

Bibliography

Cronin, John, and Robert F. Kennedy Jr., *The Riverkeepers*, 1997; Leonetti, Carol, "Robert Kennedy, Jr., A Riverkeeper Who Makes the Polluters Pay," *E Magazine*, 1995; Oliver, Joan Duncan, "A Ruffler of Feathers: Eco-Activist Robert F. Kennedy Jr. Stalks Polluters and Avenges Mother Nature," *Wildlife Conservation*, 1994; "Pace Environmental Litigation Clinic," http://www.law.pace.edu/envclinic/index.html; "Riverkeeper," http://www.riverkeeper.org/.

Kratt, Chris, and Martin Kratt

(July 19, 1969– ; December 23, 1965–)
Creators and Producers of Children's Wildlife Programs

As creators, executive producers, and stars of *Kratts' Creatures*, a television program designed for school-age children, and *Zoboomafoo*, designed for preschoolers, Chris and Martin Kratt are redefining children's programming and how future generations will view animals and the world around them. The two programs are designed not only to educate and entertain children but also to empower them to explore the world. The Kratt brothers have also written several wildlife books for children and have produced nature films. Both brothers have a wealth of travel and wildlife experience as "creature adventurers," having filmed all over the world. Together, they formed the Kratt Brothers' Creature Heroes, a not-for-profit organization to stimulate children to become interested and involved in wildlife conservation. With the funds raised, they purchased Grizzly Gulch, a wildlife refuge in Montana, to protect the endangered Great Plains grizzly bear.

Martin Kratt was born December 23, 1965, in Warren, New Jersey. He completed a B.S. in zoology at Duke University in 1990 and was awarded a Richard H. Jenrette Fellowship to the University of North Carolina M.B.A. program. After college, Kratt began his filmmaking career, when he went (with his brother, Chris) to Costa Rica to film wildlife. He later served as an intern at public television station Thirteen/WNET in New York, where he worked on the popular PBS series, *Nature*. Martin Kratt has been involved in several field training projects involving wildlife, including a training program in breeding endangered species in captivity at the Jersey Wildlife Preservation Trust in England.

Christopher Kratt was born July 19, 1969, in Warren, New Jersey. He received a B.A. in biology from Carleton College in 1992. He has conducted several field studies in ecology and ecological evolution funded by grants from the Explorers Club and the National Science Foundation. In 1990, he worked as an intern at Conservation International in Washington, D.C., and a year later founded the Carleton Organization for Biodiversity, a group dedicated to increasing public awareness of conservation issues. Upon his graduation from Carleton, he received a Thomas J. Watson Fellowship, which he used to develop wildlife documentaries for children.

Both Kratt brothers are dedicated to teaching young people about wild creatures and working for the preservation of endangered species. To that end, in 1990 they founded the Earth Creatures Company, which specializes in wildlife entertainment. In 1996, *Kratts' Creatures*, the first series created by the Kratt Brothers, premiered on PBS. The wildlife program is designed for school-age children; it has attracted enthusiastic viewers of all ages and has received numerous awards—including a 1996 Parents' Choice Award—and widespread critical acclaim; it is seen in 34 countries. Recognizing that there were no wildlife shows specifically designed for preschoolers, the Kratts created and starred in *Zoboomafoo* (zah-BOO-mah-foo), a show based around a playful leaping lemur. Each episode is based on a separate theme and introduces all kinds of creatures ranging from household pets to camels, tigers, and lesser-known creatures from all over the world. The goal of the program is for young children to befriend animals and to respect and care for the creatures with whom they share the world.

In addition to their television work, the Kratts have also produced, hosted, and filmed 12 "Earth Creature Reports" for PBS's children's television show, *Real News for Kids*, and produced segments for Nickelodeon's *Video Pen Pals Show*. They have also written several wildlife books for children, which include *Creatures in Crisis*, *Where're the Bears?*, *To Be a Chimpanzee*, and *Going Baboony*.

In 1999, they founded the Kratt Brothers' Creature Heroes, a not-for-profit organization to include and encourage children's involvement in wildlife conservation. Their first land purchase on behalf of Creature Heroes was at the end of 1999, when they acquired 1,222 acres of Grizzly Gulch, located in Montana on the fringe of the plains and the Rocky Mountains, adjacent to the Bob Marshall Wilderness Area. Of the $1.3 million purchase price, $250,000 was raised by children across North America. The Kratts hope to raise the additional $1 million during 2000 and 2001 to protect the rest of the area, which is habitat for the endangered Great Plains grizzly bear, the mountain lion, and the timber wolf, among other species.

The Kratt Brothers have won numerous awards for film and children's entertainment, including "Best Children's Film" for *Amazon Adventure* at the 1993 Jackson Hole Wildlife Film Festival and "Best Children's Program" for the premiere episode of *Big Five, Little Five* at the 1996 International Wildlife Film Festival. In September 1998, Secretary of the Interior BRUCE BABBITT presented the Kratt Brothers an award of appreciation for their ongoing commitment and public service on behalf of conservation and environmental education. The honor was bestowed by the U.S. Department of Agriculture for Partners in Conservation Education.

The Kratts continue to be creature adventurers, traveling the world, filming exotic animals for their various wildlife projects, and including children in the many projects. Their Earth Creatures Company currently works out of Toronto.

BIBLIOGRAPHY

"Kratts' Creatures," http://www.pbs.org/kratts/; "The OFFICIAL Kratt Brothers' Creature Site," http://www.krattbrothers.com/; "Zoboomafoo," http://www.pbs.org/zoboo/.

Krupp, Fred

(March 21, 1954–)
Executive Director of Environmental Defense, Attorney

As executive director of Environmental Defense, Fred Krupp first came to national attention in 1992 when he negotiated an agreement with McDonald's Corporation to stop using "clamshell" styrofoam containers made with ozone-destroying chlorofluorocarbons (CFCs). Krupp played a key role in the congressional passage of the Clean Air Act of 1990, and his innovative plan for a national trading system for pollutant emission "credits" was incorporated into the act. He has been instrumental in defusing the adversarial relations between environmentalists and corporations, leading to a "third wave" of environmentalism that calls for market-based solutions to environmental problems.

Frederic D. Krupp was born on March 21, 1954, in Mineola, New York, and grew up in Verona, New Jersey. His mother was a high school history teacher, and his father ran a business that processed waste rags that were eventually made into roofing materials. Krupp majored in combined sciences at Yale, where he became interested in finding solutions to environmental problems. He earned his B.S. in 1975. He received his law degree from the University of Michigan in 1978. Krupp formed the law firm Albis & Krupp in 1978 and remained with the partnership until 1984, when he joined Cooper, Whitney, Cochran & Krupp. He married Laurie Devitt, a public health nutritionist, in 1982.

Krupp helped found the Connecticut Fund for the Environment (CFE) in 1978 and served as general counsel for the fund until 1984. In 1979, when he was just 25 years old, Krupp led the CFE in bringing a $1.5 million lawsuit against the Upjohn Corporation for discharging toxic chemicals into the Quinnipiac River in Connecticut. The skillful handling of this bold legal action brought the young lawyer to the attention of the Environmental Defense Fund (EDF, which in 2000 shortened its name to Environmental Defense, ED), and in 1984, he was offered the position of executive director. A group of concerned citizens in Long Island had founded the Environmental Defense Fund in 1967; inspired by RACHEL CARSON, they sought to block the use of dichlordiphenyltrichlor (DDT) by challenging chemical polluters in court. Krupp's initial goal was to expand the organization's national presence. In the 16 years he has served as executive director, ED has grown from 35,000 to 300,000 members. In addition to its innovative, environmentally friendly headquarters in Manhattan, designed by green architect BILL MCDONOUGH, ED now has offices located in California, Colorado, North Carolina, Texas, and Washington and has affiliated operations in Massachusetts and the Pacific Northwest. The regional offices have been successful in raising ED's profile throughout the nation; they also effect change on the local level by encouraging local governments, businesses, and citizens to seek solutions to regional environmental problems. ED's staff has tripled in size since 1984; the multidisciplinary staff includes scientists, attorneys, and economists. The organization's budget has grown from $3 million to $23 million. In the last decade, ED has moved beyond the national scene, assuming a prominent position on the international environmental scene, particularly in the area of global warming.

Krupp is widely credited with breaking the congressional stalemate over the 1990 Clean Air Act. His innovative plan to reduce acid rain by trading emission "credits" was incorporated into the act. Under Krupp's plan, corporations that reduced sulfur dioxide emissions would be given "credits" that could then be sold to other companies. He achieved further recognition in 1992, negotiating an agreement with McDonalds to substitute cardboard containers for the styrofoam cases the company had been using. A large grassroots consumer campaign led by Penny Newman and Lois Gibbs had been pushing McDonalds for several years to discontinue using its styrofoam "clamshell" containers, whose manufacture required ozone-destroying CFCs. McDonalds called upon Krupp to negotiate an agreement in 1992. Krupp catapulted into the national spotlight, and many in the environmental movement claimed this as a symbolic victory. (Newman and Gibbs were disappointed not only because the new bleached paper containers contained little or no recycled content but also because EDF took full credit for the agreement, without acknowledging the gains of the consumer campaign.)

Krupp's willingness to cooperate with major corporations—a tendency that has been called the third wave of environmentalism—has been viewed with suspicion by many environmentalists. Corporations, accustomed to being lambasted by environmentalists, have been wary of Krupp as well. But Environmental Defense holds that it can often effect change more quickly by working with corporations than by challenging them in court. By 1994, litigation accounted for only about 5 percent of the group's activity. Throughout the early 1990s, Krupp carried ED's message to corporate executives and members of congressional subcommittees. Many of his speeches to such groups as the Aspen Institute Energy Policy Issue Forum and the Environmental Marketing Communications Forum have been published in *Vital Speeches of the Day*. During the remainder of the 1990s, Krupp forged amicable working relations between ED and global corporations such as Monsanto, Pacific Gas & Electric, and Merck. ED has been careful, according to Mark Dowie in *Losing Ground* (1995), to screen out corporate donors to exclude any that "would compromise its negotiating positions or damage its reputation as an independent advocate."

Environmental Defense reformulated its goals in 1996. The group retained its original emphasis on protecting human health from exposure to toxic chemicals but broadened its work to include safeguarding oceans and rivers from overfishing, pollution, and loss of biodiversity. The number-one priority became the reduction of emissions of greenhouse gases to stabilize global warming. Krupp embarked on a campaign to develop an international system of emissions credits similar to the one he had crafted for the Clean Air Act.

On the local level, ED has championed the public's right-to-know about environmental risks in their communities. The Chemical Scorecard web site encourages the public to learn about local pollution sources and their possible health risks to the community. The web site was launched in 1998 and currently includes information on 17,000 facilities, 2,000 counties, and all 50 states.

Krupp is a member of the President's Commission on Sustainable Development and the President's Advisory Committee for Trade Policy and Negotiations. He is a board member of the H. John Heinz III Center for Science and Economics and the Environment and an advisory board member of both the Environmental Media Association and Earth Communications Office. He serves on the Keystone Commission, a panel of environmentalists and industrialists formed to work toward Superfund reform.

Krupp lives in New Canaan, Connecticut, with his wife, Laurie, and sons Alex, Jackson, and Zack.

Bibliography

"Chemical Scorecard," www.scorecard.org; Dowie, Mark, *Losing Ground*, 1995; "Environmental Defense," www.edf.org; Krupp, Fred, "Business and the Third Wave: Saving the Environment," *Vital Speeches of the Day*, 1992; Miller, William H., "Fred Krupp: A Different Kind of Environmentalist," *Industry Week*, 1994; Reed, Susan, "Environmentalist Fred Krupp Helps Crush the Ubiquitous Fast-food Clamshell," *People Weekly*, 1991.